Green Food

Green
Food

An A-to-Z Guide

The SAGE Reference Series on
Green Society
Toward a Sustainable Future

DUSTIN MULVANEY, GENERAL EDITOR
University of California, Berkeley

PAUL ROBBINS, SERIES EDITOR
University of Arizona

SAGE | reference

Los Angeles | London | New Delhi
Singapore | Washington DC

Los Angeles | London | New Delhi
Singapore | Washington DC

FOR INFORMATION:

SAGE Publications, Inc.
2455 Teller Road
Thousand Oaks, California 91320
E-mail: order@sagepub.com

SAGE Publications Ltd.
1 Oliver's Yard
55 City Road
London EC1Y 1SP
United Kingdom

SAGE Publications India Pvt. Ltd.
B 1/I 1 Mohan Cooperative Industrial Area
Mathura Road, New Delhi 110 044
India

SAGE Publications Asia-Pacific Pte. Ltd.
33 Pekin Street #02-01
Far East Square
Singapore 048763

Publisher: Rolf A. Janke
Assistant to the Publisher: Michele Thompson
Senior Editor: Jim Brace-Thompson
Production Editors: Kate Schroeder, Tracy Buyan
Reference Systems Manager: Leticia Gutierrez
Reference Systems Coordinator: Laura Notton
Typesetter: C&M Digitals (P) Ltd.
Proofreader: Christina West
Indexer: Virgil Diodato
Cover Designer: Gail Buschman
Marketing Manager: Kristi Ward

Golson Media
President and Editor: J. Geoffrey Golson
Author Manager: Ellen Ingber
Editor: Jill Coleman, Mary Jo Scibetta
Copy Editors: Anne Hicks, Barbara Paris

Copyright © 2011 by SAGE Publications, Inc.

Printed in the United States of America

Library of Congress Cataloging-in-Publication Data

Green food : an A-to-Z guide / general editor, Dustin Mulvaney.

p. cm. — (The Sage reference series on green society: toward a sustainable future)

Includes bibliographical references and index.

ISBN 978-1-4129-9680-8 (cloth)—ISBN 978-1-4129-7187-4 (ebk)

1. Food industry and trade—Environmental aspects.
2. Produce trade—Environmental aspects.
3. Agriculture—Environmental aspects.
4. Environmental responsibility. 5. Sustainability.
I. Mulvaney, Dustin.

TD195.F57G74 2010 338.1'9—dc22 2011002868

11 12 13 14 15 10 9 8 7 6 5 4 3 2 1

Contents

About the Editors

Green Series Editor: Paul Robbins

Paul Robbins is a professor and the director of the University of Arizona School of Geography and Development. He earned his Ph.D. in Geography in 1996 from Clark University. He is General Editor of the *Encyclopedia of Environment and Society* (2007) and author of several books, including *Environment and Society: A Critical Introduction* (2010), *Lawn People: How Grasses, Weeds, and Chemicals Make Us Who We Are* (2007), and *Political Ecology: A Critical Introduction* (2004).

Robbins's research centers on the relationships between individuals (homeowners, hunters, professional foresters), environmental actors (lawns, elk, mesquite trees), and the institutions that connect them. He and his students seek to explain human environmental practices and knowledge, the influence nonhumans have on human behavior and organization, and the implications these interactions hold for ecosystem health, local community, and social justice. Past projects have examined chemical use in the suburban United States, elk management in Montana, forest product collection in New England, and wolf conservation in India.

Green Food General Editor: Dustin Mulvaney

Dustin Mulvaney is a Science, Technology, and Society postdoctoral scholar at the University of California, Berkeley, in the Department of Environmental Science, Policy, and Management. His current research focuses on the construction metrics that characterize the life cycle impacts of emerging renewable energy technologies. He is interested in how life cycle assessments focus on material and energy flows and exclude people from the analysis, and how these metrics are used to influence investment, policy, and social resistance. Building off his work with the Silicon Valley Toxics Coalition's "just and sustainable solar industry" campaign, he is looking at how risks from the use of nanotechnology are addressed within the solar photovoltaic industry. Mulvaney also draws on his dissertation research on agricultural biotechnology governance to inform how policies to mitigate risks of genetically engineered biofuels are shaped by investors, policy-makers, scientists, and social movements.

Mulvaney holds a Ph.D. in Environmental Studies from the University of California, Santa Cruz, and a Master of Science in Environmental Policy, and a Bachelor's Degree in Chemical Engineering, both from the New Jersey Institute of Technology. Mulvaney's previous work experience includes time with a Fortune 500 chemical company working on sulfur dioxide emissions reduction, and for a bioremediation startup that developed technology to clean groundwater pollutants like benzene and MTBE.

Introduction

Our food and agricultural systems have undergone a tremendous change in the 20th century. Less common now are the pastoral ways in which we have come to imagine agriculture, such as the images found on a milk carton. Agriculture and the food system have become thoroughly industrialized and increasingly globalized. A plentiful supply of cheap fossil fuels has helped to power mechanization, produce the fertilizers and pesticides, and lengthen distance from farm to plate. The result is a carbon-intensive food system that keeps food prices cheap, while large retailers and processors continue to extract value from farmers at increasing margins. The result of this transformation is an agrifood system with an enormous productive capacity, but one that causes considerable environmental burdens and that has exacerbated problems with poverty and food distribution.

The extent to which our agricultural system has become dependent on fossil fuel–based fertilizers; highly mechanized planting, harvesting, and processing; and high-tech seed has changed significantly the systems on which past agricultural practices depended. Our existing food and agricultural production systems have a long list of environmental and social impacts that suggest the need for a more sustainable agrifood system. Not only are there these externalities cause by agricultural production, but the resource intensity of the industrial food system overall could be undermining its future prosperity. For example, most agricultural soils, once rotated out of production, and once enriched with nearby farm animals, are now treated with chemicals intending to kill all of the living matter in the soil. The soil, far from its previous state as a collection of carbon and living organisms, and as a reservoir of essential nutrients, is now a sterile environment for controlled growth and pest control. Water from conventional farms now contains high levels of nitrogen pollution, salts, and other fertilizer and pesticide runoff, preventing the water's reuse, and creating anoxic zones where rivers drain into the sea, such as the infamous Mississippi River "dead zone."

The changes to our food system have not just affected how food is produced, but they have also considerably changed how the agrifood system is organized. Industrial concentration has marked the 20th century in agriculture as food and agriculture companies have vertically and horizontally integrated to dominate entire sectors; grain transportation, meat packing, agro-chemical manufacturing, and seed production are just some examples of agricultural sectors that are dominated by only a handful of large, multinational firms. This in and of itself has implications as the decisions made in several small boardrooms of multinational corporations to pursue one technology or another can have a considerable and lasting effect on the landscape and on human health.

These changes have shaped consumer expectations of the food system. Consumers demand that their food purchases defy the logic of seasons. They want tomatoes and strawberries in the cold of winter. They want consistent-tasting fast foods. This is happening while consumers have become more distant from the food they eat, and less aware of what it takes to get a food from the field to the factory to the plate.

The move toward green food is part of a reaction to the degradation and violence of industrial agriculture. The development has many roots in animal rights movements, appropriate technology movements, and back-to-the-land movements, among many others, each with its own motivations for action. The slow food movement, for example, emerges out of the hope to combat the spread of fast-food, reintroducing the cultural rituals of eating that require more time spent at the table in conversation, as opposed to the fast-paced meals eaten in the car. Evidence for the popularity of this new green food movement can be seen in the growth of organic agriculture and the constantly evoked statistic of its 20 percent annual growth in sales and revenues. The popularity and growth of farmers markets also shows evidence of this burgeoning agrifood movement. Food consumption is central to any individual's daily routine. So in many ways the environmental and social impacts of food production are closely tied to our individual choices as consumers. Hence, there is political space for those who want to see more food labeled and food's content disclosed where it currently is not.

Often lacking from the mainstream discussion of green foods, but critical to the question of environmental justice, is the treatment of agricultural workers. The improper treatment of agricultural workers has a long storied past—one could point to John Steinbeck's *The Grapes of Wrath* for an earlier account—but it continues today. Many agricultural and food system workers are minorities or immigrants, many of whom lack legal rights. The story of agricultural labor in the United States and Mexico has an interesting twist. Some of the cheap labor that has always come through the United States was recently driven to migrate from Mexico's corn-growing regions after the North American Free Trade Agreement helped the United States dump corn into the Mexican market at very low prices, undercutting local producers. Those farmers without a market for corn had little to sell but their labor, and their seasonal migrations result in remittances back to their communities. The story shows how food production in many industrialized countries is strongly influenced by everything from government policies on trade to consumer fads. We hope this volume speaks to the numerous issues and challenges we face in order to change our approach to eating.

These entries help lay out the contours of the field of agrifood studies. They on scholars working in the fields of political ecology, rural sociology, geography, and environmental studies to paint a picture of agriculture and food's past, present, and future. They look to provide the reader with a basic understanding of the institutions, practices, and concepts to help identify what is and is not a green food. Because food is so intimately connected to our daily lives, it could be that the food system offers the most promise to make changes in a sustainable direction.

What constitutes an actual sustainable and green food system is still an open question. There are many unresolved questions about what it should look like, what policies would help get it there, and what kinds of tradeoffs we face in deciding which path to choose. This volume should provide people interested in food and agricultural systems with the basic analytical and conceptual ideas that help explain why our food system looks the way it does, and what can be done to change it.

Dustin Mulvaney
General Editor

Reader's Guide

Food Challenges

Animal Welfare
Beyond Organic
Cheap Food Policy
Crop Genetic Diversity
DDT
Debt Crisis
Disappearing Middle
Export Dependency
Famine
Farm Crisis
Fast Food
Food Processing Industry
Food Safety
Food Security
Genetically Modified Organisms
Grain-Fed Beef
High-Fructose Corn Syrup
Integrated Pest Management
Irradiation
Mad Cow Disease
Malthusianism
Mechanization
Millennium Development Goals
Modernization
Nitrogen Fixation
Organochlorines
Origin Labeling
Peasant
Pesticide
Productionism
Proletarianization
Recombinant Bovine Growth Hormone
Roundup Ready Crops
Salmonella

Sewage Sludge
Soil Erosion
Sustainable Agriculture
Swidden Agriculture
Weed Management

Food Economics and Trade

Adoption-Diffusion
Appropriationism
Bracero Program
Cash Crop
Commodity Chain
Comparative Advantage
Concentration
Fisheries
Foodshed
Horizontal Integration
Just-in-Time
Labor
Land Reform
Land Tenure
Monsanto
Retail Sector
Supermarket Chains
Supply Chain
Trade Liberalization
Vertical Integration

Food Farm and Industry

Agrarianism
Agrarian Question
Agribusiness
Agricultural Commodity Programs
Agricultural Extension

Food Laws, Agreements, and Organizations

Foods and Lifestyle

List of Articles

List of Contributors

Abbott, J. Anthony
Stetson University

Adams, Rachel
Stanford University

Alkon, Alison Hope
University of the Pacific

Allen, Patricia
*University of California,
Santa Cruz*

Ambinakudige, Shrinidhi
Mississippi State University

Bacon, Christopher M.
University of California, Berkeley

Boslaugh, Sarah
Washington University in St. Louis

Cantor, Alida
California Institute for Rural Studies

Chaplin-Kramer, Rebecca
University of California, Berkeley

Coffman, Jennifer Ellen
James Madison University

Cohen, Nevin
*Eugene Lang College
The New School for Liberal Art*

Corfield, Justin
Geelong Grammar School, Australia

Darby, Kate
Arizona State University

Davis, Jason
*University of California,
   Santa Barbara*

de Souza, Lester
Independent Scholar

Donnet, M. Laura
Austral University

Edwards, Ferne
Australian National University

Evans, Tina Lynn
Fort Lewis College

Finan, Ann
Whitman College

Galt, Ryan E.
University of California, Davis

Gifford, Richard B.
*University of Arkansas Community
College at Morrilton*

Gillon, Sean
University of California, Santa Cruz

Golden, Elizabeth L.
University of Cincinnati

Graddy, Garrett
University of Kentucky

Harris, Edmund M.
Clark University

Hilimire, Kathleen Elizabeth
University of California, Santa Cruz

Howard, Philip H.
Michigan State University

Iles, Alastair
University of California, Berkeley

Ingram, Mrill
University of Wisconsin, Madison

Islam, Md. Saidul
Nanyang Technological University

Jain, Priyanka
University of Kentucky

Johnson, Paul Henry
University of Durham

Kearns, Carol Ann
Santa Clara University

Kramer, Daniel
Independent Scholar

Kte'pi, Bill
Independent Scholar

Lopez, Anna Carla
San Diego State University

Marshall, D. Jones
University of Kentucky

Matthews, Todd L.
University of West Georgia

McAfee, Kathleen
San Francisco State University

McClintock, Nathan C.
University of California, Berkeley

McCullen, Christie Grace
University of California, Davis

McKendry, Corina
University of California, Santa Cruz

Melcarek, Hilary
University of California, Santa Cruz

Minkoff-Zern, Laura-Anne
University of California, Berkeley

Monsen, Katie L.
University of California, Santa Cruz

Neff, Mark
Arizona State University

Neo, Harvey
National University of Singapore

Okamoto, Karen
John Jay College of Criminal Justice

Packwood Freeman, Carrie
Georgia State University

Panda, Sudhanshu Sekhar
Gainesville State College

Phadke, Roopali
Macalester College

Phillips, Catherine
Bishop's University

Pisani Gareau, Tara
University of California, Santa Cruz

Plec, Emily
Western Oregon University

Pokrant, Bob
Curtin University of Technology

Poyyamoli, Gopalsamy
Pondicherry University

Putnam, Heather R.
University of Kansas

Rausch, Lisa
University of Kansas

Roka, Krishna
Penn State University

Sampson, Devon
University of California, Santa Cruz

Schmook, Birgit
*El Colegio de la Frontera Sur
(ECOSUR)*

Smith, Susan L
Willamette University College of Law

Soria, Carlos Antonio Martin
*Universidad Nacional Agraria
   La Molina
Instituto del Bien Común*

Tan, November Peng Ting
National University of Singapore

Thomas, Marcia
New York University

Trevino, Marcella Bush
Barry University

Tyman, Shannon
University of Oregon

Vachta, Kerry E.
Wayne State University

Vynne, Stacy
University of Oregon

Waskey, Andrew Jackson
Dalton State College

Weissman, Evan
Syracuse University

Whalen, Ken
American University of Afghanistan

Williams, Akan Bassey
Covenant University

Yuhas, Stephanie
University of Denver

Green Food Chronology

12,000–6,000 B.C.E.: During the Neolithic Revolution, early humans learn to domesticate plants and animals, developing agriculture and the beginnings of settlements in the Fertile Crescent. Previously gathered plants, including grains and legumes, are sowed and harvested, and wild sheep, goats, pigs, and cattle are herded instead of hunted. The earliest forms of storage create food surpluses, which ensure the growth of fledgling cities.

4000 B.C.E.: The Egyptians discover how to make bread using yeast. The Chinese discover how to use lactic acid bacteria to make yoghurt, molds to produce cheese, and fermentation to make vinegar, soy sauce, and wine.

4000–3000 B.C.E.: In a seemingly simultaneous innovation, civilizations in Europe and the Middle East use oxen to pull sledges and plow fields.

3200 B.C.E.: The wheel is used in Ancient Mesopotamia.

2500 B.C.E.: Ancient Sumerians make use of the natural occurring element sulfur as the first known use of an insecticide, applying it to their crops.

1202: King John of England proclaims the first English food law, the Assize of Bread, which prohibited adulteration of bread with such ingredients as ground peas or beans.

1673: Spanish scientist Francesco Redi compares two competing theories to explain why maggots appear on rotting meat. He observes that meat covered to exclude flies does not develop maggots, whereas uncovered meat did.

1700: The British Agricultural Revolution begins at the start of the century and describes a period of the next hundred years in which significant increases in agricultural production support an unprecedented population surge.

1724: Anton van Leeuwenhoek uses his microscopes to make discoveries in microbiology. He is the first scientist to describe protozoa and bacteria and to recognize that microorganisms might play a role in fermentation.

1801: American pioneer John Chapman, better known as Johnny Appleseed, wanders the American West, planting orchards of apples. He becomes a legend in his own time for his generous ways and passion for nature.

1820–1975: Over this period, worldwide agriculture productions double four times.

1870: Forty-seven percent of gainfully employed Americans are engaged in agriculture.

1888: Vedalia beetles are imported from Australia to control fluted scale on citrus, marking the first successful biological control program of a crop pest.

1898: The Association of Agricultural Chemists establishes a Committee on Food Standards headed by Dr. Harvey Wiley, who is later the main proponent of the Pure Food and Drugs Act.

1902: Congress appropriates $5,000 for the Bureau of Chemistry to study chemical preservatives and coloring and their effects on digestion and health. Dr. Wiley's studies draw widespread attention to the problem of food adulteration. Public support for passage of a federal food and drug law grows.

1906: U.S. President Theodore Roosevelt signs the Pure Food and Drugs Act, effectively creating the Food and Drug Administration (FDA). The Center for Food Safety and Applied Nutrition (CFSAN) is the specific branch that oversees food. Now a part of the Department of Health and Human Services, the FDA regulates the safety of foods, dietary supplements, drugs, vaccines, cosmetics, and other products. The CFSAN is responsible for about $240 billion worth of domestic food in the United States. The passage of the Pure Food and Drug Act coincides with the Meat Inspection Act.

1910: Wild blueberries are domesticated.

1911: In *U.S. v. Johnson*, the Supreme Court rules that the 1906 Pure Food and Drugs Act does not prohibit false therapeutic claims but only false and misleading statements about ingredients.

1924: Austrian scientist Dr. Rudolf Steiner gives a series of eight lectures outlining the principles of biodynamic farming. His lectures are then published in the book *Spiritual Foundations for the Renewal of Agriculture*. Biodynamic farms use organic principles and, today, are a registered trademark of a U.S.-based corporation.

1933: The FDA recommends a complete revision of the obsolete 1906 Pure Food and Drugs Act.

1939: German scientist Paul Müller discovers that the chemical compound dichlorodiphenyltrichloroethane (DDT) is a very effective insecticide. In the decade to follow, manufacturers begin to produce large amounts of synthetic pesticides, and their use becomes an industry standard. Many years later, DDT would be at the heart of national outrage over irresponsible pesticide usage.

1940: Lord Northbourne writes *Look to the Land*, outlining the fundamental tenets of organic farming.

1943–1964: During these years, Mexico transforms its wheat industry and goes from importing half of its wheat to exporting half a million tons a year. The progress in Mexico sparks worldwide interest in new agricultural developments. U.S. Agency for International Development Director William Gaud coins the term "Green Revolution" in a speech, saying, "These and other developments in the field of agriculture contain the makings of a new revolution. It is not a violent Red Revolution like that of the Soviets, nor is it a White Revolution like that of the Shah of Iran. I call it the Green Revolution."

1950: The Delaney Committee starts congressional investigation of the safety of chemicals in foods and cosmetics, laying the groundwork for future legislation.

1954: The U.S. Congress passes the Miller Pesticide Amendment, spelling out procedures for setting safety limits for pesticide residues on raw agricultural commodities.

1954: The first large-scale radiological examination of food is carried out by the FDA when it receives reports that tuna suspected of being radioactive is being imported from Japan following atomic blasts in the Pacific. The FDA begins monitoring around the clock to meet the emergency.

1958: The Food Additives Amendment is enacted by the U.S. Congress, requiring manufacturers of new food additives to establish safety. A provision prohibits the approval of any food additive shown to induce cancer in humans or animals.

1959: Three weeks before Thanksgiving, U.S. cranberry crops are recalled for FDA tests to check for aminotriazole, a weed killer found to cause cancer in laboratory animals. Cleared berries are allowed a label stating that they had been tested and had passed FDA inspection—the only such endorsement ever allowed by the FDA on a food product.

1960: The Color Additive Amendment is enacted by the U.S. Congress, requiring manufacturers to establish the safety of color additives in foods, drugs, and cosmetics.

1962: Rachel Carson's *Silent Spring* is a national phenomenon, first in serial form in the *New Yorker* and then as a hardcover best seller. This exhaustively researched and carefully reasoned attack on the indiscriminate use of pesticides sparks a revolution in public opinion. Specifically, the book deals with the controversial chemical DDT, which is widely used to controls insects and other pests. Carson claims that DDT is digested by birds and causes them to lay thinner eggs, resulting in detrimental effects on the greater ecosystem. She also accuses chemical companies of promoting disinformation and lobbying public officials to ignore the dangers of modern farming practices. Historians of popular culture will later mark *Silent Spring*'s publication as a turning point in the consciousness of American consumers. In the decades to follow, extensive legislation is passed as a result of civil unrest over food production, especially on large, agro-intensive farms.

1962: The Consumer Bill of Rights is proclaimed by U.S. President John F. Kennedy in a message to Congress. Included are the right to safety, the right to be informed, the right to choose, and the right to be heard.

1966: A unified national system of food packaging labeling is established when the U.S. Congress requires manufactures to identify the product, the name and place of business, and the net quantity for all interstate commerce.

1969: The White House Conference on Food, Nutrition, and Health recommends systematic review of thought-to-be-safe substances in light of the FDA's ban of the artificial sweetener cyclamate. President Nixon orders FDA to review its list of approved substances.

1969: The Food and Nutrition Service (FNS) is established as a function of the U.S. Department of Agriculture (USDA) and is responsible for administering the nation's domestic nutrition assistance program. The FNS eventually creates the Food Stamp Program, National School Lunch Program, Farmers' Market Nutrition Program, Fresh Fruit and Vegetable Program, and many others.

1971: Several national and international groups convene to create a worldwide network of research centers under the same heading, the Conservative Group on International Agriculture Research. The group works to promote green revolutions, especially in developing nations.

1973: The first certified organic farming cooperative, California Certified Organic Farmers, begins with a group of 54 farmers who adhere to published standards promoting small-scale farming without harmful chemicals. By 2008, membership includes more than 1,800 farmers, controlling over 500,000 acres of land in 29 states and five other countries.

1973: DDT is officially banned. Historians will point to this moment as a watershed moment in the green food industry, as national opposition to end chemical additives in food production comes to head.

1977: The Saccharin Study and Labeling Act is passed by the U.S. Congress to stop the FDA from banning the chemical sweetener but requires a label warning that the substance has been found to cause cancer in laboratory animals.

1980: Genetic engineer Ananda Mohan Chakrabarty, working for General Electric, develops a bacterium capable of breaking down crude oil, which he proposes to use in treating oil spills. He requests a patent for the bacterium in the United States but is turned down by a patent examiner because the law dictates that living things were not patentable. In a 5–4 ruling, the Supreme Court sides in favor of Chakrabarty, citing: "A live, human-made micro-organism is patentable subject matter under [Title 35 U.S.C.] 101. Respondent's micro-organism constitutes a 'manufacture' or 'composition of matter' within that statute."

1983: Austria becomes the first country to develop official national guidelines for organic farming.

1985: First known as the Organic Foods Production Association of North America, the Organic Trade Organization (OTA) is established as a membership-based business group for the organic industry in America, Canada, Mexico and other countries. The goal of the organization is to coordinate the efforts of organic farming groups and to educate society about the benefits of organic food production. Founded in 1997 as part of the OTA, the Organic Materials Review Institute provides organic certifiers, growers, manufacturers, and suppliers with independent reviews of products intended for use in certified organic production, handling, and processing.

1985: The U.S. federal government establishes the Conservation Reserve Program as a cost-sharing program for farmers that actively protect the environments in which they work. In the following years, four other "Farm Bills" are passed with amendments.

1985: The Organic Crop Improvement Association forms as a nonprofit organization providing research and certification to organic growers, processors, and handlers across the globe. Incorporated in 1988, the group now has offices in Canada, China, Mexico and Japan, as well as several other nations.

1987: Several simultaneous developments occur with genetically modified (GM) plants. Calgene Inc. receives a patent for the tomato polygalacturonase DNA sequence, used to extend the shelf life of fruit. Advanced Genetic Sciences Inc. conducts field trial of a recombinant organism—a frost inhibitor—on a strawberry patch in the United States.

At the Waite Institute in Adelaide, Australia, scientists genetically modify a type of soil bacteria that causes crown gall (a disease that damages the roots of stone fruits). They remove the disease-causing gene and replace it with a gene that protects the plant from crown gall. The GM bacteria are successfully tested on almond seedlings.

In the United Kingdom, genes are added to potato plants to make them produce more protein and increase their nutritional value. Research into other foods includes removing allergy-causing proteins from peanuts.

1988: The Sustainable Agriculture Research Education program is implemented by the USDA to provide grants promoting environmentally friendly small-scale farmers.

1989: Low-input sustainable agriculture (LISA) becomes widespread on American farms, decreasing the use of chemical applications.

1990: The FDA is given the authority to require the modern food labeling on most foods regulated by the agency. It also requires that terms claims like "high-fiber" and "low-fat" meet standards set by the CFSAN.

1990: As retail sales of organics reach $1 billion in the United States, Congress passes the Organic Foods Production Act, requiring that the USDA develop national standards and regulations for organically produced agricultural products. More than a decade passes before the department completes the task with the National Organic Program (NOP).

1993: The U.S. Congress implements the Revised General Agreement on Tariffs and Trade and the new North American Free Trade Agreement, significantly lowering international trade barriers.

1994: A USDA study indicates that 1.13 million U.S. acres are devoted to certified organic production.

1994: U.S. state–regional trade groups sponsor the first trade mission of organic suppliers to Japan. The state–regional groups sponsor missions to Japan each year thereafter and add missions to Europe in 1999 and 2000.

1994: The first genetically engineered food product, the Flavr Savr tomato, receives FDA approval.

1996: Transgenic, or genetically modified (GM), plants are developed for various purposes: resistance to pests, herbicides, or harsh environmental conditions; improved shelf life; increased nutritional value; and many more. GM plants become widespread in America and in Europe, sparking international concern over how to legislate modified plants and animals.

1997: A sheep named Dolly is cloned from the udder cell of an adult sheep. The development sparks international controversy as popular culture recognizes the real, tangible possibility of widespread cloning in the near future.

1999: The United Kingdom doubles assistance for farmers converting to organic production.

2000: The USDA releases the second proposed NOP. The Secretary of Agriculture announces other initiatives to stimulate the U.S. organic sector: organic research, pilot projects on crop insurance, and marketing orders and market news reporting for organic fruits and vegetables. Organic industry members and consumers send over 275,000 comments to the USDA on their proposed National Organic Standards. The program unifies all organic foods under a cohesive national agenda and creates the organic seal, which appears on products that meet criteria.

2000: The USDA's Economic Research Service releases a major study on the status of organics in the United States, showing that certified organic crop land more than doubled during the previous decade and that some organic livestock sectors—eggs and dairy—grew even faster.

2000: The genetic code of the fruit fly Drosophila is published. Drosophila is the "lab rat" of the genetics world and is used in experiments to investigate genes and gene function.

2001: The USDA passes the Final Organic Rule after reinstating prohibitions on irradiation, sewage sludge, and genetically engineered seed.

2002: The U.S. Congress signs the NOP into law. It regulates all aspects of organic food production, processing, transportation, and retail sale. Any farmer or food producer with sales exceeding $5,000 must be certified by the USDA to use the word "organic" to describe their products. The USDA Certified Organic Seal begins appearing on food products in grocery stores and markets. The NOP does not regulate nonfood organic products, such as health and beauty products.

2002: The Farm Security and Rural Investment Act is passed to address water and other environmental issues. Programs include the Conservation Security Program, which created a reward system for eco-conscientious farmers.

2002: Researchers sequence the DNA of rice, the main food source for two-thirds of the world's population. It is the first crop plant to have its genome decoded.

2003: A group is established by the Commissioner of Food and Drugs, charged to develop an action plan to deal with the nation's obesity epidemic from the perspective of the FDA. In March 2004, the group releases "Calories Count: Report of the Obesity Working Group," which addresses issues connected to the food label, obesity therapeutics, research needs, the role of education, and other topics.

2003: The National Academy of Sciences releases "Scientific Criteria to Ensure Safe Food," a report commissioned by the FDA and the USDA, which invokes the need for continued efforts to make food safety a vital part of the overall public health mission.

2004: Australian researchers use gene mapping techniques to identify genes for tenderness and toughness in beef, allowing breeders to select stocks containing the "tender" gene.

2005: Lasers replace stickers by writing on fresh fruits, making individual pieces traceable and trackable.

2005: The gene sequence of the cow is published.

2006: According to a survey by *Vegetarian Journal*, 6.7 percent of Americans say they never eat meat and 2.3 percent are strict vegetarians, eliminating fish, seafood, and poultry from their diet.

2007: The FDA concludes that food and food products derived from cloned animals or their offspring are as safe to eat as those from noncloned animals. At the same time, the European Food Safety Authority concludes that antibiotic resistance marker genes in modified plants do not pose a relevant risk to human or animal health or to the environment.

2008: In July it is reported that a certified organic powdered ginger has been found to be contaminated with the banned pesticide Aldicarb. The organic ginger was certified organic by two USDA-accredited certifying agents in China. Under Chinese law, foreigners may not inspect Chinese farms.

2008: Research indicates that bisphenol A, a chemical found in many plastic beverage bottles, is believed to leach carcinogens into liquids.

2008: Twenty countries have labeling programs using shared criteria under the umbrella of Fair Trade Labeling Organizations International in Germany. In the United States, there is certification for chocolate, cocoa, coffee, flowers, rice, fresh tropical fruits, sugar, tea, and vanilla. The term Fair Trade Certified is registered and certification is administered by TransFair USA.

ADOPTION-DIFFUSION

Adoption and diffusion are processes by which an idea or product traverses from introduction to ubiquity. These processes are important in various fields of study, and generally diffusion focuses on the market, while adoption focuses on the individual's attitude toward the idea or product. Everything new begins in obscurity, rarity—even something highly anticipated is not adopted overnight by the general population. In economics and business, it is often critical to have a sense of how quickly a new consumer technology like DVD players will catch on. Some ideas and products, of course, never become ubiquitous; some barely move away from obscurity; others may shift back and forth along this imaginary spectrum, as record players traversed a path from luxury item, to household appliance, to specialist item for audiophiles, once the bulk of the market had moved on to digital media.

These processes have been described in a number of ways. In discussion of the diffusion of products, the "trickle-down effect" is often mentioned; this phenomenon occurs when an initially expensive product can at first be purchased only by the wealthy, but eventually becomes affordable for the general public. Sometimes this is because the cost of intensive research must be recouped, while actually producing the product is inexpensive in comparison; often it is because of economies of scale that make the per-unit cost of a product significantly lower when it's being made in large quantities. The trickle-down effect is especially noticeable in consumer technologies.

The two-step flow model introduced by Columbia University sociologists Paul Lazarsfeld and Elihu Katz privileged human agency over those simple economic factors, and described diffusion of information and adoption of ideas from the perspective of media influence. Information dispersed by the media travels to opinion leaders—those who pay the most attention to media outlets—and from them to their social and professional circles. Opinion leaders are those whose preferences are most likely to guide those around them, those who are perceived as being "in the know." Most people have a friend who is "the music guy" or "the computer expert," an opinion leader they count on.

The two-step flow model influenced the diffusion of innovations theory as articulated by Ohio State University sociologist Everett Rogers, who was an assistant professor when he wrote his textbook *Diffusion of Innovations*. Essentially a very good aggregator and reconciler of information, Rogers—whose book became one of the cornerstone texts of sociology—took over 500 studies of diffusion and from them generalized a deeper and more developed theory than had previously been available. Rogers focuses on human agency even more than Lazarsfeld and Katz do, dividing the populace into five categories of adopters, in order of when they are likely to adopt a new idea or product:

1. Innovators are highly social risk-takers with considerable financial, social, and scientific or commercial resources. Very few people are considered innovators (Rogers estimates them at about 2.5 percent of the population; however, his concepts are more useful than the accuracy of his bell curve).

2. Early adopters (13.5 percent) are typically young, well-educated, highly social opinion leaders. The idea of the early adopter has taken on a popularity that has outpaced Rogers' work, and our 21st century idea of the early adopter as the beta-tester of software, the first to switch to cruelty-free soap and a low-carbohydrate diet, first in line to purchase the new iPhone, is true in spirit to Rogers' early adopter, but doesn't always match Rogers' descriptions.

3. The early majority (34 percent) represents many of the rest of us, reasonably social and reasonably well educated, possibly opinion leaders. Those who are part of the early majority with respect to one idea or product may often be an early adopter with respect to others, but in general wait for the thing to prove itself, to show that it isn't a fad, that it's "here to last."

4. The late majority (34 percent) have little opinion leadership and are skeptical of innovation, often making a switch only in response to other pressures (changing diet in response to the preferences of others in the household or doctors' orders). Their financial situation may discourage innovation or chasing trends; their lower education may prevent them from being aware of the fast-paced exchange of ideas in the world.

5. Laggards are the last to adopt, and in Rogers's model, they are also the oldest group, focused on tradition, social only with established close friends and family members, with little money to burn. This is the stereotypical grandparent whose VCR's clock is still flashing 12:00, and who, for that matter, still uses a VCR.

The five types of adopters also constitute what is sometimes called the technological adoption lifecycle. Diffusion of innovations theory essentially incorporates the two-step flow model to describe how ideas are transmitted among these groups.

Rogers was also influenced by the work of Columbia University anthropologist Alfred L. Kroeber, who had studied diffusion from a cultural anthropologist's perspective and whose 1940 essay "Stimulus Diffusion" was especially well known. Diffusion had long been a topic of discussion in anthropology and history; in ancient history, the spread of the war chariot and the use of coins had been well-studied, and models of diffusion were so common in the discipline that many could not be persuaded to accept incidents of independent invention in cases where the evidence was overwhelmingly in its favor (such as the existence of pyramids in both Egypt and South America, prompting elaborate but unnecessary models explaining how pyramid-builders could have colonized the New World). Various kinds of diffusion were discussed, from cases in which ideas passed through a population on a person-to-person (word of mouth) basis, to the migration of ideas into new geographical areas. Stimulus diffusion, the topic of Kroeber's paper, occurs when an idea is adopted in a culture that is inspired by an idea introduced by another culture. Specifics may not be adopted, for whatever reason, but some part of the underlying premise is retained or adopted. The recipe for Country Captain Chicken is a simple example of stimulus diffusion; a southern dish of chicken, tomatoes, raisins, and curry powder, it was the result of Atlanta's prominence as a port in the spice trade, which brought southerners in contact with the ingredients of Indian cooking which they then used without consideration for the techniques and traditions of the cuisine itself.

Rogers also looked at the characteristics of ideas and technologies that will most influence their adoption: relative advantage, compatibility, complexity, trialability, and observability. Relative advantage is the improvement that an innovation offers, which is not always easy to identify—many hybrid tomato varieties offer an improvement in shelf life and durability, and were adopted for that reason; and yet heirloom varieties have been repopularized for their relative advantage of flavor over those hybrids. Compatibility is critical because some innovations can't be easily assimilated—locavorism (the "eat local" movement) is more easily assimilated in areas with long growing seasons and short winters, for instance. Complexity represents not the complexity of the idea, per se, but of the process of adopting it. Energy-efficient lightbulbs are more likely to be adopted than many other means of saving energy because, initial cost aside, switching to them requires no more effort than ordinary lightbulbs and no significant change of habit. Trialability is the ease of testing out the idea before committing to it. And observability is the idea's visibility to others; those energy-efficient lightbulbs won't be noticed by your neighbors and won't spark conversation with them, but a Prius might.

Various food practices of the green revolution have followed the processes of adoption and diffusion. Studies have shown that green farming practices have been incorporated into Brazil through the process Rogers describes. Early adopters are watched by their neighbors—people who aren't willing to risk changing the way they grow their crops until they've seen it work. They may perhaps observe for a few seasons to see if it will be successful. When results are yielded, they may adopt green practices in kind. Similar patterns have been observed in Iran, Indonesia, and southeast Asia, as green advocates have sought to encourage local farmers to turn away from pesticides and chemical fertilizers and have taught them how to achieve high crop yields using green methods.

See Also: Green Revolution; Holistic Management; Slow Food Movement.

Further Readings

Bell, Michael and Helen D. Gunderson. *Farming for Us All: Practical Agriculture and the Cultivation of Sustainability*. State College: Penn State Press, 2004.

Hall, Darwin C. and L. Joe Moffitt. *Adoption and Diffusion of Sustainable Food Technology and Policy*. New York: Emerald Group Publishing, 2002.

Kroeber, Alfred L. "Stimulus Diffusion." *American Anthropologist,* 42 (1940).

McIntosh, William Alex. *Sociologies of Food and Nutrition*. New York: Springer, 1996.

Rogers, Everett. *Diffusion of Innovations*. New York: Macmillan, 1962.

Sapp, Stephen G. and Peter F. Korsching. "The Social Fabric and Innovation Diffusion: Symbolic Adoption of Food." *Rural Sociology,* 69/3 (September 2004).

Bill Kte'pi
Independent Scholar

AGRARIANISM

Agrarianism refers to the philosophies and practices advocating for the continuity or revival of small-scale, sustainable agriculture (farming, gardening, herding) and the lifestyles associated with them. It encompasses both social and political perspectives and is rooted in ecological concerns regarding the industrialization of society in general, and of agriculture in particular. Central to the agrarian premise are an abiding respect for "traditional" or low–external input modes of farming and the moral rectitude and quality of life such heritages are said to engender. In addition, agrarianism can entail a revaluation of the traditional arts and crafts that have constituted subsistence or semisubsistence life. Agrarianists often posit themselves in opposition to hyperurbanized modernity, its alienating consumerism, and its industrialized and/or capitalist exploitation of natural resources. Despite elitist or anachronistic connotations, however, agrarian traditions have proven to be as diverse as they are universal and as current as they are ancient. Moreover, although seemingly innocuous and at times reactionary, agrarian movements have instigated reforms, revolts, and revolutions throughout history and across the world and remain a political if not radical voice for peasants' and farmers' rights as well as for environmental stewardship and justice.

Agrarian Diversity

Within this umbrella description, however, many varied strands of agrarianism have existed and have emerged more recently. Herein lie the multiple paradoxes of agrarianism: The term has carried a number of seemingly dichotomous associations, namely, that it fosters both strong communities but also rugged individualism, champions conservative autonomy as well as leftist populism, and was born of antigovernment (or specifically antifederalist) politics yet is also at the bedrock of various national patriotisms. It has been lauded as the pinnacle of self-reliance and independence, even as others argue that agrarianism is the humble recognition of one's dependence on the land and interdependence with neighbors. Agrarianism has been used to justify the primacy of private property as well as also the value of communal land tenure, exemplified in the persistent traditions of the Oaxacan *ejidos* to the *ayllus* of Quechua Peru and Bolivia.

Agrarianism has emerged again recently with these paradoxes renewed: It both speaks for the current gourmet of slow food as well as the rights of subsistence farmers stigmatized as "underdeveloped," "Third World," or poor. *Agrarianism* conjures extreme idealism and pragmatism, and although some have been accused of patriarchal nativism, many eco-feminists now embrace the name.

Agrarian Origins

Before and beyond agrarianism as a movement, however, was and remains the use of the term *agrarian* as an adjective. Agrarian societies are agrocentric and often practice ancient agricultural traditions indigenous to that ecosystem. Such communities have cultivated their lands according to agrarian calendars for millennia and have visual and musical artistic traditions correlating to the staple crops and local herbs. Myriad indigenous agrarian traditions around the world ground themselves in locally specific cosmovisions—each with respective intricate agriculturally oriented metaphysics.

Although thoroughly global in scope, within agrarianism's European lineages, Hesiod and Virgil remain enduring icons, inspiring the agrarian ideals of Thomas Jefferson, who integrated the yeoman farmer prototype into the founding literature and laws of the United States. Other prominent agrarians include the "twelve southerners" who in 1930 penned *I'll Take My Stand: An Agrarian Manifesto*. Writing in retaliation against what they considered the dehumanizing effects of postwar, northern industrialization, these writers condemned modernity, urbanity, Fordism, the Enlightenment, and the tyranny of science, factories, museums, and automobiles—all of which needlessly sacrificed the finer and more meaningful aspects of life such as craftsmanship, independent thought, silences, religion, the natural world, leisure, community, and the cyclical rhythms of agricultural life. Many of the arguments put forth in these essays—some precursors to current sustainability discourses—have been overshadowed by their elitist and racist context, wherein the lauded leisure and genteel landscape came at the expense of their disenfranchised, exploited, and unmentioned African American servants.

According to subsequent critiques, the value and values of agrarianism have not been able to overcome traces of this aristocratic, patriarchal privilege. Julie Guthman astutely addresses the consequences of romanticized agrarianism in many of her writings. In *Agrarian Dreams: The Paradox of Organic Farming*, she explains that in California, the romantic ideal of organic agriculture has obscured the political and economic reality of organic codification and the agribusiness it has spawned. This highly centralized and mechanized organics production system continues the notorious labor practices of its predecessors, remaining as socially and ecologically exploitative as its conventional counterparts. The organic label brands its populist, agrarian connotations even as it competes and succeeds in an aggressively neoliberal economy. In subsequent research, Guthman has shown how the agrarianist ideals also risk being compromised by rhetorical vestiges of the hypocritical moralism of their founding advocates of agrarianism, who were freed from the actual work of farming by the forced labor of their slaves. For her, the persistent questions of labor, inclusion, and social justice endure and must be the focus of current agrarian initiatives.

Agrarian Transitions

Nearly all civilizations and communities in the world could be classified as agrarian until the industrial revolutions of the 19th century concentrated the production of food and textiles, subsequently moving masses of farmers and herders to the cities to earn livelihoods in factories. The agrarian question, put forth in the late 19th century by Karl Kautsky, and later by Alexander Chayanov, pondered the fate of the smallholder farmers in the face of impending transitions toward capitalist, large-scale agricultural economies. Though they specifically were addressing the fate of Russian peasantry, their questions apply to rural communities and comparable agrarian conundrums across the world.

This "great transformation," as described by Karl Polanyi, has occurred globally in various fashions and to various degrees, though even today, agrarian communities persist around the world. Often, communities retain agrarian traditions, calendars, heirloom or native seeds, and the ecological knowledge that lies therein, even as they incorporate, employ, and are employed by aspects of modern, technological life. Nevertheless, the majority of agricultural systems around the world have undergone the transition from decentralized, small-scale, share-cropping cultivation to highly mechanized, monocrop production that is heavily dependent on what has been termed the "treadmill" of annually purchased chemical inputs.

Within this historical trajectory, the new agrarianist movements emerging around the world often invoke historical or even ancient agricultural principles but orient themselves toward current ecological crises and conundrums, arguing that the Earth's natural and human resources cannot continue fueling modern agriculture's voracious appetite for external inputs or absorbing its deleterious social and ecological externalities. Agrarianism posits that the fundamental values of a society are embodied in its relationship with agriculture. An industrialized agricultural system treats natural resources, its own people, and other societies as fungible elements of large machines. Writer and farmer Wendell Berry has become a veritable catalyst for current agrarian movements. His entire body of work exemplifies and elucidates agrarian philosophy, eloquently describing the "placeless and displacing" nature of industrialism's agriculture.

Agrarian Values

Ecologically, agrarians contend that living "off" the land allows people to live closer to natural processes, which in turn makes for a healthier life and relationship with the natural world. This correlation of environmental and human health unites sustainable agriculture's many incarnations, from agroecology to biodynamics to permaculture, which share the basic tenet that holistic agriculture is the crux of environmental balance, and thus salubrious people and communities (see Winona LaDuke, Paul Thompson, and the Land Institute among others).

Economically speaking, agrarianism aims theoretically for decentralized markets, localized or fair trade exchange, informal economies, barter systems, and semisubsistence production, while critiquing dominant concepts of the "creation of wealth" and unlimited growth. In contrast, some historic agrarian economies—feudal Europe and Asia, or the antebellum United States—were expressly inegalitarian: Those who worked the land did not own it or control the means of its production. Susan Mann and others have noted the agrarian origins of capitalism, whereas scholars like Wendy Wolford have noted how market-led agrarian reforms have privileged the agrarian elite while undermining grassroots initiatives for social and environmental justice, such as that of the Brazilian Landless Movement. Current forms of capital-based "agrarian" reforms—often World Bank policies—have been accused of fostering self-exploitation, displacement, and dispossession on the part of small-scale producers and large-scale laborers. Many note that such land redistributions, though called agrarian, belie agrarian principles: "a populist struggle for land that does not take into consideration power, social rights, and the historical struggle of small farmers and the landless could quickly become part of the neoliberal project and lead to increased political exclusiveness." Nevertheless, in current parlance, agrarianism usually conjures economic alternatives to the neoliberal world market.

This economic vision is irrevocably political. Agrarian movements such as Via Campesina contextualize sustainable agriculture in terms of food sovereignty—the right to grow and eat independent of world market prescriptions. Vandana Shiva, a globally renowned voice for eco-feminist agrarianism, contends that the monocrop, green revolution–based agricultural dictates perpetuate colonialist exploitation of local people, lands, resources, and knowledge. Gustavo Esteva has also launched a powerful critique against the rhetoric of "development" and the agricultural models embedded therein. Such advocates posit agrarian resurgences as antidotes to current environmental, political, social, and economic injustices. Though radical at times, agrarianism has also become more accepted in popular, political, and even educational spheres. This mainstreaming spans the global popularity of the Zapatistas to Yale's new Agrarian Studies Program.

Socially, agrarians endow farming with dignity and agency. Though at times essentializing the "peasant," agrarianism strives to reverse the social tendency to denigrate the labor and knowledge involved in small-scale agriculture and seeks to empower the *campesinas*. Historically, "agrarian" served as a synonym with the countryside, but current agrarian initiatives are often expressly metropolitan or suburban, such as urban garden movements, farmers markets, and community-supported agriculture programs. This expansion emerges partly from practicality—the world's urban majority—but in addition, the inherent class, cultural, and ethnic diversity of city-based agrarian projects mitigate the limitations of agrarian homogeneity and xenophobia.

Agrarianism entails a significant philosophical shift, moving farming from an unskilled occupation to a respected way of life, and agriculture itself from the periphery to the center of society and sustainability. Agrarians point to the honesty of such a transformation, as all food, drink, textiles, and medicines are directly or indirectly agricultural products. Moreover, they contend that such recognition would involve a transition in collective principles—from consumerism, exploitation, and selfishness to reciprocity, responsibility, and gratitude. Countering the stereotype of nostalgic pastoralists, agrarians around the world speak of the political, social, economic, and ecological urgency of such a shift and of agrarian reforms toward sustainable agriculture. Berry characterizes this as a momentous, critical struggle: What agrarians "have undertaken to defend is the complex accomplishment of knowledge, cultural memory, skill, self-mastery, good sense, and fundamental decency—the high and indispensable art—for which we probably can find no better name than 'good farming'."

See Also: Agrarian Question; Agribusiness; Berry, Wendell.

Further Readings

Carlson, Allen. *The New Agrarian Mind: The Movement Toward Decentralist Thought in Twentieth Century America.* New Brunswick, NJ: Transaction, 2004.

Freyfogle, Eric, ed. *The New Agrarianism: Land, Culture and the Community of Life.* New York: Island, 2001.

Rossett, P., et al., eds. *Promised Land: Competing Visions of Agrarian Reform.* New York: Food First, 2006.

Wirzba, N., ed. *The Essential Agrarian: The Future of Culture, Community, and the Land.* Lexington: University of Kentucky Press, 2003.

Garrett Graddy
University of Kentucky

AGRARIAN QUESTION

Historically, agriculture was based on family units and small-scale merchants. In the process of the modernization of European societies, the question came up about the future role of traditional agricultural production in the face of modernization. The question still remains relevant in multicultural societies with large tracts of land in the hands of small landholders.

Central to the agrarian question is the coexistence of three economic segments that articulate the flow of natural resources from remote rural areas and from the subsistence economies occupied by indigenous peoples and peasants, through the rural towns and medium cities that trade in natural resources in the mercantilist economy—where capital

Although the majority of the world's agricultural systems have transitioned to mechanized production, small-scale subsistence farming persists alongside it. These children were photographed near Cairo, Egypt, in June 2006, helping to harvest crops using traditional means.

Source: U.S. Agency for International Development

is based on social relationships, to when the goods arrive to the modern free market economy linked to the world economy through these exports.

The agrarian question can be summarized as asking what role there is for small-scale agriculture in the modernization process. Small-scale agriculture does not disappear but is relocated and reduced to the fringes of the modern valley, the rocky slopes, and the wild open nature, where modernization can reach if necessary in the form of a mine or an oil well. The free market, however, prefers to deal with the more trade-oriented actors from the mercantilist economy that operate in the subsistence economy and organize the production that is extracted from the areas occupied by societies in subsistence economies. Thus the extraction of resources is mediated by the mercantilist economic agents from rural areas and towns that gather, transport, and accumulate this small-scale production to be sold by free market actors that are linked to the world market, or to the regional and national markets.

With the increase in globalization in the 21st century, modern crops have expanded through areas of long-term occupation with populations that have progressively been incorporated into the free market and that are usually self-identified as mixed and no longer appeal to an indigenous origin. However, peasants and small farmers survive in remote areas where the market lacks the infrastructure or institutions to expand, although it can arrive if need be. Noncontacted indigenous peoples survive in more remote areas, particularly in tropical areas.

The Amazon is currently confronting the agrarian question in the face of globalization. The Accelerated Plan for Growth promoted by Brazil's National Development Bank is a $220,000 million investment in communication, transport, and energy infrastructure opening access to its neighbors into both oceans. In this case, migrants to the Amazon from recent decades will be faced with the arrival of fresh capital in search of land and resources. Most will sell and move to recently opened areas to repeat the cycle of opening a frontier.

In the face of globalization, probusiness governments aiming to attract investment usually leave no space for small land holders in their governments' development strategy. In such cases, the drive for modernization occurs at the expense of the traditional producers, who leave the land, turn to the cities, survive with their traditional production in the margins of the modern agricultural areas, or move to another region when possible.

Similarly, China has announced a prodevelopment economic package of around $500,000 million in a march into the west that aims to modernize the rural economies of this economic giant and to expand the free market economy of the coastal China.

Some actors from the subsistence economy can slowly adapt to the changing economic context, but most will suffer mainly a reduction in size and quality of their homeland.

Modern agriculture is characterized by export-oriented products that are aimed at the global market, from flowers to asparagus. Thus the younger generations of these subsistence economies, plus the migrants attracted to these areas by the availability of labor, will more easily adapt and benefit through education. However, many older peasants who cannot engage in the labor force will need to look for new land or migrate to the outskirts of towns and cities.

In many parts of the globe, small-scale agriculture feeds, educates, dresses, and provides health care for millions of rural populations—hence the relevance of the agrarian question from a public policy point of view. These populations can be classified as poor because of their low income, but these small-scale productions subsidize their low income with natural resources to provide for housing, food, energy, and health.

In the 21st century, the agrarian question also needs to be addressed from a global perspective. The recent growth of the world market has favored an increase in food consumption and diversity, while transnational corporations are playing a major role in organizing not only production but also the regulations that ensure growing profits while reducing labor force and intensifying land use. Since the 1990s, the global market has been characterized by an increase in consumption in some Least Developed Countries (LDCs) while other LDCs experience even greater food insecurity. The global food market offers opportunities for investors looking to supply to segments of the food market interested in diverse food exports. This has led to an expansion of export-oriented, capital-intensive production areas that are modernizing the agrarian landscape in growing economies. This process is aided by political support and recent free trade agreements that seek to ensure lasting viability of the terms of trade between industrialized and nonindustrialized nations.

Issues that highlights the role of transnational corporations can be observed in the process of current legal and political reforms aiming to increase opportunities for genetically modified organisms (GMOs), biofuels, and export-led agro-business. Harriet Friedman, a professor and researcher in the field of international regulation of food and agriculture, refers to the "private global regulation" that seeks to regulate agro-food conditions, and the organization of production and consumption in order to plan investment, supply, and marketing. Agrarian questions concerning the politics of capitalism, how competition will promote increased yields, how agro-food is accumulated, and how surplus is dealt with will remain highly relevant.

See Also: Agrarianism; Agribusiness; Land Reform; Peasant.

Further Readings

Goodman, David. *Globalizing Food: Agrarian Questions and Global Restructuring.* London: Routledge, 1997.
Kautsky, Karl. *The Agrarian Question.* London: Unwin Hyman, 1988.
Rossett, P., et al., eds. *Promised Land: Competing Visions of Agrarian Reform.* New York: Food First, 2006.
Saith, Ashwani. *The Agrarian Question in Socialist Transitions.* London: Routledge, 1986.

Carlos Antonio Martin Soria
Universidad Nacional Agraria La Molina
Instituto del Bien Común

AGRIBUSINESS

A portmanteau of the words agriculture and business, *agribusiness* generally refers to the range of industries and activities associated with farming, including the manufacturing, processing, marketing, storage and distribution of agriculture related commodities. However, the term *agribusiness* can have different connotations depending on the context and viewpoint. Several agribusiness philosophies and critiques from industry, environmental, Jeffersonian, Marxist, and labor perspectives are discussed here. The agribusiness industry itself promotes agribusiness as a natural progression in the development of modern agriculture. Under this definition, *agribusiness* simply means the business of agriculture. This includes inputs such as land, equipment, and labor, as well as the management, research, and development necessary to fully modernize the agricultural system by increasing efficiency and production. Environmental critics recognize flaws in the agribusiness model, as environmental concerns such as water and soil quality, ecosystems management and preservation, and biodiversity maintenance commonly are not accounted for in the large conventional farming and food processing operations that generally signify agribusiness. The Jeffersonian critique looks at agribusiness from the perspective of proponents of small-scale farms. Under the Jeffersonian definition, the term *agribusiness* signifies large-scale industrial and corporate farming and is contrasted with yeoman or self-sustaining farmers. This critique sees the development of agribusiness as destructive to the American farming ideal that includes agrarian lifestyles and values. In contrast, proponents of the Marxist critique look at agribusiness through the lens of capitalist production, examining food commoditization into the components of agricultural production processes and the consequent social division of labor. This critique focuses on the tribulations of agribusiness as profit is prioritized over consumer and worker health and the environment. A derivative of the Marxist critique, the labor movement focuses on the exploitation of workers by the agribusiness industry. Hence, the term *agribusiness* refers to the progression of the farm into a modern industrial enterprise and the positive and negative consequences of that progression.

Industry Perspective

From the perspective of the agribusiness industry itself, the term *agribusiness* refers to business-like agriculture including modern management and equipment. This definition includes the physical farm property, inputs, and the growing processes. These processes may include all stages of agriculture and food production, from the development of seed germplasm to the processing of food products. From this perspective, the assumption is that agricultural and food-processing businesses will increase in scale like any other business. The agribusiness model is seen as one of efficiency and a rational business choice. Issues of labor exploitation and land stewardship are not usually discussed in this version of the definition. Proponents of agribusiness are generally promoters of international agricultural commodity trade and agricultural agglomeration and development.

Some of the larger and better-known agribusiness companies include Monsanto and ConAgra Foods. Although these companies are multinational corporations, smaller family-owned operations may be termed agribusiness companies as well. Monsanto is representative of the production end of agribusiness, manufacturing agricultural technologies and inputs such as seeds, herbicides, and pesticides. ConAgra is a packaged food company, processing and marketing food products around the world. Both corporations market food and agriculture products under various brand names.

In this context, agribusiness is mostly a descriptive term and refers to the business involved in agricultural production. This may be very wide-ranging, including farming, seed supply, agrochemicals, farm machinery, wholesale and distribution, processing, marketing, and retail sales. Undergraduate and graduate degrees are available specifically for agribusiness at universities such as Texas Tech University and Kansas State University. The California Polytechnic State University has a Department of Agribusiness, boasting the largest agribusiness degree program in the United States. A degree in agribusiness typically includes courses in topics such as agricultural economics and business administration, combining agricultural management and technical skills with studies in business and economics.

Environmental Critique

Many environmental and local-food activists see agribusiness expansion as negative to ecological and ecosystem health. Agribusiness operators are primarily concerned with profiting from their production process. Farming ecologically, without synthetic agrochemicals, for example, is generally more labor intensive and therefore not cost-effective for many agribusiness operations. It should be noted that organic farms and input manufacturers, natural food processors, and other ecologically minded businesses are included under some definitions of agribusiness. Some agribusiness operations may find that being environmentally conscious may be profitable, as they may fit a niche market. Many environmentalists are skeptical about these businesses, however, because they may market themselves as ecologically sustainable, but because they prioritize the highest possible profit margin, they may not make sufficient efforts to curb the environmental consequences of their production.

Agribusiness operations include manufacturers and distributors of agrochemical products and genetically modified seeds, which many environmentalists see as threats to environmental and human health. Many large food-processing companies also use chemical additives to replace natural ingredients, as they are less expensive to produce. Agribusiness degree programs tend to promote the use of chemical or conventional farming techniques and food processing.

Depending on the business model and the size of the operation, some agribusiness promoters and operators may manufacture, market, and trade their agricultural commodities globally and take advantage of global agricultural markets to the best of their ability. Local-food activists conceive of global agribusiness producers and distributors as environmentally destructive because of the resource cost and pollution resulting from transporting food globally.

Jeffersonian Critique

This critique of agribusiness uses the Jeffersonian ideal of the yeoman or self-sustaining farmer. Thomas Jefferson, third president of the United States, promoted the notion that the United States would grow strong and prosperous via small family farms, whereby men cultivated their own land. He critiqued the growth of the U.S. economy through manufacturing and insisted that the country must maintain its small family farms. Those who critique agribusiness from this perspective are generally against applying a liberal economic model to farming and use the term *agribusiness* to suggest a corporate or industrial farming model. Proponents of this view of agribusiness include activists working to support small-scale farms, some small-scale farmers who have inherited farms, and back-to-the-landers—recent generations of farmers who do not come from agricultural backgrounds.

In comparison with Jeffersonian farms, farms run by agribusiness are large-scale operations. Promoters of the Jeffersonian critique see modern agribusiness operations as monopolistic entities that have destroyed small farms and rural or agrarian-based values and lifestyles. They see the development of agribusiness as a process by which small-scale family farmers are pushed out of the market by large corporations.

This critique also suggests that agribusiness operations have negative environmental and social impacts. The notion of the Jeffersonian or yeoman farmer implies that one physically lives on the land that is farmed and therefore keeps a watchful eye on the effects of his or her agricultural cultivation on the local flora and fauna to sustain their rural livelihood and community. Agribusiness owners and operators, in contrast, usually do not live on the land they own and/or manage. This also means they may not be connected to the local community and ecology. Large-scale farm owners may own many plots of land in different areas and regard the cultivations site(s) as real estate investments that can be bought and sold according to the success of production, with little care as to the long-term effects of agricultural processes on the land and local people. Agribusiness managers therefore may have less investment in the long-term viability of agriculture in the region and the environment and social effects of cultivation.

Marxist Critique

Many social scientists have understood the concept of agribusiness specifically as an accumulation of capital that includes the means of production (a land-owning class), the exploitation of wage labor, and production for profit. This is a classic Marxist analysis of the industry and complements the labor critique in that it emphasizes class relations. The vantage point of a Marxist critique views the small-scale farmer as part of a larger capitalist system, which contrasts with the Jeffersonian ideal of the independent farmer. This critique states that under an agrarian capitalist system or agribusiness model, food quality and safety may be compromised for profit. Many social scientists have contributed to this critique and the related literature, including David Goodman, Jack R. Kloppenburg, Phillip McMichael, Harriet Friedmann, Richard Walker, and Michael Watts.

Geographer Richard Walker contests the term *agribusiness* itself. Instead, he argues that the term should simply be *agrarian capitalism*, being that in modern times, agriculture and the countryside have always been dominated by the capitalist mode of production. This analysis points to the commodification of food, land, and labor as part of creating a fully capitalist agricultural system. In the agrarian capitalist system, food, land, and labor's values are measured by their ability to be bought and sold—also called their exchange value—rather than their value as a usable entity, which is called their use value. This critique points out that when the capitalist mode of production is applied to the food and agriculture system, it disrupts the human connection between land and labor and separates people from their food supply. A theme that resonates throughout these critiques of agribusiness is the separation of the worker from the land, which was termed *primitive accumulation* by Karl Marx.

Under agrarian capitalism, increasing quantity and rate of production is prioritized. This includes increasing labor productivity and using new forms of plants and equipment—also known as the industrialization of agriculture. Industrialization includes the mechanization of labor, using machinery to aid or replace human labor. This critique understands the industrialization of agriculture to be inevitable under a capitalist system.

The risk involved in the industrialization and commodification of the food and agriculture system is that the focus of food production becomes one of profit and efficiency, not one of the health of the consumer, worker, or environment, as discussed in the environmental critique. The primary purpose of food production—providing sustenance for people—also becomes obscured. This can be observed in new food processing technologies, such as the use of food additives and the substitution of chemical ingredients for natural ones. The controversial use of genetically modified organisms in food and agriculture products has been critiqued as a highly commodified form of crop production and as problematic in its effects on agricultural sustainability and consumer health.

Labor Critique

Building on the Marxist critique, the critique of agribusiness by proponents of the labor movement looks at how the scaling-up of agriculture, especially in the United States, has increased the division of labor and the exploitation of the laboring class in agricultural production. The expansion of agricultural production and the issue of exploited farm labor first brought widespread attention through John Steinbeck's work, especially his acclaimed novel *The Grapes of Wrath,* published in 1939. Authors and academics such as Carey McWilliams, Miriam Wells, and more recently Julie Guthman, among many others, have also explored how agribusiness, or what many term *industrial agriculture*, exploits the laboring class. Much of this work highlights agribusiness and labor relations in California in particular, as it is the home to the largest agricultural economy in the United States.

Such authors have worked to shatter the idealized Jeffersonian notion of farming as the core of American agriculture, pointing to the hidden social costs and consequences of the expanding industrial style of U.S. agriculture. Focusing on class relations as well as on issues of race, gender, and immigration, the agricultural labor movement and its proponents critique the way that agribusiness creates a class system in which the "grower" or owner does not engage in physical farm labor at all. In large-scale farming, the grower's job is to manage workers and their labor. The people who actually work the land are therefore not landowners or managers themselves. Wages for agricultural workers have been historically lower than in other industries, and benefits such as workers' compensation and health care have been virtually nonexistent.

Agribusiness operations have a history of hiring immigrant labor. These workers can be paid lower wages and are unable or unlikely to organize or protest against working conditions, as many are undocumented workers who fear deportation. The Bracero Program, a temporary labor program, was an exchange between the United States and Mexico from 1942 to 1964. It has been attributed as giving a "jump start" to the U.S. agricultural industry by providing around 4.5 million guest workers from Mexico to work on U.S. farms for low wages. Although Mexican immigrants had been coming to work in U.S. agriculture for decades before the program began, during this time a strong pattern of immigration was established between Mexico and the United States that exists to this day.

Conclusion

In contrast to the perspective of the agribusiness industry itself, which promotes the industrial growth and modernization of agriculture as a generally positive process, the critiques covered here point to the flaws and risks in adopting a large-scale industrial food and agriculture system. For the food and agriculture industry to grow as it has, sacrifices to rural livelihoods and communities, consumer health and safety, worker rights, and the environment will continue to be made.

See Also: Bracero Program; ConAgra; Genetically Modified Organisms; Horizontal Integration; Labor; Modernization; Monsanto; Seed Industry; Yeoman Farmer.

Further Readings

Goodman, David, et al. *From Farming to Biotechnology: A Theory of Agro-Industrial Development.* Oxford: Basil Blackwell, 1987.

McMichael, Philip. *Food and Agrarian Orders in the World Economy.* Westport, CT: Praeger, 1995.

McWilliams, Carey. *Factories in the Field: The Story of Migratory Farm Labor in California.* Berkeley: University of California Press, 1999.

Walker, Richard. *The Conquest of Bread: 150 Years of Agribusiness in California.* New York: New Press, 2004.

Laura-Anne Minkoff-Zern
University of California, Berkeley

AGRICULTURAL COMMODITY PROGRAMS

Agricultural commodity programs have taken several forms in the United States. They are usually programs to meet the problem of overproduction.

Historically, American farmers were faced with a new continent to farm. At the beginning of the republic, it took about four farmers to raise enough to feed one extra American. However, in the 1860s, several laws promoting agricultural expansion were passed. One was the Homestead Act of 1862, which gave land—a quarter-section (160 acres)—as a freehold grant to those willing to settle on it to farm or ranch it. After a few years of occupation, the land became the property of the settler. During the great expansion of the railroads following the Civil War, a checkerboard pattern of sections along the railroad was used to promote railroad expansion. One section (640 acres on 1 square mile) would belong to the railroad, and the opposite section to the government. Usually both sold some of their land to immigrants.

The Homestead Act, the use of public lands for financing railroad expansion, was matched with the Morrill Act of 1862, which was the authority for the creation of a number of land grant colleges in the last decades of the 1800s. From the investment in the agricultural and mechanical land grant colleges came many new scientific advances in agriculture. The result was a vast expansion of U.S. agriculture across the continent. However, it was really an overexpansion.

From 1900, people began to leave their farms for a more profitable life in the towns and cities. The problem was in making a living on a farm: With the improvements in agriculture available to farmers, crop yields continually increased; however, in bumper crop years, prices became depressed as an excess of cereal grains, cotton, or other agricultural products overwhelmed the markets. The depressed prices would drive people off the land and into bankruptcy.

In wartime, prices were high, but times like the 1920s were again years of overproduction. When the Great Depression arrived in 1929, there were cases of farmers burning their crops or slaughtering their herds to keep the surplus from the market. The strategy was outrageous to many because there were many people going hungry who could not afford to buy food.

The administration of President Franklin D. Roosevelt began to address the problem of farm surpluses with a program of acreage allotments. Farmers were given a fixed amount of land that they could farm for any one crop. If the allotment was for 200 acres of corn, no more could be planted or federal agricultural agents would see that the surplus was plowed under. Even if the overage was for personal consumption, the overage was illegal. However, the more the land was limited, the more intensely it was farmed, causing production to increase on smaller amounts of land.

World War II was a time during which agricultural production was encouraged to feed vast numbers of military personnel and the starving millions in war-torn areas. However, with the end of the war, the problem of agricultural surpluses returned. The federal government instituted a commodity program under which farmers were allowed to store their crop (commodity). If the price rose, the commodity could be sold at a profit; however, if the price remained static or declined, the farmer could sell the crop to the government at a commodity support price.

Eventually the government owned and was storing billions of dollars worth of commodities such as wheat, corn, barley, oats, rice, peanuts, sugar, oilseeds, feed grains, sorghum, tobacco, soybeans, milk, cheese, and butter. Many of these commodities were given to those on welfare. Some foods such as cracked wheat were not familiar to many people, who simply fed it to their chickens. Others received honey or other foods once a month in quantities that could feed them for days.

Public school lunch programs were also given such commodities as field peas and molasses. These would be served cooked together along with bread and milk. Often other surplus foods were fed to U.S. schoolchildren.

The agricultural commodity programs have lasted for the same reason that they were instituted: Farm products often lack "parity," which is a price comparable to the returns from performing urban or industrial jobs. To see that food shortages do not occur, commodity programs make political sense. However, they interfere with market forces that fix the price so that "inefficient" farmers are forced out of business. The problem with that application of a free market solution is that farm problems arise from an excess of efficiency in which a U.S. farmer today produces enough to feed well over 100 other Americans.

The administration of farm commodity programs is the responsibility of the U.S. Department of Agriculture (USDA). The basic laws are the Agricultural Adjustment Act of 1933, the Agricultural Act of 1949, and the Commodity Credit Corporation (1933). The USDA still subsidizes two dozen commodities. These include, in addition to those previously mentioned, wool and vegetable oils. The subsidy puts a floor on the price of a commodity. For the farmer, it is a guarantee that the price cannot fall below the subsidy price.

Beneficiaries of the commodity subsidies have changed over the decades. Much of the money goes to large agribusinesses. However, the United States and global consumers have benefited from cheap-to-inexpensive foods that have risen in quality as well as in quantity. The great increase in global meat consumption developed from surplus grains, which was a result of commodity programs that have kept the price of feed grains low and in abundant supply. Other beneficiaries have been nursing mothers, through the Special Supplemental Nutrition Program for Women, Infants, and Children that supplies them with milk. The poor also benefit through distributions of free surplus foods.

Commodity programs include the National School Lunch Program, Child and Adult Care Food Program, and the Summer Food Service Program. The USDA's Food Distribution Division of the Food and Nutrition Service coordinates the distribution of commodities that provide meals to needy students in nearly 100,000 public and private nonprofit schools. These programs also support U.S. agricultural producers, who receive cash for the products for the meals and other nutrition programs.

Regions of the world that are plagued with poverty, war, or civil strife are often beneficiaries of food aid drawn from both U.S. and European food surpluses. The USDA operates the Foreign Agricultural Service, which provides U.S. commodities to millions of hungry people each year. It operates the Food for Progress Program, the McGovern-Dole International Food for Education and Child Nutrition Program, and the Food for Peace Program. It also cooperates with the food programs of the United Nations and with private voluntary associations to supply food to people around the world who are in need.

Beginning in 2005 there was a rise in food prices on a global basis. One reason for this was the congressional mandate that ethanol be used in gasoline. Derived from corn, the use of ethanol diverted about a fifth of the corn crop from the global marketplace. The rise in food prices, along with the global financial crisis that began in 2007, contributed to a significant increase in the number of people both in the United States and globally in need of food aid.

The European Union supports farm subsidies, but as in the United States, they are a matter of political controversy. The opponents usually point to the costs of the programs as a waste. However, the proponents see the programs as a way to keep farmers in business and to ensure adequate supplies of food. A pertinent objection is that dumping cheap surplus commodities on Third World markets depresses prices and hinders the development of local commodity production. Efforts to develop market-government joint programs have met with some success.

One fact that is important about surplus commodities is that the surpluses are a food/commodity bank that is available in times of famine or other forms of suffering. Without food surpluses, famines will occur because there will be inadequate supplies of commodities for people or their animals. Famine conditions promote civil unrest leading to wars.

See Also: Agribusiness; Department of Agriculture, U.S.; Land Grant University; Public Law 480, Food Aid.

Further Readings

Blank, Steven C. *The Economics of American Agriculture: Evolution and Global Development.* Armonk, NY: M. E. Sharpe, 2008.

Cochrane, Willard W. *The Curse of American Agricultural Abundance: A Sustainable Solution.* Lincoln: University of Nebraska Press, 2003.

Ferris, John N. *Agricultural Prices and Commodity Market Analysis*. East Lansing: Michigan State University Press, 2005.

Hadwiger, Don F. *Federal Wheat Commodity Programs*. New York: John Wiley & Sons, 1970.

O'Donnell, Christopher. *Commodity Price Stabilisation: An Empirical Analysis*. Surrey: Ashgate, 1993.

Schnepf, Randy. *High Agricultural Commodity Prices: What Are the Issues?* Hauppauge, NY: Nova Science, 2008.

Wright, Brian D. *Reforming Agricultural Commodity Policy*. Washington, D.C.: American Enterprise Institute for Public Policy Research, 1995.

Andrew Jackson Waskey
Dalton State College

AGRICULTURAL EXTENSION

Drawing its name from the act of "extending" scientific knowledge to the general public, agricultural extension serves as the bridge between scientific research and farmers. Beginning in the late 19th century, government-funded agricultural schools and research stations developed practices and technologies that were then transferred to farmers by extensionists. This top-down model was used to disseminate new agricultural technologies and was vital to the scale-up of agriculture in the United States. A similar model of extension was used in the global South by colonial administrations and independent governments alike to increase export crop production. Cuts in government spending under structural adjustment led to the decline of government-funded extension and a growing role for nongovernmental organizations (NGOs) in many such countries. The failures of extension and new technologies to better the lives of small

Agricultural extension has grown more collaborative, as suggested by this photograph of a farmer (second from left) consulting with an extension agent, a systems analyst, and a mechanical engineer about conditions in his peanut field in Terrell County, Georgia.

Source: U.S. Department of Agriculture, Agricultural Research Service

farmers led to calls for greater farmer participation and an overhaul of the top-down extension model. Its scope was expanded to include other aspects of rural development. Although top-down transfer of research and technology—and biotechnology, in particular—persists, agricultural extension has become a more collaborative and participatory practice, where farmer knowledge and experience play a significantly larger role.

The dissemination of agricultural information to farmers is as old as civilization itself. Bountiful harvests fed growing urban populations, provided tax revenue to governments, and fueled armies and the expansion of empires; rulers thus had a vested interest in ensuring good production. Examples of cropping calendars and recommendations for improved farming practices were widespread in ancient Mesopotamia, Egypt, Greece, Phoenicia, and Rome. Agricultural research and extension began in China as early as the late Han Dynasty. In Europe, the development of printing technology led to the widespread distribution of treatises on crop and livestock husbandry during the Renaissance. During the Age of Enlightenment, agricultural clubs were founded by gentlemen farmers interested in applying scientific methods to production on their estates. By the early 19th century, itinerant agricultural teachers throughout Europe and North America were hired by landowners to educate their tenants on improved production techniques.

Birth of Modern Agricultural Extension

The financial support of the state was central to the development of modern agricultural extension. By the middle of the 19th century, most European countries had agricultural schools conducting training and research and disseminating ideas through publications and fairs. The first wholly state-funded agricultural extension service was established in France in 1879 to keep farmers up to date on the latest discoveries in agricultural science. Britain soon followed suit. In the United States, formal state-sponsored extension began with a series of federal laws establishing an infrastructure for agricultural research, education, and extension. The land grant university system, a network of state colleges teaching agriculture and mechanical science, was created with the signing of the Morrill Acts of 1862 and 1890; state agricultural research stations were established under the Hatch Act of 1887. The Smith-Lever Act of 1914 formalized cooperative extension—an organizational hierarchy linking the federal and state Departments of Agriculture and land grant universities. Under this system, a network of extension agents working from county offices disseminated agricultural research conducted at the universities and state research stations.

In the global South, agricultural extension was central to European colonial projects during this period. Colonial economies in Africa and Asia were largely based on the export of raw agricultural products destined for factories and mills in the urban centers of Europe. Agricultural schools, research stations, and extension services were established by colonial administrations to increase production of commodity crops. The primary role of extensionists was to provide farmers with highly subsidized inputs and credit. Despite being viewed by many as vestiges of colonialism, the extension infrastructure largely remained intact in many African and Asian countries following their independence in the mid-20th century. Extension was also widespread in Latin America during this period, having been initiated to support export crop production.

A top-down "transfer of technology" model defined agricultural extension worldwide for much of the 20th century. In the United States, cooperative extension played a central role in the development of industrial agriculture. Following Agriculture Secretary Ezra Benson's warning in the 1950s to farmers to "get big or get out," the scale-up of production was made possible through the spread of new technologies to farmers. Mechanization, hybridized seeds, fertilizers, and pesticides were developed by land grant universities, promoted by extension agents, and sold by agricultural supply firms. By the late 1970s, however, the land grant–cooperative extension system was criticized for playing to the interests of agribusiness at the expense of the environment and of small family farms unable to keep up with capital-intensive demands of the new technologies demanded by economies of scale. The unidirectional flow of technology ultimately privileged only the wealthiest farmers.

Technology and the Green Revolution

This same model of technology defined extension in the developing world during the green revolution. Working with researchers from a network of international agricultural research centers in Asia and Latin America (and modeled after the U.S. land grant system), extensionists disseminated technology to middle- and large-scale farmers in the highly productive "breadbaskets" of the developing world with the assumption that technologies would diffuse down to poorer farmers who were considered backward and in need of modernization. Extension was highly centralized under this top-down model and was rarely present in more isolated areas. Although the green revolution was widely successful for large-scale production on high-quality land, small-scale subsistence farmers living on marginal land reaped few of its benefits.

Changes in the global economy forced a massive transformation of agricultural extension in the global South during the late 20th century. The debt crisis of the late 1970s and the drop in commodity prices in the early 1980s undermined postcolonial agricultural export economies. Under the structural adjustment programs that followed in the 1980s and 1990s, agricultural extension services were severely gutted or decommissioned. Finally, the restructuring of global agrifood commodity chains in the 1990s saw demands for traditional export crops such as grains and fiber crops replaced by new "cool chain" markets for fruits and vegetables.

These structural changes, alongside the failure of small farmers to adopt new technologies, underscored the weaknesses of the top-down model and called for a more equitable partnership between agricultural extension and farmers. Beginning in the late 1970s, many rural development workers affiliated with peasant organizations, and NGOs were heavily influenced by the popular education theories of Paolo Freire, which underscored the primacy of local culture and knowledge. Incorporating his ideas of people-centered liberation, many NGO workers began to promote simple but effective farming techniques appropriate for local conditions, including biological soil and pest management and soil conservation. Agricultural extension slowly evolved to encompass rural development more broadly, emphasizing livelihoods, economic development, and women's empowerment. Farmer participation began to take a more prominent role in research and extension.

Agricultural extension worldwide has been revolutionized by this populist "farmer first" approach. Considerably more participatory than its top-down antecedents, extension has become more horizontal in its approach both in the global South and in the industrial North. Farmer-to-farmer exchanges and field days serve as forums for information exchange. Indeed, the role of extensionist has shifted largely from teacher to facilitator, with an emphasis placed on communication, marketing skills, and networking rather than on transfer of technology. Nevertheless, echoes of top-down extension remain with the promotion of genetically modified crops by the land grant system and the new green revolution for Africa.

See Also: Adoption-Diffusion; Agribusiness; Commodity Chain; Debt Crisis; Department of Agriculture, U.S.; Export Dependency; Green Revolution; Land Grant University; Mechanization; Peasant.

Further Readings

Scoones, Ian and John Thompson, eds. *Beyond Farmer First: Rural People's Knowledge, Agricultural Research and Extension Practice.* London: Intermediate Technology, 1994.
Swanson, Burton E., et al., eds. *Improving Agricultural Extension: A Reference Manual.* Rome: Food and Agriculture Organization of the United Nations, 1997.

Nathan C. McClintock
University of California, Berkeley

AGRODIVERSITY

Based on the principle that diverse systems are more resilient and adaptable, agrodiversity contrasts with industrial agriculture. Although the green revolution increased food harvests in terms of quantity and efficiency, critics point to the simplification of ecological systems, loss of food security, and centralization of technological control as problems that can be attenuated through approaches incorporating agrodiversity. Rather than adopting universal approaches to agricultural production, agrodiversity incorporates multiple organisms, landscapes, technologies, and management strategies to ensure farm livelihood. Components of agrodiversity include agrobiodiversity, agroecosystem management, and farm organization.

Agrobiodiversity refers to the ongoing and simultaneous cultivation and harvest of multiple organisms. Agrobiodiversity is a scalable term applied to differences among and within species. An agrodiverse system may have many different organisms—various species of field crop interplanted with trees accompanied by livestock production—or may have broad genetic variation within a single species—for example, an orchard with many varieties of apple.

Agroecosystem management refers to various cultivation strategies tailored to distinct situations of soil fertility, moisture, temperature, and pests. Mountain agriculture provides many examples of microclimatic cropping; for example, big-seeded corn in valley bottoms and small-seeded, drought- and frost-tolerant corn on valley slopes. Intermingling different species and varieties of plants slows the spread of pests. There are innumerable ways to manage soil fertility and moisture using terraces, channels, mulches, and green manures.

Farm organization refers to the partitioning of property and labor resources. Land might be held communally, corporately, or privately and might be worked through a variety of labor arrangements such as sharecropping, reciprocal exchange, and wage or contract farming. Labor and land resources can be organized for multiple small harvests to satisfy household needs or for a single large harvest destined for market. Agricultural production may be the basis of farm livelihood or merely supplemental to wage income.

Ecological Systems

Monocropping—cultivating one variety of plant across an entire field—is fundamental to modern agriculture and contrary to agrodiversity in its simplification of biotic systems. The uniformity of a monocrop makes management of labor and chemical inputs simple but raises concerns for long-term soil health related to biota and nutrient availability.

Nonpest and beneficial organisms are collateral casualties of monocropping. Because such organisms play an important role in nutrient cycling and natural food webs, their loss undermines overall ecosystem sustainability. Mounting evidence indicates that diverse agricultural systems maintain soil health better than simplified ones.

Agrodiversity advocates point to contradictions in the systemic viability of modern agriculture. An industrial crop variety is typically productive for less than 10 years before pests adapt to make commercial production unprofitable. New crop varieties for industrial agriculture are developed using genetic material from unrefined, domestic, and wild strains. This reliance on genetic diversity for breeding simultaneous with homogenization of farming systems is inconsistent with sustainable resource management.

Food Security and Risk Management

Agrodiversity is associated with risk management rather than with maximum economic yield. In monocropping, a narrowly prescribed regime of fertilizer and pesticide application generally creates a bounteous harvest; however, pests can move easily through the crop, increasing the risk of catastrophic failure. A dramatic example of this is the U.S. corn leaf blight epidemic of 1970. An agrodiverse system contains a mosaic of genetic and management adaptations. Although a certain organism or cultivation strategy may not be successful for a given production cycle, many within a diverse system will be, thus ensuring a harvest of some type. Single-commodity farms are vulnerable to changing markets; spreading labor across various production activities ameliorates economic shortfalls. Variation in the farming system ameliorates negative outcomes rooted in dynamic ecological and economic conditions.

Technology Concentration

Modern and commercial agricultural research focuses on the model of monocropping across vast acreage. This technology, emphasizing hybrid and transgenic plants, is created mainly by Western research institutions, which hold the patents for new plant varieties, fertilizers, and pesticides. The concentration of technology in economically developed countries and its distribution to farmers engaged with global markets makes it difficult for small-scale farmers in less-developed countries to compete economically.

Before the green revolution, farmers, rather than research scientists, accomplished most advances in plant breeding. The advent of advanced hybridization, continuing today with transgenic technology, shifted the loci of agricultural innovation from numerous farmers' fields to a few institutional research sites. The collapse of in-field innovation at multiple sites was accompanied by a universalizing approach to agricultural production that was dependent on irrigation, petrochemical fertilizers, and pesticide inputs. As in-field innovation on numerous farms declines, opportunities to capture the beneficial diversity resulting from coevolution among crops and pests also decline.

Barriers to Agrodiversity

Barriers to adopting agrodiversity for modern agricultural systems are legion. Agrodiverse systems are not well suited to mechanization or industrial food processing. Interplanting, varying crop maturation periods across a field, and diverging input requirements are labor- and management-intensive. Diverse foods are difficult to market and incorporate into industrial food processing systems, which makes them irrational for short-term economic production and development strategies. As a result, there is relatively little institutional or commercial research dedicated to agrodiversity as a farming process or model; diversity conservation paradigms in agriculture emphasize seed and gene banks to preserve dwindling genetic diversity rather than looking to fields as centers of innovation.

See Also: Agroecology; Agrofood System (Agrifood); Green Revolution.

Further Readings

Brookfield, Harold. *Exploring Agrodiversity*. New York: Columbia University Press, 2001.
Brookfield, H. C., et al. *Agrodiversity: Learning from Farmers Across the World*. Tokyo: United Nations University Press, 2003.
Zimmerer, Karl. *Changing Fortunes*. Berkeley: University of California Press, 1996.

J. Anthony Abbott
Stetson University

AGROECOLOGY

Agroecology is the study of agriculture through an ecological lens; that is, by treating agricultural systems as ecosystems managed by humans for food, fiber, and fuel production. Ecosystems have both biotic and abiotic components: They are composed both of organisms and the elements of the environment in which they live. Agroecosystems include organisms such as crop plants, livestock, weeds, vertebrate and invertebrate pests, bacteria, viruses, and fungi, as well as inorganic components such as water, minerals, air, light, energy, and the nonliving soil matrix. Each agroecosystem also has a specific geographical, economic, and social context; that is, it is located in a particular landscape, economy, and set of communities ranging from local to global. To underscore the importance of these nonagronomic features and the breadth of scope, some agroecologists define their field as the ecology of the food system.

The field of agroecology stems from the disciplines of ecology and agronomy. It is similar to other types of applied ecology, such as restoration ecology, in its work to use ecological principles to improve managed ecosystems, as well as to use managed ecosystems as experimental sites to test ecological principles. This focus on application may make it seem to closely resemble its agronomic roots; however, agroecology differs from agronomy in at least two ways. First, it more purposefully includes the human component of agroecosystems, reaching across disciplinary boundaries to include sociological, anthropological, and economic analysis of agricultural problems and potential solutions. Second, where agronomy tends toward reductionism, agroecology approaches farming systems from a holistic perspective. Instead of confronting a particular management challenge, such as a weed, an insect pest, or the need for a particular nutrient concentration in the soil to support a potential maximum yield, with a single mechanistic solution, such as an herbicide, insecticide, or fertilizer, and then trying to minimize the solution's negative effects, agroecologists look for strategies that treat the problem as part of the system. Such strategies might imitate natural ecosystems in some way or draw on knowledge from or approaches used in traditional agroecosystems. They also might be multifaceted, trying to meet multiple goals such as improving soil fertility while disrupting several pest cycles. For some agroecologists, the ultimate goal is to create agricultural systems that have cycles and processes that imitate natural systems with reduced human intervention. This might include systemic regulation of pest populations or of nutrient cycling.

There is no single group of agroecological production methods. Agroecology is not synonymous with organic, biodynamic, or other specific sets of codified or culturally defined practices but is a way to evaluate one or more practices within a given agricultural system. For example, there are several interacting components to consider when evaluating shade-grown coffee production from an agroecological perspective. Researchers have found that growing coffee in shade can minimize disease and support noncrop biodiversity, including species richness of birds, ants, and native trees. Increased abundance and richness of bird and ant species can in turn provide ecosystem services such as a reduction in invertebrate pest damage or outbreaks. Reduced pest damage can increase the yield of saleable fruits, and sales of noncoffee products from the shade vegetation, such as fuel wood and construction materials, can provide additional farm income. However, certification costs (e.g., for marketing under a bird-friendly or organic production label) can reduce the final net income, and certification schemes themselves may be susceptible to market forces. There are also questions as to whether certification programs directed to geographically distant consumers can fundamentally transform the relationships with and vulnerability of farmers and thus the biodiversity associated with the shade production system. Agroecologists may examine both these natural and social factors of shade-coffee agroecosystems and work with people along the commodity chain, from farmers to consumers, to look beyond short-term yield and economics and to develop more sustainable systems appropriate to each location.

Overall, agroecologists view their discipline as an agent of social change with its sights set on sustainability. It is a very activist-oriented field in which researchers are working with farmers and others in the food system to better understand the system's many natural and social components and processes and thus create a way to supply adequate amounts of wholesome food that is justly produced and distributed with minimal negative effects on the environment.

History

The initial development of the field of agroecology is commonly traced to agronomist Karl Klages, whose 1928 article concerning the training of agricultural scientists described what he called "ecological crop geography." The root of "ecology" is the Greek *oikos*, meaning "house," and so the work of ecologists is ultimately to study organisms in their natural homes. In this sense, Klages pointed out, most agronomic studies draw on ecological principles as researchers work to encourage optimal growth of crops in the field. However, even by the 1920s, Klages saw agronomy becoming focused primarily in the natural sciences and suggested ecological crop geography as a way to consider the relationships of crop plants to their physiological environment as well as to what he later termed the *social environment*, including economic, political, historical, technological, and social forces. He was particularly interested in identifying which crop plants are best suited to a given region and then using that knowledge to develop government policies that support the production of best-adapted plants and curtail or eliminate production of crops that are grown in "artificial" environments, relying on high external support.

Although Klages introduced the core concepts of this field, the term *agroecology* first appeared in print in a 1928 book by agronomist Basil Bensin. The Russian-born Bensin, who also did agronomic work in Czechoslovakia and in the United States, adapted the term from the Czechoslovak Botanical Society. Similar to Klages, he was interested in the adaptations of crop plants to different agricultural regions—what he called "agrochora." Bensin wanted to compile data for each crop type grown in each agrochora throughout the world. The data would include crop characteristics such as the growing period and heat units required for fruiting and flowering, along with local climatic data, what varieties of the crop are used in the region, and specific cultivation practices; for example, the date and method of planting, the soil fertility practices used, and how harvesting is done. Bensin, however, did not mention social or other nonphysiological questions.

The development of agroecology proceeded slowly from the 1930s through the 1950s. Ecological Society of America President Herbert Hanson called on his disciplinary colleagues to conduct research on agricultural systems in his 1938 address but saw this work as separate from, though related to, that of agronomists, geographers, sociologists, and economists and actually disagreed with the use of the term *agroecologist*, favoring simply *ecologist* instead. Three influential books appeared during this period: a crop ecology compendium by Argentinian agronomist Juan Papadakis in 1938, Klages's work on ecological crop geography in 1942, and a book on agricultural ecology by Italian meteorologist Girolamo Azzi in 1956. Meanwhile, the work of many agronomists became increasingly focused on production, as the chemical fertilizer and pesticide by-products of the world wars transformed agriculture in the United States and other developed countries.

Agroecology's growth gained momentum in the 1960s and 1970s. Agronomists had produced most of the field's early writings, but a growing interest in systems-level approaches and concern about the negative environmental impacts of industrialized agriculture brought ecologists back to the table. The new leadership from this community was evidenced in part by the publication of a book on agroecology by German ecologist Wolfgang Tischler, who also taught courses on the ecology of crop plants, and the foundation of the journal *Agro-Ecosystems* in 1974, with British ecologist John Harper as its editor.

In the opening editorial of *Agro-Ecosystems*, Harper discussed several themes that remain important in agroecological research. He called for studies on the effects of high fertilizer application rates, and particularly the effects of the leaching of excess nutrients into neighboring freshwater ecosystems. He also called for research into the effects of pesticides on natural ecosystems, especially on wildlife, and on the relationship between diversity and stability, especially given the recently reduced germplasm diversity of the major cereal crops. Reflecting the interest in systems-level approaches, Harper noted that research must be conducted on the whole of a system as well as on its individual components. To do so, he wrote, requires a multidisciplinary approach.

Two other system-level agricultural journals were founded in the 1970s. *Agriculture and the Environment* (1974) published papers on the interdependence of food production and the environment, with the goal of adding to knowledge that would help achieve a balance between the two spheres. *Agricultural Systems* (1976) was created as a place for papers that combined economics and social sciences with biology, addressing field scale through global problems. British agricultural scientist Colin Spedding, editor of the latter, described agricultural systems research as a bridge-building subject that reaches across disciplines and challenges traditional ideas about how to conduct research.

Most of the early work in agroecology came from the United States, Britain, and Europe, but in the 1970s and early 1980s, several scientists turned their attention to tropical agroecosystems, particularly ones in Latin America. Ecologist Dan Janzen wrote one of the first articles on tropical agroecology in 1973, based in part on what he observed while conducting biodiversity research in Costa Rica. Janzen called to task agronomists who primarily sought technological fixes to any and all problems in tropical agriculture. He argued that this reductionistic and parochial approach assumed ignorance on the part of tropical farmers and failed to take into account both the ecological and socioeconomic contexts in which farmers make decisions. Tropical agriculture poses a great number of challenges to farmers, from low nutrient-reserve capacities of soil to governmental promotion of cash cropping, and understanding these different, interacting components well enough to apply general principles at a local scale—for example, through holistic research done at regional experimental stations—could allow for the development of more sustainable agroecosystems.

Agroecologist Steve Gliessman was also heavily influenced by time spent in Costa Rica. While completing his dissertation research on the ecology of an invasive weed that plagued Costa Rican farmers, he realized that he wanted his research to do more than add to the body of academic literature: he wanted his work to actually help local farmers. Shortly after completing his graduate studies, Gliessman spent a couple of years growing vegetables and coffee at a Costa Rican farm, Finca Loma Linda. During this time he shared ideas on agriculture and ecology with faculty and students from the Organization of Tropical Studies, who used Finca Loma Linda as a field site. One of those engaged in the discussions with Gliessman was John Vandermeer, who has developed long-term research with Ivette Perfecto on traditional shaded coffee agroecosystems in southern Mexico and made substantial contributions to tropical agroecology.

Gliessman was also was influenced by his time spent as a faculty member at the Colegio Superior de Agricultura Tropical in Mexico, where he and his colleagues worked on combining agriculture and ecology in the study of *agroecología*. A key component of their work was sharing knowledge in a two-way exchange with local farmers, learning traditional methods of farming without reliance on modern inputs, from mechanical labor to synthetic fertilizers and pesticides. Gliessman continued to develop his understanding of the field as part of the faculty of the University of California Santa Cruz, working with undergraduate and graduate students in one of the longest-running agroecology programs in the United States and publishing a widely used textbook and laboratory manual.

Another agroecologist whose work draws on his experiences in Latin America is Miguel Altieri, a contemporary and colleague of Gliessman. The Chilean-born Altieri completed graduate work in Colombia and the United States, focusing on entomology, before accepting a professorship at the University of California, Berkeley, in 1980. Three years later, he published a book on agroecology that became a standard for the rapidly expanding field. In the book, as in much of his subsequent work, Altieri made a case for studying traditional farming systems to understand the ecological basis of their practices. Because traditional agroecosystems are a nature–social hybrid that coevolved with humans over time, they are typically adapted to their place and allow people to meet food needs without dependence on modern fuel- and chemical-intensive methods. This adaptation to place has come about through centuries of experimentation by farmers, who have found ways to optimize productivity on a long-term basis, maintaining their social and natural resource bases in the process. Talking with and learning from traditional farmers thus could demonstrate ways of optimizing pre–green revolution methods to increase yields for poor farmers while providing other benefits, from empowering local communities to increasing biodiversity. Traditional agroecosystems also may give clues for how to remedy some of the problems faced in modern agriculture, both in developed and developing countries, as it presses against limits of fuel usage, on- and off-site environmental damage, and increasingly resistant pest populations. However, research into traditional methods must come soon, Altieri wrote, before the agroecosystems themselves are lost.

Altieri recognized structural problems well beyond the plot level that threaten the survival of traditional agroecosystems and make adaptation of more sustainable practices difficult. Barriers such as the socioeconomic structure of corporate agriculture, the availability of funding for research into alternative production methods, the structure of land tenure, and subsidies for pesticides and fertilizers create substantial challenges to developing more sustainable agroecosystems. As a result, the solutions proposed from agroecological research must ultimately be part of larger societal changes at multiple loci in the food system. Altieri specifically included the importance of the role of the consumer in his 1983 book, an early reference to what would become an important movement within agroecology in the following decades. Altieri continues to challenge agronomists and government leaders with his "voice from the South" through his writings, both within and outside the academic press, and worldwide guest lectures.

In the 1990s and 2000s, the field of agroecology continued to develop in many aspects. The number of publications using "agroecology" or "agroecological" as a title or keyword began to rise in the late 1980s and exploded by the late 2000s. More researchers are involved in the discipline, including researchers from previously unrepresented countries, such as Nigeria, China, India, and Brazil, as well as women researchers. For example, the Scottish geographer Joy Tivy published a text on agricultural ecology in 1990, and in 1997 University of California, Davis, researcher Louise Jackson—who truly bridges agronomy and ecology in her work—edited a volume covering topics from the level of plant physiology up though biotic interactions and ecosystem process. The scope of research has grown from adaptations of individual crop species to studies at the level of the food system, and research has become both more interdisciplinary and more participatory, directly involving farmers and others in determining which problems are important enough to merit attention, and in some cases collaborating on the study design, data collection, and interpretation of results. Agroecology has begun to be institutionalized, both in professional societies such as the Ecological Society of America, which now has an active Agroecology Section, and in graduate degree programs in locations such as the United States, Mexico, Brazil, Germany, Norway, and Uganda.

Challenges in Agroecology

Although many researchers share the overarching goal of studying and understanding agroecosystems holistically, carrying out this work is difficult. Not all researchers who identify as agroecologists work across natural and social science realms, even in teams. Many natural scientists maintain a field-scale approach focused on production, using economic measures as the primary nonbiological or chemical estimates of agroecosystem sustainability. Even within the natural sciences, agroecologists may study topics as diverse as birds, insects, crop genetics, soil nutrients, pathogenic bacteria, or beneficial fungi, each with its own language, literature, and methodologies. Overcoming these differences to understand problems at the level of the food system, and thus propose solutions that are socially, economically, and environmentally sound, is a substantial challenge. Further, because the field is both specific and broad—a characteristic recognized by Klages and repeated throughout the agroecological literature—management practices developed at a research station may not be suitable for other farms within a heterogenous environmental and social landscape.

As agroecologists try to figure out how to conduct interdisciplinary research, they also are trying to teach the next generation of researchers. Faculty face challenges such as the structure of universities, their own limited breadth of knowledge as graduates of those universities, and the tension between providing students with a broad understanding of the field and adequate depth in appropriate subfields. Ecologists tend to focus on the relationships between organisms and their environment, whereas agronomists tend to focus on production details, and the students themselves come with their own expectations and interests. Processes outside the farm field, from the work required to develop and maintain seed lines or to produce synthetic or organic fertilizers to the actual steps of processing, marketing, and consumption of food receive little attention. Creative redesign of undergraduate and graduate programs may be required to train future agroecologists who can work across disciplines to develop sustainable agroecosystems both now and under uncertain future climatologic, population, and resource scenarios.

See Also: Agrofood System (Agrifood); Holistic Management; Integrated Pest Management; Soil Nutrient Cycling; Weed Management.

Further Readings

Altieri, Miguel. *Agroecology: The Science of Sustainable Agriculture.* Boulder, CO: Westview Press, 1995.

Bensin, Basil. *Agroecological Characteristics Description and Classification of the Local Corn Varieties-Chorotypes.* Prague: 1928.

Francis, Charles, et al. "Agroecology: The Ecology of Food Systems." *Journal of Sustainable Agriculture*, 22/3 (2003).

Gliessman, Stephen. *Agroecology: The Ecology of Sustainable Food Systems*, 2nd ed. Boca Raton, FL: CRC Press, 2007.

Harper, John. "Agricultural Ecosystems." *Agro-Ecosystems*, 1/1 (1974).

Jackson, Louise, ed. *Ecology in Agriculture.* San Diego, CA: Academic Press, 1997.

Janzen, Daniel. "Tropical Agroecosystems." *Science*, 182/4118 (1973).

Klages, Karl. "Crop Ecology and Ecological Crop Geography in the Agronomic Curriculum." *Journal of the American Society of Agronomy*, 20/4 (1928).

Wezel, Alexander and Virginie Soldat. "A Quantitative and Qualitative Historical Analysis of the Scientific Discipline of Agroecology." *International Journal of Agricultural Sustainability*, 7/1 (2009).

Katie L. Monsen
University of California, Santa Cruz

AGROFOOD SYSTEM (AGRIFOOD)

A food system is the processes of and links between production, distribution, consumption, and waste that in total supply food for a population. Production includes farming, ranching, or fishing; labor; processing, packing, and/or preserving; and all forms of preparation to create an edible product. Distribution can occur between different production activities and between production and consumption; it includes the transportation, labeling, certification, and marketing required to deliver food products to the consumer. Consumption often involves the sale of the food item and results in eating the food product, whether at a restaurant or at home. These first three parts of the food system essentially describe a chain that links food from the field or water where it started to the end-consumer's fork. One simple chain in a food system could be eating the beans you grew in your backyard; at the more complex extreme, a food chain could entail the production of corn distributed and sold to cattle growers, the sale and distribution of beef to a restaurant, the preparation of a steak at that restaurant, and the eventual consumption of the steak by the diner. Waste is also a factor in each of these chains, and the extent to which that waste is cycled back into the system is an important consideration for the sustainability of the food system. The totality of these interlinked chains and cycles that feed a community make up a food system. People are involved in all aspects of a food system and will ultimately determine the stability of the system.

It is common in food systems thought to compare and contrast conventional food systems with alternative or sustainable food systems. In reality, it is more of a gradient from more to less sustainable practices, but for the purposes of illustration, it is helpful to consider the extremes. Conventional versus sustainable delineations can be found in every aspect of the food system, through production, distribution, consumption, and waste, including how people—from producers to consumers—are treated in these processes. It is also important to draw a distinction between "conventional" and "traditional" food systems. By their most basic definitions, these two words are somewhat synonymous. In the food and agriculture world, however, "conventional" means the industrialized and often globalized food systems that have existed since World War II, and "traditional" refers to the premodern food systems that dominated agriculture before the mid-20th century (and still exist in some developing countries). Alternative food systems seek to reclaim many aspects of traditional food systems while recognizing that the reality of our situation has undergone some dramatic changes in terms of population, land use, and urbanization.

From the production perspective, conventional food systems tend to rely on external inputs and often produce unwanted outputs in addition to the desired food harvest. The green revolution, which resulted from new technologies discovered in World War II, introduced dozens of synthetic inputs to improve crop production. The Haber-Bosch process enabled the synthesis of fertilizers from the fixation of gaseous nitrogen so that agricultural production was no longer constrained by preexisting nitrogen in the system. This newly "created" nitrogen boosted crop yields substantially but meant a new reliance on fossil fuels (because nitrogen fixation is an energy-intensive process) and caused a nearly threefold increase in the global nitrogen supply. Long term, this increase in total global nitrogen has not contributed to more productive soils but is, unfortunately, instead, washing away into our rivers and ultimately into our oceans, causing massive dead zones such as the one found in the Gulf of Mexico.

The Growth of Pesticides

Along with synthetic fertilizers came synthetic pesticides, developed as weapons during World War II and now employed as weapons against pests. Though they seemed a panacea at first, these chemicals often resulted in pest rebounds or secondary outbreaks when previously benign herbivores became pests through the removal of their natural predators. Sadly, these chemical pesticides also have a serious effect on the humans who apply them to crops, in many cases causing respiratory distress, cancer, learning disabilities in children, and in severe cases, death. Furthermore, after the pesticide toxins have done their work on the insect populations, they remain in the soil or wash into our waterways, poisoning our drinking water and threatening wildlife. Rachel Carson's *Silent Spring* catapulted pesticides to the forefront of national attention, but although some restrictions have been placed on these chemicals, they are still a cornerstone of production in conventional food systems. Production in conventional food systems is often imagined as "factory farms," because the farms are indeed treated as factories, and food as something that can be manufactured rather than something that should be cultivated with an understanding of the system's ecological web.

Alternative food systems, in contrast, include organic, beyond organic, biodynamic, or otherwise low-input agriculture that employ nonsynthetic and reduced or at least less harmful inputs, whose nonfood outputs can thus be more easily integrated back into the system. These food systems are characterized by more holistic management, from nutrient production to pest control to waste reduction, recognizing the links between these processes and focusing on managing all simultaneously. Waste can be essentially eliminated if it is used as an input for another process, and this is often the aim of alternative food systems. Alternative food systems are seen as more of a cycle than a linear chain, using animal or crop waste as fertilizer and closing the loop so that nitrogen does not leave the system. Integrated pest management, using such methods as biological control instead of pesticides, helps reduce crop damage in a more sustainable way, and other farming techniques help keep the crops from attracting so many pests in the first place. Rather than the vast monocultures characterizing conventional food systems, alternative food systems use polyculture or permaculture to thwart pest species, to spread the risk of loss to any one crop, and to take advantage of synergies between different crops and animals (e.g., natural fertilization by legumes, pest management by chickens, etc.).

The production of meat deserves a little special attention. In conventional food systems, confined animal feeding operations (CAFOs) create waste products that must be treated as pollutants because they are so far removed from the field or range where they would provide nutrients instead. These CAFOs are inhumane to the animals and are generally not healthful for the people who consume them, either. Grain-fed beef is higher in saturated fats, less nutritious, and more at risk for contamination by *Escherichia coli* and mad cow disease than grass-fed beef. The high densities of animals in CAFOs also present disease risks, as many livestock-borne epidemics (which can threaten humans as well, as seen in bird flu and swine flu) are more likely to originate in CAFOs than in smaller, more diffuse, or more genetically diverse animal operations. Feeding grain to animals requires the use of prime farmland that could be used to grow food directly for people, whereas grazing or "pasturing" (in the case of animals other than cattle) occurs on land generally unfit for farming and thus does not compete with other food crops. Grazing has its own host of problems, and plenty of grazing practices would belong in our idea of a conventional food system, even if the animals were never fed grain. However, grazing can be done in a way that stimulates grass growth and promotes diversity, allowing for a coexistence between food production and nature that is not as feasible in even the most sustainable cropping systems. At smaller scales, animals can even be integrated into the cropping system, to the benefit of both.

The Role of Food Processing

A final consideration for production is the food processing industry, whose role in conventional food systems has increased dramatically over the past few decades. Initially seen as a modern convenience, freeing up time otherwise spent preparing meals, processed foods are now seen by many health experts as a major cause of diabetes, heart disease, and stroke. Simple sugars, such as those making up high-fructose corn syrup (found in nearly all conventional processed foods), are so easy for our bodies to process that we quickly accumulate a surplus, which is exacerbated by the quantities in which they are found in processed foods. The nutritional quality of these foods is also lower than that of their whole-food counterparts. Still, processed foods dominate the supermarket and our diets, resulting in a decreasing amount of our food dollars returning to the farmer (19 percent in 2000 compared with 40 percent as recently as 1973). Alternative food systems, in their tendency to support local economies as discussed below, use more whole-food ingredients, and thus fewer processed goods. It is important to note that preserving food (e.g., canning, freezing, drying) is a form of processing, so any food system is likely to include processing as a production method to make food available outside the growing season. Alternative food systems simply use few or no synthesized ingredients in their processing, rendering these foods still more "whole" than foods processed in conventional food systems.

Because the production processes associated with alternative food systems often incorporate external costs into food production and food producers that are more conventional are allowed to ignore these externalities, our current economic system often forces the more-sustainable producer to charge more for a similar product (e.g., an organic peach costs more than a conventional peach). Third-party certification, or eco-labeling, of production processes has emerged as a tool for food marketers to quickly communicate to the consumer the extra value of the product and thereby explain to the consumer the reason for the extra cost. "Certified Organic" has become a common label used on fruits and vegetables to tell the consumer that no chemical pesticides or fertilizers were used in the production of that item. Organic certification is carried out by neutral third-party certification organizations, such as California Certified Organic Farmers. Because of the complexity of the food system, removing artificial chemical inputs does not necessarily make a food "sustainable." Therefore, the certification of other aspects of food production has become increasingly common, such as the treatment of food workers (Fair Labor, or the Union label of the United Farm Workers) and of animals (Certified Humane).

The Higher Cost of Eating Well

The higher cost of more-sustainable food that is a result of the externalization of downstream effects of unsustainable production practices, combined with the unhealthful qualities of cheap, conventional, and/or factory-processed foods, means that people of lesser means eat less well than wealthy people. In addition, the conventional food distribution methodology sells most of America's calories through supermarket chains or big box stores, which are rarely built in poor neighborhoods. Many people in poorer, urban areas live in "food deserts" that do not provide access to nutritious sustainably produced food and are forced to rely on small convenience stores for conventionally produced, heavily processed food that can cause long-term health problems. One common approach to solving this problem of food security/food justice is through education and outreach programs that promote backyard homegardens and other urban agriculture endeavors that allow urban residents greater food sovereignty and a modest amount of independence from industrially controlled food systems. Through these programs, members of marginalized communities and those that are most at risk for hunger are empowered to reclaim spaces (often empty lots) in their neighborhoods to produce local, sustainable, nutritious foods. These community gardens often preserve traditional production practices and provide culturally appropriate foods available to at-risk populations. Because the food is locally sourced and embedded within the urban landscape, it is easily accessible to the consumers who benefit the most from this healthful, nutritious food.

The creation of regional food systems takes this concept of local sourcing and applies it to a larger metropolitan area instead of just the urban core. This conceptual approach attempts to define and manage a "foodshed" for a city; that is, the geographic area of fisheries, rangeland, and agriculturally productive lands that provide the urban population with its food. Each city has a unique foodshed that is influenced by the size of the city, the types of land and water surrounding it, and the types of food that can be produced from these areas. Consumers who interact with their food system through this approach often refer to themselves as "locavores" and sometimes intentionally limit their food purchasing to within a certain radius from their home. More commonly, consumers who wish to support a regional food system with their purchasing power will buy from a local farmers market or subscribe to a community-supported agriculture program, in which a farmer delivers weekly "shares" of his/her product to neighborhood drop spots in return for monthly, quarterly, or annual subscriptions. Many regional food system approaches encourage greater contact between the producer and consumer and allow a larger share of each food dollar to bypass producers and agribusinesses to reach the producer directly. The additional benefit to a regional food system is the reduced fossil fuel footprint from transporting the food. In conventional food systems, the average food item travels more than 1,500 miles from farm to plate. As fuel grows scarcer and the threat of global climate change grows more extreme, these local food movements may begin to look less "alternative" and more "just common sense."

See Also: Beyond Organic; Biodynamic Agriculture; Biological Control; California Certified Organic Farmers; Certified Humane; Certified Organic; Commodity Chain; Community Gardens; Community-Supported Agriculture; Confined Animal Feeding Operation; Eco-Labeling; Factory Farm; Fair Labor Association; Family Farm; Fertilizer; Food Processing Industry; Food Security; Foodshed; Food Sovereignty; Grain-Fed Beef; Grazing; Green Revolution; High-Fructose Corn Syrup; Holistic Management; Homegardens; Integrated Pest Management; Locavore; Low-Input Agriculture; Supermarket Chains; Sustainable Agriculture; United Farm Workers; Urban Agriculture.

Further Readings

American Farmland Trust. "San Francisco Foodshed Report." http://www.farmland.org/programs/states/ca/Feature%20Stories/documents/ThinkGloballyEatLocally-FinalReport8-23-08.pdf (Accessed May 2009).

Eat Wild. "Grass-fed Basics." http://www.eatwild.com/basics.html (Accessed May 2009).

Pollan, Michael. *In Defense of Food: An Eater's Manifesto.* New York: Penguin, 2008.

Pollan, Michael. *The Omnivore's Dilemma.* New York: Penguin, 2006.

Savory, Alan. *Holistic Resource Management: A Model for a Healthy Planet.* Washington, D.C.: Island, 1988.

World Resource Institute. "Nutrient Overload: Unbalancing the Global Nitrogen Cycle." http://www.wri.org/publication/content/8398 (Accessed May 2009).

Rebecca Chaplin-Kramer
University of California, Berkeley

Daniel Kramer
Independent Scholar

ANIMAL WELFARE

Concerns about the welfare of animals has grown since animal farming emerged with sedentary civilization. Many argue that animals should be treated humanely because they are sentient beings, meaning that they experience pleasure and pain. In the last quarter century, the arguments have become more vocal, particularly in industrialized countries, where the vast majority of land animals and certain sea creatures are raised in confined animal feeding operations (CAFOs), which are considered by many to be an inhumane system of animal productions. CAFOs are highly industrialized systems that strive for maximum efficiency, where animals are produced as objects according to principles of mass production, warranting the label "factory farming." The problem of factory farming and CAFOs motivate some consumers to purchase animal products that are produced by smaller farms, where animals may be unconfined or "free range," or by avoiding meat products altogether by becoming vegetarian or vegan.

Among other issues, animal welfare reformers focus on reducing confined animal feeding operations while promoting free-range farms that allow animals like these Angus cattle to feed by traditional grazing.

Source: U.S. Department of Agriculture, Agricultural Research Service/Scott Bauer

Development of Farmed Animal Welfare Attitudes

Ethical and spiritual concerns for animals have motivated vegetarianism for several millennia. Ancient vegetarian proponents include Eastern religious leaders Buddha, Mahavira, and Lao Tzu, as well as prominent ancient Greeks such as Pythagoras, Plato, Porphyry, and Plutarch, who proposed the sentience, rationality, and kinship of other animals. Plutarch critiqued cruel farming methods practiced to improve taste, which suggests that even traditional farming involved suffering.

Vegetarianism waned in Western culture during the reign of Christianity, which relies on the human/animal dichotomy. Animal welfare writings resurfaced in the 18th century, largely in resistance to "Cartesianism"—Descartes's scientific viewpoint suggesting that nonhuman animals were automata with little consciousness or feeling—an instrumental attitude serving to justify animal use amid growing concerns over animal suffering in science and agriculture. In contrast, utilitarian philosophers Jeremy Bentham and John Stuart Mill acknowledged animal sentience and promoted restricting animal use to only necessary food and research. Popular sentiment dictated that causing wanton animal suffering led to inhumanity toward other humans. This anthropocentric concern for creating a peaceful civilization was common in 19th-century vegetarian writers, such as Leo Tolstoy, Mahatma Gandhi, and Albert Schweitzer. Henry Salt and J. Howard Moore even promoted an early idea of extending human rights to other animals, often referring to animal cruelty as criminal.

These ideas anticipated philosophies that ushered in the late-20th-century animal rights movement, largely credited to Tom Regan and Peter Singer. Regan considered it morally inconsistent to take away the life of a nonhuman animal—a fellow subject of a life—when one would not take a human life, such as for food. Singer argued that animal agribusiness, whether free range or intensive, is a speciesist practice, meaning it discriminates based on one's species, sacrificing their major interests (life) to satisfy our minor interests (taste).

Agricultural Practices

Historically, a largely rural society supported animal farming because they witnessed animals leading wholesome, natural lives. However, over the previous quarter-century, most family farmers were forced out of business by larger corporate farms. Today's agricultural status quo severely deviates from the bucolic ideal of the pastoral "Old McDonald's Farms," as industry has largely confined the animals behind closed doors. Pigs and birds are the most intensively confined, cows raised for dairy are semiconfined, and cattle are the least confined until sent to a feedlot.

Animal protection campaigns tend to emphasize the worst cruelties, specifically the extreme confinement systems of hen battery cages, pig gestation crates, and veal crates. The industry must fight nature and manipulate animals to fit the unnatural CAFO conditions and improve profitability, either through genetic engineering or painful medical procedures such as debeaking, branding, castration, dehorning, toe clipping, ear and tail docking, and teeth clipping. Although the industry argues they must care about animal welfare to be profitable, animal protection organizations cite the high mortality rates in farms and transport as evidence of the animals' poor living conditions and lack of individual veterinary care. Undercover videos reveal corpses rotting among the living; workers beating to death sick animals or runts; and male chicks—useless to the egg industry—suffocating in trash bags. Slaughterhouse videos show sick or lame animals being dragged to slaughter and a portion of the animals, particularly many birds, being conscious through throat-slitting—sometimes up to the point of experiencing scalding tanks and dismemberment.

Cartesian views have proven outdated as research reveals that mammals, fish, and birds in agriculture endure both physical and emotional pain during their lives and slaughter. The frustration of increased competition for food, restricted movement, and pressure on social relationships, including separation of mothers from their young, result in animals experiencing painful emotions such as fear and depression.

Annual Death Toll

Over 50 billion land animals are slaughtered worldwide, and likely a larger number of sea animals. The United States kills more than 10 billion land animals, including millions of farm mortalities and millions of male chicks at egg hatcheries. According to 2007 U.S. Department of Agriculture statistics, animals slaughtered include approximately 9.4 billion broiler chickens, 450 million laying hens, 317 million turkeys, 121 million pigs, 39 million cows, and millions of ducks, rabbits, sheep, and goats. The sea animal lives taken (whether in aquaculture CAFOs or commercial fishing) are recorded by weight, not by individual animal, but Americans eat an estimated 17 billion aquatic animals.

Humane Farming Reforms

In the 19th century, the first Societies for the Prevention of Cruelty to Animals, formed in England and the United States, initially prioritized farmed animal welfare and slaughterhouse reforms. In fact, the American Societies for the Prevention of Cruelty to Animals's first cruelty conviction was of a butcher. America's first federal anticruelty statute, the "twenty-eight hour law" passed in 1871, regulated the care of animals transported to slaughter. It was not until 1958 that the U.S. Congress passed another farmed animal statute, the Humane Slaughter Act. It required rendering land animals unconscious before slaughtering but left fish and birds unprotected.

Although many European nations have recently banned the worst intensive confinement systems, as have some U.S. states and municipal legislatures, the U.S. government has not instituted federal humane regulations on the farm, despite public support. So, in recent years, animal protection organizations started using statewide referenda (in Florida, Arizona, and California) to circumvent a strong federal agricultural lobby and let voters directly ban crates. Even major fast-food companies, such as McDonald's have voluntarily instituted stricter welfare regulations for suppliers, and college campuses are switching to cage-free eggs.

This interest in reform has sparked debate between animal protection organizations over whether to prioritize vegan campaigns or legislative reforms. Welfarists argue that absolutist campaigns, such as veganism, are less effective and impractical at saving billions of animals, whereas farming reforms are pragmatic, incremental steps that reduce suffering now. Rightists argue that reforms ultimately ease industry and consumer conscience more than they actually ease animal suffering. Rightists support veganism because, with global free trade, if humane reforms drive up costs, a continued demand for inexpensive animal products will simply send factory farms to developing countries, exporting the environmental, health, and welfare problems overseas.

Welfare and environmental concerns in agriculture have recently taken center stage in books and slow food movements, promoting a return to local, organic, free-range, sustainable farming. Conscientious consumers are boycotting hormones, drugs, pesticides, and genetic engineering and demanding cage-free eggs and free-range, grass-fed meats. Some, like food journalist Michael Pollan, suggest a more sustainable diet consists of eating mostly plants, boycotting factory farms, and occasionally hunting or fishing for meat. He argues that pastoral animal farming is an evolved symbiotic relationship that ecologically harnesses solar power to naturally fertilize soil and grow food. However, Pollan and pro-animal groups critique some new welfare claims, such as "animal care certified" or "free-range," as sheer marketing ploys meant to mislead consumers into believing that animals lead a bucolic existence; in actuality, many are raised in warehouses, and even cages, with little outdoor access. This led Singer to claim that veganism is the easiest solution to animal cruelty and unsustainable farming, as it provides greater moral clarity than meat-eaters' confusing attempts at conscientious consumerism. In this post-Cartesian era of factory farming, consumers are starting to demand greater ethical and ecological accountability from food producers.

See Also: Confined Animal Feeding Operation; Factory Farm; Vegan; Vegetarian.

Further Readings

Beers, Diane. *For the Prevention of Cruelty: The History and Legacy of Animal Rights Activism in the U.S.* Athens: Ohio University Press, 2006.
Compassion in World Farming. http://www.ciwf.org.uk (Accessed February 2009).
Curtis, S. E. and W. R. Stricklin. "The Importance of Animal Cognition in Agricultural Animal Production Systems." *Journal of Animal Science*, 69 (1991).
Farm Sanctuary. "Summary of the Scientific Evidence Establishing Sentience in Farmed Animals." http://www.sentientbeings.org/SB_report_web.pdf (Accessed February 2009).
Fraser, David. "Farm Animal Production: Changing Agriculture in a Changing Culture." *Journal of Applied Animal Welfare Science*, 4/3 (2001).
Linzey, Andrew and Paul Clarke. *Animal Rights: A Historical Anthology.* New York: Columbia University Press, 2004.
Pollan, Michael. *The Omnivore's Dilemma: A Natural History of Four Meals.* New York: Penguin, 2006.
Regan, Tom. *The Case for Animal Rights.* Berkeley: University of California Press, 1983.
Singer, Peter. *Animal Liberation.* London: Random House, 1990.
Singer, Peter and James Mason. *The Ethics of What We Eat: Why Our Food Choices Matter.* Emmaus, PA: Rodale, 2006.
U.S. Factory Farming. http://www.hsus.org/farm or http://www.factoryfarming.com (Accessed February 2009).
Walters, Kerry S. and Lisa Portmess. *Ethical Vegetarianism: From Pythagoras to Peter Singer.* New York: SUNY Press, 1999.

Carrie Packwood Freeman
Georgia State University

APPROPRIATIONISM

Appropriationism describes the transformation of agricultural production processes into industrial activities that minimize constraints to profit accumulation imposed by agriculture's basis in biological cycles. As a consequence, appropriationism also entails the reincorporation of the production processes taken over by industry back into agricultural production as purchased inputs. Examples of appropriationism include the use of fossil fuel–powered machinery in place of human and animal farm labor and the use of industrially produced fertilizer in place of farm-based nutrient management systems such as crop and livestock rotations that use farm-based methods of crop fertilization.

David Goodman, Bernardo Sorj, and John Wilkinson developed the parallel concepts of appropriationism and substitutionism in their theory of agro-industrial development described in *From Farming to Biotechnology*. Together, these concepts explain the industrialization of agriculture through the application of science and capital investments to discrete segments of the rural labor and biological processes in agricultural production. Although *appropriationism* refers primarily to the industrial replacement of biological or farm-based production processes, *substitutionism* refers to industrial intervention in postharvest, downstream agricultural product processing.

Agriculture is different from manufacturing or industrial sectors in its reliance on biological processes, such as plant growth and animal gestation. These processes take time and make agriculture a less-profitable investment, relative to other forms of industrial production, which can often be accelerated or made more efficient through technological innovation or labor reorganization alone. Agriculture also requires larger amounts of land than most forms of industrial production. The spatially extensive land requirements of agriculture limit farm size, given a fixed supply of labor and no industrial intervention to replace animal and human labor with fossil fuel–powered machinery.

Appropriationism, then, refers to industrial sectors' attempts to reduce these natural barriers to accumulation by, for example, accelerating plant growth rates with industrially produced nitrogen fertilizer or increasing harvesting efficiency with farm machinery. The industrial sector invests in agricultural technologies to make a profit by selling manufactured products to farms as purchased inputs that perform the duties once fulfilled by rural labor and biological processes.

Industry, however, has been unable to completely transform agricultural production processes. The biological cycles inherent in agricultural production cannot be performed entirely by industry—plants need sunlight and soil for growth, and animal growth rates cannot be reduced to zero. Given that industrial appropriation of agricultural production is only partial, the historical processes of appropriationism have taken several discrete forms. Among them are mechanical, chemical, and genetic appropriations of rural labor and biological production processes. These are discussed in turn here, drawing on the United States' case as an example.

Appropriationism first emerged through the mechanization of rural agricultural production. Between the mid-1830s and early 1850s, U.S. crop harvesting was revolutionized by the development of animal-powered mechanical reapers that replaced human-powered harvesting tools like scythes or cradles. Increased harvesting efficiency led to reductions in the quantity of labor needed at harvest time and made it possible for a single farmer to plant more acres. Thus, the prevalence of mechanical harvesters reduced the constraint that land as space had on possibilities for accumulation in agriculture. Farmers could cover more ground more quickly, and farm-implement manufacturing industries were born.

A second major appropriationist moment came with the development of the gasoline-powered internal combustion engine. In the United States in the 1920s and 1930s, these engines powered combine harvesters and their diffusion. The use of gasoline-powered tractors also rose at this time. The gasoline engine, because of its mobility, was better fit to confront the spatial extensiveness of land-based agricultural production than other customary 19th-century energy sources—animal power, the cumbersome steam engine, or electricity. Aided by rising incomes associated with a World War II–based agricultural boom, postwar U.S. farmers were nearly fully transitioned from animal to mechanical and electric power as the energy basis of rural production processes. That is, industrial capitals had appropriated the rural production of energy (animal power fed by crops and pasture) with fossil fuel and the manufacture of agricultural implements: the tractor, gasoline engine, and the electrical motor. This appropriation helped remove the constraint that land, as space, imposed on the profitability of agriculture.

A third turning point in the historical process of appropriationism is marked by the development of mineral-based and industrially produced agricultural fertilizers. Mined minerals and the Haber-Bosch process for manufacturing nitrogen fertilizer supplemented what typically limited plant productivity—nitrogen supply—with industrially produced nutrients. Instead of relying on historical methods of crop fertilization, like crop-animal rotations and manure composting and spreading, industrial activities allowed the purchase of these inputs, as well as the spatial separation of crops and livestock, which became less necessary for fertilizing crop fields. The fertilizer industry, developed in the 1920s, was stagnant through the 1930s until the development of high-yielding seed varieties that could more efficiently convert nitrogen into plant matter. Declining crop yields associated with mechanization created the desire for intervention in the biological determinants of crop yields through research in plant breeding.

Although previous rounds of appropriationism transformed the rural labor process, the first real appropriation of natural production processes occurred through developments in plant genetics and crop hybridization techniques. High-yielding crop varieties were often more responsive to soil nutrients and sunlight and often matured sooner than other varieties. Public research laid the foundations for developing new seed varieties, but appropriation by industry was made possible with the introduction of hybrid seeds. Their inability to produce viable offspring, as well as the biophysical variation of agricultural regions, made multiple seed lines necessary. In other words, farmers must purchase high-yielding hybrid seeds as inputs each year instead of saving portions of their crop harvest to plant as seed the subsequent year. Thus, the application of science appropriated the biological processes of plant growth as a center for capital accumulation via increased plant productivity and seed research and sales.

These innovations in plant breeding became the pivot of subsequent agro-industrial development as the chemical and farm equipment sectors abandoned their more independent strategies of capital accumulation and converged toward the biological innovations in seed production. Plant breeding techniques permitted the complete mechanization of crop cultivation as varieties that fit requirements of mechanization were developed. An oft-cited example of controversial public agricultural research serves here: the University of California's development of the VF-145 tomato plant was designed to bear fruit that would ripen simultaneously and be sufficiently durable for mechanical harvest by its companion tomato harvester, developed near the same time. Similar advancements in breeding for uniform height, ripening timing, and other characteristics that make crops amenable to mechanical picking, processing, or transport occurred for sugar beets, cotton, and corn, among others.

Appropriationism continues through biotechnologies that, for example, replace farm-based methods of pest and weed control with industrially produced crop varieties engineered to be poisonous to pests or resistant to herbicides that kill surrounding weeds. Although productivity gains have been realized, industrially appropriated agriculture has led to significant environmental costs, including nutrient pollution from plant and animal growth intensification as well as soil erosion and biodiversity loss resulting from large-scale monocrop cultivation.

See Also: Agribusiness; Substitutionism.

Further Readings

Goodman, David and Michael Redclift. *Refashioning Nature: Food Ecology and Culture.* London: Routledge, 1991.

Goodman, David, et al. *From Farming to Biotechnology: A Theory of Agro-Industrial Development.* Oxford: Basil Blackwell, 1987.

Kloppenburg, Jack. *First the Seed: The Political Economy of Plant Biotechnology, 1492–2000.* Madison: University of Wisconsin Press, 2004 (1988).

Mann, Susan and James Dickinson. "Obstacles to the Development of a Capitalist Agriculture." *Journal of Peasant Studies,* 5 (1978).

Sean Gillon
University of California, Santa Cruz

AQUACULTURE

Aquaculture refers to the growing of aquatic floral and faunal organisms under controlled conditions. These range from preindustrial forms of stock enhancement to systematic intervention in the entire life cycle of the organisms most commonly found in today's semi-intensive and intensive forms of commercial aquaculture. Attention here is confined to faunal aquaculture; that is, finfish, mollusks, crustaceans, and other organisms. In the face of declining or static global stocks of open capture fisheries, aquaculture has been promoted by government, business groups such as the Global Aquaculture Alliance, and global governance agencies such as the World Bank as an efficient and environmentally sustainable way of providing animal protein for humans, particularly for fish-eating populations in developing countries. It has also been criticized by academics, national and international nongovernmental organizations (NGOs), local communities, and some policymakers as ecologically damaging, socially and economically regressive, and unsustainable in the long term. Particular criticism has been directed at tropical-water shrimp producers who supply elite consumer markets in the developed world and at salmon farmers in developed and developing countries. In response to such criticism and to the growing global demand for seafood, efforts are being made to move toward more sustainable and socially equitable forms of aquaculture production.

Aquaculture has expanded greatly in the past three decades and accounts for some 43 percent of global fish supply. It is dominated by freshwater finfish such as carp and other cyprinids (carp and minnows), mollusks (oysters, clams, cockles, ark shells), and crustaceans (shrimp). Freshwater aquaculture contributes the highest volume and value of the sector, but marine and diadromous (salt- to freshwater migratory) species add higher monetary value relative to production volume. Aquaculture is an important export industry in many developing countries, providing direct employment for over 12 million people and earning valuable foreign exchange. Asia—and China in particular—accounts for over 90 percent of the volume and over 80 percent of the value of global aquaculture production and has seven of the top 10 export countries. It is a good example of the globalization of production, exchange, and consumption in which hatcheries, farmers, processors, exporters, transporters, retailers, and consumers are linked through vertical and lateral networks of trade. These trading networks are increasingly dominated by corporate retailers in the developed world who exert control over the supply chains linking consumers to producers and processors located in developing countries. It is such "industrial aquaculture" that has attracted most attention and criticism. However, many developing countries also farm aquatic products under a variety of economic, social, and environmental conditions serving large, ethnically diverse, and economically differentiated local and domestic markets.

Types of Aquaculture

There are several types of aquaculture including freshwater and brackish water aquaculture, mariculture, open capture aquaculture, pond and tank culture, pen and cage aquaculture, and sea ranching. Farming systems vary from traditional extensive through modified extensive and semi-intensive to intensive and hyperintensive. Traditional extensive, modified extensive, and semi-intensive farming systems are found mainly in developing countries, whereas highly intensive production systems are more common in developed countries such as Canada and Norway. These systems are distinguished by levels and types of capital investment, use of commercial feeds, stocking densities, use of water exchange, aeration systems and pumps, and antibiotic and chemical usage. The more intensive the system, the greater the degree of technical and organizational

A manager at a catfish farm in Columbus, Mississippi, loads 2,000 pounds of fish onto a truck. Aquaculture facilities employ over 12 million people worldwide and now provide as much as 43 percent of the world's supply of fish.

Source: U.S. Department of Agriculture Agricultural Research Service/Stephen Ausmus

control of the production process, the higher the capital costs, and the more disembedded the system from surrounding environments and communities.

Debates about aquaculture have centered on how to increase production and productivity while at the same time ensuring greater environmental sustainability, improved distribution of the economic benefits of the industry, better working conditions along the supply chains, and enhanced food security for the poor of the developing world. Initially, much criticism was directed at the local effect of export-oriented brackish water aquaculture and mariculture on land and water use and the unequal distribution of the economic benefits of fish and shrimp farming at the grassroots level. For example, studies show that in several brackish water shrimp farming areas, local farmers have seen rice yields decline because of increased soil and water salinization and from pressures placed by wealthy aquaculture investors on local farmers to shift to shrimp production. Artisanal fishers have seen fish stocks diminish as access to common pool resources is restricted, and estuarine, coastal, and inland water bodies are privatized for shrimp farming. Some farming areas have witnessed a reduction in grazing land and feeding sites for household poultry as fish and shrimp farms expand. More recently, there has been a greater academic and political focus on the ways in which the entire value chain from extraction and production to retailing and consumption affects small farmers, rural workers, women, and children. For example, there are ongoing campaigns by NGOs and others against poor working conditions in the industry, especially the use of child labor, gendered inequalities in pay, and the plight of smallholder fish and shrimp farmers. Critics argue that reforms to the international seafood trading system have been top-down with little if any participation by local people engaged in aquaculture. Thus, certification schemes to ensure quality products are costly to implement and favor larger producers and processors rather than small-scale fish farmers.

Pros and Cons of Aquaculture

Environmental criticisms of the industry range from its effect on local ecological systems to its global carbon footprint. Aquaculture is said to reduce wild fish stocks as a result of the use of fish as feed for cultured carnivorous fish species, high bycatch rates of nontargeted species from wild seed collections, fish death from fish waste and chemical use, and escape of cultured species into the wild. Aquaculture is criticized for destroying or degrading local ecological systems and their biodiversity, thereby reducing the ecological services provided by, inter alia, mangroves and salt marshes. Aquaculture is blamed for introducing new pathogens harmful to both wild and farmed fish and crustaceans and for disrupting feeding links between species. A more recent concern is over the creation of genetically engineered fish and their effect on other fish and human health. Finally, it is argued that export aquaculture has wider negative effects, as it has made fish too expensive for local people and reduced their intake of fish protein. Thus, the claim made by industry proponents that aquaculture can "feed the world" is rejected because over 90 percent of the products of export aquaculture are consumed by relatively wealthy consumers in Europe, North America, and Japan and increasingly by the new middle classes of China, India, and Brazil.

In response to their critics, proponents of the sector such as governments, industry representatives, and international development agencies recognize that aquaculture has had negative effects in the past. However, they argue that much ecological and social damage has been a result of poor planning and management and can be reduced through improved culture practices, proper land zoning, financial and technical support for small farmers, and other measures. Organizations such as the Global Aquaculture Alliance question claims that mangrove destruction has been largely caused by aquaculture, pointing out that agriculture, forestry, urbanization, and a range of other activities have played a significant role. Regarding fish food and bycatch, it is argued that the use of fish feed can be lowered by improved methods of processing and a shift to nonfish meal sources and to the raising of herbivorous species. Bycatch volumes can be reduced through such policies and also through improved catching methods. Industry proponents point out that increasingly the industry is required to meet new regulatory standards and certification schemes applicable throughout the supply chain to ensure improved quality and safety of the product, better working conditions, and more environmentally sensitive aquaculture practices. On the question of the declining affordability of cultured fish for local peoples, it is argued that the growing scarcity of finfish and crustaceans is the result of a wide range of environmental and economic factors and not simply to the high export prices for farmed fish. Also, reference is made to the culture of low-value fish, which is often cheaper than wild-caught fish. Industry proponents argue that many fish farmers have benefited economically compared with those who remain in rice and cereal production. However, they also recognize that this has sometimes been at the expense of other members of the community, particularly landless laborers and artisanal fishers. Attempts to redress this imbalance are central to pro-poor aquaculture programs. With regard to wild fish stocks, they argue that properly managed aquaculture can both help to relieve pressure on wild stocks and supplement those wild stocks under threat. The industry's waste products can be used ecologically and contribute to reducing harmful land use practices. Genetically modified organisms can contribute to producing fish that grow faster, have greater tolerance for environmental variability, and are more efficient feeders. On the broader question of environmental sustainability, some aquaculture scientists, environmentalists, and industry representatives support moves toward closed system aquaculture and recirculating aquaculture systems to ensure biosecurity and environmental protection.

Future of Aquaculture

Although there have been calls from some industry critics for dismantling the export-oriented aquaculture sector, this is a minority view. At present there is some evidence of a cutback in export production as a result of the global financial downturn, but over the longer term, domestic and export aquaculture will continue to increase in Asia and in Latin America and sub-Saharan Africa. Governments, the corporate seafood sector, the World Bank and other international agencies, and prominent NGOs such as the World Wildlife Fund officially support shifts toward more sustainable aquaculture practices. For example, the World Wildlife Fund is to co-found an Aquaculture Stewardship Council, which will provide third-party certification according to guidelines laid down by the highly respected International Social and Environmental Accreditation and Labeling. In Europe and North America, there are pressures to move toward closed systems aquaculture, in which the culture of aquatic organisms is separated from the natural environment as a means of producing high-quality products with limited environmental impact.

Aquaculture is now the focus of pro-poor development programs in several countries such as India, Bangladesh, Cambodia, and Vietnam aimed at raising incomes and enhancing food security through the greater involvement of marginalized populations in domestic and export-oriented aquaculture programs. Poor people also participate directly and indirectly in urban and peri-urban wastewater aquaculture in India, Bangladesh, and Vietnam. More integrated and polycultural systems of aquaculture are being developed for small farmers and rural workers that combine fish, shrimp, and rice production in ways designed to protect natural ecosystems and complement seasonal agricultural cycles. Co-management schemes are being extended to aquaculturalists to improve their competitive position in global seafood markets, to ensure greater local participation in management, and to allow a more equitable distribution of the economic returns from aquaculture. Aquaculture is being incorporated into integrated coastal zone management schemes that promote ecosystem-management approaches to the management of diverse species and habitats. The use of saltwater ponds for fish production in arid lands is being trialed as part of dry land livelihood diversification schemes. In several countries, unused ponds are being converted for fish farming to supply local populations. These types of programs face several problems, including elite capture of the institutions serving projects, loss of commons lands through project enclosures, conflicts over land use, consumer fears over contaminated products, and concerns over long-term funding to ensure continued activity. However, they are now mainstream components of development planning and are likely to grow in importance in coming years.

See Also: Commons; Fisheries; Food Safety; Supply Chain.

Further Readings

Costa-Pierce, Barry A. *Ecological Aquaculture: The Evolution of the Blue Revolution.* Malden, MA: Blackwell Science, 2002.
Dina, James S. "Aquaculture Production and Biodiversity Conservation." *BioScience,* 59/1:27–38 (2009).
Naylor, Rosamond L., et al. "Effect of Aquaculture on World Fish Supplies." *Nature,* 405/29 (June 2000).
World Bank. *Changing the Face of the Waters: The Promise and Challenge of Sustainable Aquaculture.* Washington, D.C.: World Bank, 2007.

Bob Pokrant
Curtin University of Technology

ARCHER DANIELS MIDLAND

Archer Daniels Midland (ADM), once self-titled "supermarket to the world," is an Illinois-based agribusiness and the largest supplier of ethanol in the United States. With more than 230 factories and 27,000 employees in over 60 countries, ADM is also one of the world's largest food processors.

ADM operates an extensive agricultural sourcing, storage, processing, and transportation network. Each day, ADM converts 91,000 metric tons of oilseeds, 50,000 metric tons of corn, 27,000 metric tons of wheat, and approximately 15 percent of the world's cocoa crop into a wide variety of products. Corn alone becomes some two dozen products, including food and beverage ingredients (e.g., high-fructose corn syrup that may end up in soft drinks), animal feed ingredients (e.g., lysine, bulk vitamins), and industrial products.

Some highlights in ADM's history of expansion include its beginnings in 1902 as a linseed-crushing business under partners George A. Archer and John W. Daniels, the 1923 acquisition of Midland Linseed Products Company, continued takeovers of Midwest oilseed processors throughout the 1920s, expansion into flour milling and the discovery of how to produce lecithin from soybean oil in the 1930s, post–World War II growth into overseas locales, mass-marketing of textured soy protein in the 1960s, acquisition of Corn Sweeteners (high-fructose syrups and glutens) and Tabor (grains) in the 1970s, takeover of Colombian Peanut and the formation of a grain-marketing joint venture with GROWMARK in the 1980s, and the 1989 entry into the lysine market (an amino-acid feed additive primarily for livestock), as well as the realm of biotechnology. Shortly thereafter, ADM also became known as the biggest recipient of "corporate welfare" in the United States.

A federal investigation was launched in 1992 charging that ADM conspired with its supposed competitors to fix prices and allocate sales of lysine and citric acid. In 1995, the FBI joined the investigation, building its case on information provided by ADM executive

Mark Whitacre. ADM pleaded guilty in 1996 to violating the Sherman Antitrust Act and paid $100 million in penalties, which was then a record for a U.S. antitrust case. Interestingly, Whitacre himself was later convicted of defrauding ADM of $9 million, so he lost his immunity and, along with two other ADM executives, was sentenced to prison in 1999. ADM's expansion hardly slowed during this period. In fact, ADM continued to benefit from federal and state subsidies for furthering corn-based ethanol production. The Government Accounting Office estimated subsidies to the ethanol industry at $11 billion for the years 1980–2000, and as the largest ethanol producer in the United States, ADM received the greatest portion of them.

ADM continues to venture into new, yet related, partnerships. For instance, in 2008, ADM entered a biofuels collaborative research program with Monsanto and Deere & Company. ADM currently owns or co-owns biodiesel production facilities in Brazil, Germany, India, Indonesia, and the United States (specifically, in Missouri and North Dakota). ADM also has partnered with multiple companies and other institutions to develop bio-based plasticizers and other crop-derived alternatives to petroleum-based plastics and chemicals.

In its mere century of existence, ADM has greatly influenced the very ways in which food and energy are produced.

See Also: Agribusiness; Corn; Monsanto.

Further Readings

Archer Daniels Midland. http://www.adm.com (Accessed February 2009).
Eichenwald, Kurt. *The Informant: A True Story*. New York: Broadway, 2000.
Lieber, James. *Rats in the Grain: The Dirty Tricks and Trials of Archer Daniels Midland*.
   New York: Four Walls Eight Windows, 2000.

Jennifer Ellen Coffman
James Madison University

B

Berry, Wendell

Wendell Berry (1934–) is an American writer and farmer whose work is steeped in the South, rural America, and the world of agriculture. His publications and awards span the three major modes of literature—fiction, nonfiction, and poetry. The grounding of his work in agrarian themes and a strongly developed sense of place and community, as well as his focus on the fictional Kentucky town of Port William, have invited comparisons to William Faulkner.

The oldest of the four children of lawyer John Berry and his wife Virginia—tobacco farmers who came from families that have farmed Henry County, Kentucky, for five generations—Berry attended the University of Kentucky, where he earned his bachelor's and master's degrees in English, completing his master's in 1957. He married Tanya Amyx that same year and enrolled in the creative writing program at Stanford University, where he studied under Wallace Stegner in a seminar that included a future "who's who" of American literary talent: Edward Abbey (*The Monkey Wrench Gang*), Nobel winner Ernest J. Gaines (*The Autobiography of Miss Jane Pitman*), Ken Kesey (*One Flew Over the Cuckoo's Nest*), Pulitzer winner Larry McMurtry (*Lonesome Dove*), Tillie Olsen (*Tell Me A Riddle*), and Robert Stone (*Dog Soldiers*). Berry's first novel, *Nathan Coulter*, followed in 1960, and after some travel to Europe on a Guggenheim fellowship and a brief stint at New York University, he took a position as professor of creative writing at the University of Kentucky, where he taught from 1964 to 1977, and again from 1987 to 1993.

Shortly after taking the University of Kentucky job, Berry and his family moved to Lane's Landing, a farm on the banks of the Kentucky River just outside Port Royal, Kentucky, near the birthplaces of both his parents. The farm has expanded to 125 acres, where Berry grows corn, grain, and tobacco.

Berry has published 11 volumes of fiction (novels and short story collections), 16 of nonfiction, and 25 of poetry. His nonfiction makes explicit the ideas that are "behind the scenes" in much of his other work. Berry prizes an informed agricultural/rural life, in which people are well connected to the community that surrounds them and to nature and the interconnectedness of life, the economy is primarily locally driven, agriculture is sustainable, food is fresh and good, and work is done well and responsibly. In essence, Berry prizes the yeoman farmer who represented the ideal American for the agrarians of the 18th and 19th centuries. Opposed to this life are threats to the environment, both locally and globally;

large-scale industrial farming and corporate agribusinesses; greed and ignorance and indifference; a lack of respect for life and nature and others; and the increasing globalist focus of economics. Many of his essays, even when addressing contemporary concerns that are now 30 or 40 years old, evince a general philosophy that is applicable beyond that timely context.

One of Berry's most significant and influential works of nonfiction is 1977's *The Unsettling of America: Culture and Agriculture*, which in Berry's words "describes and opposes the abuses of farmland and farming people," and by extension the abuse and deterioration of America's agrarian, yeoman heritage. The book discusses the loss of farmland to urban and suburban development, the rise of agribusinesses (which then operated nationally and now have globalized—something Berry strongly opposes), the disastrous rate of soil erosion and the rampant problems of soil and water pollution, and in general, the losing battle fought by agrarianism against industrialism. In a follow-up essay, "The Agrarian Standard," written 25 years later (and pointing out that the conditions described in *Unsettling* grew worse in the interim), Berry describes his role as an agrarian writer as "writing essays and speeches that one would prefer not to write, that one wishes would prove unnecessary, that one hopes nobody will have any need for in twenty-five years . . . but I have never doubted for a minute the importance of the hope I have tried to serve: the hope that we might become a healthy people in a healthy land."

Berry's fiction chronicles the world and "membership" (population) of Port William, Kentucky, over the course of eight novels and 28 short stories. Through the history of Port William, we see the transformation of the public sphere as the Depression, the New Deal, and the rampant industrialization after World War II change the nature of American agriculture and the relationship between farm and town; we also see private dramas of alcoholism, murder, marriage, and unrequited love. Most of Berry's stories are set in the recent past—the early and middle decades of the 20th century. Again and again we see the alienation of generations from one another as the family farm diminishes in priority and the younger members of the family pursue life in the city. Marriage and family are frequent topics of exploration, and more generally the bonds between community members are demonstrated throughout the course of Berry's work.

Berry's first volume of poetry was a single elegiac poem, *November Twenty Six Nineteen Hundred Sixty Three*, published in 1964 to commemorate the recent death of President John F. Kennedy. His first full-length collection, *The Broken Ground*, followed later that year and developed Berry's pastoral themes of a sense of place and the cycle of life and death, grounded in imagery of the Kentucky River and the farming communities of north-central Kentucky. Sometimes elegiac, often pastoral, often ecologic, Berry's poetry is didactic, concerned primarily with conveying its values.

See Also: Agrarianism; Agribusiness; Yeoman Farmer.

Further Readings

Angyal, Andrew. *Wendell Berry*. New York: Twayne, 1995.
Berry, Wendell. "The Agrarian Standard." *Orion* (Summer 2002).
Berry, Wendell. *Standing by Words*. San Francisco: North Point, 1983.
Berry, Wendell. *The Unsettling of America: Culture and Agriculture*. San Francisco: Sierra Club, 1977.

Bill Kte'pi
Independent Scholar

BEYOND ORGANIC

The phrase *beyond organic* is used by a constituency of growers and consumers that are not satisfied with the current organic standards and want to see a significant change in the food system. Since the organic standards have been narrowed to focus only on the input used in organic agriculture, those who want to go beyond organic aim to include broader considerations about impacts of growing, processing and distributing products, or "whole systems" thinking. They believe that organic agriculture is just as capable of being industrialized as conventional agriculture. While the number of those looking to go beyond organic remains small in the United States, it is growing as food producers and consumers become more conscious of how their choices affect the Earth and human health.

The demands of the U.S. Department of Agriculture organic certification program do not satisfy all growers, producers, or consumers within the sustainable agriculture movement. Current regulations allow for longer-distance, and even global, marketing with legal assurance of certain standards but do little to ensure the protection of small farms, local production, the health of rural communities, social justice, and farm workers' rights. Many criticize the lack of emphasis on issues such as biological diversity, renewable energy, and environmentally aware land stewardship. A counter-effort by small growers, activists, chefs, and consumers is creating a diverse movement that pushes for the inclusion of these concerns in food system politics. These ambiguous, diverse, and nuanced arguments, movements, and individuals all represent beyond organic. From a beyond organic perspective, it is not enough to eliminate only chemical fertilizers and pesticides.

The organic movement began small in reaction to the environmental and social impacts of industrial agriculture, but it has since made its way into the mainstream. Organic food (produced without using most conventional pesticides, fertilizers made with synthetic ingredients or sewage sludge, bioengineering [genetic modification], or ionizing radiation) production is now a $4-billion-a-year industry, and in many cases large organic growers produce thousands of acres of genetically similar monocultures. For example, Safeway, a large national supermarket retailer, has its own organic brand, and Wal-Mart has expressed ambitions to become the largest organics retailer in the country. Today, the fastest-growing processed food on the market is organic packaged food. The organics industry has not escaped agricultural consolidation either. For example, according to research done at the University of California, Davis, 2 percent of California's organic farms accounted for half of the state's organic sales in the mid-to-late 1990s. Lobbyists, generally representing large-scale organics, have recently been pushing to weaken the definition of organic, and to a certain degree they have been successful. For example, products with preservatives that allow for a longer shelf life are now considered certified organic.

Those whose ideals can be considered beyond organic, in contrast, see the need for a fundamental change in the food system and have not been satisfied with the progress made through the organics movement—notably, certified organic foods. Specific concerns of the movement include the increasing size of farming operations, lack of labor standards, corporate ownership of the organics industry, and increasingly lax organic regulations. The beyond organics movement claims to be represented by people who value socially just, economically viable, and ecologically friendly foods and practices over those that seek higher profits and efficiency. For beyond organic farmers, growing food the right way, whatever that might mean to the individual in question, means selling an idea, not just healthy food.

Locally grown food is an important platform of beyond organic politics. Factory farms are designed to efficiently produce a vast quantity of food that then gets shipped (an average of 1,500 miles) to supermarkets around the world. The local food movement encourages people to buy from within their local community. In doing so, the argument is, these "locavores"—defined by the 2007 *Oxford American Dictionary* as someone who eats food only from a 100-mile radius of their home—strengthen personal ties in the community, support their local economy, and become more aware of, and therefore engaged with, how their food is grown and processed.

Although the definition of beyond organic varies by individual, there are some commonalities in addition to eating locally that many of the movement's constituents encourage and that are not requirements under the U.S. Department of Agriculture's organic certification program. Some of these are:

- Food sovereignty, or assuring food security. This may mean growing food for oneself in a community garden plot or empowering others to provide for the community.
- Reintroducing the social element of food that, many argue, has been lost via the predominant supermarket mentality.
- Direct marketing practices such as community supported agriculture, where customers pay a set price for, in general, a weekly box of farm goods.
- The slow-food movement, which encourages people to savor their food and funds the reintroduction of heirloom poultry and livestock breeds.
- Biodynamic agriculture, which incorporates natural rhythms into agricultural practices.
- The use of just wages and farming practices that do not put the worker at risk, including the fair trade movement.
- Conservation of habitat for endangered species and responsible water use, neither of which are addressed by organic certification. Programs such as the Oregon-based Pacific Rivers Council Salmon Safe program, Earth Island Institute's Dolphin Safe Tuna certification, and the Growers' Wool Cooperative's Predator Friendly Wool certification program go beyond the expectations of organic certification and highlight the connection between food production and wildlife preservation.

The changes to the agricultural sector envisioned by the beyond organic movement cannot simply be encompassed by a single certification stamp. Rather, a radical rethinking of the entire agricultural system is sought. Some argue that land reform is needed, and that farm machinery must be redesigned to accommodate polycultures (agriculture using multiple crops in the same space). The agrarian and permaculture movements support new concepts of land stewardship, and the empowerment of peasant communities is the focus of activists in the developing world, such as Vandana Shiva. One term used by many in the beyond organic movement to represent food that is grown locally and sustainably is *real food*. Real food incorporates an ethical responsibility into food consumption and asks that people know who grew their food, where they grew it, and what techniques they used. Community supported agriculture, farmers markets, and co-ops are important venues for real food.

Ultimately, beyond organic argues that the organic movement, as it stands, does not challenge the social, political, cultural, and economic structures underlying modern agriculture. Beyond organic suggests that prevailing agricultural paradigms must be questioned. The philosophical, political, ethical, and scientific tensions between the various positions of the movement represent its strength rather than its weakness and together compose a powerful and diverse force that is attempting to form a creative new vision for food systems in the 21st century.

See Also: Biodynamic Agriculture; Community-Supported Agriculture; Locavore; Permaculture; Slow Food Movement.

Further Readings

Ecological Farming Association. http://www.eco-farm.org (Accessed January 2009).
Imhoff, Daniel. "Food for Thought: Beyond Organic—Farming With Salmon, Coyotes, and Wolves." *Sierra Magazine,* 2000. http://www.sierraclub.org/sierra/199901/food.asp (Accessed January 2009).
Lappé, Francis Moore. *Diet for a Small Planet.* New York: Ballantine, 1971.
Pollan, Michael. *The Omnivore's Dilemma: A Natural History of Four Meals.* New York: Penguin, 2006.
Salatin, Joel. *Everything I Want to Do Is Illegal: War Stories From the Local Food Front.* Swoope, VA: Polyface, 2007.

Shannon Tyman
Stacy Vynne
University of Oregon

Biodynamic Agriculture

Biodynamic agriculture, or biodynamics, is a worldwide movement that advocates a "farm-as-organism" approach to agriculture. Biodynamics predates the organic agricultural movement and is sometimes described as "organic plus metaphysical." Proponents of biodynamics argue that the improvement and maintenance of soil life is fundamental to the care of landscapes, production of food crops, and management of livestock. Each of these elements must be understood in relation to the other, and farms will thrive if these elements can be true to the individual nature of each farm-organism. Farming success also requires that biodynamic practitioners embrace their own inner development and connectedness to the unique landscapes in which they farm.

Background

The principles of biodynamic agriculture were laid out in 1924 through a series of eight lectures given by Austrian scientist Dr. Rudolf Steiner (1862–1925) at Schloss Koberwitz in then Silesia, Germany (now Poland). Steiner delivered the lectures at the request of farmers concerned about soil degradation and declining crop and livestock health resulting from industrial farming techniques. Called the "agricultural course," the eight lectures were then published as the book *Spiritual Foundations for the Renewal of Agriculture,* and they form the intellectual basis of the biodynamic agriculture movement.

Steiner proposed the biodynamic method, in which each agricultural practice or measure related to life in its totality. Steiner, who was at the time renowned for his work on Johann Wolfgang von Goethe's scientific writings, had also developed innovative and holistic approaches that would lead to transformations in philosophy (anthroposophy), medicine (anthroposophical medicine), education (the Waldorf schools), special education (the Camphill movement), the arts (eurythmy), and more.

Steiner drew on his experiences researching the forces that regulate life and growth to propose a spiritually integrated approach to agriculture that nurtures the relationship between the ecology of the "Earth-organism" and the cosmos. Toward this end, Steiner proposed such "common-sense practices" as structuring activities to be in rhythm with nature; becoming self-sufficient in terms of energy, fertilizers, and the reproduction of plants and animals; building farm health by cultivating diverse plant and animal species; and maintaining oneself in concert with the farm by being reliable, orderly, and attuned to one's own—and the farm's—needs. Steiner and subsequent practitioners of biodynamic agriculture recognize that any farm is open to elements from beyond the farm's boundary, but the notion of farm-as-organism means that a healthy farm should be self-nourishing by reducing losses of organic matter and nutrients. Although Steiner noted that every farm should operate as self-sufficiently as possible, he acknowledged that such an ideal can rarely be completely attained. Accordingly, Steiner noted that materials brought in from outside the farm—for example, nonsynthetic fertilizers and compost preparations—should be considered as medicine for an ailing farm.

Steiner's lectures conveyed complex, dynamic relationships between farmers and farms, and he also provided practical measures to support these relationships. In the 1930s, Dr. Ehrenfried Pfeiffer, who had worked with Steiner, brought biodynamic agricultural concepts to the United States, leading to the founding of the U.S.-based Biodynamic Farming and Gardening Association in 1938 (http://www.biodynamics.com).

Method

Soil life is key to biodynamic agriculture and can be improved through proper humus management, crop rotation, intercropping, and other protective measures. Biodynamic principles suggest that farm manure and compost are the most valuable fertilizers, and soil health depends on applying sufficient organic manure and compost that have fermented according to prescribed processes. On the basis of his understanding of the material and spiritual worlds, Steiner outlined specific soil preparations, crop treatments, and other methods to enable "biologically dynamic"—or biodynamic—agriculture. These include fermented herbal and mineral preparations as compost additives and field sprays, methods for controlling weeds and pests, and an astronomical calendar to guide the timing of farming activities.

Two main types of "dynamically effective" substances are the field preparations Horn-manure (also called preparation 500) and Horn-silica (preparation 501), incorporated into a biodynamic farm by being sprayed directly on the soil or on growing plants. To prepare these compounds, cattle manure (for preparation 500) or silica from finely ground quartz (for preparation 501) is packed into cattle horns. Horn-manure preparations are then buried in autumn and retrieved in the spring, whereas Horn-silica preparations are buried in the spring and retrieved in autumn. During the period in which the cattle horns are buried, the horns serve as antennae to receive and concentrate cosmic forces into the materials inside. Small quantities of the contents of the horns then undergo "dynamization"— being stirred into water for an hour (amounts vary per preparation and field size) in such a way as to create a vortex that infuses the compounds and the water with the fundamental principle of life. The very processes of mixing and applying are prescribed to maximize the efficacy of the ingredients and the connections of the preparer with other natural forces. Meant to stimulate soil life, Horn-manure should be applied when cultivating the soil to make a seedbed or sprinkled on moderately moist soil during the late afternoon. Horn-silica is sprayed directly on plant leaves, as well as on the developing parts of the plant meant for harvest or use, during the morning of a sunny day.

The other key biodynamic compounds are compost preparations 502–8, made of herbs and other plant derivatives commonly used in homeopathic medicines. Preparations 502–6 include packing the plant materials (yarrow flowers, chamomile flowers, stinging nettle plants, oak bark, or dandelion flowers) into the skulls or organs of particular animals or into peat or manure, burying them in the Earth for particular periods of time, and then retrieving, diluting, and applying the preparations to compost. Preparation 507, which contains valerian flower extract, is also applied to compost, but without fermentation. Finally, preparation 508 is a tea made of horsetail plant used as a spray to counter fungal diseases. As with the Horn-manure and Horn-silica preparations, the creation and application of these preparations are regulated by an astronomical calendar.

Certified Biodynamic

Originally founded in 1928, the Demeter Association (http://www.demeter.net) of biodynamic farmers exists to regulate farming and processing standards for biodynamically produced foods. Demeter has developed a worldwide network of individual certification organizations, and since 1997, Demeter International has overseen certification organizations and ensured cooperation in the legal, economic, and spiritual spheres that resonate with biodynamic agricultural concepts. Demeter International currently represents approximately 4,200 Demeter-certified producers in 43 countries.

To be certified Biodynamic, farms must prohibit synthetic fertilizers, herbicides, and pesticides; genetically modified plants or animals; and synthetic additives used in processing. Demeter certification also requires specific measures to strengthen the life processes in soil and foodstuffs, as described above. Such preparations are a "non-replaceable element of biodynamic agriculture" for Demeter certification.

See Also: Holistic Management; Organic Farming; Slow Food Movement; Soil Nutrient Cycling.

Further Readings

Klett, Manfred. *Principles of Biodynamic Spray and Compost Preparations*. Edinburgh, UK: Floris, 2005.
Koepf, Herbert. *Bio-Dynamic Agriculture*. Herndon, VA: SteinerBooks, 1976.
Lampkin, N. H. and S. Padel, eds. *The Economics of Organic Farming*. Wallingford, UK: CAB, 1994.
Steiner, Rudolf. *Agriculture Course: The Birth of the Biodynamic Method*. London: Rudolf Steiner, 2004 (1924).

Jennifer Ellen Coffman
James Madison University

BIOLOGICAL CONTROL

Biological control is the use of an organism to control the growth rate of a pest population through predation, parasitism, herbivory, or competition. Optimally, biological control reduces a pest population to a level that is below economic injury to the crop, but

This U.S. Department of Agriculture Agricultural Research Service staffer is working with an insect called *Encarsia formosa*, which may be beneficial in controlling the whitefly pest.

Source: U.S. Department of Agriculture Agricultural Research Service/Scott Bauer

high enough to sustain the population of the beneficial organism. Because biological control relies on ecological processes, it is considered an essential tactic in integrated pest management programs.

Most animal populations experience fluctuations in numbers as a result of abiotic disturbances, such as drought, floods, and hurricanes, and biological factors, such as natural enemies, disease, and food quantity and quality. When resources are abundant, favorable environmental conditions exist, and natural enemies are scarce, population growth rates can be high. Conversely, when resources are limiting, environmental conditions are unfavorable, and activity from an antagonist organism increases, the animal population can decline.

Most plants possess physical (e.g., spines, waxy coatings, tough cell walls) or chemical (e.g., plant secondary metabolites) defenses against plant-feeding organisms. In agricultural crops, breeding has mainly focused on characteristics such as uniform growth, durability during shipping, and palatability. In the process, the plant's natural defenses against herbivores may be reduced or lost, increasing their susceptibility to damage from pests. When susceptible crops are planted in large "monocultures," they can be more apparent to and therefore more easily located by herbivores than in natural systems, which tend to be more diverse. Hence, in agroecosystems, food resources are generally not limiting for herbivores, and their populations may increase to levels that reduce crop yield or marketability, thereby becoming a "pest." Furthermore, intensively managed monocultures may lack the diversity and stability of resources to support a diverse natural enemy community. Thus, the idea behind biological control is to increase the population numbers of natural enemies of pests by either introducing or attracting them to fields or conserving those that may already be present.

Biocontrol agents can be herbivores, predators, parasitoids, or pathogens. Within those functional groups, biocontrol agents may be generalists (use a broad number of species) or specialists (use a limited number of species). When their numbers are sufficiently high, specialist natural enemies can be particularly effective at quickly and efficiently reducing the population of a single host species or genus. On contrast, the collective activity of generalists, which may be common in the farm landscape, can prevent a host population from reaching pest status.

Weed populations are controlled by herbivores and plant pathogens, which use above-ground (stems, leaves, flowers, and seeds) or below-ground (roots and tubers) plant parts.

Ground beetles (Coleoptera: Carabidae) have been widely documented for their ability to consistently consume weed seeds present in the soil (the seed bank), thereby reducing the following generation of weed seedlings. The combined granivory (seed-feeding) of crickets, earthworms, slugs, rodents, and birds can also reduce the seed bank of several economically important weed species, such as jimsonweed, pigweed, and common lamb's-quarters, by as much as 90 percent.

Arthropod pests are controlled by invertebrate and vertebrate predators, parasitoids, and pathogens. Ladybird beetles (Coleoptera: Coccinellidae), minute pirate bugs (Hemiptera: Anthocoridae), green lacewings (Neuroptera: Chrysopidae), and syrphid flies (Diptera: Syrphidae) are common predators of aphids, thrips, small caterpillars, and other soft-bodied insects. Parasitoids are wasps (Hymenoptera) or flies (Diptera) that lay their eggs in or on the egg, larva, or pupa of a host insect. The parasitoid larva that hatches from the egg consumes its host and eventually emerges as an adult. Trichogramma wasps are minute egg parasitoids of lepidopteran species and one of the most commonly released parasitoids in orchards and row-crop agriculture.

Just like other animals, insects contract diseases. Pathogenic microorganisms, including bacteria, viruses, protozoa, and fungi, can exert significant pressure on arthropod populations by interfering with reproduction, curbing growth, or killing them outright. The most widely used microbial insecticide is the spore-forming bacterium *Bacillus thuringiensis*, used mostly for caterpillar pests, but also for beetles. *Bacillus thuringiensis* is a crystalliferous bacterium that, when ingested, kills its host by dissolving in the midgut and causing gut paralysis.

There are three approaches to biological control: classical, augmentative, and conservation. Classical biological control involves the importation of an exotic control agent to reduce the population of an exotic (introduced) pest. Augmentative biological control is the periodic release of native or non-native natural enemies to boost local populations. Conservation biological control aims to preserve and enhance endemic or naturalized populations of natural enemies by reducing their mortality and maintaining habitat that provides necessary resources, such as alternate food and shelter.

Exotic organisms can become pests because they escape their natural enemies that occurred in their native range. The first step in developing a classical biological control program is to identify the origin of the exotic pest and to collect any natural enemies that are associated with that pest in the original range. These natural enemies are screened for their ability to reduce populations of the exotic pest, their host specificity, and potential effects on nontarget organisms. The quarantine process can take several years before a natural enemy species is identified as being effective and ecologically benign. Even then there is no guarantee that the natural enemy will successfully establish in a new geographic region and control the pest population. Successful programs depend on favorable environmental conditions and methods for rearing and delivering the natural enemy. One of the earliest and most successful classical biological programs was for cottony cushion scale (*Icerya purchase*) that was devastating citrus production in California in the late 1800s. The Vedalia beetle (*Rodolia cardinalis*) was introduced from Australia and effectively controlled the scale to the point where it is rarely a pest today. Classical biological control programs occur on a regional scale to achieve widespread extermination of a pest and for this reason often require significant institutional and government support.

Augmentative biological control differs from classical biological control in two ways: first, the natural enemy is typically present in the region, but not in numbers sufficient to regulate the pest population, and second, releases are periodic and occur at the scale of the field or farm. Examples of augmentative biological control include release of ladybeetles to control aphids, Trichogramma wasps to control lepidopteran pests, *Phytoseiulus persimilis* to control two-spotted spider mite, and minute pirate bugs to control thrips in greenhouses. Classical biological control and augmentative biological control are similar approaches in that they are commodified inputs that figure into the costs of production.

Conservation biological control is a preventive strategy that aims to build up local populations of arthropod natural enemies in farmlands by modifying production practices that cause mortality of beneficial insects, such as tillage and pesticide applications, and creating or conserving habitat that provides necessary resources, such as nectar and pollen, alternate prey, and stable sites for overwintering and reproduction. Perennial hedgerows, commercial annual flower mixes, and perennial grass buffer strips or "beetle banks" are examples of different types of habitat that a producer may establish in field borders or as strips within fields to enhance the activity of natural enemies. The area of influence of beneficial habitat on biological control depends on the extent of movement of the natural enemy. In general, large-bodied insects, such as syrphid flies, ladybird beetles, and ichnuemonid wasps, can travel longer distances (in the range of miles) than small-bodied insects, such as mites, small parasitoids, and minute pirate bugs (in the range of yards). Much of the interest in conservation biological control grew out of the 1965 classical study by R. L. Doutt and J. Nakata that documented an increase in parasitism rates of a pest (the grape leafhopper) in vineyards closer to riparian habitat. During the winter, the riparian habitat supported an alternate insect host for the leafhopper's key parasitoid, *Anagrus epos* (Hymenoptera: Mymaridae). Then, in the spring, *A. epos* emerged from its alternate host, dispersed into adjacent vineyards, and parasitized the grape leafhopper.

Habitat modifications increase the complexity of ecological interactions on the farm, which presents a challenge for developing predictive models of natural enemy and pest response in conservation biological control programs. Even so, conservation biological control can be the most self-sustaining and cost-effective of the approaches because the natural enemies are already present in the landscape.

Biological control has several constraints. First, through selective pressure by the natural enemy, resistant new strains of the pest can evolve that can then elude the natural enemy. Second, lack of host specificity can reduce the efficacy of the biocontrol agent against a specific pest and can be a problem in weed management if an introduced herbivore begins using a crop or a native plant as a resource. Finally, biological control does not achieve 100 percent pest control, nor does it offer an immediate solution to a pest problem. Because the increase in natural enemy populations depends on a robust pest population, some pest presence and activity in the crop field is expected. For high-value produce crops that have low economic thresholds, or for insects that transmit plant disease, biological control alone may be insufficient to prevent economic injury to a crop. There is also a time lag between the introduction of a biological control agent and the decline in its host pest numbers. For these reasons, biological control is most effective when used in a broader IPM program.

See Also: Agroecology; Integrated Pest Management; Organic Farming.

Further Readings

Barbercheck, Mary. "Biological Control of Insect Pests." http://www.extension.org/article/ 18931 (Accessed March 2009).

Bellows, Thomas S. and T. W. Fisher. *Handbook of Biological Control: Principles and Applications of Biological Control*. San Diego, CA: Academic, 1999.

Doutt, R. L. and J. Nakata. "Overwintering Refuge of *Anagrus Epos* (Hymenoptera: Mymaridae)." *Journal of Economic Entomology*, 58/3 (1965).

Norris, Robert F., et al. *Concepts in Integrated Pest Management*. Upper Saddle River, NJ: Prentice Hall, 2003.

Schonbeck, Mark. "Promoting Weed Seed Predation and Decay." http://www.extension.org/article/18544 (Accessed March 2009).

Tara Pisani Gareau
University of California, Santa Cruz

BRACERO PROGRAM

Bracero is a Spanish word that means "hired hand" and comes from the word *brazo* or "arm." The term *bracero* was generally used to refer to Mexican contract laborers working in the United States. Beginning on August 4, 1942, and authorized by Congress shortly thereafter, the U.S. government formally operated the Bracero Program, a federal guest-worker program designed to provide low-wage, temporary immigrant labor for agriculturalists and for the railroads—two industries that had grown accustomed to Depression-era wages and had a history of reliance on migrant and immigrant workers. Although the program was intended to be a temporary measure that would be ended after the war, strong lobbying on the part of California agriculture, combined with U.S. government permissiveness and the weakening of the Mexican government after the war, contributed to the program's continuation for more than two decades. Roughly 5 million braceros were employed under the program over the 22 years of its operation. The Bracero Program was, and continues to be, heavily criticized by immigrant and human rights groups as well as by labor organizations. Its demise can be linked, in part, with the 1962 founding of the United Farm Workers by Dolores Huerta and César Chávez and the activism of its predecessors, the National Farm Labor Union and the National Agricultural Workers Union. As recently as 2007, critics of President George W. Bush's immigration legislation proposals referred to the temporary guest worker provisions as a modern-day version of the Bracero Program.

Following a wave of deportations during the Great Depression, gradual economic recovery led to increased immigration from Mexico. Some of the roughly 500,000 Mexican workers who had been deported returned, along with many others. Workers were also recruited from the Caribbean, with Florida's cane fields being their primary destination. In the case of agriculture, the wartime economy and deployment of citizen workers, especially in states such as California, Texas, and Florida, allowed many farm laborers to find year-round employment in shipyards and factories. The resulting competition for workers among industries and agriculture presented a unique opportunity for both farm workers and labor organizations to demand higher wages and worker protections. In response, agribusiness lobbied Congress, arguing that labor shortages were causing economic ruin, particularly in the Southwest. Farm employers had been requesting permission from the Immigration and Naturalization Service (INS) to import Mexican contract labor for years without success. In the spring of 1942, however, the INS formed a committee to consider the question of labor importation. The result of their efforts was the drafting and signing of the bilateral agreement known as the Bracero Accords. The agreement between the United States and Mexico allowed the importation of Mexican temporary workers under U.S. government supervision, with the Mexican government bargaining on behalf of contract workers. Labor standards for bracero contracts were established by the accords, but they were not routinely enforced or observed, especially when control of the program was given over to the employers.

The Bracero Accords were revised in the months following agreement, and a final version was released in April 1943. According to the formal agreement, wage standards for braceros were to be equivalent to the wages earned by domestic workers in similar employment. The minimum wage was set at 30 cents per hour, and a subsistence wage of $3 per day was to be paid if braceros were unemployed for more than a quarter of the contract period. The U.S. and Mexican governments agreed that Mexican contract workers would not engage in U.S. military service or be employed to replace or reduce the wages of domestic workers. Ironically, many of the provisions of the agreement promised laborers the same sanitary and housing conditions enjoyed by domestic farm workers, who had been routinely omitted from U.S. labor legislation guaranteeing basic standards and protections on such matters. In addition to wage and workplace guarantees, the Mexican negotiators initially required that employers in Texas be excluded from contracting braceros because of the history of discrimination against Mexican workers in Texas. In fact, one of the general provisions of the revised bracero agreement stated that Mexicans entering the United States as Bracero Program workers, in accordance with an executive order, shall not suffer discriminatory acts of any kind. Additional guarantees included transportation and repatriation, as well as a prohibition against the employment of minors under the age of 14 years. Regular violations and abuses of workers' rights and guarantees, however, quickly tarnished the program's reputation.

In addition, the mechanization of much factory and farm labor led to increased work hazards and production demands while at the same time forcing a decline in available employment opportunities. Before 1949, many bracero contracts expressly forbade workers from operating machinery. If they did so, they were to be paid at a higher rate. The accords explicitly acknowledged that these higher-paying jobs should be reserved for domestics and citizens. Migrant field workers, who were exempted from the 1935 National Labor Relations Act along with domestic workers, faced significant health hazards as a result of the physical arduousness, pesticide and sun exposure, water contamination, and mechanical failures that often characterized their workplaces.

From 1947 until 1964, when the program was ended by the federal government, bracero contracts were handled directly by growers who set the terms of contracts, which often fell short of the requirements of the original agreement. Before this time, the U.S. Farm Security Administration functioned as the "employer" in the Bracero Accords, with farm-labor contractors listed as subemployers. Before 1950, fewer than 70,000 workers were brought into the United States annually on bracero contracts, but after the growers became directly involved in the bracero contracts in 1947, and as demand for agricultural workers in the United States grew, the number of migrant bracero workers increased dramatically, regularly exceeding 400,000. During the later periods of the Bracero Program, before it was officially terminated in 1964, Mexican migration to Illinois, in addition to California, increased substantially. In many cases, undocumented workers who were already residing in the country were contracted as braceros as well, adding to the overall numbers of immigrant workers in the fields. The practice of paroling undocumented immigrants into bracero contracts, thus legalizing their presence as workers in the United States, led to a widespread perception among Mexican workers that it was easier to cross illegally into the United States and then be hired as a bracero than it was to wait to be contracted from within Mexico. This perception contributed to the problem of illegal immigration, along with the widespread practice of hiring undocumented workers who would accept wages below those guaranteed by the Bracero Program.

The Bracero Program ended as a federal program in 1964 and was replaced by maquiladora agreements, which allowed U.S. companies to move factory production and other business to the border to benefit from the low-wage workforce in Mexico. Labor organizations, religious groups, and community organizations protesting the often horrific treatment of immigrant workers in migrant camps drew widespread media attention that resulted in a series of hearings and committee meetings to examine migrant labor in the United States. Additional pressure to end the program came from an oversupply of immigrant labor in the United States, in part as a result of farm and factory mechanization and to high unemployment rates among working-class citizens.

See Also: Agribusiness; Contract Farming; Fair Labor Association; Labor; Pesticide; United Farm Workers.

Further Readings

Calavitas, Kitty. *Inside the State: The Bracero Program, Immigration and the I.N.S.* New York: Routledge, 1992.

Galarza, Ernesto. *Merchants of Labor: The Mexican Bracero Story.* Santa Barbara, CA: McNally & Loftin, 1964.

Gamboa, Erasmo. *Mexican Labor and World War II: Braceros in the Pacific Northwest 1942–1947.* Seattle: University of Washington Press, 2000.

Nevins, Joseph. *Operation Gatekeeper: The Rise of the "Illegal Alien" and the Making of the U.S.-Mexico Boundary.* New York: Routledge, 2002.

U.S. Department of State. "Temporary Migration of Mexican Agricultural Workers." Executive Agreement Series No. 278. Washington, D.C.: Government Printing Office, 1943.

Emily Plec
Western Oregon University

Bt

Bt, or *Bacillus thuringiensis*, is a species of spore-forming bacteria that occurs naturally in most soils around the world, as well as in some plants. During sporulation, Bt produces proteins that aggregate into crystals, some of which operate as toxins against certain species considered agricultural pests, such as those within the orders of Lepidoptera (moths and butterflies), Coleoptera (beetles), and Diptera (flies and mosquitoes). There are hundreds of different Bt strains, most of which can produce several different crystal proteins, and the same protein may be found in multiple different strains. Because of their insecticidal properties and purported safety to humans and many beneficial insects, different Bt strains have been used in so-called traditional agriculture, and Bt is even approved as an organic pesticide. More recently and controversially, Bt has been incorporated into trans-genic/biotech crops.

Bt Crystals as Insecticides

Bt's appeal as a pesticide is the specificity with which it works. Most Bt strains produce different combinations of insecticidal crystal proteins (ICPs), and different ICPs prove toxic to different types of insects. This is so because the toxins bind only to particular receptors in midgut epithelial cells.

Basically, insects must ingest Bt crystals and spores to be affected. Once ingested, the crystal toxins dissolve if the conditions in the gut are suitable, and then the toxins bind to receptors and erode the gut lining, which allows the Bt spores and gut bacteria to spread out of the gut and eventually kill the host. Although this entire process may take days, the insect will stop feeding within hours of initial ingestion.

Because not all insects possess the same gut receptors, different ICPs—from different strains of Bt—may be developed to target specific insect species. Bt is considered nontoxic or even harmless to vertebrates, including humans, and many other arthropods, including a wide range of "beneficial" insects, which do not possess receptors as described.

Bt in Commercial Products

In 1901, Japanese bacteriologist Shigetane Ishiwatari first isolated what is now known as Bt as the cause of *sotto,* or sudden-collapse disease, which was decimating silkworm populations and threatening Japan's silk industry. Given the negative effects of the bacterium, there was little desire by Ishiwatari to expand its use. But in 1911, German scientist Ernst Berliner rediscovered it as he examined dead flour moth caterpillars. Because of the bacterium's effectiveness in killing a "pest" species, Berliner detailed the bacterium as a potentially valuable pesticide. He published a scientific description and assigned the name *Bacillus thuringiensis* in 1915. Efforts to cultivate Bt as a pesticide then began in earnest.

French insecticide industries began to commercially produce Bt-based formulas, such as Sporeine in 1938, as powders containing sporulated cells and their toxic crystals. Over the decades, additional formulations of Bt-based products targeting a variety of insects were developed and marketed in Europe and in the United States. By 1995, the U.S. Environmental Protection Agency listed 182 registered Bt products, but they accounted for only 2 percent of the money spent globally on insecticides.

Bt insecticides are mixed as sprays and applied to leaves or other surfaces where the targeted insect larvae feed. Bt toxins demonstrate great specificity, degrade when exposed to ultraviolet light, and are easily washed away by rain. These qualities attracted environmentalists to Bt as an appealing alternative to dangerous and bioaccumulating organochlorides, such as DDT. In fact, Bt even received an endorsement from Rachel Carson in her 1962 best seller *Silent Spring.* However, those same qualities led many large-scale farmers to dismiss Bt spray applications: They complained that Bt had a narrow spectrum of activity and was less toxic compared with broader-range and more potent pesticides that killed multiple insect species (e.g., organophosphate pesticides), and Bt had to be sprayed more often and at higher prices than readily available synthetic pesticides. So, although external applications of Bt eventually became approved for use on certified organic farms, Bt did not become a pesticide of serious renown until the mid-1990s, when the biotech industry began to genetically engineer Bt crops.

Biotech Bt

Since the mid-1990s, a variety of crop plants have been genetically modified to express the ICPs that Bt produces, with the goal being to create crops that could generate ICPs and thus protect themselves throughout their lifetimes from insects and without need for externally applied pesticides. Over 100 Bt toxin genes have been cloned and sequenced. They have been spliced into the genomes of multiple major food crops, including corn, cotton, potatoes, soybeans, broccoli, and cabbage. Other Bt crops being developed include apples, canola/rapeseed, eggplants, peanuts, rice, sunflowers, tobacco, tomatoes, and more.

By 2007, an estimated 115 million hectares of land worldwide grew genetically modified crops. Today, about half of the corn grown within the United States is Bt corn. Clearly, this technology has been widely adopted, but serious concerns remain about insect resistance to Bt, unintended genetic crossover, negative effects on nontarget species, changes in the nutrient composition of transgenic plants, and potential toxicity and allergenicity of the introduced proteins to humans and livestock.

For example, because some insects have demonstrated resistance to Bt sprays, it seems reasonable to expect that some insects will also develop resistance to certain Bt proteins expressed in transgenic crops. In an attempt to slow such resistance in the United States, the U.S. Environmental Protection Agency mandated refuges of non-Bt host plants to surround fields of Bt crops. These refuges are meant to enable susceptible target insects to survive so that they can breed with resistant individuals. Known resistance mechanisms to Bt are recessive, so this process of interbreeding resistant and nonresistant individuals is meant to diminish resistance genes. It is too early to tell how effective this method will be.

In addition to concerns over unintended biological consequences, questions about social, political, and legal/regulatory issues have arisen. One of the most common Bt cash crops grown in so-called developing countries is cotton, but reports on the economic benefits have been highly variable over the years. Activists like Vandana Shiva and members of the Coalition for GM-Free India charge that the current emphasis on biotechnology has diverted efforts from more sustainable agricultural practices and forced Indian cotton farmers into debt. Others draw attention to the growing illegal economies in India and elsewhere, in which people sell and trade unapproved transgenic cotton hybrids to avoid the high prices charged by Monsanto.

Monsanto dominates the biotech industry in Bt crops. Now engaged in a collaborative research program on biofuels with Archer Daniels Midland and Deere & Company, Monsanto continues to find a variety of outlets through which it can extend its Bt corn and other products. Because biotechnology and the creation of transgenic plants are relatively new phenomena, long-term effects of Bt crops on non–genetically modified crops and food production systems remain to be seen.

See Also: Agribusiness; Cash Crop; Genetically Modified Organisms; Monsanto.

Further Readings

Aroian Laboratory at the University of California, San Diego. http://www.bt.ucsd.edu (Accessed February 2009).

Shelton, Anthony M. "Considerations on the Use of Transgenic Crops for Insect Control." *Journal of Development Studies*, 43/5:890–900 (2007).

Shiva, Vandana. *Earth Democracy: Justice, Sustainability and Peace*. Cambridge, MA: South End, 2005.

Smale, M., et al. "Bales and Balance: A Review of the Methods Used to Assess the Economic Impact of Bt Cotton on Farmers in Developing Economies." *AgBioForum, 9/3:*195–212 (2006).

Jennifer Ellen Coffman
James Madison University

CALIFORNIA CERTIFIED ORGANIC FARMERS

Started in 1973, California Certified Organic Farmers (CCOF) was one of the first organizations to certify organic farm products. For most of this history, CCOF has worked through regional chapters to maintain standards that were agreed on by the growers: peer-review inspections took place where the farmers mutually certified each other's adherence to the standards. However, because the U.S. Department of Agriculture (USDA) passed federal standards in 2002 under the National Organic Program, CCOF now serves as a third-party certifier for the USDA label. It is currently the largest organic certifier in the country, certifying nearly 80 percent of the organic farmland in California and over 11 percent of the organic businesses in the United States.

CCOF Structure

CCOF is the only third-party certifier that provides both organic certification and a trade association to market organic products. These are two of the three components in the structure of the not-for-profit CCOF organization, which also includes a foundation:

- *CCOF Trade Association.* The CCOF Trade Association promotes organic food in the marketplace. It tries to increase consumer demand and public support through outreach programs, including education, public relations, marketing strategies, and political advocacy. The association is divided into 15 chapters representing 13 regions of California, an At-Large chapter, and a Processors/Handlers chapter. In addition to the 2,100 certified members in the trade association, there are an additional 350 supporting members—individuals and businesses, both certified and not—who seek to engage in the marketplace of organic food.
- *CCOF Certification Services.* A subsidiary of the Trade Association, CCOF Certification Services is a distinct entity because of the USDA's requirement that third-party agents must be free of conflict to ensure certification, thus separating the certifiers from those who promote those products in the marketplace. CCOF Certification Services is accredited by the USDA National Organic Program and maintains an international program to allow for overseas shipping, including to Europe, Japan, and Canada. CCOF currently certifies farms and products in 35 states and three foreign countries.
- *CCOF Foundation.* The CCOF Foundation is a nonprofit organization created to develop education and outreach activities. Working with growers and the general public, it aims to foster growth of organic farming through activities such as training and tours of organic farms.

CCOF certifies nearly 600,000 acres of farmland and over 1,300 different crops, products, and services. Crops include fruit, nuts, nursery flowers, vegetables, herbs, beans, and grains. Its sign is a common fixture for booths at farmers markets around the state of California. With a stated goal of creating a "farm-to-fork" certification process for organic farming, however, CCOF certifies products beyond agriculture, including livestock, food processors and retailers, private labelers, and restaurants. For example, CCOF certified the Ukiah Brewing Company in 2001, making it the first certified organic brewpub in the United States.

CCOF Chapters

CCOF is organized into the following chapters:

- Big Valley: Contra Costa, Merced, San Joaquin, and Stanislaus counties
- Central Coast: Alameda, Monterey, San Benito, San Mateo, Santa Clara, Santa Cruz, and San Francisco counties
- Kern: Kern County
- Fresno-Tulare: Fresno, Kings, Madera, and Tulare counties
- Humboldt-Trinity: Humboldt, Del Norte, and Trinity counties
- North Coast: Marin, Napa, and Sonoma counties
- Mendocino: Lake and Mendocino counties
- San Luis Obispo: San Luis Obispo County
- South Coast: Santa Barbara and Ventura counties
- Pacific Southwest: Imperial, Los Angeles, Orange, Riverside, and San Diego counties
- North Valley: Butte, Glenn, Lassen, Modoc, Plumas, Shasta, Sierra, Siskiyou, Tehama, and Yuba counties
- Sierra Gold: Amador, Calaveras, El Dorado, Placer, and Tuolumne counties
- Yolo: Colusa, Nevada, Sacramento, Solano, Sutter, and Yolo counties
- At-Large: Certified members outside established CCOF chapters, including international members
- Processor/Handler: Processors, handlers, packers, retailers

History

The green revolution of the 1950s allowed for the conversion of food production from small farms to larger operations, often called agribusiness. This movement led to an increase in reliance on chemicals. CCOF, based in Santa Cruz, California, was one of the first groups to certify food as organic, after 54 farmers agreed on standards and conducted their own enforcement of these standards, which forbade artificial pesticides and fertilizers. "Standards developed by the farmers themselves added credibility to the young movement." explained Peggy Miars, executive director and chief executive officer of CCOF; this model would become an example for the entire country. This farmer-centered structure required that organization be decentralized: regional chapters are largely autonomous in marketing and certification but are also still connected to the parent organization. Because of this structure, the quarterly newsletter *Certified Organic* became an important component of the development of the organization.

From the outset, CCOF also involved itself in policy and advocacy. CCOF helped draft and pass legislation in California—the California Organic Food Act of 1979—again allowing the organic farmers themselves to control standards. Enforcement of the organic standards was not included in the law, and this left private groups such as the CCOF to sue the state, seeking enforcement. The first big case was against Pacific Organics in 1988: The company had been repacking conventional carrots as organic, and CCOF's campaign and publicity pushed the state to enforce the law. CCOF returned to Sacramento in 1990 to help draft the updated California Organic Food Act. Also in the 1990s, CCOF founded two organizations: the Organic Farming Research Foundation and, along with the Oregon Tilth, the Organic Materials Review Institute, which is the country's large clearinghouse for organic materials.

CCOF later would help develop national standards. In fact, the organic industry first approached the USDA to seek federal standards, which took 12 years to finalize. Dating from 2002, when these national standards were implemented, the USDA accredits all organic certifiers, including CCOF; CCOF now works as a third-party certifier. Changes made to the National Organic Program and its standards are made through the USDA. CCOF is one of hundreds of organizations contributing to the continually updated standards, although as one of the oldest voices in the field, CCOF enjoys high credibility in the organic community. CCOF continues to grow, adding more staff at the headquarters as member farmers add more acreage in the field. Expansion of their practice is a stated goal of CCOF. Miars states the organization's vision: "We want organic to become more prevalent in the world. It will create a better world for everyone—consumers, farmers, and farm workers."

Criticisms

Perhaps the greatest criticism of organic farming in general is the cost of certification. Small farms and processes can get certified for $400–$1,000, whereas large ones average about $30,000. Fees paid to CCOF cover an annual fee and the inspection process. However, proponents counter that organic practices themselves can be more expensive—for example, using manual weeding to replace chemical application—and that to retain the credibility of the organic label, certification needs to be specific and regulated. An additional dilemma has been in the accreditation of particular products: For example, a fertilizer had been considered organic and was used by farmers, but later the practices of the fertilizer's manufacturer were questioned. The manufacturer reached a private settlement with the U.S. government. Because the fertilizer was deemed organic at the time of use by the certified-organic farmers, however, those farms were not penalized.

See Also: Agribusiness; Certified Organic; Farmers Market; Green Revolution; National Organic Program; Northeast Organic Farming Association; Organic Farming.

Further Readings

California Certified Organic Farmers. "Organic Certification, Trade Association, Education & Outreach, Political Advocacy." http://www.ccof.org (Accessed May 2009).
Proctor, Keith L. "CCOF History 1973–1979/1980–1990/1990–2000/2000–2003." *CCOF Certified Organic* (Spring 2003/Summer 2003/Fall 2003/Winter2003).

Rachel Adams
Stanford University

CASH CROP

Cash crops are grown for direct sale in the market, rather than for family consumption or to feed livestock. Coffee, cocoa, tea, sugarcane, cotton, and spices are some examples of cash crops. Food crops such as rice, wheat, and corn are also grown as cash crops to meet the global food demand. Production of many of these crops is a form of export-oriented production, a development associated with European colonization in the topics and subtropics. Europeans developed cash crops in large-scale, capital-intensive, and export-oriented plantation agricultural systems. Today, smallholder farmers in developing countries grow most of the cash crops, such as coffee.

A man spreads coffee beans out to dry in May 2008 in Kwanza Sul Province, Angola, where the coffee industry has only recently recovered after decades of war. For some countries, coffee brings in 75 percent or more of all export earnings.

Source: U.S. Agency for International Development

Cultivation of cash crops was not always associated with colonization. Precolonial societies also engaged in growing crops for sale to balance their dual foci of risk and subsistence security. Colonization brought another dimension to cash crop production by introducing new crops, technology, and scale of production.

Colonization in the countries of Africa, Asia, and South America has transformed peasant households into commodity producers. Cultivation of cash crops has brought structural changes in the peasant livelihoods in developing countries as a result of the rising cost of production, increased in- and out-migration, and dependency on the global market. Largeholder farmers in developing countries often devote a larger share of their land to cash crops compared with that of smallholder farmers. During most of the 20th century, many countries intervened in primary cash crop markets. In the late 1980s and early 1990s, commodity agreements, such as the International Coffee Agreement, failed, and commodity parastatals became financially strained.

Cultivation of export-oriented cash crops has increased the peasant farmers' vulnerability to fluctuating global market prices. With rising costs of land and labor—the increased cost of inputs without guarantees of higher output for cash crops—farmers in developing countries are forced to intensify their cash crop production, to reduce their level of consumption, or both. Despite this disadvantage, cultivation of cash crops affords a means for smallholder farmers to improve their economic conditions.

Coffee is an important cash crop, grown mostly in developing countries of Asia, Africa, and the Americas. Coffee grows as a red cherry on a bush that grows to about 1.5 meters in diameter. Coffee cultivation was expanded to Africa, Asia, and Latin America by the European colonial powers. Coffee, produced in more than 60 countries, is one of the world's most heavily traded commodities. For many countries, coffee exports account for more than 75 percent of their total export earnings. The coffee market was highly regulated until 1989, when the international coffee agreement between producing countries and consuming countries collapsed. The 1990s and 2000s experienced coffee crises as global coffee prices plummeted and livelihood conditions of millions of smallholder farmers throughout the coffee-growing countries deteriorated.

Tea remains an important cash crop worldwide. The tea plant is a large evergreen shrub. Ideally, the tip of the new shoot, including two leaves and a bud, is used in making tea. Most of the tea consumed in the world is produced on large plantations. There are also many small tea gardens. The world tea production is about 3.15 million tons annually. The largest producers are India and China, followed by Kenya and Sri Lanka.

Cotton is also grown throughout the world as a cash crop. China, India, and the United States are the three major cotton producers in the world. The United States is the top exporter of cotton, followed by Uzbekistan and India. Cotton was domesticated in India, and archeological evidence shows that cotton cloths were used in Indus Valley civilizations. Cotton played a major role in Britain's economy during the Industrial Revolution. In the beginning of the Industrial Revolution, cotton was supplied from India to the factories in Manchester, England. Eventually, the United States became the major source of cotton to Britain's textile industry. Cotton became an important cash crop in the southern United States, where it was directly tied to slavery. Cotton changed U.S agricultural history and literally caused the enslavement of hundreds of thousands of men and women until the mid-19th century.

Sugarcane can also be considered a cash crop, and most countries in the world grow it. Brazil is the world's largest producer of sugarcane, followed by India. Sugarcane is used in the production of sugar, molasses, alcoholic beverages, and ethanol for fuel. Sugarcane was spread by the colonial empires to different parts of the world from India and Africa and was one of the first cash crops of early colonial America. It grew plentifully in the southern states and was a major source of income for many plantations.

Cocoa is grown mainly by smallholder farmers in developing countries as a cash crop. The cocoa plant originated in the upper Amazon basin. A cocoa plant can grow up to 50 feet tall. Cocoa trees produce pods containing the seeds that will become cocoa beans, which are used in making chocolate. The top three countries that grow cocoa are Côte d'Ivoire, Ghana, and Indonesia.

Spices were one of earliest cash crops grown in south Asia. India accounts for more than 85 percent of the world's spice production. China and Bangladesh also produce a significant amount of spices, which have played a major role in history as cash crops. The spice trade linked the civilizations of Asia, Europe, and the Middle East.

Ecuador, Costa Rica, the Philippines, and Colombia contribute more than two-thirds of the world banana exports. The export of bananas and coffee account for most of the food trade in Central American countries in the 20th century. The United Fruit Company controlled most of the banana trade in Central America in the 19th and 20th centuries. The company is currently operating under the name Chiquita Brands International. Costa Rica, Honduras, and Panama were once known as "banana republics" because the banana trade dominated their economies.

Several vegetables, fruits, and flowers also fall under the term *cash crops*. Perishable cash crops such as vegetables and fruits are grown closer to cities or consumers. Other cash crops worth mentioning are oilseeds, rice wheat, corn, and chilies.

See Also: Green Revolution; Rice.

Further Readings

Akiyama, T., et al. "Commodity Market Reforms: Lessons of Two Decades." Washington, D.C.: World Bank (2001).

Bernstein, H. *Agrarian Classes in Capitalist Development.* London, Routledge, 1994.

Dictionary of the English Language, 4th ed. Boston: Houghton Mifflin, 2006.

Fafchamps, M. "Cash Crop Production, Food Price Volatility, and Rural Market Integration in the Third World." *American Journal of Agricultural Economics,* 74/1 (1992).

Food and Agriculture Organization of the United Nations. "Major Food and Agricultural Commodities and Producers." http://www.fao.org/es/ess/top/commodity.html?lang=en&item=156&year=2005 (Accessed January 2009).

International Cocoa Organization. "Production." http://www.icco.org/economics/promotion.aspx (Accessed January 2009).

Singh, J. and S. S. Dhillon. *Agricultural Geography.* New Delhi, India: Tata McGraw-Hill, 2004.

Talbot, J. *Grounds for Agreement: The Political Economy of the Coffee Commodity Chain.* Lanham, MD: Rowman & Littlefield, 2004.

Watts, M. *Silent Violence.* Berkeley: University of California Press, 1983.

West, Jean. "King Cotton: The Fiber of Slavery." http://www.slaveryinamerica.org/history/hs_es_cotton.htm (Accessed January 2009).

Shrinidhi Ambinakudige
Mississippi State University

CERTIFIED HUMANE

The Certified Humane Raised and Handled Label is a consumer certification and labeling program identifying egg, dairy, meat, and poultry products that are produced in a manner that considers the welfare of the animals from birth to death. The program came into being because of the concern among animal welfare groups and consumers that farm animals were living in inhumane conditions before their slaughter. Certified Humane programs are growing, both in the United States and internationally, as consumers recognize the perceived benefits and show more interest in knowing how their food was raised and processed.

Animals certified as humanely raised must be allowed to engage in natural behaviors such as flapping wings and dust bathing for chickens and rooting for pigs; have sufficient space, shelter, and handling so as to eliminate stress; be provided with access to ample amounts of clean water; and be fed a natural diet free of growth hormones and antibiotics. All caretakers must also be trained in animal husbandry and welfare. Operators typically pay a fee (e.g., $50) for the certification application as well as a fee for operation inspection (e.g., $500–$600/day/inspector).

Requirements for the raising, handling, and processing or slaughter of animals are provided by different certification programs. Standards vary depending on the animal but typically are provided for beef cattle, dairy cows, pigs, sheep, broiler chickens, laying chickens, turkeys, goats, and young dairy beef. Certified Humane does not necessarily mean that the products are organic or free range, as the program is more concerned about the welfare of the animals than the type of farming system.

To obtain humane certification, facilities must meet precise, objective standards set by nonprofit organizations and local, state, and federal governments and be certified annually by an inspector. For example, meat processors must comply with the American Meat Institute Standards, which works to ensure optimal care of animals with minimal suffering. Although the government has some standards for the welfare of farm animals, such as the Animals Welfare Act and Humane Methods of Slaughter Act passed by Congress, these laws do not set certification requirements, nor are they applied to all types of livestock. Many animal welfare groups have accused the U.S. federal government of not doing enough to safeguard farm animals, leading to the creation of nongovernmental humane certification programs.

There are two major Humane Certification Programs in the United States: the Humane Farm Animal Care Certified Humane Raised and Handled Label and the American Humane Association American Humane Certified program. The Humane Farm Animal Care, a national nonprofit organization, worked with a team of animal scientists and veterinarians to develop the standards for Certified Humane Raised and Handled Label in 2003. The standards are based on a review of existing programs in Europe and on scientific review and are updated as new information becomes available. The program has grown significantly since it started: in 2003, 143,000 animals were raised under its standards; by the end of 2006, there were more than 9 million certified animals.

Humane Farm Animal Care is audited annually by the U.S. Department of Agriculture, as well as the International Organization for Standardization. Producers that are Certified Humane are also audited annually. Failure to meet any of the standards will result in revocation of certification.

The American Humane Certified program is a third-party, voluntary, independent verification program that certifies the producer's care and handling of farm animals to ensure that their treatment meets the animal welfare standards of the American Humane Association. The American Humane Association has worked since 1877 to improve the living, handling, and processing conditions for cattle, hogs, sheep, and poultry. The American Humane Certified standards are based on the standards of the Royal Society for the Prevention of Cruelty to Animals and rely on scientific reviews by animal science experts, veterinarians, and animal husbandry specialists.

There are programs outside the United States that are similar to the Certified Humane program. In the United Kingdom, the Royal Society for the Prevention of Cruelty to Animals monitors the Freedom Food label, which offers certification for farmers, haulers, abattoirs, processors, and packers to ensure well-managed free-range, organic, and indoor farms. The welfare standards are based on scientific research, veterinary advice, and input from farmers and undergo constant review to ensure they incorporate the latest information on animal welfare. Operators must meet standards for rearing, handling, transport, and slaughter of farm animals that are broadly based on the "Five Freedoms," as defined by the Farm Animal Welfare Council: (1) freedom from hunger and thirst; (2) freedom from discomfort; (3) freedom from pain, injury, or disease; (4) freedom to express normal behavior; and (5) freedom from fear and distress. Standards have been developed for beef cattle, chickens, turkeys, dairy cattle, ducks, laying hens, pigs, salmon, and sheep.

Demand for Certified Humane is growing within the United States as well as internationally, even though the cost of certified products is slightly higher than the cost of conventionally raised products. As consumers have become more aware of the negative implications confined spaces have on the animals they eat, demand for products that are humanely raised has risen. The benefits of humane treatment of farm animals that have been used to develop laws and guide nonprofit certification processes include reducing physical and mental harm to animals (which, in addition to the pain incurred by the animal, may affect the health of the meat) and reducing the spread of disease among animals. Some claim that humanely raised food tastes better because animals have been able to perform natural behaviors. Purchase of Certified Humane products has also grown with recognition of conditions beyond the animals: the overuse of antibiotics in animal products that has led to more strains of drug-resistant bacteria, the reduced earnings for independent farm operators as a result of the growing number of factory farms, and pollution in waterways from factory farm runoff. Others argue that the use of Certified Humane programs is simply to allow humans to feel less guilty about slaughtering animals. Regardless of the various arguments over the benefits of Certified Humane labeling, by identifying which meat products have been humanely raised, consumers are given more purchasing power.

See Also: Animal Welfare; Certified Organic; Confined Animal Feeding Operation; Factory Farm; Recombinant Bovine Growth Hormone.

Further Readings

American Humane Certified. http://www.thehumanetouch.org (Accessed December 2008).
Certified Humane. http://www.certifiedhumane.com (Accessed December 2008).
Factory Farm. http://www.factoryfarm.org/?page_id=48 (Accessed December 2008).
Farm Animal Welfare Council. http://www.fawc.org.uk (Accessed January 2009).
Humane Farm Animal Care. "Policy Manual." August 15, 2008. http://certifiedhumane.com/documents/Pol08.v8A.pdf (Accessed January 2009).
Royal Society for the Prevention of Cruelty to Animals. "Freedom Food Label." http://www.rspca.org.uk/servlet/Satellite?pagename=RSPCA/RSPCARedirect&pg=FreedomFoodHomepage (Accessed December 2008).
Royal Society for the Prevention of Cruelty to Animals. "RSPCA Freedom Food Label." http://www.rspca.org.uk/servlet/Satellite?pagename=RSPCA/RSPCARedirect&pg=FreedomFoodHomepage (Accessed January 2009).

Stacy Vynne
Shannon Tyman
University of Oregon

CERTIFIED ORGANIC

Certified Organic is an agricultural certification process that refers to the production, processing, and handling of food and agricultural products in a way that reduces harm to humans, animals, and soil life. To become Certified Organic, food producers must meet strict standards set by the governments in the countries where the food is grown and imported. Although pricing of organic foods remains higher than that for conventional foods, prices are coming down as demand for organics increases.

The term *organic* refers to the methods used in the growth of agricultural products, as well as the processing and handling of these products. Farms growing organic foods vary in size: they may be small family farms that sell products only at local farmers markets, or mass-production farms that distribute products in supermarket chains across the country and the world. Compared with conventionally grown foods, organic food products are produced without using most conventional pesticides, fertilizers made with synthetic ingredients or sewage sludge, bioengineering (genetic modification), or ionizing radiation. Organics are typically minimally processed, without the addition of artificial or synthetic ingredients. The processing equipment and storage and shipping containers must be verified to ensure that they do not contain synthetic fungicides, preservatives, or fumigants. Growers must also prove that no prohibited substances have been used on the land for at least three years.

Animal care and manure composting techniques must be done in a manner to reduce the risk of transmitting food-borne diseases. For meat, poultry, eggs, and dairy to be certified organic, the animals must be uncaged inside barns or warehouses, with some access to outdoor areas. They must be fed an organic, all-vegetarian diet containing no antibiotics, pesticides, or growth hormones. To become certified, a producer must select an approved third-party organic certifier, maintain records of production and all materials used, and agree to an annual inspection by a certifier.

With population growth following the Industrial Revolution, the use of pesticides and synthetic chemicals in food production expanded to grow more food with less hands-on management. The 1962 publication of Rachel Carson's *Silent Spring* awakened the world to the harm that pesticides cause to humans and to the environment. The benefits of growing and consuming organic foods are debated. However, research on organic farms done over several decades has shown that organic farming reduces soil erosion, lowers fossil fuel consumption, lessens leaching of nitrates into the soil, improves carbon sequestration, and eliminates pesticide use, which is beneficial to humans and the environment. Some also argue that organically grown food simply tastes better.

Although communities across the world have grown organic food since the beginning of agriculture, it was not until 1990, when the Organic Foods Production Act—part of the 1990 Farm Bill—authorized the secretary of agriculture to appoint a National Organic Standards Board, which was done in 1992. The role of the board is to develop the standards for organic food production and handling (i.e., what can and cannot be used) as well as to implement the national organic program. The program is responsible for developing, implementing, and overseeing the labeling standards for organic products, as well as for authorizing agents who inspect organic production and handling to ensure that they meet the U.S. Department of Agriculture (USDA) organic standards. For example, nonsynthetic materials are allowed unless specifically prohibited, but no synthetic materials are allowed unless specifically approved for use postharvest, such as biopesticides, chlorine dioxide, citric acid, and waxes made of natural materials.

Before the federal government set the standards for organics in the 1990s, individual states set requirements for food grown or processed in the state or imported from other countries. For instance, Oregon and Washington were early implementers in the organic movement, with the establishment of the Oregon-Washington Tilth Organic Producers Association in the early 1970s.

Four types of organic labeling are approved by the government and are currently in use in the United States: "100% Organic" is the only label that receives the USDA Organic seal and means that everything in the packaged product, or a single-ingredient product, is certified organic; "Organic" is the label used when 95 percent of the product (by weight) is organic, if the 5 percent of nonorganic products are not available commercially in organic form; "Made with organic ingredients" is used if the product contains between 70 percent and 95 percent organic ingredients; and "This product has some organic ingredients" is used if less than 70 percent of the ingredients are organic and they are listed on the packaging.

Farmers must not only meet USDA requirements, but must also pay a fee to apply for certification. Although these costs are already inaccessible for some small-scale farmers, according to the National Sustainable Agriculture Information Service, costs may continue to rise, depending on the fees charged by certification agents, the size and complexity of the farm operation, inspection, and other factors.

It has been argued that organic certification is driving independent or small organic farmers out of business because of the expenses associated with gaining certification. To offset the costs of organic certification, the government has instituted a cost-share program for a select number of states. This has not yet been implemented in some of the major organic growing states, such as Oregon, Washington, or California, and there is concern about farmers being able to afford the costs of becoming USDA-certified organic.

The process for receiving organic certification, as well as who does the certification, varies by country. In some cases, the government publishes legislation for organic standards, but even so, certification is often overseen by nonprofit or public companies. Products in Europe must be certified by agencies complying with international organic standards (some of which the United States does not meet). At the international level, the International Federation of Organic Agriculture Movements is working to set global organic certification standards. Because standards for organic labeling vary by country, to ensure consistency with their own standards, some countries only import products that are certified by the International Federation of Organic Agriculture Movements under the International Organic Accreditation Service.

Organic foods historically have been more expensive to purchase than conventionally grown or raised food (e.g., in 2000, consumers paid up to 50 percent more for organic tomatoes, poultry, and dairy compared with their conventional counterparts). However, prices are soon expected to become more competitive with conventional products as demand increases and production expands.

See Also: Beyond Organic; California Certified Organic Farmers; Certified Humane; National Organic Program; Organic Farming.

Further Readings

Fernau, Karen. "Shopping Tips Help Cut Cost of Going Organic." *The Arizona Republic* (September 27, 2006). http://www.azcentral.com/arizonarepublic/food/articles/0927organic 0927cheap.html (Accessed November 2008).

Humane Society of the United States. "A Brief Guide to Egg Carton Labels and Their Relevance to Animal Welfare." http://www.hsus.org/farm/resources/pubs/animal_welfare_ claims_on_egg_cartons.html (Accessed November 2008).

International Federation of Organic Agriculture Movement. http://www.ifoam.org (Accessed November 2008).

Kuepper, George. "Organic Farm Certification and the National Organic Program." National Sustainable Agriculture Information Service. http://attra.ncat.org/attra-pub/organcert .html#steps (Accessed November 2008).

Kuepper, George and Lance Gegne. "Organic Crop Production Overview." National Sustainable Agriculture Information Service. 2004. http://attra.ncat.org/attra-pub/organiccrop.html (Accessed November 2008).

U.S. Department of Agriculture. "National Organic Program Cost Share Program." http://www.ams.usda.gov/nop/StatePrograms/CostShare.html (Accessed November 2008).

U.S. Department of Agriculture. "Organic Foods Standards and Labels." http://www.ams.usda.gov/nop/Consumers/brochure.html) (Accessed November 2008).

World Organic Certifiers. http://www.organic.com.au/certify (Accessed November 2008).

Stacy Vynne
Shannon Tyman
University of Oregon

CHEAP FOOD POLICY

Cheap food policy refers to the suite of policies, most prevalent in the industrialized world, that are designed to keep urbanites' spending on food a relatively small percentage of their total expenses. Cheap food policy as it stands at the beginning of the 21st century emerged in response to three historical drivers: (1) urbanization associated with industrialization and, in the case of the United States especially, immigration; (2) the Great Depression, which triggered the development of centralized food and agricultural policies; and (3) World War II, when Europe's agricultural productivity and infrastructure was mightily affected, resulting in food shortages and an increased value placed on reliable access to food, which in part is guaranteed through lower food costs.

There are two primary policy mechanisms for ensuring cheap food for consumers in industrialized countries, First, trade policies favoring imported food produced at comparative advantage at the expense of domestic producers (i.e., "free trade" policies) are often touted as a boon for food consumers, in that cheaper imported food becomes available at market prices. Second, government-funded subsidies and payments to producers are designed to appeal to multiple publics. They are meant to control costs of production, decrease producer risk, and increase production, resulting in larger supply—and thereby lower food prices. Though not as associated with the political rhetoric of cheap food, a third policy category—investment in agricultural research and technological development—can also be included in the discussion of these policies that are influential in decreasing the relative cost of food production. The cost of food relative to other household expenses in the developed world has certainly fallen over the last several decades, in the United States going from about one-third of a household's expenses in the 1960s to only about 11 percent in the early 2000s.

Historically, the repeal of the Corn Laws in Britain in the mid-19th century is perhaps the first major instance of the free trade strategy used in the pursuit of the politically expedient cheap food policy. Corn Laws were a suite of trade policies offering protection to domestic grain producers. Their repeal ushered in an era of free trade in the agricultural arena, opening the British market to imported food from its colonies and North America, which were able to out-compete British-produced food based on lower cost of production. Although this policy was detrimental for domestic farmers, and therefore politically quite contentious, it was advantageous for urban populations, who were engaged in what was seen as the new economic engine—manufacturing. Industrialization and concomitant urbanization drove a demographic shift in populations dominated by rural people engaged in agricultural production to those dominated by urban people engaged in manufacturing; this change in demographic characteristics was paralleled by a shift in political priorities. Arguably, cheap food policy most benefited owners and managers of industry, rather than primarily laborers, because it allowed manufacturing wages to remain low. Urbanization and industrialization continue to drive cheap food policy in both the developed and developing worlds. As fewer people make their living through agriculture, and as more people are concentrated in urban areas, consumer concerns gain increased political importance.

Fluctuating Prices and Overproduction

In the United States, the Great Depression of the late 1920s and 1930s was presaged by dramatically fluctuating and decreasing farm prices, starting in the 1920s. These were at least partly a result of agricultural overproduction, related to the shrinking of foreign markets, and the resulting price decline during the previous decade and were complicated by the eco-social phenomenon of the Dust Bowl. At first, U.S. policy focused mostly on controlling overproduction, but, in large part in response to urbanites' protest over the perceived effects on food prices, additional food and agricultural policy was developed to control food prices. The ongoing balancing of these rural and urban interests—or producer and consumer interests—has shaped U.S. agricultural and food policy ever since.

In most of the time since the Great Depression, U.S. agricultural policy has balanced payments to farmers, encouraging production with those controlling production, with the goal of stabilizing both food prices for consumers and farm gate prices for farmers (although farm prices continued to fluctuate in response to a variety of market expansions and contractions). However, in the mid-1990s, the production control mechanisms in farm subsidies were removed from U.S. farm policy, leading to a contention that the current iteration of cheap food policy in the United States goes overboard on encouraging production. Critics argue that this production boom results in an oversupply of cheap U.S.-produced food on the international market, with negative consequences for consumers, as well as international competitors.

In Europe, the effect of the two world wars on the food system was devastating, especially that of World War II. Widespread undernutrition during and after the war added fuel to the cheap food policy of the region. Opening markets to imports, at that time especially from North America, was one of the primary mechanisms used to address food shortages and affordability.

However, as at the time of the repeal of the Corn Laws, in the early 21st century, reducing protectionism remains an important piece of both implemented trade policy as well as political rhetoric for countries interested in maintaining inexpensive food for consumers. Recent decades have seen the rise of international trade bodies with the mandate to encourage the lifting of all sorts of trade restrictions—a goal consistent with much of the cheap food policy rhetoric. The World Trade Organization has prohibited trade barriers such as import tariffs and taxes among member nations on most agricultural goods. As a result, many food-exporting nations of the developing world have been able to access markets in the industrialized nations, often out-competing domestic producers. Cheaper agricultural labor, land, and inputs often result in an overall lower cost of production in less-developed nations, even when storage and transportation costs are accounted for. Having effectively addressed traditional trade barriers of taxes and import duties on agricultural goods, recent efforts regarding agricultural trade rules within the World Trade Organization have focused on reducing and regulating production side subsidies.

Controlling Production Costs

Although free trade policies can be seen to encourage cheap food on the consumption side, through comparative advantage, many countries also have a suite of policies aimed at controlling the cost of production of food crops. In much of the developed world, agricultural subsidies are employed in the name of cheap food policy, especially those that encourage production. For example, the United States' cheap food policy emerged after the Great Depression. Farm subsidies were seen as aiding both farmers and consumers, in that they provided a sort of guaranteed income to farmers, reducing their cost of production and economic risk, and thereby resulting in cheaper food prices for urban consumers. Related to this second set of policies, agricultural policies that invest in agricultural research and development and technical assistance to farmers also tend to result in abundant production and lower food prices for urban consumers.

Developing countries may also engage in a version of cheap food policy. Historically, many developing nations have capped grain prices to maintain food at an affordable cost for urban consumers. This type of policy is, of course, less expensive (at least in the short term) for the government to implement than subsidy programs; however, it often results in increasing rural poverty as farmers' incomes are concomitantly reduced. As a result, it can fuel rapid urbanization of many developing world cities, which is associated with a wide array of social problems.

There are many critiques of these policies, leveled by academics, politicians, and farmer and nutrition activists. Both free trade and production subsidies are controversial, though for different reasons. The assumption that a cheap food policy in general is good for a society is also being questioned, especially in the developed world. Critics argue that some of the effects of these policies can negatively affect farmers, national food security, environmental sustainability, and human health. Finally, some question whether these policies actually result in lower food prices for consumers or whether other players in the food and agriculture value chain are the beneficiaries of the policies.

Free trade policies sacrifice domestic producer needs in the interest of lower overall food prices. This can result in hardship for farmers, eventual loss of farmers, and arguably decreased food security at the national level. There is also the concern that exporting countries may drop off production or export in the face of political unrest, natural disasters, and so on, perhaps leading to shortages in importing countries whose agricultural infrastructure has already been affected. These issues are more of a concern in developing countries that lack the extensive subsidy programs of the industrialized countries.

One of the most significant critiques of cheap food policies is the effect that overproduction has on environmental quality. Critics argue that policy incentives to produce ever-greater amounts of food encourage farmers to use natural resources at an unsustainable rate, having negative effects on water supplies, wild biodiversity, and land use. Environmentalists are also critical of the use of pesticides and artificial fertilizers that can cause pollution and harm to both human and nonhuman life.

The production-side subsidies of developed nations have been implicated in the rising levels of obesity and other nutrition-related health problems observed in both wealthy and poorer nations. For example, two-thirds of American adults are overweight or obese—a steep increase over previous generations. Critics argue that an excess of cheap grain has lead to excessive consumption of sweeteners (especially corn sweeteners) and high-fat animal products (fed cheap grain). Critics of the cheap food policy in the industrialized world also contend that the policies compromise the quality of food grown, both in terms of diversity of crops (i.e., encouraging grain production at the expense of fruits and vegetables) and in terms of varieties within crops (i.e., encouraging the most productive varieties without regard to nutritional or other quality attributes).

Whether these policies labeled "cheap food policy" actually result in significantly cheaper food is itself a contentious issue. Many agricultural economists argue that subsidies that were instigated as "cheap food policy" today actually have very little effect on the price that consumers pay in the supermarket because the proportion of the consumer price that is paid to the producer is very small, usually under 10 percent. It is argued that once-cheap grains are now directed toward products like biofuels and animal products.

Cheap food policies were developed in response to population shifts from rural to urban living, and the parallel economic shifts from an economy dominated by agricultural interests to one dominated by industrial interests. These policies take the form of free trade policies and of production-side subsidies. Critics of cheap food policy worry about the effects of overproduction on the environment and human health. As the global economy changes in the 21st century, we may see a change in how food policies are implemented in the industrialized world.

See Also: Agribusiness; Comparative Advantage; Food Security.

Further Readings

Eggert, R. J. "Advantages and disadvantages of direct payments with special emphasis on marketing considerations." *Journal of Farm Economics* 29, 250–55 (1947).
Knutson, R. D., et al. *Agricultural and Food Policy*, 4th ed. Upper Saddle River, NJ: Prentice Hall, 1998.
Moore, A. *The Farmer and the Rest of Us*. Boston: Little, Brown, and Company, 1945.

Ann Finan
Australian National University

CODEX ALIMENTARIUS

The Codex Alimentarius (Latin for *book of food*) collects international standards, practices, processes, guidelines, and recommendations related to food, food production, and food safety. In 1963, the Food and Agriculture Organization and the World Health Organization—two United Nations agencies—established the Codex Alimentarius Commission with the goal of establishing a set of standards and practices for the international food commodities trade that would protect consumers. The World Trade Organization has since recognized the Codex as the relevant authoritative work when resolving international disputes pertaining to food safety and consumer health.

The Codex is inspired in name and aim by the Codex Alimentarius Austriacus (CAA), a collection of standards and product descriptions pertaining to food used by the Austro-Hungarian Empire, and later Austria. Work on the CAA began in 1891, with portions of it circulating informally in the next two decades until it was collected in three volumes and published between 1910 and 1917. Although the CAA was produced principally by universities and the food industry itself, and primarily established the identity and proper treatment of various foods, it was relied on by the court system as an authoritative reference. In 1975, it was finally formally incorporated into the body of Austrian law, by which time its spiritual descendent, the Codex Alimentarius, had been formulated. During the early years of the European Economic Community, leading up to the European Union, the idea of a pan-European Codex Alimentarius had been explored and eventually helped lead to the UN-sponsored Codex.

The Codex is published in English, French, Spanish, Chinese, and Arabic and is updated periodically to keep up with changes in the world and new safety-related information. Officially, it is meant to cover all foods, but in the course of its development it has come to focus principally on food that is sold to the consumer (flour and bread and dry pasta, for instance, rather than unprocessed wheat).

Specific standards and practices documents are produced for meat products (including processed or cured meats and the handling of frozen meat products); fish and other seafood products; milk and dairy products; vegetables, fruits, fruit juices, and processed products composed thereof; cereals and dried legumes and products derived therefrom; fats and products derived therefrom; and special miscellaneous products that do not fit a broad category but require special mention (such as bottled water; sugar, honey, and other sweeteners; chocolate—an important food commodity; baby food; and baby formula).

Furthermore, in addition to those "vertical" topics, documents cover "horizontal" topics that are not limited to one type of food: food labeling (including nutrition information, health benefit claims, ingredients, and the language used in ingredients, such as what is meant by "artificial coloring" or "natural flavors"), food additives (including not only flavorings and food coloring but also acidity regulators, anticaking agents, antifoaming agents, antioxidants, bulking agents, emulsifiers, humectants, stabilizers, thickeners, and, of course, preservatives), the prevention of and appropriate response to food contaminants (including specific guidelines for radionuclides, aflatoxins, and mycotoxins), maximum residue limits for pesticides, food hygiene (specific practices designed to limit the risk in the handling, preparation, and packaging of food), and biotechnology-derived foods such as DNA-modified crops.

The Codex is formulated as a reference document, not a code of law, and the commission is not empowered to enforce these standards on any nation or entity within that nation. However, just as the CAA was used as an authoritative source by the Austro-Hungarian courts, so too has the modern Codex been recognized by the World Trade Organization, which is, in fact, empowered to enforce standards on entities engaged in international trade. Though this recognition does not preclude the authority of other information sources, the Codex has been criticized for essentially having the force of law without—so the critics would argue—being composed by a commission run as rigorously as a legislative body would be—one that is not influenced by the legislators or electorate of the countries its standards affect.

Furthermore, many—particularly alternative health enthusiasts—find issue with the Codex's handling of vitamins, minerals, and other health supplements. There is no international consensus on the handling of nonpharmaceutical health supplements, whether vitamin C pills or herbal supplements that claim to prevent or treat cancer. In some countries, the relevant factor is whether any health claims are made by the seller or on the label; in others, this is not important. In some countries, such products are treated as over-the-counter drugs; in others, they are not. A recent question in U.S. law is whether food items traditionally sold for consumption—as opposed to vitamin pills that have no nontherapeutic purpose—can be advertised with reference to their health benefits, and in such cases, what claims may be made before the product is classified as a drug. (Cheerios cereal crossed what is by general estimation a vague line when its boxes were labeled with new language claiming "in six weeks, Cheerios can reduce bad cholesterol by an average of 4 percent," which the FDA deemed was a pharmaceutical claim made by a nonpharmaceutical product.) U.S. alternative health enthusiasts fear that when international bodies have jurisdiction over such matters, their rights to dietary supplements (as outlined in the 1994 Dietary Supplement Health and Education Act) are abridged. Often, allegations are made that the commission is unduly influenced—or even "in the pocket of"—the pharmaceuticals industry, and that it is therefore motivated to discount the efficacy of—and even to criminalize—any remedies or treatments other than those of that industry.

See Also: Food and Agriculture Organization; Food Safety; Meats; Pesticide.

Further Readings

Livermore, Michael A. "Authority and Legitimacy in Global Governance: Deliberation, Institutional Differentiation, and the Codex Alimentarius." *New York University Law Review*, 81 (2006).
World Health Organization. *Understanding the Codex Alimentarius*. Rome: World Health Organization Publishing Management Service Information Division, 2005.

Bill Kte'pi
Independent Scholar

Commodity Chain

A commodity chain is the process or series of steps that describes the life span of a commodity before it arrives at the consumer. The process of planting an orange tree, picking the orange, waxing or otherwise treating it, packaging it, and delivering it to a retail store is one commodity chain, and sourcing tomatoes, peppers, chiles, onions, garlic, lime, and salt for salsa; heating it and jarring it; packaging it; shipping it to restaurants; and serving it with a bowl of chips is another. The chain is not just a list of steps, though—commodity chain theory values and emphasizes the contributions of the actors driving the chain: the workers planting the tree and harvesting the fruit, the people pasting labels to jars, or driving the trucks to the warehouse. The dependence of those actors is a frequent topic of discussion in commodity chain theory. Commodity chains can be extremely complex, but even the most complicated ones are becoming more transparent in the 21st century.

Commodity chains are highly affected by the market. Each step of the chain involves a business that is concerned with making money, with the regulations governing its activity, and with competing with the other businesses that perform its function. Such steps include not only production and processing but also advertising, marketing, and research. Ideally, healthy and successful food commodity chains provide affordable food products that are still safe, without sacrifices in quality, while constantly expanding into new markets. Prices remain low because of the high competition and the economies of scale that reduce per unit costs at each stage, as well as, in some cases, the ability to benefit from comparative advantages. Although it is only recently that it has become normal in a developed country for a family's meal to consist primarily of food that did not originate locally, food commodity chains have always spanned great distances—early trade routes, even in prehistory, included food commodities, and diplomatic relationships between countries ever since have usually included some trade in food. The spice trade was the most lucrative food commodity chain, because of the long shelf-life of its products and the high ratio of value to volume relative to things like grain or fruit. The Industrial Revolution brought refrigeration, fast transport, and better, faster methods of canning and otherwise preserving food, whereas the 20th century introduced cheap air transport, preservatives, and a greater degree of automation in factory work. As a result, food exports soared from 4 million tons (worldwide) around the time of the Civil War to 40 million tons when World War I started. Government regulations quickly became a normal fact of life for those in the food industry, as one law after another was passed in response to food safety concerns and crises brought about by this expanded activity.

Commodity chains demonstrate the interconnectedness of the global economy, as Illinois college students serving espresso in the student union are bound up in the same chain as Tanzanian plantation workers. Tracing a commodity chain tells you more than just the life of the commodity itself. It tells you who holds the power, who profits from the commodity, and who influences its life. Since the 1990s, for instance, there has been a great deal of discussion about the power that large bookstore chains have on the publishing industry, particularly with respect to fiction and popular nonfiction. Because those chains like to have broadly similar inventories in all their stores, their purchase orders represent a sort of "voting bloc" that outnumbers any other smaller store or even regional chains. Although publishing is not a democracy, it is market driven, and those agents elsewhere in the book commodity chain—publishers, editors, literary agents, authors, illustrators, and the employees and staff of all the above—are arguably swayed by the knowledge that a given book will or will not appeal to that voting bloc. Similarly, we can see the ways that supermarket chains have a great deal of control over fresh vegetables, and that coffee roasters have more control over the coffee market than growers do.

The same examination of commodity chains implies that food grain chains (corn, wheat, rice, etc.) are driven not by multinational corporations or retail chains but by national governments and agencies like the European Union. This has been the case since about the middle of the 20th century. Regulation of grain trade and grain production was characteristic of the response to the Great Depression throughout the West—not just in the United States—often with the attendant price supports on domestic grains and restrictions on foreign grain imports. Ironically, perhaps, the West adopted stricter and more specific regulations on agriculture at the same time that the Communist states of China and the Soviet Union were refining their centrally planned state-controlled agricultural programs for entirely different reasons. Regardless of motive, however, control over the means of production—whether direct or indirect—and over foreign imports put the power of the food grain commodity chains in the hands of the state, around the world.

Consumers generally do not want to know about the commodity chain when it comes to food. There is a perception, at least, that people are happier shopping in supermarkets where the apparent produce selection is largely the same year round, with more blueberries in the summer and more pumpkins in the fall, and seasonal varieties like blood oranges functioning as an edible liturgy. Where those fruits and vegetables come from—for that matter, where the meat comes from, and the rice, and the nuts—varies according to the season, with an extraordinary amount of activity occurring below the surface to preserve the appearance of stasis above. There have always been exceptions to this, particularly when the region of origin becomes used like a brand name—as with Vidalia onions, Washington apples, Idaho potatoes, or Vermont maple syrup. Increasingly, origin may be used in supermarket signage when it is a selling point, such as local strawberries in the summer.

Safety concerns, more than concerns with the quality of taste, have made food commodity chains more transparent. Food poisoning crises may not be more common in the 21st century, but they seem to affect wider areas when they do occur, because of the vast coverage of distribution networks. When meat from a certain packager or produce from a certain country comes into question, consumers previously happy purchasing anonymous skinless chicken breasts and bagged spinach suddenly demand to know what they are buying and where it came from. The widespread suspicion of Chinese goods after an unfortunate succession of poisoning crises caused by various Chinese products led many supermarket chains to note which of their produce goods were American and which were imported.

See Also: Export Dependency; Food Safety; Supermarket Chains.

Further Readings

Allen, Gary J., et al., eds. *The Business of Food: Encyclopedia of the Food and Drink Industries.* New York: Greenwood, 2007.

Bair, Jennifer, ed. *Frontiers of Commodity Chain Research.* Stanford, CA: Stanford University Press, 2008.

Djurfeldt, Goran, et al. *The African Food Crisis: Lessons From the Asian Green Revolution.* New York: CABI, 2005.

Jackson, Peter, et al. "Mobilizing the Commodity Chain Concept in the Politics of Food and Farming." *Journal of Rural Studies*, 22/2 (April 2006).

Pollan, Michael. *In Defense of Food.* New York: Penguin, 2009.

Pollan, Michael. *The Omnivore's Dilemma: A Natural History of Four Meals.* New York: Penguin, 2007.

Bill Kte'pi
Independent Scholar

Commons

Lands that are controlled by owners but over which other people exercise rights of use or other rights are known as *common land,* or more colloquially, "the commons." This type of relationship between humans and land was developed in England, though it is has been used throughout different parts of the world, including in the United States. Modern understanding of the commons derives largely from the influential 1968 essay by Garrett Hardin titled "The Tragedy of the Commons," in which Hardin documents the potential threat to common land that exists when individuals act independently and in their own self-interest.

However, evidence is available that demonstrates how the commons often are managed rather effectively, and thus possess a rather high conservation value.

Traditional rights to the commons in England and Wales were generally held by those who occupied the land at a certain point in time. However, common rights could be extended to more than simply pasture or land use. Common rights were also granted for the right to fish estuaries (piscary), the right to take sods of turf for medieval fueling purposes (turbary), the right to take sand and gravel (common in the soil), the right to turn out pigs to eat certain nuts (mast or pannage), and the right to take wood for the erection of the commoner's home (estovers). These common rights of use were a central part of the largely communal organization of medieval agriculture.

The rights of common use existed before the development of established law and were largely unregulated until the 15th century and the emergence of the enclosure movement, which irrevocably altered the common lands. This movement has been called "the revolution of the rich against the poor" by many historians, as it involved a series of political and legal acts that literally enclosed publicly used land. Enclosures typically involved the placement of hedges, ditches, or other barriers that limit the free movement of individuals and animals across the land. It placed the land under private control and forced peasants off the lands that they and their ancestors had used for subsistence for generations. Commentators have noted that the enclosure movement may have been the critical development that altered the relationship between humanity and the natural world, as it converted land from something people belong to into a commodity people possess and control for profit. By the 19th century, the enclosure movement was largely complete, even though common lands remained. These lands have since been largely neglected, as most commoners made the move to urban areas in the search for a better quality of life.

Today, references to the commons derive inspiration from the aforementioned Hardin essay on the tragedy of the commons. The general point that Hardin makes is that commonly used resources will eventually be overused and degraded by individuals seeking to rationally maximize their own gains from the resource (thus the "tragedy" of the commons). This problem is heightened by upward population pressures, which lead more and more users to deplete resources that are largely fixed or in balance. Hardin's essay has been incredibly influential in many circles and has drawn sharp rebukes from others. Maintaining or protecting the commons has taken on a broader context following Hardin, as it now is typically used to refer to the stock of natural resources that are generally seen as indispensable to the survival of humanity (food, water, energy resources). Significant theoretical and empirical efforts have emerged in recent years in response to this work, broadly held together under the umbrella of environmental resource management.

Debate continues in this arena over the most efficient way to manage and protect these commons, with most arguing either for government intervention or for further privatization. Others have argued for a third approach, which involves resource users constructing flexible, collective social institutions in which management responsibility is invested in the users themselves. Research and theorizing continues on this critical issue, though few successful applications of commons management have been found to date, except for those that are primarily localized and rather small in scale, like localized or community-level agricultural activities.

The issue of the commons continues to vex human societies, as it has for centuries. The enclosure of the commons was a critical development for humanity—one that certainly changed the relationship of humanity to nature, and in many ways to each other. Although the enclosure movement was largely successful in its primary purpose, the commons remain and are increasingly important to the future of the ecological health of the planet.

See Also: Community Gardens; Community-Supported Agriculture; Land Tenure.

Further Readings

Hardin, Garrett. "The Tragedy of the Commons." *Science*, 162:1243–48 (1968).
Ostrom, Elinor. *Governing the Commons: The Evolution of Institutions for Collective Action*. Cambridge: Cambridge University Press, 1990.
Rifkin, Jeremy. *The Biotech Century: Harnessing the Gene and Remaking the World*. New York: Tarcher/Putnam, 1998.

Todd L. Matthews
University of West Georgia

COMMUNITY GARDENS

Community gardens are pieces of land that are gardened by a group of people. Community gardens improve food security by providing access to land on which to grow fresh produce. They also create green space and beautify neighborhoods, build a sense of community and connection to the environment, and provide a venue for satisfying labor and exercise. In some cases, researchers have found that urban neighborhoods with community gardens experience less crime and vandalism, probably as a result of increased neighborhood activity. Community gardens often exist in poor, inner-city neighborhoods with high proportions of apartment dwellers and other residents without space to grow their own food, but they can exist in rural areas as well. Aside from private gardens in people's yards, community gardens are perhaps the most familiar type of urban agriculture in the United States.

Community garden land can be owned either publicly or privately. In the United States, it is common for community gardening advocates to clean up abandoned vacant lots and turn them into productive gardens. In these cases, community garden land may or may not be used with owners' permission. Alternatively, community garden land can be owned by public agencies such as parks and recreation districts, city departments, nonprofit organizations, or land trusts. Community gardens are sometimes included in public parks as health or recreational amenities. Access to land and security of land tenure remains a major challenge for community gardeners worldwide, as in most cases the gardeners themselves do not own the land directly. Especially in high-density urban areas, land for community gardens can be expensive to obtain and maintain.

The term *community garden* is the preferred term for collective gardening ventures in the United States and Canada. The most common types of community garden found in northern North America are areas of land divided into smaller plots that are tended by individuals or households. Individual garden plots are generally small but can range in size from as small as 5 feet by 5 feet to as large as 50 feet by 50 feet. Garden members are usually asked to follow a set of rules and rent plots by the year for a nominal fee. Waiting lists for plots can be quite long—sometimes a year or more. Many community garden programs also ask members to participate in garden workdays, fundraisers, and social gatherings. Some community gardens exist to provide access to gardening to those who otherwise could not have a garden, such as the elderly, recent immigrants, or the homeless. Although the primary goal of these types of gardens is to provide access to land on which to grow produce, community garden organizers typically say that "growing community" is just as important. According to the American Community Gardening Association, the primary advocacy group for community gardening in the United States and Canada, "in community gardening 'community' comes first." In Europe, "allotment gardens" are closely related to this type of community garden but are usually much larger and can have dozens of large plots that have been rented by the same family for generations. Some community gardens in the United States resemble European allotment gardens in terms of their size and layout. These gardens often originated as Victory Gardens during World War II. Although not usually termed *community gardens*, communally held land used for vegetable production is also common in the developing world in both urban and rural areas.

Sometimes, community gardens are set up as "common gardens" or "community farms," meaning that instead of consisting

This New York City community garden is divided into small plots to allow residents to cultivate their own plants and flowers in an urban environment.

Source: National Science Foundation/Mark Whitmore

of individually tended plots, they are set up as one garden that is tended collectively by employees, community members, or volunteers. If vegetables are grown in these types of gardens, members distribute produce equitably among themselves or donate some or all of it to the surrounding community, neighborhood food banks and soup kitchens, or other charitable organizations.

Although community gardens broken up into individual plots and used primarily for fruit and vegetable production are the most common form, the practice of community gardening in northern North America includes a variety of styles and approaches. Some gardens are primarily gathering places for community members and serve as demonstration gardens that showcase art and organic or ecological techniques. Examples include the Clinton Street Garden in Manhattan and the Peralta Garden in North Berkeley—gardens that produce food, but only as part of a much larger vision of urban environmental sustainability. Other community gardens represent reclamation of space and cultural heritage. For example, *casita* gardens in the Lower East Side of Manhattan are developed by Puerto Rican immigrants to re-create a familiar, cultural space in a new environment. In New York, these gardens replicate the small wooden structures surrounded by vegetable gardens commonly seen in Puerto Rican shantytowns and are used primarily for community gatherings. Still other community gardens are entirely devoted to creating ecological green space or habitat for urban wildlife, especially in neighborhoods with few public parks. Others exist mainly to offer educational opportunities, job skill training, and transitional employment to disenfranchised members of society. For example, the Homeless Garden Project in Santa Cruz, California, is an organic community garden that employs homeless men and women in their community-supported agriculture program.

Although community gardening exists all across northern North America, the practice is strongest in the northeastern states. Literally thousands of community gardens exist in New York City, Philadelphia, and Boston. This abundance is in large part a result of prominent, long-standing community garden organizations in these regions—for example, Green Thumb and Green Guerillas in New York City; the Food Project in Lincoln, Massachusetts; and the Neighborhood Gardens Association in Philadelphia. Community garden programs are also prominent along the West Coast, especially in Seattle, Washington, and in British Columbia, and in Midwestern cities such as Chicago and Minneapolis.

Permanency of community gardens is conditional on economic and real estate interests, especially when land tenure is not secure. This vulnerability was seen on a large scale during the New York City urban garden crisis of the late 1990s when Mayor Rudy Giuliani announced plans to auction gardens located on city-owned properties to private bidders. Sales were halted with the combined efforts and funds of community garden activists, the Trust for Public Land, and actress Bette Midler. Although many New York City gardens had a happy ending, countless other community gardens throughout the United States have been bulldozed due to rising land prices and the push for development.

See Also: Food Security; Homegardens; Urban Agriculture.

Further Readings

American Community Gardening Association. http://www.communitygarden.org (Accessed February 2009).

Hynes, H. Patricia. *A Patch of Eden: America's Inner-City Gardens*. White River Junction, VT: Chelsea Green, 1996.

Hilary Melcarek
University of California, Santa Cruz

Community-Supported Agriculture

Community-supported agriculture (CSA) is a form of direct marketing in which consumers purchase "shares" in a farm and then receive a regular—often weekly—installment of fresh, seasonal produce. On one level, CSA is a marketing strategy, allowing farmers to receive preseason financing to help cover the costs of seeds, fuel, salary, and other expenses and providing consumers with fresh, locally grown produce. On another level, however, it is a cooperative partnership formed as an alternate response to an increasingly corporatized, decentralized food system.

History

The CSA movement in the United States began in the mid-1980s with two separate but nearly simultaneous projects. In 1985, Jan Vandertuin approached Robyn Van En of Indian Line Farm in Massachusetts with the idea of recreating a form of agriculture he had seen in Switzerland. Vandertuin and Van En worked with others to create the name and structure of this first CSA project, and shares were sold for that year's apple harvest. Many families who participated in the apple shares also signed up for vegetable shares for the next year. Similarly, in 1986, Trauger Groh drew on his experience at a German farm to start Temple-Wilton Community Farm in New Hampshire. From these two farms the movement grew. In 2005, a U.S. Department of Agriculture database listed 1,150 CSAs; others now place the estimate closer to 2,000 operations.

Literature about CSAs often includes the phrase "food with the farmer's face on it." The phrase is a philosophical translation of *teiki*, the name for partnerships that Japanese women householders started with organic farmers in the 1970s in response to concerns about pesticide use, the loss of farmland, and increased food imports. Although this narrative is frequently mentioned in the history of U.S. CSAs, it appears that the U.S. movement was not directly inspired by *teiki*, but that these simply were parallel paths.

Structure

CSAs begin in two main ways, either as consumer-initiated projects that arise when a group of people with similar concerns hires a farmer, or as farmer-initiated projects. Approximately three-quarters of CSAs are farmer-initiated, in which case CSA is often one of multiple marketing strategies; for example, complementing farmers market or restaurant sales. Some CSAs are run by nonprofits, such as universities, churches, and farms, that train people who are homeless or who have mental or physical disabilities.

Production methods at CSA farms are typically organic or biodynamic; this is not a structural requirement of the projects but tends to be a preference of the consumers and farmers involved. Vegetables tend to be the mainstay of CSA production, sometimes supplemented by fruit (e.g., strawberries, apples) in the week's delivery, but other types of projects have also appeared. Many CSAs add eggs, herbs, honey, or flowers to the share for an extra fee, and others offer grass-fed beef, baked goods, local specialty crops (e.g., Kona coffee or macadamia nuts), or foods made from farm produce in commercial kitchens (e.g., pesto, jam, or goat milk soap and cheese). A few specialized CSAs supply members with whole and processed grains, and CSA members of a farm in Washington State receive organic, pastured broiler hens as well as first priority in reserving Thanksgiving turkeys. Most CSA products are edible, although the Martha's Vineyard Fiber Farm offers a yarn CSA, membership in which includes an invitation to the annual Shearing Day Celebration along with a share of the spring "yarn harvest."

Most CSA projects offer weekly installments during the main summer growing season, which can vary from 12 to 20 weeks in the Midwest and northeastern United States to 36 weeks in California, and year-round in Hawaii. Some CSAs offer winter shares, such as a one-time box of winter squashes and root vegetables to supply the Thanksgiving and winter holiday tables, or monthly installments of greens produced in greenhouses. Wishing

Stone Farm in Rhode Island offers an alternative to a weekly produce box: CSA members can pay up-front and then pick up whatever produce they choose at the farm's stand at one of several farmers markets. The funds from this debit-CSA must be used by the end of the season; any unused credits are kept by the farm.

CSA share pick-up sites vary but include urban offices, churches, retirement centers, farmers markets, stores, or volunteer members' homes. Some farmers prefer the connection that comes from consumers directly picking up produce from the farm; however, others favor the reduced carbon footprint of driving a large quantity of produce into an urban area in one trip instead of multiple trips being made by CSA members out to the farm. Freewheelin' Farm in California bypasses driving altogether by delivering shares via bicycle, and in 2008 the Creston Grain CSA in British Columbia delivered the year's harvest of wheat, spelt, and oats by sailboat to a community up-lake.

The involvement of members in CSAs varies greatly. Most farms require no labor from their members but may have special volunteer days for labor-intensive projects. However, members of some CSA projects are required to work a certain number of hours per season; other CSAs offer reduced prices or free "working shares." Work can range from assisting with mailings to helping with harvest or distribution. Most CSAs are run fully by the farmer, but about one-quarter have a "core group"—a team of members who make decisions regarding CSA operation. Although production decisions are typically left to the farmer, the core group may work with the farmer to determine what should be grown, what the share price should be, and what distribution methods should be used. The core group members may also take charge of CSA jobs, including managing the money, organizing distribution, recruiting and retaining members, or scheduling member workdays or festivities.

Benefits

One key benefit of CSAs for farmers is prepayment by the consumers for part or all of the season's produce. This provides the farmer with capital for covering the season's expenses, from seed to labor, as well as a guaranteed market. This also theoretically places part of the inherent risk in growing produce on the consumer: If part of the crop fails, then the consumer receives a smaller share. Some farms clearly state this as part of the seasonal membership application. However, some farmers decide not to spread the risk so greatly and try to supplement weeks of low harvest with purchased fruit or vegetables from other farms.

Another benefit of CSAs is the direct exchange between shareholder and farmer, with funds going directly to the farm instead of being split with intermediate buyers. This also benefits shareholders, as the products are generally less expensive than if purchased separately at farmers markets or grocery stores.

CSA produce tends to be harvested very shortly before packing and delivery, providing consumers with fresh, seasonal produce. The short travel time allows farmers to include varieties that do not ship well, including heirloom varieties that may be difficult to acquire elsewhere.

Another frequently cited benefit of CSAs is the partnership between the member and the farm or farmer. Members can talk with "their" farmer about the farm and production methods and either visit or participate in the work of the farm, strengthening their connection to and understanding of how their food—and food in general—is produced. Many CSA farmers try to develop means of connecting the consumers with the rhythm of the production season and their particular piece of land, especially those consumers who do not participate in regular farm labor. One popular option is a newsletter with stories about planting, weeding, and the crops themselves (often answering the particular question, "When will tomatoes be ready?"), as well as recipes to help members know what to do with the produce. Another method for encouraging member–farm connection is field days, including special member u-pick tomato days, pumpkin patches, or harvest celebration potlucks.

Many members report positive attitudinal and behavioral changes with their participation in a CSA. Attitudinal shifts include a new appreciation for the challenges of farming or for the seasonality of produce, which can affect members' choice of restaurant when dining out or off-season grocery store purchases. Others say that an increased proportion of vegetables in their diet helps them develop healthier eating habits, including a reduced dependence on meat. Some are pleased that such changes affect their whole family, particularly their children. Other members appreciate a new spiritual or cosmic connection to their food. However, these changes do not appear to spread to nonfood areas of consumption.

Challenges

Although preseason payment for shares benefits the farmer financially, the large single sum can be difficult for consumers to pay and thus may act as a barrier to acquiring members. Some CSAs divide the price into installments, with partial payment due up-front and the rest due during the growing season. Some researchers suggest that CSA farmers could generally set higher prices for shares but that their pricing decisions are affected by altruistic feelings toward the shareholders.

The production of a box filled with a variety of fresh vegetables, fruits, and herbs each week can be challenging for farmers. Consumers generally desire variety as well as relatively small quantities of any one kind of produce. Variety is important not just in a single week's offerings but across weeks as well. This can require the farmer to plant and manage 30–60 crops—a greater diversity than may be needed for other marketing avenues. Some farms overcome this challenge by partnering with other farms in different microclimates, using the produce that grows best at each site to fill the week's delivery.

Another substantial challenge for farmers is member turnover. Consumers join CSAs for a variety of reasons, the most common of which are concern for the environment and eating fresh, seasonal, local produce. Once involved, however, members face the challenge of integrating their ideology with their daily life as they have to adapt their purchasing, processing, and eating to the production and distribution schedule of the farm. This adjustment to cooking and eating based on the farm's production and distribution schedule is sometimes known as "supermarket withdrawal." Customer loss is often a result of this withdrawal, particularly the inconvenience of receiving a weekly box on a particular day, followed by receiving too much produce and having too little choice. Customers who are unaccustomed to cooking with Swiss chard, for example, may be daunted to receive it several weeks in a row, and vegetables can accumulate in the refrigerator if a family goes on vacation during the production season. One way some farms help customers determine whether they can make the lifestyle changes necessary without a full season's commitment is by offering short-term (e.g., four-week) trial shares. Recipes and cooking tips in the weekly newsletters also help with the transition.

Critiques of CSA

CSAs, along with other alternative forms of marketing such as farmers markets, have disproportionately low participation by ethnic minority and low-income populations. On the basis of perceptions of CSA managers in California, approximately 90 percent of CSA members have middle or high incomes. CSA managers attribute higher participation by affluent people primarily to the ability of that population to afford produce. They also attribute it to better education, particularly in regard to the societal and environmental costs of food production. Many CSAs use one or more strategies to help improve affordability and accessibility of produce, including offering low-priced or sliding-scale shares, covering the costs of shares through donations (including subsidies from other CSA members), gleaning at the farm, or donating food left after distribution. Although many CSA managers indicate a desire to support increased food security through the use of this and other alternative forms of marketing, they also point to the importance of maintaining fair wages for the farmer and any farmworkers. This tension between food and farm security may limit the ability of CSA to fundamentally change the food system and improve food security, particularly for those most at risk of being food insecure. Extending government programs such as the Farmers Market Nutrition Program to CSA could help resolve part of this tension.

CSA projects tend to be dominated by women: women often begin CSA projects, even when they are family run; women very commonly initiate CSA memberships and take the lead in lifestyle changes that make CSA work as part of their family's food procurement; women shareholders spend more time working at the farm and make up greater proportions of core groups; and families with a full-time woman householder are most likely to participate in farm events. In a case study of a CSA in Michigan, these differences in activities and responsibilities—although noted by both male and female members—were not used as fuel for a larger conversation about gender reform. Women members did not see feminist identity as a reason behind joining the CSA, nor was such an identity developed through their participation. Although this lack of critical engagement can be seen as a lost opportunity, the work carried out by women in the social and physical space of the CSA can also be considered a form of quiet activism or a subtle feminine empowerment.

Some researchers have raised questions as to how much "community" really is in CSAs. About half the consumers who took part in a study of eight CSA projects in the Twin Cities region of Minnesota ranked the community aspect low among reasons for CSA participation; this low ranking even held true for members of CSAs with high member participation and low turnover. However, the actions of those members say that community is valued: Many CSA members make lifestyle adjustments to accommodate this new way of cooking and eating and return as members in subsequent years. Although this low level of participation may be seen as a "community of interest" instead of a "community of relationships," there have been examples of CSAs in which members have become integral to the continued existence of the farm, even in for-profit farmer-initiated CSA projects. Several CSAs have appealed to members for additional pledges as part of capital campaigns, raising substantial funds for special projects, land, or upgrades. Members of Angelic Organics in Illinois invested in a limited liability company to purchase additional nearby acreage and then lease it back to the farm, increasing the farm's stable land base without adding a mortgage (the farm's story, including this particular chapter, is told in the film "The Real Dirt" on Farmer John).

See Also: Family Farm; Farmers Market; Food Security; Urban Agriculture.

Further Readings

Cone, Cynthia and Andrea Myhre. "Community-Supported Agriculture: A Sustainable Alternative to Industrial Agriculture?" *Human Organization,* 59/2 (2000).

DeLind, Laura and Anne Ferguson. "Is This a Women's Movement? The Relationship of Gender to Community-Supported Agriculture in Michigan." *Human Organization,* 58/2 (1999).

Guthman, Julie, et al. "Squaring Farm Security and Food Security in Two Types of Alternative Food Institutions." *Rural Sociology,* 71/4 (2006).

Henderson, Elizabeth and Robyn Van En. *Sharing the Harvest: A Citizen's Guide to Community Supported Agriculture,* revised and expanded edition. White River Junction, VT: Chelsea Green, 2007.

Schnell, Steven. "Food With a Farmer's Face: Community-Supported Agriculture in the United States." *Geographical Review,* 97/4 (2007).

Katie L. Monsen
University of California, Santa Cruz

COMPARATIVE ADVANTAGE

Comparative advantage is one of the principal ideas used in economics to explain the potential for gains from trade between countries. The theory of comparative advantage—developed by 19th-century economist Robert Torrens but usually attributed to David Ricardo—asserts that a country should focus on producing those goods it can make most efficiently while importing goods which the country can make relatively less efficiently. This is true even if one country has an absolute advantage in producing all goods in question (i.e., if it can produce all goods using fewer resources than other countries). The assumed benefits of comparative advantage are challenged from a number of perspectives, including a national security and food self-sufficiency perspective—a critique that emphasizes diminishing terms of trade, particularly for agricultural goods, and another that highlights the politics of the implementation of comparative advantage in global trade agreements.

An example will be useful in explaining how comparative advantage works: Country A and country B both manufacture steel and grow corn. They do not trade with each other, but each consumes all of their own corn and steel. Country A makes more steel and grows more corn than country B. However, country A is only marginally more efficient than country B at making steel and significantly more efficient at growing corn. In other words, the cost of growing a bushel of corn in country A is much less than in country B, but the cost of making a ton of steel is only slightly less. Country A has a greater comparative advantage in growing corn than it does in manufacturing steel. Country B has a comparative advantage in making steel relative to growing corn. Therefore, if country A puts more of its resources into growing corn, at which it is significantly more efficient than it is at manufacturing steel, and country B puts more of its resources into making steel, at which it is more efficient than it is at growing corn and only slightly less efficient than country A, between the two countries more steel will be manufactured and more corn will be grown overall. The countries can then trade, country A selling its corn to country B for that country's steel and vice versa. By putting more of the nation's resources into producing the good in which they hold a comparative advantage, both countries can have more of each good less expensively than they would have without trade, making both countries wealthier. According to this argument, even countries with weak economies will benefit from such trade because, by focusing on their comparative advantage, they will be using their available resources as efficiently as possible.

One major critique of comparative advantage is from a national security and self-sufficiency perspective. Such critics assert that there are some goods that a country should produce domestically to ensure access to them at all times. These may include crucial foodstuffs, military technology, or key raw materials for a nation's main industries. Even if the country produces these goods inefficiently, if a war or other event blocks trade, the well-being of a country will not be threatened by loss of access to these important items. Particularly in regard to food security, if staple goods are produced domestically, a country's basic food supply will not be threatened by price fluctuations or shortages in the world market. In terms of military security, others argue that it may not be in a country's strategic interest to trade with enemy countries, even if it would be economically beneficial to do so.

Another major critique argues that for poor countries whose economies are based on the production of agricultural commodities, trade based on their comparative advantage is unlikely to lead to significant economic development. The first reason for this is that agricultural goods are inexpensive compared with machinery, technology, and other more processed goods. Therefore, if a country holds a comparative advantage in growing bananas or cotton and focuses all of its resources on growing this product and selling it internationally to import technology, food, and other necessities, it will not be able to import as much of these more expensive goods as the society might need. As the price of bananas is unlikely to rise as fast as the price of rapidly advancing technology, this situation will tend to worsen over time. Second, specializing in agriculture creates few spin-off industries and requires little advanced skills. It will therefore neither spur broader economic growth through the creation of new businesses nor advance the overall education and skill level of the country's population. Proponents of this set of critiques often assert that if a country wishes to industrialize, it must temporarily protect its initially inefficient infant industries until they are able to compete in the world market.

A third set of critiques of comparative advantage asserts that as logical as the theory is, it ignores the real-world politics of trade. First of all, critics assert, shifting production from one set of products to another to promote a country's comparative advantage is likely to be met with stiff resistance from the groups in society who will bear the personal costs of this transition in the loss of jobs and livelihoods. This may lead to social unrest and other forms of resistance to such policies. Second, critics highlight the inconsistency with which developed nations embrace the principles of comparative advantage and free trade. This is particularly true for agriculture. Despite being active proponents of free trade in areas at which they are strong, such as technology and manufactured goods, developed countries have largely refused to consider eliminating subsidies and other protections for their agricultural industries. Developing countries—the comparative advantage of many of which is in agricultural goods—have decried this stance as hypocritical and argue that wealthy countries only support free trade to the extent that it benefits their national interests. Though not responding directly to the theoretical claims of comparative advantage, critics who emphasize the importance of politics in trade argue that the theory cannot be examined separately from these real-world issues.

Despite these and other critiques, the theory of comparative advantage has withstood the test of time. It is broadly accepted by most economists and continues to be the basis of much international trade policy and theory.

See Also: Cash Crop; Doha Round, World Trade Organization; Export Dependency; Food Sovereignty; Trade Liberalization.

Further Readings

Lipsey, Richard G., et al. *Economics*, 12th ed. Reading, MA: Addison-Wesley, 1999.
O'Brien, Robert and Marc Williams. *Global Political Economy: Evolution and Dynamics*, 2nd ed. New York: Palgrave Macmillan, 2007.
World Bank. *World Development Report 2008: Agriculture for Development*. Washington, D.C.: International Bank for Reconstruction and Development/World Bank, 2007.

Corina McKendry
University of California, Santa Cruz

COMPOSTING

Cured or finished compost is organic plant or animal matter that has decomposed into a stable soil, usually dark brown or black, called humus. All organic matter eventually decays, but composting is defined as using techniques that create an ideal environment that speeds the process up to reduce waste and provide easily accessible nutrients for growing plants. Composting is a key component of organic gardening and farming.

Composting is essentially the recycling of soil nutrients by decomposers. Decomposers are organisms that break down dead animals and plants; they include fungi, bacteria, worms, enzymes, and aerobic soil microbes and are cultivated to break down organic matter. The rate of decomposition depends on the ratio of carbon to nitrogen. Carbon-rich material is called "brown" matter and includes leaves, wood chips, straw, and manure. "Green" matter is nitrogen heavy and includes food scraps, coffee grounds, and grass clippings. The ideal brown-to-green ratio—when the material will decompose the fastest and produce the richest humus—is 25:1. Too much carbon will slow the decomposition process, and too much nitrogen will cause odor. The rate of decomposition slows as outside temperatures drop and accelerates in the heat of summer. Because they are slow to decompose and thus create odor and attract rodents, animal products (meat, eggs, dairy, fat), cooking oil, and other grease should not be put in a home compost heap or bin.

There are many methods of composting; the two broadest categories are passive and active. Passive composting takes much longer and relies heavily on natural forces. Throwing food scraps and lawn waste into a large pile or bin is considered passive composting. This form of composting reduces waste but is not ideal for creating compost that can be used on farms or gardens as a source of nutrients.

Sheet composting is another example of passive composting whereby organic materials, which can include manure, are spread over the soil, dug in, and allowed to decompose

The ideal mixture for composting is 25 parts of carbon-rich material for every one part of nitrogen-rich material. Compost that is made correctly can have over 6 billion microbes per teaspoon, which is very beneficial for crops.

Source: iStockphoto.com

slowly. The soil should not be planted until the materials are fully broken down, which can take three months or more. U.S. Department of Agriculture organic standards regulate the use of manure, which must be allowed to reach a sterilizing temperature. If raw animal manure is used, 120 days must pass before the crop is harvested.

Active compost involves more planning and labor but "finishes" more quickly and creates a more reliable and nutritious product. There is not one right way to actively compost. Early organic pioneer Sir Albert Howard, for instance, developed the Indore method, a type of aerated windrow composting that involves the introduction of animal residues to the compost process to activate decomposition.

Composting can be done commercially or by individual households. Most home composters build a bin or buy a composting tub, of which many shapes and varieties are available. The ideal size for a backyard or garden compost bin is approximately three cubic yards. Temperature should be monitored frequently and can reach above 150 degrees Fahrenheit. Compost should be moist, but not wet—the consistency of a damp sponge. Contrary to common belief, compost piles should not smell. If they do, carbon sources should be added and the compost turned or aerated.

As sustainability becomes an increasing concern, urban composting has become more popular and is often supported by city governments. Many cities offer composting classes, and some give away composting bins of various sizes and styles. Nonprofit organizations also offer helpful resources and sometimes starter kits for urban composters. Master Composter courses are also offered at low fees through Cooperative Extension offices across the country. Vermiculture, or composting with worms, is the easiest way to accomplish successful and speedy decomposition in a small apartment setting. Redworms are the most commonly used for vermicomposting and can break down an incredible amount of kitchen scraps quickly. The compost can then be placed on potted houseplants or in window boxes.

Composting is one of the cornerstones of organic farming. Proper compost is an incredibly nutritious soil amendment and is thus often referred to as "black gold" by organic farmers and home gardeners alike. Compost is chemical free and serves as a cheaper, more environmentally friendly alternative to commercial fertilizers. Chemical fertilizers, used by conventional agriculture, break down almost immediately, whereas compost decomposes more slowly, and therefore fewer excess nutrients leach out into nearby water systems.

Nitrogen, phosphorous, potassium, and other macro- and micronutrients needed for healthy growth are made more accessible to plants via the composting process and encourage vigorous root systems, which in turn cool the soil surface and mitigate erosion. In addition, compost buffers the soil pH and retains moisture and oxygen in the soil. Unfinished or hot compost can destroy garden plants, though, so it is important to be sure the compost has completely broken down and the temperature dropped before applying it.

Compost also improves soil tilth, or texture, by breaking up the structure of heavy clay soils and retaining water in sandy soils. Compost feeds bacterial and fungal growth beneficial to plant health and is thus key to a holistic approach to farming or gardening. A teaspoon of healthy compost contains more than 6 billion microbes. Certain microbes produce antibiotics that help plants resist disease. Simultaneously, compost decreases the presence of harmful pathogens. The U.S. Department of Agriculture Agricultural Research Service's Environmental Microbial Safety Laboratory estimates that proper composting reduces *Escherichia coli* and *Salmonellae* by 99.999 percent.

There are other innovative uses of compost. For example, it has been used as landfill cover and to aid in the bioremediation of contaminated sites. Studies have shown that plants grown in composted soil absorb fewer toxins such as heavy metals.

Siting, permitting, and management of organic material are regulated at the state level except for biosolids and manure. It is possible to compost human waste, which is done in many countries, but it must be done safely, and there are strong regulations for doing so in the United States under the Clean Water Act.

The U.S. Environmental Protection Agency estimates that 24 percent of solid waste is composed of yard trimmings and food waste. Therefore, eliminating this waste by composting it onsite diverts a significant portion of daily waste from landfills. Composting is key to keeping large-scale events sustainable and potentially zero waste. Many disposable plates, cups, and silverware are now compostable, and organizers of large-scale events and college campuses, to name a few relevant venues, often require that all vendors sell only such materials.

See Also: Department of Agriculture, U.S.; Organic Farming; Soil Nutrient Cycling.

Further Readings

Blum, Barton. "Composting and the Roots of Sustainable Agriculture." In *The History of Agriculture and the Environment,* edited by Douglas E. Bowers and Douglas Helms. Washington, D.C.: Agricultural History Society, 1993.

Gershuny, Grace and Deborah L. Martin. *The Rodale Book of Composting: Easy Methods for Every Gardener.* Emmaus, PA: Rodale/St. Martin's, 1992.

Jenkins, Joseph. *The Humanure Handbook: A Guide to Composting Human Manure.* Grove City, PA: Jenkins, 1999.

U.S. Environmental Protection Agency. "Wastes—Resource Conservation—Reduce, Reuse, Recycle—Composting." http://www.epa.gov/epawaste/conserve/rrr/composting/index.htm (Accessed January 2009).

Shannon Tyman
Stacy Vynne
University of Oregon

ConAgra

ConAgra Foods, Incorporated, is an international foods company with global operations. Its headquarters are at 1 ConAgra Drive in Omaha, Nebraska. It is listed on the New York Stock Exchange (symbol: CAG). Its vision statement is "One company growing by nourishing lives and finding a better way today . . . one bite at a time." Its goal is to provide consumers with food that they can trust to be safe, tasty, and nutritious at a fair price.

The company began September 29, 1919, when Alva Kinney combined four grain mills in south-central Nebraska into a single company, the Nebraska Consolidated Mills Company (NCM), headquartered in Grand Island, Nebraska. With bumper crops of wheat arriving at the storage elevators, Kinney soon added a fifth mill in Omaha to mill the abundant cereal grains. In 1922, the corporate headquarters moved to Omaha, where Kinney profitably ran the company until 1936, when he retired.

Kinney's successor, R. S. Dickinson, continued the company's successful milling operations but, similar to other companies, began to expand. In 1942, a flour mill and animal feed mill was opened in Alabama. The new flour mill was also a research center that developed new kinds of prepared foods such as the company's Duncan Hines cake mixes, which were put on the market in the 1950s.

As a result of competition from General Mills, Pillsbury, and other companies that were dominating the market, it was decided that a different corporate strategy was needed. So NCM sold its Duncan Hines cake mix brand to Procter & Gamble, and the new president of the company, J. Allan Mactier, used the proceeds of the sale to expand milling operations. A new mill was opened in Puerto Rico and operated by a subsidiary, Caribe Company. After successfully supplying flour, the subsidiary began a new company, Molinos de Puerto Rico, to supply animal feeds, which aided Puerto Rico's nascent beef industry. The company also aided the growth of the dairy industry on the island.

By the 1960s, the global demand for flour had leveled off, and more prosperous consumers were seeking new types of foods. NCM, along with other milling companies, began to diversify. In NCM's case, growth was sought in the animal feed business in the U.S. South, where the company supplied chicken feed to the growing number of chicken producers in Georgia, Florida, Alabama, and elsewhere.

With consumers seeking higher-quality meats that required more feed, operations in the Midwest were expanded. The company also expanded into the European market as a partner of a Spanish company.

In 1971, corporate officials decided to change the name from Nebraska Consolidated Mills Company, which reflected its original base in Nebraska, to a new name that reflected its growing operations in many places. The new name chosen was ConAgra, which means "in partnership with the land." In 1973, its stock was listed on the New York Stock Exchange. However, the new name was not followed by financial success because many of its expansion acquisitions were not highly profitable.

Charles Harper, former CEO of Pillsbury, was hired to restore profitability. He successfully prevented bankruptcy and enabled the company to regain its success. However, in 1980, the purchase of Banquet Foods brought a return to the processed food market via an increase in its chicken sales. Other moves were to expand into agricultural chemicals. This was a move designed to aid ConAgra's commodity business, which was afflicted with cyclical booms and busts.

The 1981 decision to move into prepared foods as an antidote to the cyclical commodity market was successful because of the stringent financial regulations instituted by Harper. That move would eventually make ConAgra the second-largest food company. Among its acquisitions were seafood companies: Singleton Seafood, Sea-Alaska Products, Trident Seafoods, O'Donnell-Usen Fisheries (Taste O'Sea frozen foods), and others. The moves positioned ConAgra against Mrs. Paul's Kitchens, Gorton's, and Van de Kamp's.

In 1982, ConAgra became dominant in the poultry business with the formation of Country Poultry. The next year it delivered over a billion pounds of chicken to the market. In 1986, it formed ConAgra Turkey Company, which was followed by the purchase of Longmont Foods. Another important acquisition was Banquet. ConAgra was then able to move into areas of chicken sales well beyond the basic broiler in the meat cooler.

The purchase of Armour Food, a processor of red meat, added the frozen food line Dinner Classics. The purchase came as Armour was experiencing a down cycle that caused it to eliminate about 40 plants. Despite reorganization, it was difficult to profit from the move, with the exception of the Dinner Classics line. The company also continued to expand in the frozen foods market with the purchase of the Morton, Patio, and Chun King lines. It also made more acquisitions in the red meat industry.

Because consumers were seeking leaner beef as part of a growing health consciousness, ConAgra began developing leaner beef. In 1987, it also acquired a stake in Swift Independent Packing Company, and it used that capacity as part of its expansion into foreign markets. Part of the expansion was the development of ConAgra's won financial services business to aid foreign money movements.

By 1990, ConAgra was selling food products across the food chain. The 1990 purchase of Beatrice foods brought Hunt's Tomato Paste and Butterball Turkey into its lines of business. The purchase was designed to mitigate commodity market swings. The company also purchased RJR Nabisco. The purchases also at times led to sales of some parts of the acquisitions that did not match the goals ConAgra had established.

Similar to other food processing companies, ConAgra's brands have experienced food safety issues. The recall of Peter Pan peanut butter and other product recalls have hurt the company. Legal problems have involved labor union issues in several plants.

Today ConAgra Foods is one of the largest food companies in North America. It has organized its operation into two businesses: Consumer Foods and Commercial Products. It is one of the largest packaged-food companies, and its products sell in supermarkets, restaurants, and food service operations. Among its brands are Chef Boyardee, Egg Beaters, Hebrew National, Hunt's, Orville Redenbacher's, and PAM. It suffered a major setback on June 10, 2009, when an explosion in its Slim Jim factory at Garner, North Carolina, killed three workers and injured others. The explosion closed the plant, creating a shortage soon afterward of Slim Jims—a popular beef stick snack.

New genetically modified foods have posed difficulties for ConAgra. Opposition to genetically modified foods led it to purchase websites both supporting and opposed to genetic modification. It is using a two-part strategy: sell genetically modified products when not controversial and avoid them otherwise. The food industry is dynamic because of consumer demands, new food safety issues, and other biological issues not previously imagined.

See Also: Agribusiness; Food Processing Industry; Food Safety; Genetically Modified Organisms.

Further Readings

Andreas, Carol. *Meatpackers and Beef Barons: A Company Town in a Global Economy.* Boulder: University Press of Colorado, 1994.

Bonanno, Alessandro, et al., eds. *From Columbus to ConAgra: The Globalization of Agriculture and Food.* Lawrence: University Press of Kansas, 1994.

Elliot, Jack. *Agribusiness: Decisions and Dollars.* Florence, KY: Cengage Learning, 2008.

Hamilton, Bill, et al. *Agribusiness: An Entrepreneurial Approach.* Florence, KY: Cengage Learning, 1991.

Jansen, Kees and Sietze Vellema, eds. *Agribusiness and Society: Corporate Responses to Environmentalism, Market Opportunities and Public Regulation.* London: Zed, 2004.

Limprecht, Jane E. *ConAgra Who?: The Story of ConAgra's First 70 Years.* Omaha, NE: ConAgra, Inc., 1990.

Andrew Jackson Waskey
Dalton State College

CONCENTRATION

Concentration is the process through which control over resources and economic decision making is placed in the hands of ever-fewer economic units (e.g., firms, farm households, etc.). Concentration is one of the central "laws" of capitalism, as identified by Karl Marx and as understood by the new rural sociology. In a capitalist political economy, concentration occurs through; (1) firms and farm households being in direct competition with one another through the production of commodities for a self-regulating market; and (2) differential success in capital accumulation between these units. Units that are outcompeted are often integrated into other units, so that over time there are fewer—and larger—units. In agrofood systems, concentration can be found in all sectors, from agricultural inputs and production to food processing and distribution. Data on concentration in food processing and distribution are generally more difficult to gather.

Excessive concentration can prevent competition through oligopolistic or monopolistic control over markets. Competition, according to both neoclassical economics and Marxian political economy, is the mechanism through which productivity increases in a capitalist economy. Collusion—illegal in North America and Europe because of antitrust laws—can become especially problematic in highly concentrated sectors, as it becomes easier to do and harder to detect and prove. James Lieber has documented a famous case of price setting between Archer Daniels Midland and other large grain purchasers and processors.

Concentration in Agrifood Sectors

Social scientists have pointed to farm losses and the resulting larger farms as an outcome of competition under capitalist political economy, with Willard Cochrane's "treadmill of production" featuring prominently in explanations. The treadmill of production refers to the pressure on farmers to adopt new technologies, as they increase efficiency relative to factors of production. Early adopters of new technologies—such as tractors, fertilizers, pesticides, and hybrid seeds—have a competitive edge over farmers who have not adopted.

If nonadopters are to remain competitive, they need to adopt the new technology as well. When all farmers do so, production increases and, everything else being equal, farm gate prices drop. The main ways in which farmers can maintain their incomes with declining prices is to produce more, or to produce with greater efficiency. Thus, new technologies increase productivity but also lead to intense competitive pressures on farmers, many of whom will end up being outcompeted by other farmers.

In the United States, the treadmill of production has taken the form of substituting capital and management for labor and land. The number of farms has dropped by two-thirds from 1910 to 1997. The social changes set in motion by the green revolution in many areas, and the "farm crisis" in the United States in the 1980s, are important examples of concentration among farms. Concentration of production among an ever-smaller number of producers has resulted in the decline of population, community, and economic activity in rural areas where agriculture is a prominent activity. Walter Goldschmidt's work, and subsequent research along similar lines, has shown that rural communities surrounded by fewer, larger-scale farms have lower indices of community health than communities surrounded by a larger number of small-scale farms.

Although the data are harder to find, it is also evident that concentration in food processing and distribution in the United States has been increasing over time. Concentration in processing offers farmers fewer outlets for their agricultural commodities and increases the possibility of collusion among purchasers, including price setting and price manipulation. Concentration in the retail sector means increased power for retailers, who increasingly determine the conditions of production, and enables them to become price setters rather than price takers. Many commentators note that concentration in the U.S. food system is extreme and is growing rapidly. Indeed, Thomas Lyson and Annalisa Raymer found that only 138 individuals sit on the boards of directors of the 10 firms responsible for more than half of the U.S. food supply. Similar to in the conventional sector, the organic foods sector has also experienced significant concentration in processing and retail.

Responses

Most social scientists agree that government regulation and intervention are required to maintain competitive markets. In contrast, neoliberalism, the hegemonic political discourse from the 1980s to the present, maintains that government intervention in the economy is harmful (critics maintain that such intervention serves largely the interests of economic elites). The dominance of neoliberalism in the United States has allowed unprecedented concentration to occur in many economic sectors, similar to classical liberalism at the turn of the 19th century, which led to strong grievances and eventual "trust busting" by President Theodore Roosevelt. Whether a similar type of mandated deconcentration will occur in agro-food systems will depend on political will and popular movements.

See Also: Archer Daniels Midland; Disappearing Middle; Farm Crisis; Food Processing Industry; Green Revolution; Supermarket Chains; Wal-Mart.

Further Readings

Cochrane, W. W. *The Development of American Agriculture: A Historical Analysis.* Minneapolis: University of Minnesota Press, 1979.

Goldschmidt, W. R. *As You Sow: Three Studies in the Social Consequences of Agribusiness.* Montclair, NJ: Allanheld, 1978 (1947).

Heffernan, W. D., et al. "Consolidation in the Food and Agriculture System." *Sustainable Agriculture,* 12/2:12–13 (2000).

Hendrickson, M. and W. D. Heffernan. *Concentration of Agricultural Markets.* Columbia: Department of Rural Sociology, University of Missouri, 2007.

Lieber, J. B. *Rats in the Grain: The Dirty Tricks and Trials of Archer Daniels Midland.* New York: Four Walls Eight Windows, 2000.

Lyson, T. A. *Civic Agriculture: Reconnecting Farm, Food, and Community.* Medford, MA: Tufts University Press, 2004.

Lyson, T. A. and A. L. Raymer. "Stalking the Wily Multinational: Power and Control in the U.S. Food System." *Agriculture and Human Values,* 17:199–208 (2000).

Marx, K. *Capital: A Critique of Political Economy.* New York: Penguin/New Left Review, 1990.

Ryan E. Galt
University of California, Davis

CONFINED ANIMAL FEEDING OPERATION

Commonly abbreviated to CAFO, a confined (or concentrated) animal feeding operation is a category of animal-raising operation defined by particular quantities of livestock. The term *CAFO* originated as a regulatory category through which federal and state laws regarding waste management and environmental protection are implemented. CAFOs are a central element of the industrial approach to agriculture known as factory farming and are a controversial presence in rural landscapes because of the potential for pollution and damage to environments and to human health.

These Missouri cattle being fed together outdoors have a flush tank and lagoon system in place to ensure a cleaner feeding area. Liquid manure from the lagoon behind them will later be used on nearby pastures.

Source: U.S. Department of Agriculture, Natural Resources Conservation Service/Ken Hammond

Regulatory Context

The U.S. Environmental Protection Agency (EPA) is responsible for protecting the natural environment and regulates the emissions from industrial farming operations. The Clean Water Act (1972) addressed the need to restore water quality, and the EPA identified livestock "feedlots" as "point sources" that discharge waste into water systems. The National Pollutant Discharge

Elimination System (NPDES) was established to regulate such emissions and required CAFOs to prevent the discharge of manure-contaminated and process-generated wastewater into U.S. waters. In 1976, the EPA defined more clearly the poultry and livestock operations that were subject to NPDES regulations. *Animal Feeding Operations* were defined as operations that fed or maintained animals for 45 or more days of the year in a confined area marked by the absence of vegetation. Within this larger category, some operations were further designated as CAFOs, and thus as "point sources," and were subject to the NPDES regulations. The EPA now defines CAFOs as animal feeding operations that "stable or confine as many or more" of certain threshold numbers of "animal units" (e.g., "large CAFO" threshold values include 700 dairy cows, 2,500 swine weighing over 55 pounds, and 125,000 chickens) and that "discharge pollutants into the waters of the United States." The EPA also defines "medium" and "small" CAFOs that are subject to lesser requirements.

Although CAFOs are subject to federal regulation, many states have set stricter standards in response to local conditions. The regulation of CAFOs is a dynamic element within agricultural policy. In response to pressure from environmental and community groups, the EPA updated the regulations in 2003 to require CAFOs to produce a Comprehensive Nutrient Management Plan that documents strategies for handling waste storage and discharge, disposal of animal carcasses, handling chemicals, testing soil and manure, and diverting clean water. The focus remains, however, on ensuring water quality, and the regulations still pay little attention to ground and air pollution issues. Following a lawsuit initiated by environmental groups (*Waterkeeper v. EPA*), the EPA revised the regulations further, as many CAFOs were storing liquefied manure in open pits called "lagoons" and then spraying the liquid onto surrounding land, causing nutrient overload and runoff. The EPA issued further revisions in 2006 and 2008, making it the responsibility of all CAFOs holding NPDES permits to keep detailed records of discharges. The situation may change again, however, following the arrival of the Obama administration, which has indicated that CAFO regulations will be under scrutiny once again.

The Role of CAFOs in the Food System

CAFOs are an integral part of industrial animal production that emerged during the 1930s, when the first mechanized swine slaughterhouses began operation. Several factors, chiefly the availability of cheap feed crops, have encouraged the growth of CAFOs in a way that epitomizes the trend in U.S. agriculture toward consolidation, simplification, and specialization. A few large corporations now control the majority of CAFOs, and swine and poultry production chains are vertically integrated to a high degree. As such, growers contracted to provide animals to slaughterhouses are responsible for the disposal of animal waste and the carcasses of animals that die before shipment. As a result of the high degree of consolidation that exists in industrial food production, slaughter, and processing, growers rarely have the market power to negotiate prices high enough to cover the costs of waste management that arise when such high numbers of livestock are raised in small spaces.

Supporters of the CAFO approach to animal production argue that the high degree of concentration and consolidation in the industry is the result of market forces and point to the better supply chain management and reduced transaction costs possible through an integrated meat production system. Advocates also point to high consumer demand for cheap meat and argue that CAFOs represent the most efficient production system and avoid the need to increase the area of land in agricultural production. Others have pointed out, however, that CAFOs have emerged during a period in which agricultural commodity policies and weak environmental regulation gave large-scale producers a strong price advantage over smaller-scale, diversified operations. Researchers Elanor Starmer and Timothy Wise have estimated that in a climate of stronger environmental regulation and full-cost (unsubsidized) feed, hog CAFOs' operating costs would rise by between 17.4 percent and 25.7 percent.

The price of animal feed represents the largest single operating cost for a CAFO. When changes to U.S. agricultural policy in the 1990s brought commodity prices below their production cost, CAFOs experienced dramatic reductions in feed costs. Rather than reinstating production controls to combat depressed prices, feed producers were provided with emergency subsidies to combat the developing crisis in farm incomes. As such, farm policy in recent years has supported the CAFO model of animal production and subsidized feed crop production. Detractors have accused corporate agribusiness of exerting inappropriate influence and pressure over agricultural policymaking through networks of networks of agriculture commodity groups, scientists funded by the industry, political lobbyists, and sympathetic politicians.

Concerns About CAFOs

CAFOs have serious negative external effects in three broad areas: environmental degradation and pollution, risk to human health, and damage to rural communities. The detrimental effects of CAFOs on water systems, air quality, and soils result from the large quantities of waste produced by animals grown at high stocking densities. The volumes of manure produced by a typical hog CAFO are equivalent to the human waste produced by a small town. Unlike in a human settlement, however, CAFO waste is not processed through a sewage treatment plant. Traditional solutions such as spreading the manure on nearby land are environmentally damaging at such high volumes, creating artificially high nutrient levels that contaminate water resources. CAFO waste contains a range of chemical contaminants including pesticides, heavy metals used as dietary supplements, and antibiotics. Campaigners argue that because CAFO operators are not obliged to bear the full cost of disposing of waste safely, these costs are externalized by the industry, resulting in the widespread availability of artificially cheap processed food.

The potential for pathogen transfer and the development of infectious diseases within CAFOs pose risks to human health. Researchers are concerned about the potential for the development of new viruses within large and highly concentrated animal populations, from which pathogens can be transmitted either by agricultural workers or by untreated waste spread onto surrounding land. The use of antibiotics and growth hormones in animal feeds has resulted in growing antimicrobial resistance in livestock and the emergence of antibiotic-resistant bacteria populations. CAFOs also affect the health of agricultural workers and local communities. Toxic gases generated by industrial animal production cause respiratory irritation and can lead to chronic illness. Communities suffer from dust and air pollution, and the health effects of living close to CAFOs have increasingly been the focus of epidemiological research exploring respiratory health and neurobehavioral issues. Rural communities also suffer as industrial-scale farming pushes smaller locally owned farms out of business. Research in such areas reveals higher poverty rates, property devaluation, and persistently low wages.

The conditions for animal welfare within CAFOs have been broadly criticized. Pew Commission research examining industrial farm animal production has described most intensive confinement systems, such as restrictive cages for veal and battery cages for poultry, as constituting inhumane treatment because of the restriction of movement. Public demand for higher animal welfare standards is increasing, and it is being met by both regulatory change and consumer pressure for retailers and producers to change suppliers and production practices.

The future for CAFOs is highly dependent on agricultural policy and on the price of feed crops. New trends, such as corn production for ethanol, have increased feed crop prices, affecting the operating costs for CAFOs and threatening their operating cost advantage over smaller, diversified operations. Although regulation of agribusiness is likely to increase, CAFOs are likely to remain a central part of the industrial food system. Corporate agribusiness is also expanding aggressively around the world, particularly in countries where cheap land and labor are available and where demand for meat is growing as incomes increase.

See Also: Agribusiness; Cheap Food Policy; Concentration; Factory Farm; Sewage Sludge; Vertical Integration.

Further Readings

"EPA Administered Permit Programs: The National Pollutant Discharge Elimination System," 40 CFR §122.23. http://www.gpoaccess.gov/cfr/index.html (Accessed January 2009).

Pew Commission on Industrial Farm Animal Production. "Putting Meat on the Table: Industrial Farm Animal Production in America." http://www.ncifap.org (Accessed January 2009).

Starmer, Elanor and Timothy A. Wise. *Living High on the Hog: Factory Farms, Federal Policy, and the Structural Transformation of Swine Production.* Working Paper 07-04. Medford, MA: Global Development and Environment Institute, Tufts University, 2007.

Edmund M. Harris
Clark University

Contract Farming

Contract farming is farming under an agreement made between producers (farmers) and buyers. Rather than making all their decisions independently (albeit informed by the market and past observed behavior), as farmers did by default for thousands of years, in a contract farming situation, the farm is hired to produce a specific quantity of agricultural product, meeting certain specifications (fruit of a particular quality, for instance, or eggs of a certain size, though often the specifications will be much more specific than this). Often an acceptable time range of delivery is specified. In return, the buyer commits to purchasing the product and may pay some part of the price up front or commit to providing other support, such as technical information or advice in meeting the required specifications. An agricultural corporation may want carrots of a certain sweetness level or may provide contract farmers with information on the planting and care of their new hybrid crop. The contract may stipulate other requirements such as acceptable and unacceptable pesticides and other practices, organic certification, and so on. More involved than the relationship between producer and purchaser, the relationship between the contract farmer and the purchaser can be like that between a nanny and a parent. Contract farming is used for all types of agricultural products, including both those that will be sold as is and those that will be processed in some way.

Ideally, both parties benefit from a contract farming arrangement, which guarantees (barring calamities like natural disaster or bankruptcy) that supply and demand will be met at prices amenable to both. This provides a predictability to an industry that depends on historically unpredictable factors like crop yields. The assurance of the contract reduces risk for both parties, with the comparatively minor trade-off that one might miss out on an unusual opportunity—the farmer may harvest his crop at a time of scarcity, when he could have sold it for more than was agreed on, or the purchaser may pay at a time of plenty, when he could have found the same goods elsewhere for a much lower price. So long as the price agreed on is fair and profits both parties, the predictability and assurance offsets this risk, much as the security of low-risk investments offsets the lost opportunity of high returns.

Contract farming has become a common arrangement between farmers in developing nations or in poor areas of developed nations and buyers in developed nations. There are various benefits to both sides. The buyer may find that his dollar will go farther in developing nations than in developed ones—more than offsetting the cost of transport and the transaction fees of overseas business. The farmer benefits from improvements that the contract brings about, ranging from better irrigation facilities and other equipment to infrastructure improvements to the community, as the buyer builds or funds the building of new roads to accommodate transporting the crops. There are further transaction costs and expenses that are also reduced for a farmer who can deal in volume—selling mainly to one buyer, or to a handful of buyers, instead of to smaller local stores or directly to consumers.

Contracts and Risk

All contractual arrangements entail some risk as to the ability of both parties to agree that the terms have been met. Buyers may claim that the goods are a lower quality than agreed, and that a lower price should thus be paid; farmers may try to present substandard goods as proper quality. One side or the other may try to get out of the contract if circumstances in the market or their business have changed. Improvements the buyer agreed to fund or arrange may not be adequately funded or arranged, leaving the farmer responsible for delivering a product he or she is not properly equipped to deliver.

Of course, any working arrangement between rich and poor, between developed and developing nations, brings with it risks of exploitation on the one hand and corruption or poor performance on the other. These two sides of the world are often characterized as the Global North—the developed world, though this includes Australia and New Zealand, which are located in the geographic south—and the Global South, or developing world. Given the general industrial/agricultural divide between the north and south in the United States, the model is especially resonant from an American perspective, but it was actually popularized by German Chancellor Willy Brandt in 1980.

Individual buyers can easily attain a position of monopsony over farmers, and a complaint is often made that the Global North has collectively assumed this position over the Global South. *Monopsony*, a term coined by British economist Joan Robinson in 1933, is the converse condition of monopoly: a circumstance in which there is one buyer and many potential sellers, resulting in imperfect competition, as the buyer is able to dictate the market terms in the same way that a monopolizing seller is able to in a monopoly. As retail outlets like Wal-Mart and Barnes & Noble become bigger and bigger, industry observers often complain that they enjoy monopsonistic privileges, acting as the "one buyer" before reselling to consumers who have little control over the selection offered to them. In a farming scenario, when one buyer has such power, he or she can use the situation to his advantage and to the detriment of the farmers; desired products may be changed at a whim, forcing major readjustments at the farm and potentially affecting the job market. Situations may also develop similar to those of "company towns" in the 19th century, in which all the workers in town worked for the same company, which also ran a company store to sell them the goods they needed. The buyer in a contract farming arrangement may charge for the improvements made to the farm that enable the farmer to enter into the contract, and although the farmer presumably makes a profit beyond the cost of those improvements, he or she is made that much more dependent on the buyer by turning his or her farm into a more specialized endeavor that requires the continued business provided by the contract.

Contracts and the Law

Some countries have laws specific to contract farming; in most, the existing body of contract law covers such arrangements sufficiently. The clauses of the contract typically cover the responsibilities of the contracting partners; specify in as much detail as possible the product or products the contract concerns; detail the means and methods of production, including the specific seed varieties to be used, the methods of cultivating and preparing the soil and of preventing disease and animal trouble, the transportation processes in place or to be developed, and storage facilities; the conditions of purchase, including price, timing of delivery, and method of delivery (such as who is responsible for transporting the goods from the place of production or the storage facility to their final destination); the jurisdiction applicable to the contract; and some description of a dispute settlement mechanism, should either party feel the conditions of the contract have not been met.

Contract farming has become a significant source of agricultural revenue for farmers in India and Thailand, where conditions are excellent for agriculture and the local economies include significant natural resources and skilled labor, as well as the ability to benefit from the guarantees and economies of scale of contract farming arrangements. However, contract farming is certainly not limited to the interaction between the Global North and the Global South. Throughout the southeastern United States, contract farming is becoming more common among grain farmers, who are provided with the seed, chemicals, and fertilizers the grain company wants used with the product they purchase, and who need only the land, skill, and labor to bring the grain to harvest. In other cases, grain farmers contract with hog and poultry farms, who contract for a specific amount of feed grain delivered at a specific time or at specific points in the season. In the Midwest, corn farmers contract with ethanol producers. Skittishness toward the futures market may increase the prevalence of contract farming among U.S. farmers in the coming decade.

See Also: Agrarianism; Commodity Chain; Cooperative.

Further Readings

Grossman, Lawrence S. *The Political Ecology of Bananas: Contract Farming, Peasants, and Agrarian Change in the Eastern Caribbean.* Chapel Hill: University of North Carolina Press, 1997.

Little, Peter D. *Living Under Contract: Contract Farming and Agrarian Transformation in Sub-Saharan Africa*. Madison: University of Wisconsin Press, 1994.

Rehber, Erkan. *Contract Farming: Theory and Practice*. Hyderabad, India: Icfai University Press, 2007.

Bill Kte'pi
Independent Scholar

COOPERATIVE

A cooperative is a social enterprise, an association of individuals with both economic and social goals and a shared philosophy, which has become increasingly common since the 1990s, when there was a concerted push to popularize the cooperative movement—then about a century and a half old. The Statement on the Cooperative Identity, drafted by the Geneva-based International Cooperative Alliance in 1995, defines a cooperative (often shortened to "co-op") as "an autonomous association of persons united voluntarily to meet their common economic, social, and cultural needs and aspirations through a jointly-owned and democratically-controlled enterprise . . . based on the values of self-help, self-responsibility, democracy, equality, equity, and solidarity. In the tradition of cooperative founders, cooperative members believe in the ethical values of honesty, openness, social responsibility, and caring for others." Those clauses of the statement allude to the Rochdale Principles—first formulated by the Rochdale Society of Equitable Pioneers, the consumer cooperative that founded the modern cooperative movement in 1844—and have been updated by the International Cooperative Alliance to retain the original spirit but reflect the modern economic climate. The Rochdale Principles is a set of ideals or values that guide the activities and function of cooperatives.

Voluntary and open membership is a critical requirement of cooperatives, which must be open "to all persons able to use their services and willing to accept the responsibilities of membership, without gender, social, racial, political, or religious discrimination." This does not prevent the adoption of rules such as limiting the membership to citizens of a particular geographic area, or those who belong to a certain industry (farmers, for instance) or who have specific experience in that industry; nor does it prevent the charging of membership dues and thus turning away those who cannot pay.

Democratic member control means that cooperatives function as democratic groups controlled by the members, who make decisions and set policies; most cooperatives give all members equal voting rights, though some may weight votes according to factors particular to their circumstance (much as shares of stock provide weighted votes to stockholders in proportion to the number of shares they own). Generally, elections are held for representatives who are responsible for the day-to-day operations of the coop.

Member economic participation says that members contribute to their co-op's capital, at least a portion of which becomes the co-op's common property, used to develop the co-op, set up reserves, benefit members in proportion to their participation, and support other operations jointly approved by the co-op's membership. Generally speaking, members will put money into the co-op for its activities, and when a profit is made, that profit will be paid back to the members in proportion to what they put in, with surpluses used to further expand or fund co-op activities. Depending on the nature of their activities, some co-ops may pay dividends to their members, the way stock companies do.

Autonomy and independence require that whatever relationship a co-op enters into with some other body must not be one that compromises the co-op's independence. A co-op cannot be bought out by some corporation and still remain a co-op, for instance.

Education, training, and information are the ideal principles that encourage co-ops to provide education and training for members and to make information available to the general public, "particularly young people and opinion leaders," about the benefits of cooperatives.

Cooperation among cooperatives calls for co-ops to consider themselves as part of the cooperative movement that has been pursuing and promoting these objectives since the 19th century.

Concern for community calls for cooperatives to be part of the sustainable development of their local community.

The legal existence of cooperatives varies by jurisdiction. In the United States, most co-ops are noncapital stock corporations organized under and beholden to state co-op laws. They may sometimes be formed as limited liability corporations instead or under other forms of organization, depending on what the co-op wants to do and how a co-op is defined by local law. There are many kinds of co-ops, such as the rural utility cooperatives—public utilities owned by their customers—which were formed to provide electricity and telephone service throughout the rural United States; retail cooperatives, the members of which are businesses, rather than individuals; banking cooperatives (such as credit unions); and agricultural cooperatives, among many others.

Agricultural Cooperatives

There are two sorts of agricultural cooperatives, or farmers' co-ops. An agricultural production cooperative, in which farmers pool their resources to create a megafarm and enjoy economies of scale, is a fairly rare thing outside of the kibbutzes of Israel and the collective farms that were attempted in the Soviet Union. More common is the agricultural service cooperative, created so that its members can buy their supplies at bulk discounts or market their crops and livestock as a joint entity. This latter sort, the agricultural marketing cooperative, is so common in the United States that Americans who live outside farming communities, and who may not give much thought to where their food comes from, are likely unaware of the everyday regional and national brands used by these co-ops:

- Blue Diamond Growers, a California-based co-op of almond growers.
- Cabot Creamery, a dairy co-op with 1,350 members in Vermont and upstate New York, which has been marketing its cheese internationally since 2007.
- Emerald Nuts, a brand of Diamond Foods, which was a co-op until it was taken public in 2005.
- Dairylea, a New York dairy co-op—its very name a contraction of the Dairymen's League. In the 1920s, the league represented more than 100,000 farms; today it represents 9,500, a sign of the United States' agricultural contraction and mergers.
- Florida's Natural Growers, owned by 1,100 citrus growers to market fruit snacks and citrus juices.
- Land O' Lakes, a dairy co-op based in Minnesota.

- Ocean Spray, a Massachusetts cranberry and grapefruit co-op.
- Sunkist Growers, a California and Arizona citrus growers' co-op with 6,000 members. Sunkist has made good use of its brand name, marketing it for use by snacks, candies, nuts, sodas, and juices in 45 countries, including Sunkist soda, the most popular orange soda in the United States and one of the few with caffeine.
- Sun-Maid Growers of California, a raisin growers co-op.
- Sunsweet Growers, operators of the largest dried fruit plant in the world, best known for their prunes and prune juice.

See Also: Agribusiness; Community-Supported Agriculture.

Further Readings

Chayanov, Alexander. *The Theory of Peasant Cooperatives.* Athens: Ohio State University Press, 1991 (Originally Moscow, 1927).
Cobia, David W. *Cooperatives in Agriculture.* New York: Prentice Hall, 1988.
Lyson, Thomas A. *Civic Agriculture: Reconnecting Farm, Food, and Community.* Medford, MA: Tufts University Press, 2004.
Pollan, Michael. *The Omnivore's Dilemma: A Natural History of Four Meals.* New York: Penguin, 2007.

Bill Kte'pi
Independent Scholar

CORN

Corn is one of the most widely grown crops in the world. In recent decades, modern chemistry and other industrial processes have changed corn from something people ate at the dinner table to something in almost everything people eat. In fact, it is in much of what they use as well: Corn is now a part of an industrial food chain, as well as the source in part or in whole of a myriad of products.

The Taínos were pre-Columbian inhabitants of the Bahamas and other Caribbean islands. When the Spanish first encountered them, they adopted corn and used the word *maize* as the Spanish form of the Taínos word *maiz*. Maize is used fairly globally today; however, corn, the English general name for a cereal grain, is still used in English-speaking North America. In the United States and Canada, the term *Indian corn* is used to describe a variety of multicolored field cultivars. They range in color from red to blue to black and may be sold as an autumn decoration or as a popcorn specialty item.

Today, corn is a globally important food crop. It originated in the Americas, specifically in central Mexico, from a plant the Aztecs called *teosinte* in their Nahuatl language, meaning "the mother of corn." It is believed to have originated on a single mountain in southern Mexico. *Teosinte* still grows wild in southern Mexico and northern Central America; however, it is not corn. It took thousands of years for the plant to be developed into modern corn. As it became corn, it was able to support the Mayan and later Aztec civilizations, becoming a god (and some times a goddess) that they worshipped as their corn culture became central to their survival.

The development of corn as a food is one of the greatest of human achievements. From this miraculous agricultural achievement has come a major food source for millions of descendants of the Indians who first produced corn, as well as hundreds of millions of people and countless animals around the world.

Corn is a hybrid that seems to have originated from genetic mutations of some *teosinte* plants about 9,000 years ago. Since then, corn has been a hybrid plant dependent on humans for its continued existence. It is an existence that, since Christopher Columbus, has spread from the Americas around the world to most of the places humans farm. The only cereal to outrank corn in production as a cereal grain is wheat. Rice is in third place.

Organically grown corn like this is surrounded by more weeds than conventional corn because only nonchemical weed management techniques are used, but the amount of corn yielded is similar.

Source: U.S. Department of Agriculture, Agricultural Research Service/Bob Nichols

As a food commodity, more than half of the world's corn production is in the Corn Belt of the midwestern United States, with Iowa as the center of production. Corn is really a tropical or semitropical plant. To grow well, corn needs hot rainy summers that turn dry when the corn is nearly ready for harvesting. Large quantities of corn are also grown in the southern United States, where cornbread in a variety of forms—including as hush puppies—is a favorite food.

Corn has a multitude of uses in its natural state. It is widely used as fodder to feed animals or is stored green and fermented to make silage. Until the advent of indoor plumbing, corn cobs were used as toilet paper in outhouses. It was also important as the mash from which corn whiskey was made, the taxing of which sparked the Whiskey Rebellion in the early history of the United States. Today, large quantities of corn mash are used to make alcohol (ethanol) as a renewable resource to reduce the consumption of petroleum in gasoline and other fuels.

Historically, the preferred use of corn was as a food. The kernels of corn can be eaten directly off of the cob (the inedible center of the ear of corn) after boiling, steaming, or roasting. Or the kernels can be cut off, cooked, and eaten while still fresh. Corn on the cob can be eaten with butter and salted for flavor. Street vendors can be seen selling corn on the cob on the streets in the United States, in China, and in other places in the world, where it is eaten like a popsicle on a stick. In Mexico, corn on the cob—or *elote* (a Nahuatl word)—may be smeared with mayonnaise and then coated with shredded Cotija cheese. Other condiments may also be used. Corn is also popular when preserved frozen or canned.

However, for use out of season, corn can be allowed to dry in the fields. After harvesting, the ears of corn are shucked and the corn kernels are shelled (removed from the cob). Whole dried corn keeps well when stored properly—it can be easily kept for a year. This allows users to grind the kernels of corn fresh before use as corn meal, which is used in many foods including corn bread, tortillas, and griddle cakes, as well as an ingredient in many other dishes. Some varieties can also be popped as free kernels or popped on the cob and eaten from the cob.

A Global Food

In Europe and the Orient, pizzas may be topped with corn. It is also widely eaten in salads as kibbled kernels. Sweet corn can also be cooked with milk or cream sauce to make creamed corn. Sweet corn picked very young may be processed or pickled and eaten as baby corn, usually in salads.

In Latin America, corn has traditionally been eaten with beans—this combination balances the diet. If confined to either beans or corn alone, the diet would be deficient in an essential amino acid, which would lead to dietary disease. In addition, traditionally in Latin America corn is soaked in limestone, which opens the kernels to make vitamin B available.

Sweet corn and popcorn are grown for human consumption. Sweet corn can be a hybrid or, in some traditional seeds, a natural mutation of the original corn first grown in the Americas. The Iroquois are credited with the first use of sweet corn as a vegetable.

Field corn is grown for fodder and animal feed. In some northern climes such as Vermont, corn is usually grown as fodder because of the short growing season; however, agribusiness companies are constantly working to develop hybrids that will produce the best yields in the soil and climate conditions of a particular locality. In other places, corn is grown for the corn seed, which is ground into a coarse feed for poultry, hogs, or fish. All these ways in which corn has been used as food for humans and animals have radically changed in recent decades. The change began in 1852, when the Chicago Board of Trade invented a grading system for corn. Before the grading system, farmers sold their corn in sacks embroidered with their names, as a kind of branding. However, with grading, corn became a commodity that was the same for every grade, regardless of where it had been grown. It was not so much a farm crop thereafter but an industrial agriculture resource that would emerge in the coming decades.

During the presidency of Richard Nixon, several events occurred that radically changed the way in which corn was produced. The Arab Oil Embargo of 1973 instituted by the Oil Producing and Exporting Countries and its allies created oil shortages and, ultimately, inflation. The embargo had been preceded by several disastrous harvests in the Soviet Union. The Soviets thus came looking to buy grain in the United States and bought a huge amount of corn. This was to be the impetus for a change in the farm subsidy laws from a system of loans and other devices to direct payments. The change promoted the development of corn as a widely grown monocrop planted from fencerow to fencerow.

The change to monoculture in corn was fostered by the green revolution. New hybrids were planted that produced greater yields. Since the 1970s, the use of fertilizers, hybrids targeted to specific climates and soils, pesticides, and farm machinery that plants ever-larger areas radically increased the yield of bushels of corn per acre. It also meant that there was no longer a need for farm animals, for crop rotation, or—ultimately—for as many people on the farm as had previously been there. One of the consequences of the change to corn as a monoculture has been a population decline in the Corn Belt.

More Than a Food

As an ever-greater supply of corn has been grown, it has produced a commodity that is foundational to industrially processed food and products. Much of the corn grown is used as animal feed. It is fed to cattle to tenderize the beef with fat marbling. The fat comes from the weight gained by cattle raised on grass but fed grain in feed lots. Production levels in 2007 reached 13.308 billion bushels of corn in the United States alone. Yields per acre were on average 180 bushels; however, using intensive farming methods—which are more expensive—yields can exceed 220 bushels per acre.

Most of the huge volume of corn produced in the United States is consumed or processed into products for other uses. Only about 2.25 billion bushels are exported. The size of the 2008 and 2009 crops were down from 2007 because of adverse weather in the Corn Belt. Farmers in 2008 survived flood conditions through crop insurance and government direct payments.

An increasing volume of corn production in the United States, as much as 25 percent of the total crop, is being diverted into making ethanol. The increased demand is a boon for farmers, a concern for automobile owners who believe it rots engine parts, and a cause of higher food prices on the global market that could foster famine in some places.

The increase in corn production in the United States has been matched by increases in Brazil and other places such as Europe and China. However, global warming is seen as a threat to global corn production because it increases climatic conditions favorable to insects and diseases that attack corn. Other expected effects of global warming would be longer growing seasons and increased carbon dioxide for plant consumption. However, stronger storms, more weeds, and other adverse effects could harm corn production. The need will be for increases in green technologies such as nonchemical weed management techniques, precise irrigation techniques, and other methods for sustaining crop yields.

Processing corn into a food began when John Harvey Kellogg (1852–1943), a physician, and his brother Will Keith Kellogg (1860–1951) invented cornflakes as a food for the patients in the Battle Creek, Michigan, sanitarium that Dr. John Harvey Kellogg was running. Influenced by Seventh-Day Adventist Church teachings, the Kelloggs were vegetarians. Cornflakes as a breakfast food was followed by other processed corn products.

Products that are made from corn after processing and are used in other foods include cereal grits for cereals; brewer's grits; polenta grits used in hot cereals, as hominy grits, and in pizza; and bakery applications. Fine grits are used in extruded cereals and processed pet foods. Other products include corn oil, cornstarch, liquid glucose, dextrin, malt dextrin, sorbitol, and dextrose monohydrate. Cornstarch is the source of corn syrup. The use of corn syrups, including high-fructose corn syrup in soft drinks, ice creams, and a great many other foods, goes far beyond its simple use as a sweetener for pancakes or other foods. It is used to sweeten ketchup, preserved meats, canned fruits and vegetables, soups, beers, and enchilada sauce. It also sweetens the taste of sealable envelopes, stamps, and aspirins.

Nonfood products made from corn number in the thousands. They include plastics that are biodegradable (food wrapping), paper and plastic cups, textile fibers, binders for polymer sinks, and a binder in rat poison.

See Also: Agribusiness; Crop Rotation; Green Revolution; High-Fructose Corn Syrup.

Further Readings

Diamond, Jared. *Guns, Germs, and Steel: The Fates of Human Societies.* New York: W. W. Norton, 1999.

Fussell, Betty. *The Story of Corn.* New York: Knopf, 1994.

Hardeman, Nicholas P. *Shucks, Shocks and Hominy Blocks: Corn as a Way of Life in Pioneer America.* Baton Rouge: Louisiana State University Press, 1981.

Mangelsdorf, Paul C. *Corn: Its Origins, Evolution, and Improvement.* Cambridge, MA: Harvard University Press, 1974.

Pollan, Michael. *The Omnivore's Dilemma: A Natural History of Four Meals.* New York: Penguin Books, 2006.

Smith, C. Wayne, et al., eds. *Corn: Origin, History, Technology, and Production.* New York: John Wiley & Sons, 2004.

Thompson, Gare. *Corn: An American Indian Gift.* New York: Harcourt Achieve, 1998.

Walden, Howard T. *Native Inheritance: The Story of Corn in America.* New York: Harper, 1966.

Warman, Arturo. *Corn & Capitalism: How a Botanical Bastard Grew to Global Dominance.* Trans. Nancy L. Westrate. Chapel Hill: University of North Carolina Press, 2003.

Andrew Jackson Waskey
Dalton State College

COVER CROPPING

Cover cropping is the planting of crops specifically for one or more of several non-revenue-generating purposes, most commonly for protecting against soil erosion, improving soil fertility, and minimizing soil nutrient loss. Cover crops are used in both annual and perennial and organic and conventional agroecosystems, varying in management and species composition according to the system and purpose.

Cover Crop Benefits

Many annual cash crops in temperate regions are planted in the spring and harvested in the fall, leaving the soil bare, and thus open to the erosional forces of wind and water for several months each year. Quick-growing cover crops, particularly cereals such as barley, oats, rye, and winter wheat, as well as mustards, can take advantage of fall weather and put on substantial biomass to minimize soil loss caused by wind and rain. These crops can also take up—or immobilize—nutrients, such as nitrate, that are susceptible to loss down through the soil. This maintains the nutrients in the rooting zone so they are accessible to the following cash crop and reduces the input of nutrients to groundwater or surface water, the latter of which can lead to toxic algal blooms or anoxic zones. Reduced leaching by such cover crops, also known as "catch crops," has helped improve water quality in locations such as Chesapeake Bay. Cover crops can also protect against soil and nutrient loss in perennial agroecosystems, such as orchards and vineyards, where they are planted between the rows.

In addition to maintaining soil nutrients, cover crops can add nitrogen to the soil. Legume cover crops, such as clovers, cowpeas, fava beans, field peas, and vetches, host bacteria that convert atmospheric nitrogen into forms used by the plant. When the crop is mowed, rolled, or undercut and incorporated, soil microbes mineralize the organic nitrogen, making it available for the following cash crop. These nitrogen-fixing legumes, also known as "green manures," are particularly important to organic or other low-input agroecosystems, in which fertilizer derived from the Haber-Bosch process is not or only minimally used. Legume cover crops are also recommended by some researchers as a strategy for developing agricultural sufficiencies in sub-Saharan Africa and other food-insecure regions.

A relatively new use of cover crops is "biodrilling," particularly in systems with reduced tillage. Fall-planted cover crops form root channels in soft, wet soil. Without cultivation, these root channels, some as small as 1 millimeter in diameter, remain and allow the cash crop to penetrate compacted zones and reach deep soil moisture that would otherwise be inaccessible. This can help reduce water stress in the summer.

Cover crops can also provide pest control in agroecosystems, depending on both the pest species and cover crop species present. Cover crops may compete with weeds for resources or suppress their growth through allelopathic chemicals. They may also provide a "biofumigation" function, reducing populations of plant parasitic nematodes or the incidences of diseases (e.g., club root or Verticillium wilt) or pathogens (e.g., pea root rot). Cover crops may also indirectly control pest species, such as by providing favorable habitat that maintains or increases populations of beneficial nematodes or beneficial insect predators and parasitoids.

These low cover crops growing between rows of trees in a California orchard help prevent erosion from rain and wind.

Source: U.S. Department of Agriculture Natural Resources Conservation Service/Gary Kramer

Cover crops may be grown as monocultures or in mixes, with different plants performing different services. A common example is combining cereals (to rapidly immobilize soil nutrients and provide cover in fall) with legumes (to add nitrogen to the system).

Although fall-planted cover crops are most common in temperate climates, summer cover crops can be used before overwintering cash crops. Cover crops may also be grown as perennials, such as between rows of trees or vines.

Cover Crop Challenges

Using cover crops can present several challenges to growers. One of the main challenges involves the timing of operations. Stands of cover crops need to be mowed, rolled, or undercut and then incorporated into the soil, so they can begin to break down before the planting of the cash crop. The timing of spring planting is often weather dependent, with growers beginning fieldwork as soon as soil and climatic conditions allow. Growers who have to include field activities to manage the cover crops then have to delay planting, compared with growers who can immediately till and plant cash crops. The nutrients immobilized by the cover crops are also not immediately available to the cash crop, as time is needed for microbes to decompose the cover crop biomass and release the nutrients. This can have negative effects on the cash crop, particularly if the nutrient demand of its seedlings or some other early life stage is high.

Another challenge of cover crops is their effect on soil water content. If overwintering cover crops have a substantial growth period before they are incorporated into the soil in spring, they may deplete the soil water by transpiring it to the atmosphere. This preemptive competition can have a deleterious effect on the subsequent cash crop if the depleted soil water is not replaced by rains or high percolation from groundwater.

In reviewing data from field trials and demonstrations, researchers at the University of Michigan found that nitrogen provided by legume cover crops could be enough to replace synthetically derived nitrogen used in agroecosystems globally. Others have disputed these findings, saying that biological means are inadequate to provide nutrients to support agriculture for a growing population. The debate over the review offers a window into competing discourses of how to grow food for a future world.

The pest control provided by cover crops varies: in some cases, susceptible cover crops have increased parasitic nematode populations or the incidence of pathogens. Further, recent concerns over the potential for cover crops to harbor vertebrate pests that could carry pathogens such as *Escherichia coli* O157:H7 led to a leafy greens marketing agreement that discouraged noncash crop vegetation in buffers surrounding crop fields.

Cover crops can also provide opportunities for weed growth and reproduction. Cover crops frequently are not managed for weeds, yet depending on the growth form and timing of germination and canopy development, weeds may establish a foothold and set seed, creating problems later in the cropping cycle. Increased seeding rates may hold promise for earlier canopy cover, and hence weed suppression. However, some persistent cover crops may actually become weeds themselves in subsequent seasons. In addition to suppressing weed growth, the allelopathic chemicals produced by cover crops can also have a negative effect on the establishment and growth of the following cash crop.

The use of cover crops in resource-poor regions is not a simple panacea. Extra labor may be necessary to implement cover crops, particularly woody species that do well in tropical areas; such labor may not be available for food-insecure households, so subsidies may be necessary to help cover the cost of starting such systems.

See Also: Agroecology; Biological Control; Intercropping; Legume Crops; Low-Input Agriculture; Organic Farming; Soil Erosion; Soil Nutrient Cycling.

Further Readings

Badgley, Catherine and Ivette Perfecto. "Can Organic Agriculture Feed the World?" *Renewable Agriculture and Food Systems*, 22/2 (2007).

Clark, Andy, ed. *Managing Cover Crops Profitably*, 3rd ed. Beltsville, MD: Sustainable Agriculture Network, 2007.

Thorup-Kristensen, Kristian, et al. "Catch Crops and Green Manures as Biological Tools in Nitrogen Management in Temperate Zones." *Advances in Agronomy*, 79 (2003).

Weil, Ray and Amy Kremen. "Thinking Across and Beyond Disciplines to Make Cover Crops Pay." *Journal of the Science of Food and Agriculture*, 87/4 (2007).

Katie L. Monsen
University of California, Santa Cruz

CROP GENETIC DIVERSITY

Genetic diversity in crops provides the raw material that allows breeders to improve yield, increase protein and vitamin content, and augment drought and pest resistance in food plants.

The increasing food needs of a growing human population, combined with global climate change, regional water shortages, and altered environments, make it imperative to preserve crop genetic diversity. Loss of diversity has been severe in recent decades, as the number of crops grown has declined, and large-scale planting of genetically uniform, high-yield varieties has replaced traditional, locally grown crops.

Crops originated from wild plants. Through selective breeding, farmers have produced higher-yielding, more-nutritious forms of these plants. As people migrated around the globe, they brought their crop plants with them and selected forms best adapted to the new environmental conditions. Regionally adapted variants of crop plants, known as

The National Plant Germplasm System from the U.S. Department of Agriculture's Agricultural Research Service stores unusual varieties of Latin American maize like these, which may be used to enhance the genetic diversity of the U.S. corn crop.

Source: U.S. Department of Agriculture, Agricultural Research Service/Keith Weller

landraces, are the result of hundreds of years of local breeding and selection. The size, shape, color, and taste of different landraces result from genetic differences that have been selected in response to local tastes, customs, and markets. Selection in response to local soil characteristics, water availability, and cultivation practices has resulted in different varieties in different areas. Often, traditional farmers maintain many varieties that can be planted under different conditions and for different tastes.

The green revolution in the mid-1900s dramatically increased crop yields for a rapidly growing human population. Scientific principles were applied to crop breeding: Better irrigation systems, mechanized cultivation, and the expanded use of pesticides, herbicides, and fertilizers quickly changed the nature of agriculture. Although landraces persisted in subsistence agriculture, modern agriculture no longer focused on local markets but provided large quantities of fewer crops to international markets. Now, of 7,000 crop plants that are grown globally, only 150 are widely cultivated, and only three—rice, wheat, and corn—provide the bulk of the world's calories. Genetically homogeneous, high-yield crops, cultivated in standardized ways, are planted over large expanses of land. However, such homogeneous crops are genetically vulnerable in that a disease, pest epidemic, or environmental change can decimate a crop. If high-yield varieties succumb to pest resistance, yields are significantly reduced within about 5 years. It takes 8 to 11 years to breed new varieties, using genes from wild relatives, landraces, and stock exposed to mutagens to induce variation. However, the loss of wild relatives through habitat destruction and the decline in cultivation of landraces have resulted in gene erosion, limiting options for change.

A few examples from the United Nations Food and Agriculture Organization (FAO) illustrate the extent of gene erosion:

- 96 percent of the 7,098 U.S. apple varieties cultivated before 1904 are extinct
- 95 percent of U.S. cabbage and 81 percent of tomato varieties are gone
- Only 10 percent of the 10,000 wheat varieties grown in China before 1940 remain
- 80 percent of Mexican maize varieties grown in the 1930s are lost

In addition to these types of gene erosion, researchers from Bioversity predict that global climate change will result in the extinction of 16–22 percent of the wild relatives of food plants.

In the 1970s, several UN programs, including the FAO, realized the need to conserve genetic diversity in crops. An international body, the Consultative Group on International Agricultural Research (CGIAR), was formed, and CGIAR research centers around the world established gene banks to store seeds and propagules. At this time, there are over 1,300 gene banks, some affiliated with CGIAR and others representing national or regional efforts. These gene banks house 6.1 million accessions. The largest, the National Plant Germplasm System (NPGS), is housed in several facilities across the United States and maintains 450,000 accessions. Plant materials are provided to breeders and researchers free of charge. Most materials are requested from within the United States, but over 191 countries have used these resources for breeding and, more recently, for genetic engineering of crops. The value of these genetic resources is apparent from many examples:

- Researchers at Cornell University have bred a nonedible wild potato with an edible potato to produce plants covered with small hairs that trap feeding and ovipositing insects.
- The Russian wheat aphid causes $300 million damage to U.S. wheat crops annually. Crossing wheat with Turkish landraces has produced an aphid-resistant form.
- Stored beans intended for food cannot be fumigated and are often decimated by weevils. Breeders have introduced genes from a wild bean that produces a chemical preventing weevil reproduction.

Although the NPGS has been catalogued and digitized, many other gene banks are less accessible. In the 1970s and 1980s, plant materials were collected with an urgency formed by scenarios of an impending crisis. Many accessions were not "characterized," meaning that information concerning yield, nutritional value, drought resistance, soil preferences, and other features are not available. Even some NPGS materials suffer from lack of characterization. In addition, gene banks are costly to staff and operate. Seed viability declines with time, and seeds must periodically be grown into plants to secure fresh seeds, which are then dried and stored at low temperatures. Plants that do not store well as seeds must be maintained as tubers or cryopreserved tissues.

In addition to gene banks, in situ conservation programs, sponsored by governments, international organizations, and nongovernmental organizations, influence farmers to grow landraces. Major programs exist in Ethiopia, the Philippines, and Mexico, where crops continue to evolve under cultivation. Some countries have set aside preserves for wild relatives of fruit trees (Brazil, Germany, and Sri Lanka), wheat (Israel and Turkey), and other crops.

At this time, crop genetic resources are protected through two major international treaties: the Convention on Biological Diversity and the FAO's International Treaty on Plant Genetic Resources for Food and Agriculture. These treaties mandate conservation and sustainable use, facilitate access to genetic resources, and ensure equitable benefits to source nations and nations developing new varieties.

See Also: Agribusiness; Green Revolution; Seed Industry.

Further Readings

Consultative Group on International Agricultural Research. "Who We Are." (2007–08). http://www.cgiar.org/index.html (Accessed December 2008).

International Treaty on Plant Genetic Resources. "The Treaty." http://www.planttreaty.org (Accessed December 2008).

U.S. Department of Agriculture, Agricultural Research Services. "National Plant Germplasm System" (November 2008). http://www.ars-grin.gov/npgs/index.html (Accessed December 2008).

Carol Ann Kearns
Santa Clara University

CROP ROTATION

Crop rotation is a farming technique that has been used for centuries to maintain agriculturally productive land in a manner that strives for ecological sustainability. This practice helps maintain soil fertility and reduces the occurrence of pests, diseases, and erosion. Crop rotation is a form of polyculture—the practice of growing different crops in the same space. Using the technique of crop rotation, farmers plant a series of different crops in seasonal sequence over several years on the same piece of land. For example, on a single field over a period of three to six years, a farmer may plant a rotation of cereal crops, followed by a "break crop" of potatoes or soybeans, alternating grass each year. Crop rotation is distinct from monoculture—the continuous planting of a single variety of crop on the same piece of land. Monoculture, or continuous cropping, is a technique that is commonly practiced in American commercial agriculture. Following a brief history of crop rotation, this article discusses the benefits of crop rotation, as well as the reasons for and consequences of the switch to monocropping in the United States, concluding with a look at contemporary trends in crop rotation farm-management techniques.

Historical Background

Dating back to some of the earliest eras of organized cultivation, crop rotation is a traditional farming technique that has been used at one time or another in many different parts of the world. Detailed records show that the technique was used to improve agricultural production in China during the Han Dynasty from the 3rd century B.C.E. to the 3rd century

Sunflowers growing next to millet as part of a U.S. Department of Agriculture experiment in crop rotation. One of the benefits of crop rotation is that it may reduce the use of manufactured fertilizers, pesticides, and herbicides.

Source: U.S. Department of Agriculture, Agricultural Research Service/David Nielsen

C.E. In the Near East, crop rotation was scientifically developed during the Islamic Golden Age, as crop rotation techniques were combined with advances in irrigation technology. A three-year crop rotation pattern was followed in Europe during the Middle Ages, in which a wheat crop was followed by a year of oats or barley, broken by a year-long period in which the field lay fallow; that is, unplanted. The technique was developed further in Britain with the four-field crop rotation pattern, in which a different crop was planted in three of the four fields, with the fourth allowed to lay fallow, rotating each year. Polycultural farming practices were used by indigenous peoples in the Americas before European colonization. Settlers adapted European agricultural practices to the North American environment, borrowing indigenous crops and techniques. The benefits of crop rotation, and the consequences of failing to rotate crops, were so well understood that by the 19th century, it was common for leases on agricultural land in the United States to include a stipulation in which the tenant agreed to rotate crops annually. The traditional corn-oats-clover crop rotation was routinely practiced in U.S. agriculture until the late 1950s, when it was phased out with the advent of new developments in mechanized farming. Inorganic fertilizer production and the breeding of high-yield crop varieties, as well as other economic factors, further contributed to this trend. By the mid-1970s, monocropping had almost completely replaced crop rotation as the dominant agricultural practice in the United States.

Benefits of Crop Rotation

There are several important benefits of crop rotation that have made this agricultural technique such a long-standing tradition. The most significant benefit of crop rotation is its effectiveness at controlling weeds, pests, and diseases, as well as promoting soil fertility, eliminating the need to use chemical pesticides, herbicides, or fertilizers. Pathogens such as fungi, bacteria, viruses, and pests all have specific life cycles and feed off certain crops. By rotating crops—that is, by taking away the food of a particular pest for a period of time—crop rotation cuts off the regeneration of that pest. By introducing another plant variety in its place, competition among pathogens is introduced, thereby lessening their negative effect. For example, a particular pest associated with wheat will be mitigated by the planting of a root crop, such as potatoes, following the wheat harvest.

Along with keeping weeds, pests, and diseases associated with a particular crop at bay, crop rotation has the added benefit of maintaining and enhancing soil fertility. Crop rotation achieves this through the sequencing of crops according to their nutrient needs, the introduction of organic matter, and the fixing of nitrogen in the soil. Planting crops according to their nutrient needs involves planting deep-rooted plants, followed by more-shallow-rooted plants, so that the nutrients at different soil depths are used and then allowed to regenerate. Letting a field lay fallow for a year, and using it as grazing land for livestock, allows for the buildup of a nutrient-rich mixture of manure and grass roots. Meanwhile, soil fauna, like earthworms, unhindered by the use of pesticides, contribute to the buildup of organic matter in the soil. Allowing fields to lay fallow also aids in the fixation of nitrogen in the soil, which occurs through rainfall and the growth of leguminous (nitrogen-fixing) crops or weeds. Nitrogen is an essential element of life on Earth; however, plants and animals cannot directly use atmospheric nitrogen, and instead use nitrogen compounds, or "fixed" nitrogen. Legumes have a special symbiotic relationship with root-borne bacteria called rhizobia, which fix nitrogen. Thus, the planting of a legume crop, such as alfalfa or soy, helps to replenish the nitrogen in the soil that may have been depleted by the nonlegume crop, like wheat or corn.

In addition to the agricultural benefits of crop rotation, the practice is also believed to benefit the surrounding ecosystem. Polyculture in general, including sequential crop rotations as well as intercropping—the practice of growing two or more crops at the same space and time—tends to benefit species richness in an agricultural ecosystem. Growing a variety of different plants naturally attracts a variety of different insect and bird species. By eliminating the use of pesticides, crop rotation allows invertebrates to fully develop, and this in turn attracts birds. Another benefit to crop rotation is that it reduces soil erosion. Allowing a field to lay fallow ensures that a field is not overtilled, meaning more soil—and thus more water—is retained in the field. Erosion is also reduced through the use of perennial crops, which eliminate the need for tilling to reseed. As crop rotation reduces the effect of farming on the land, it likewise allows for a broader distribution of manual and instrument labor.

Overall, crop rotation is a crop management system that strives for energy efficiency. Crop rotation lessens the need for manufactured fertilizers, made from fossil fuels, as well as for chemical pesticides and herbicides. In addition, crop rotation that spaces out the growing cycles of different plants throughout the year (by planting a winter cover crop, for instance) enables a more efficient use of solar energy, as opposed to growing a single crop of corn on a field, which may only absorb three months of solar energy.

Crop Rotation Versus Continuous Cropping

The late 1950s saw a decline in the use of crop rotation as a crop-system management technique and the adoption of monocropping methods instead. The practice of polyculture gave way to monoculture, with most farms producing either a single crop, or using a single rotation, such as crop/grass or corn/soy. This was made possible by the increased availability of mechanized farming implements, as well as industrial herbicides, pesticides, and fertilizers. At this time, there was also an increased need for cereal grains to be used as animal feed, with increased prices giving farmers an incentive to specialize in grains as a cash crop. Monocropping was also more profitable, because the production of a single crop meant that machine and building costs were kept at a minimum, as the production of just one variety of crop requires just one type of farming implement and storage facility. Likewise, monocropping allowed farmers to gain expertise in farming a single crop, contributing to further specialization.

The predominance of monocropping in U.S. agriculture has not been without its negative consequences. Crop specialization has led to the proliferation of weeds and fungal diseases. Herbicides are used to control weeds, but this has caused the added problem of herbicide-resistant weeds, and a similar situation exists with the use of pesticides. Crops that are genetically modified to resist different diseases and pests are being designed, but this is often costly and time consuming. In addition, the use of large machines on expansive monocrop fields has taken a toll on the soil, leading to increased rates of erosion. This in turn leads to more water runoff from the fields, overburdening local streams and rivers. The runoff carries synthetic fertilizers into lakes and streams, leading to a phenomenon known as algal bloom—an increase in the algae population—that significantly affects aquatic ecosystems. Mechanized farming and the use of petroleum-based synthetic fertilizers not only has had serious effects on local ecosystems but also has contributed to the depletion of nonrenewable natural resources, as well as the release of greenhouse gases.

Contemporary Trends in the Use of Crop Rotation

Crop rotation is being adopted by a growing number of organic farmers. Farmers raising organic beef, for instance, rotate grazing land and several different kinds of feed crops. However, nonorganic farms are also returning to crop rotation techniques. The rotation of noncereal commercial crops, such as beans, peas, and potatoes, is compatible with intense, mechanized agricultural production and is beginning to be implemented through the use of computerized planting sequencing tools. Finally, the practice of crop rotation continues as a popular technique in home vegetable gardening.

See Also: Cover Cropping; Intercropping; Legume Crops; Nitrogen Fixation; Organic Farming.

Further Readings

Jackson, Dana L. and Laura L. Jackson. *The Farm as Natural Habitat: Reconnecting Food Systems with Ecosystems.* Washington, D.C.: Island, 2002.
Loomis, R. S. and D. J. Connor. *Crop Ecology: Productivity and Management in Agricultural Systems.* New York: Cambridge University Press, 2004.
Magdoff, Fred and Ray R. Weld, eds. *Soil Organic Matter in Sustainable Agriculture.* Boca Raton, FL: CRC Press, 2004.
Thorne, D. Wynne and Marlowe D. Thorne. *Soil, Water and Crop Production.* Westport, CT: AVI, 1979.
Watson, Andrew M. *Agricultural Innovation in the Early Islamic World: The Diffusion of Crops and Farming Techniques, 700–1100.* New York: Cambridge University Press, 1983.

D. Jones Marshall
University of Kentucky

Dairy

Dairy products in most parts of the world have traditionally been milk and foodstuffs produced from milk that came from cows. However, in recent times there has been an increasing availability in the West of dairy products from milk from goats, sheep, camels, yaks, horses, and water buffalo, although all these have been used by people around the world since ancient times.

Traditionally, milk from cows and other animals was consumed soon after its production or was quickly turned into other products such as butter, cheese, or yogurt. This was because milk quickly deteriorated, especially in hot or warm temperatures. However, with the invention of homogenization and pasteurization in the 19th century, milk could be stored more easily and for longer periods. This ability to store dairy products has increased with the introduction of refrigeration, also in the 19th century.

Essentially, for dairy products to be "green," it is necessary for the animals concerned to have been raised naturally. The rearing and breeding of particular types of cattle such as Jersey cows; or Guernsey cows, which produce richer milk; or Friesian cows, which produce more milk and live longer, are still a part of the natural process and have happened, in the case of other species, for several thousand years. The artificial insemination of cattle and the injection of cattle with artificial hormones affects the cattle, and hence the milk. Therefore, some customers do not regard dairy products as being "green" if the breeding of the cattle involves artificial processes.

Then it is necessary for the cattle or other animals to be able to graze naturally, as far as is possible. For Jersey cows, used for the production of creamy milk in the United Kingdom, it is traditional to graze them with short chains, with the cattle moved every few hours. Other breeds have been found to benefit from different treatments. Obviously, in winter in many parts of the world it is impossible for cows or other animals to find enough naturally growing grass to eat, so hay is used instead. This does not affect the quality of the dairy product. However, to feed the animals foodstuffs and food supplements that are man-made or not natural has an effect on the animal, and therefore by definition, on the milk produced. There is also some question about inoculating animals against diseases, as this too can have an effect on the milk produced.

Although some people prefer milking to be done by hand, in practice this would raise the price of milk so considerably that most people recognize that the milking process needs to be done with machines. This has the added benefit of ensuring both little waste of milk and that the milk itself is quickly stored in hygienic conditions.

The pasteurization process was developed in the 1860s by Louis Pasteur, who worked initially with wine and beer, showing that abnormal fermentation could be stopped by heating the liquids to 57 degrees C (135 degrees F) for several minutes. It was not long before it was found that the process could be repeated with milk that was raised to a temperature of 63 degrees C (145 degrees F) for 30 minutes. By experimentation, it was soon shown that heating the milk to 72 degrees C (162 degrees F) for as little as 15 seconds produced the same result.

There were soon more experiments, involving heating the milk to a higher temperature; ultrapasteurization involved heating milk to 138 degrees C (280 degrees F) for two seconds. This in turn was to lead to the introduction of ultra-high-temperature, or UHT, milk, whereby the milk is heated to 150 degrees C (302 degrees F) for one or two seconds. As a result, UHT milk—providing it is packaged in sterile and sealed containers—can last unrefrigerated for several months. It is accepted by most people that pasteurization by heating the milk does not remove its ability to be "green," although radiation pasteurization, in which beta or gamma rays are applied to foods, definitely removes the ability to market a product as "green." The next process developed for the marketing of milk was homogenization, which prevented the fat or "cream" from separating, providing a more even consistency in the milk. This involves passing the milk through one or two valves under high pressure. Homogenized milk is still able to be marketed as "green," because the process does not involve treatment with any chemicals.

With other dairy products such as butter, cheese, and yogurt, the process by which these are made determines whether or not these products are "green." Butter made from milk fat that is created by churning cream can easily be a natural product. However, butter made this way tends to have a very high calorie level and also contains a considerable amount of added salt, and it can also become very hard and difficult to spread. As a result, many manufacturers add other products to make butter; these products can include vegetable oils, such as olive oil or canola oil, and emulsifiers, such as soy lecithin. The issue arising from this is therefore the source of the olive oil or canola oil—specifically, whether it comes from genetically modified canola, which would obviously cause any such product not to qualify as "green."

The making of yogurt is a natural process in which milk is fermented by the *Streptococcus salivarius* species or other bacteria, and there are many types of yogurts available in stores that are marketed as "natural," using live yogurt cultures. These are often known as "plain" yogurt. Most of the different types of yogurts for sale in stores are flavored. Of these, some advertise that they are made with natural products such as fruit, and others have artificial flavoring and added products that can sometimes preclude marketing them as "green."

Cheese has also been produced for thousands of years; it is made by coagulating the milk, separating the whey, and then letting the cheese ripen. Again, it is possible for this to be a natural and "green" product, and there are also cream cheeses and cottage cheeses that can be "green." However, there are also a range of additives for cheese such as firming agents, lactic starter cultures, and vegetable rennet, some of which are artificial and some are natural. Often, to extend the shelf-life of cheeses, artificial products are added, and radiation pasteurization can also be used. Both of these mean that the product produced cannot be regarded as "green."

In addition to natural cheeses, there are also processed cheeses, the most famous of which was developed by James L. Kraft (1874–1953). Kraft came up with his patented processed cheese during World War I, when he had a contract to supply the U.S. Army. This process involved grinding down natural cheese, blending it, and then pasteurizing it. Although some processed cheeses could be regarded as being "green," most are usually not, as the modern technology used in their production involves the addition of artificial products to the process. Ice cream also can be made from solely natural products, but its shelf life is not long. As a result, most ice cream manufacturers include many additives either for flavor or as preservatives. These can include maltodextrin from maize, emulsifiers, vegetable gums, flavoring, and coloring. In other types of ice cream, crushed animal bones or whale blubber may be added. Technically, the latter products are "natural," but to be added to ice cream, they have to be so heavily processed that the ensuing product is not generally regarded as "green."

There are also a range of other dairy products. Ghee, common in India, is made from clarified butter—butter that is heated gently to remove solid matter. There are also casein and buttermilk, the latter of which is often used as a food additive for animals. There are also some classifications that include eggs as dairy products because they are produced by animals, but eating them does not involve eating any part of the animal.

See Also: Animal Welfare; Factory Farm; Grazing; Recombinant Bovine Growth Hormone.

Further Readings

Dahr, Tirtha and Jeremy D. Foltz. "Milk By Any Other Name ... Consumer Benefits From Labeled Milk." *American Journal of Agricultural Economics,* 87/1:214–28 (February 2005).
Jukes, Thomas H. "BST and Milk Production." *Science,* 265/5169:170 (July 8, 1994).
"Keep Milk Fresh." *Science News-Letter,* 44/15:227–28 (October 9, 1943).
Rubin, Andrew L. and Mark Goodman. "Milk Safety." *Science,* 264/5161:889–90 (13 May 1994).

Justin Corfield
Geelong Grammar School, Australia

DDT

Dichlorodiphenyltrichloroethane (DDT) is one of the best-known synthetic pesticides. It was first synthesized by German chemist Othmar Zeidler in 1874. However, it remained a laboratory curiosity until Swiss chemist Paul Hermann Müller discovered its insecticidal properties in 1939. DDT was adopted for widespread use in public health programs. It was the most effective agent known at eradicating diseases that are transmitted by insects and was the first synthetic organic substance used in large quantities for insect control. As a consequence, the undesirable side effects of chlorinated hydrocarbons were first discovered along with it. During World War II, DDT was used by the military and civilians to control the spread of malaria and typhus by mosquitoes and lice, respectively. As a result of his discovery of DDT's insecticidal properties, Müller, of Geigy Pharmaceutical, was awarded the 1948 Nobel Prize in Physiology and Medicine.

After the war, DDT was used as an agricultural insecticide. The first reports of ecological problems related to the use of DDT were published in 1950. These reports were followed by the publication in 1962 of Rachel Carson's *Silent Spring*, which documented the chemical's devastating effect on the ecosystem and launched the modern environmental movement. The American biologist highlighted the environmental impacts of the indiscriminate spraying of DDT in the United States and questioned the release of the chemical into the environment without analyzing its effects on ecology and human health. The book suggested that DDT and other pesticides could cause cancer, explaining that their agricultural use was a threat to wildlife. *Silent Spring* resulted in an extensive public outcry that led to the ban of DDT in the United States. DDT was first banned in Hungary in 1968, Norway and Sweden in 1970, and the United States in 1972, but it was not banned in the United Kingdom until 1984. It was subsequently banned for agricultural use worldwide under the Stockholm Convention, though it is still being used in some underdeveloped countries for disease vector control. The Stockholm Convention was signed by 98 countries and is endorsed by most environmental groups.

Available epidemiological and experimental data indicate that the presence and persistence of DDT and its metabolites worldwide are still problems of great relevance to public health. The use of DDT in vector control has been largely replaced by less persistent insecticides. The large-scale manufacture and distribution of organochlorines took place after the accidental discovery of DDT. It is an organochlorine insecticide commonly used for the control of disease-bearing insects and on a variety of food crops. Although DDT has been banned in most nations because of its potential for human toxicity and severe ecological effects, it is still found worldwide as a contaminant because of its extreme persistence and high mobility in the environment.

DDT has the molecular formula $C_{14}H_9C_{l5}$ and molar mass of 354.49 g/mol. Its density is 0.99 g/cm^3, and its boiling point is 109 degrees C. Its structure is similar to the pesticides dicofol and methoxychlor. It is a highly hydrophobic, colorless, crystalline solid with a weak odor. It is soluble in most organic solvents, fats, and oils and is produced by the reaction of chloral (CCl_3CHO) with chlorobenzene (C_6H_5Cl) in the presence of sulfuric acid, which acts as a catalyst. Trade names that DDT has been marketed under include Anofex, Cezarex, Chlorophenothane, Clofenotane, Dicophane, Dinocide, Gesarol, Guesapon, Guesarol, Gyron, Ixodex, Neocid, Neocidol, and Zerdane. Commercial DDT is a mixture of some related compounds. The components include *p,p*-DDT isomer (77 percent), *o,p*'-DDT (15 percent), dichlorodiphenyldichloroethylene (DDE), and dichlorodiphenyldichloroethane (DDD). DDE and DDD are also the major metabolites and breakdown products of DDT in the environment and in the body. Total DDT in a sample, therefore, refers to the sum of all DDT congeners (*p, p*-DDT, *o, p*-DDT, DDE, and DDD).

The most common route of DDT exposure is through diet, particularly fatty foods such as fish, meat, and dairy products. It is known to adversely affect the nervous system, and there is particular concern that DDT and its metabolites, DDE and DDD, may interfere with normal reproduction and development as a result of its endocrine-disrupting properties. The continued use of DDT in some countries further contributes to worldwide environmental contamination, as it can accumulate to high levels in soil, sediment, plants, animals, fish, and humans. Analysis of blood and urine are the most common methods for detecting DDT exposure. It can also be measured in fatty tissues and breast milk. DDE has the shortest biological half-life, followed by DDT, and then DDD. It is the persistence of DDT and its breakdown products that leads to its bioaccumulation and bioconcentration in the food chain.

The problems with DDT are related to its molecular structure. DDT is the common name for 1,1,1-trichloro-2,2-bis(p-chlorophenyl)ethane. The microbial degradation of DDT is fairly well understood. It is a hydrocarbon and, hence, does not dissolve in water but does dissolve in oils and fats. It is not easily degraded to other substances, and it therefore remains in the soil long after it has been applied to agricultural crops. All chlorinated hydrocarbons share these properties. The fat solubility of DDT results in its being concentrated throughout the food chain. The result is that organisms at the top of the food chain have excessively high DDT levels in their body fat—levels that are much higher than those in soil or water. DDT has been associated with many health problems in humans including cancer. Banning its use in developed countries removed a major health threat for much of the world's population, even though underdeveloped countries still use it.

DDT is moderately toxic, with a rat LD50 of 113 mg/kg. It has potent insecticidal properties: It kills by opening sodium ion channels in the neurons, causing them to fire spontaneously, leading to spasms and eventual death. Insects with certain mutations in their sodium channel gene are, however, resistant to it and other similar insecticides. DDT is classified as "moderately toxic" by the U.S. National Toxicological Program and "moderately hazardous" by the World Health Organization (WHO), based on the rat oral LD50 of 113 mg/kg. The U.S. National Toxicology Program classified DDT as "reasonably anticipated to be a human carcinogen," and the U.S. Environmental Protection Agency classified DDT, DDE, and DDD as class B2 "probable" human carcinogens. These evaluations and classifications are based mainly on the results for animal studies. DDT is a persistent organic pollutant that is extremely hydrophobic and strongly absorbed by soils. Its soil half-life ranges from 22 days to 30 years. Its breakdown metabolites, DDE and DDD, are also highly persistent and have similar chemical and physical properties. DDT, DDE, and DDD magnify through the food chain. They are stored mainly in body fat. DDT is toxic to a wide range of animals, insects, aquatic life, and many species of fish.

See Also: Organochlorines; Pesticide.

Further Readings

Pesticide Action Network. http://www.Pesticideinfo.org (Accessed February 2009).

"Toxicological Profile: for DDT, DDE, and DDE." Atlanta, GA: Agency for Toxic Substances and Disease Registry, 2002.

University of Minnesota. "Lecture 25. Microbial Degradation of Pesticides." http://jan.ucc .nau.edu/~doetqp-p/courses/env440/env440_2/lectures/lec25/lec25.html (Accessed February 2009).

World Health Organization. *Environmental Health Criteria 9: DDT and Its Derivatives.* Geneva: World Health Organization, 1979.

World Health Organization. *The WHO Recommended Classification of Pesticides by Hazard.* Geneva: World Health Organization, 2005.

Akan Bassey Williams
Covenant University

Debt Crisis

The Third World debt crisis was a consequence of the inability of debtor nations to service their external debts. Their burgeoning debts dramatically increased in the 1970s and 1980s and were a result of legacies of colonialism coupled with structural factors in the postwar world economy and historically specific circumstances related to fossil fuel–dependent development and the oil price shocks of the 1970s. Seeds of the crisis were sown at the 1945 United Nations Monetary and Financial Conference, commonly referred to as Bretton Woods. The International Bank for Reconstruction and Development (World Bank) and the International Monetary Fund (IMF) were created at Bretton Woods, and plans were laid for creating the General Agreement on Tariffs and Trade. These institutions have played major roles in creating structural conditions for unpayable Third World debt within the neoliberal economic paradigm of globalization. Many Third World nations continue to service unpayable debts to the present day, and these debts create leverage points for the domination and exploitation of both people and environments by transnational corporate interests. The effects of social and environmental exploitation that derive from the economic vulnerability of nations in debt crisis are exacerbated by socially and environmentally destabilizing legacies of the postwar green revolution in agriculture that forced many subsistence producers from the land and into megacity slums and resulted in massive losses of topsoil and declines in soil fertility. During the debt crisis, Third World nations experienced rapidly increasing needs for social investment and environmental protection at the same time that their debts and the neoliberal policies of international banking interests prevented them from addressing these needs, thereby intensifying poverty and environmental damage.

The Bretton Woods conference was a forum for Allied nations to devise strategies and institutions to aid in financing the rebuilding of Europe and to prevent future economic depressions. At Bretton Woods, the stage was set for a postwar world system that carried forward in time inequities imposed during the period of European colonization. The United States led in the creation of a system that would allow it to maintain its trade surpluses, thereby increasing the global concentration of wealth and power and intensifying the economic dependency of newly independent nations. The United States also advocated conducting international trade in a free market, using national currencies. Nations running trade surpluses would be under no obligation to expend their surplus earnings by purchasing exports of debtor nations. The United States had seen its productive capacity rise as a result of the war, and its infrastructure and factories remained intact. The American delegation was concerned with maintaining and expanding U.S. trade surpluses as an outlet for American productive capacity and a vehicle for avoiding a postwar recession. Many former colonies that had not modernized and industrialized would find themselves at a competitive disadvantage as they imported high-priced manufactured goods from the First World and sold comparatively lower-priced raw materials and agricultural produce in a competitive global market.

World Bank, IMF, and Third World Development

The World Bank and the IMF have set the conditions for Third World development since Bretton Woods. The IMF was created to provide an international financial pool of funds on which member countries could draw to help resolve a temporary balance of payment deficits that threatened the stability of their currencies. Any nation that experienced a negative balance of trade that threatened to upset its economy could borrow from the fund on a short-term basis to avert an economic downturn or currency crisis. Stabilization provided by the IMF would also prevent future global depressions because IMF loans would provide the liquidity necessary to maintain aggregate consumer demand in the global economy. IMF loans would also encourage countries to maintain employment during an economic downturn so as not to compound existing problems. The World Bank was charged with financing development and the rebuilding of economies shattered by the war. The General Agreement on Tariffs and Trade served as a postwar platform for removing tariffs and subsidies deemed to be barriers to free trade.

The postwar global monetary system was also established at Bretton Woods. Under this system, nations running trade surpluses could exchange financial surpluses for gold held by debtor nations. Gold was valued in dollars, and all other currencies were valued relative to the dollar—a system known as the gold exchange standard. Even though the United States unilaterally ended the gold exchange standard in 1971 by ending exchange of dollars for U.S. stocks of gold, the Bretton Woods monetary system had established the dollar as a world reserve currency that would be held in the reserves of foreign central banks. Loans made by the World Bank, the IMF, and many commercial banks would be denominated in dollars, and the dollar would serve as the currency of choice for many international business transactions. Dollar hegemony in international business and finance bestowed economic and political advantages on the United States by strengthening the value of the dollar during the postwar era, even as the United States became the world's foremost debtor nation. Developing nations experienced no such advantage during and since the debt crisis.

The oil price shocks of the 1970s created conditions for the lending of vast sums of dollars throughout the world while simultaneously triggering a worldwide recession that would make it difficult for nations to repay their debts. Third World nations pursuing oil-dependent development faced unexpected costs for a commodity that had become essential to their economies. Hunger and poverty increased sharply in the Third World, and some regions experienced absolute declines in food grain consumption. Globally, an economic recession was triggered in part by the high price of oil. The balance of international trade for low-income countries turned sharply negative, and the debt burden for developing countries skyrocketed. During this period, members of the Organization of Petroleum Exporting Countries that were capable of exporting large amounts of oil accumulated immense surpluses of dollars because their oil sales were (and continue to be) transacted solely in U.S. dollars. Members of the Organization of Petroleum Exporting Countries invested many of these dollars in the United States, and these funds were then lent by First World banking interests to developing countries that were in desperate need of dollars to finance oil purchases and other imports. This process greatly exacerbated the debt situation of developing nations. To make matters worse, interest rates were increased sharply in the United States in 1979 to prop up the value of the dollar, which had been falling during the 1970s, in part as a result of the negative balance of trade in the United States during the Vietnam War. Debtor nations saw interest rates on their dollar-denominated loans rise as a result, and this rise increased their debt burden at the same time that economic stagnation and inflation in the United States created deteriorating conditions for Third World countries to earn dollars through exporting to the United States. Defaults on external debts followed.

Neoliberal Philosophy and Third World Debt

Since the 1980s, the IMF and the World Bank have advanced neoliberal philosophy in the process of renegotiating Third World debt. Neoliberal ideologues adhere to extreme market fundamentalism. They believe in small government and the ability of a self-regulating market to serve effectively as the ultimate arbiter of economies and all social life. Neoliberals promote privatization of public industries, services, and functions; capital market liberalization; and the removal of barriers to trade. All of these strategies have been packaged in various combinations as "structural adjustment programs" and are required by the IMF and the World Bank for renegotiating loans. Structural adjustment program strategies typically also include fiscal austerity (reduced government spending) and high interest rates for domestic lending—policies that make it difficult to impossible to develop the national economy and repay external debt. In the manufacturing sector, free trade has created a race to the bottom in wages and environmental protection as transnational corporations move production facilities around the globe, seeking production cost advantages. Although a large measure of Third World debt is commercial, the IMF and the World Bank can leverage economic policy in Third World nations because loan agreements with these institutions are often preconditions for commercial bank lending.

Under neoliberalism, structural inequalities built into the Bretton Woods paradigm contributed to creating a world system in which the economic and political positions of Third World nations approximates that of colonies with regard to First World corporate interests. The perpetual dependency of many Third World nations on external economic assistance, which paradoxically weakens their relative economic position, exacerbates their dependent status with regard to institutions, corporations, and nations that strip them of infrastructural and business assets, jobs, decision-making power, natural resources, and social support that could serve as bases for creating freer, more self-sufficient, and more sustainable societies. Thus, neoliberal policies enforced through the IMF and the World Bank often circumscribe the resilience and sustainability of communities and nation-states. The oil shocks of the 1970s occurred for historically specific reasons but triggered a debt crisis in the world system that threatens the socioecological sustainability of Third World debtor nations.

See Also: Export Dependency; Fair Trade; Green Revolution; Trade Liberalization.

Further Readings

Kaplinsky, Raphael. *Globalization, Poverty and Inequality: Between a Rock and a Hard Place.* Malden, MA: Polity, 2005.
Stiglitz, J. E. *Globalization and Its Discontents.* New York: W. W. Norton, 2002.

Tina Evans
Fort Lewis College

DEPARTMENT OF AGRICULTURE, U.S.

The U.S. Department of Agriculture (USDA) is a department of the executive branch of government of the United States. Its multipronged mission is to maintain an adequate supply of food in the country, to promote agricultural research, to promote the marketing of U.S. farm products at home and abroad, and to seek fair prices for consumers and for farmers both at home and abroad. The USDA is headed by the Secretary of Agriculture, who is a member of the President's cabinet. The Secretary of Agriculture is appointed by the President and confirmed with the advice and consent of the Senate.

Historically, the farm vote was very significant so it was common for the appointment to the post of Secretary of Agriculture to be a reward for farm vote support from members of the President's party. Today, farmers in the United States number fewer than three million, so the vote is less important than it was historically. It is still important, however, because the United States is—by its geographic and geologic nature—an enormous agricultural country, and the appointment is still a reward for farm interests. Many of these farm interests today are giant agribusinesses that seek to use all manner of science and technology to provide food for global markets. Green interests are also a part of farm interests, from the small farmer to giant farms and food corporations.

As long as people need to eat, agriculture will be a subject of permanent legislative interest for the U.S. Congress. Its interest is in feeding an ever-growing number of U.S. citizens, along with selling huge volumes of agricultural products to the world. To deal with the many concerns posed by agriculture, Congress began to deal with these interests early in U.S. history. In 1836, Henry Leavitt Ellsworth became the Commissioner of Plants—a position within the Department of State.

The U.S. Department of Agriculture actively promotes the sale or donation of American agricultural surpluses abroad. These men are covering an enormous pile of food aid for the African country of Malawi that was donated through the U.S. Agency for International Development.

Source: U.S. Agency for International Development

Ellsworth began collecting seeds and plants from around the country through members of Congress and agricultural societies, as well as from foreign sources via people serving overseas. The goal was to improve agriculture by identifying and using those plants that could adapt to U.S. agricultural conditions and that would produce useful harvests. Because this work involved property claims, the Agricultural Division was created as part of the Patent Office in 1839. Ellsworth produced annual reports, prepared agricultural statistics, and applied chemistry to agriculture.

In 1849, the Patent Office, along with the Division of Agriculture, was transferred to the Department of the Interior. However, both within the Division of Agriculture and from agricultural interests there was agitation for a separate department of agriculture. On May 15, 1862, Congress authorized the Department of Agriculture, headed by the Commissioner of Agriculture as its chief operating officer; however, the department did not have cabinet status. A separate agency was needed because about 60 percent of Americans were farmers. They were in constant need of good seed, information, and other help to maximize their productivity.

In the 1880s, business, labor, and farm interests sought cabinet-level departments. Eventually, the departments of Commerce, Labor, and Agriculture were created. Agriculture reached cabinet status on February 9, 1889, when President Grover Cleveland signed a bill into law elevating the Department of Agriculture as the eighth cabinet-level department. A cabinet-level department was needed to manage the growing agricultural programs that were creating an extensive agricultural infrastructure. The 1887 Hatch Act had funded agricultural experiment stations in each state, and the government was actively employing etymologists to deal with the locust problems in the West.

In 1914, the Smith-Lever Act funded cooperative extension services in each state. This act was educational, providing instruction in better agriculture, home economics for farm families, and a wide range of other subjects. The programs still continue to provide services that would have been unavailable to people living in remote areas. In times of poverty like the Great Depression, the support for farm families was vital for their stability and success.

Today the USDA is an elaborate bureaucracy presided over by a secretary for whom the deputy secretary serves as chief operating officer and principal adviser. There are seven undersecretaries for the following divisions: (1) Farm and Foreign Agricultural Services; (2) Marketing and Regulatory Programs; (3) Rural Development; (4) Food, Nutrition, and Consumer Services; (5) Natural Resources and Environment; (6) Research, Education, and Economics; and (7) Food Safety. These undersecretaries run many different antihunger programs for the poor in the United States: food stamps, school lunches, school breakfasts, and the Special Supplemental Nutrition Program for Women, Infants, and Children.

Eating a proper diet is vital to good health, so the USDA spends millions every year on nutrition research. Using knowledge acquired through nutrition research, its promotion programs encourage balanced meals and proper diets for good health for all Americans. The food pyramid developed by the USDA is regularly modified to better educate the public on what is in a healthy diet.

Other USDA programs for rural areas have promoted rural housing, electrification, and safe drinking water. These programs are aiding the development of rural infrastructure, including the Housing and Community Facilities Programs for rural development.

The United States has some of the best and most extensive areas of farmland in the world. The country has been a major food, fiber, and livestock exporter since its earliest beginnings. Today the USDA promotes sales of U.S. agricultural products abroad. With at least two billion of the world's population either poor or impoverished, U.S. agricultural surpluses provide food for millions who may be suffering dislocation as a result of war, civil strife, famine, or other calamities.

Food aid, administered through the U.S. Agency for International Development, is made available to foreign governments or to international bodies such as the World Food Program or to nongovernmental organizations. The various food programs were authorized by Congress in the Agricultural Act of 1949 and the Agricultural Trade Development and Assistance Act of 1954 (Public Law 480 or the Food for Peace Program).

The U.S. Forest Service is an agency of the USDA. It manages about 200 million acres of national forests and rangelands, practicing a mixed-use policy that promotes conservation and timber management that is renewable (replanting after timber lands are harvested). It also promotes the growth of wildlife and recreation. Hunting and fishing on national forest lands are important conservation programs that promote sustainable herds of animals or fish stockings. Many of the Forest Service's conservation programs promote environmental projects on privately owned lands. These projects are scientifically designed to increase soil and water conservation, as well as to protect wildlife.

Food safety is a major area of USDA activity. It provides inspectors for meat, poultry, and eggs consumed by Americans. The inspection programs seek to protect consumers from animal diseases while also helping farmers and ranchers find ways to keep their flocks and herds free from diseases, which is to their economic benefit and helps ensure the public's safety. The USDA also inspects grain supplies to prevent the sale and consumption of grains that have diseases that can infect humans and animals.

To aid its work, the USDA conducts agricultural research directly or cooperates with colleges and universities to develop new strains of plants and animals that are more productive and disease free. It is currently engaged in programs to identify and to combat newly emerging diseases or pests. It also has promoted the adoption of new crops such as soybeans. The USDA is still seeking new crops that will diversify and enrich U.S. agriculture to the benefit of farmers and consumers.

Food safety is also concerned with natural disasters that affect crops and with terrorist threats. Without this protection, markets at home and abroad could be damaged or destroyed and innocent people hurt.

See Also: Agribusiness; Agricultural Commodity Programs; Agricultural Extension; Food Safety; Public Law 480, Food Aid.

Further Readings

Espy, Mike. *Reinventing the U.S. Department of Agriculture: 1994 Annual Report of the Secretary of Agriculture.* Darby, PA: Diane, 1998.

National Research Council. *Investing in the National Research Initiative: An Update of the Competitive Grants Program in the U.S. Department of Agriculture.* Washington, D.C.: National Academies, 1995.

Outlaw, J., et al., eds. *Agriculture as a Producer and Consumer of Energy.* Oxford: CABI, 2005.

Rosaler, Maxine. *Department of Agriculture.* New York: Rosen, 2005.

Shames, Lisa. "U.S. Department of Agriculture: Improved Management Controls Can Enhance Effectiveness of Key Conservation Programs: Congressional Testimony." Darby, PA: Diane, 2009.

USA IBP, Global Investment and Business Center, Inc. *U.S. Department of Agriculture Business Opportunities Handbook, Vol. 13.* Washington, D.C.: USA IBP, 2005.

Andrew Jackson Waskey
Dalton State College

DIAMOND V. CHAKRABARTY

On June 16, 1980, the U.S. Supreme Court decided the case of *Diamond v. Chakrabarty* (*Chakrabarty*), in which it affirmed the patentability of a live bacterium modified in a laboratory by design. The 5–4 decision affirming the patentability is of interest because of the subject of the patent itself as well as its novelty. The details and the reasons for the dissent when compared with the majority reasons shed light on some of the issues and implications of the decision.

The case involved a submission for three types of patents, of which two were granted at first instance without dissent. It is the third type that was the issue in the appeals. This third type of patent claim was for the bacterium itself. The inventor, a microbiologist named Ananda Mohan Chakrabarty, used a bacterium from the genus *Pseudomonas* and engineered it to carry at least two stable energy-generating plasmids, each of which provide a separate pathway to degrade hydrocarbons.

Plasmids carry DNA, and when introduced into the bacterium, the proteins in the DNA activate and override the native DNA, enabling the bacterium to operate in ways consistent with the DNA in the plasmid, which may be different than the ways in which the bacterium would usually operate. To stabilize the presence of the plasmid in the bacterium and prevent rejection, an antibiotic is usually required. When the plasmid is removed or rejected, the bacterium resumes operating as it would usually. Although the bacterium is a host to the plasmid(s), they have a symbiotic relationship.

Footnote 1 in the majority decision states:

> Plasmids are hereditary units physically separate from the chromosomes of the cell. In prior research, Chakrabarty and an associate discovered that plasmids control the oil degradation abilities of certain bacteria. In particular, the two researchers discovered plasmids capable of degrading camphor and octane, two components of crude oil. In the work represented by the patent application at issue here, Chakrabarty discovered a process by which four different plasmids, capable of degrading four different oil components, could be transferred to and maintained stably in a single *Pseudomonas* bacterium, which itself has no capacity for degrading oil.

Two points are worth noting. One is that the rights to the patent were assigned to General Electric and the other is that similar modifications to living plant cells such as yeast are patentable. The material difference here is that the patent sought is for a microorganism and not a plant cell.

The reasons supplied by the previous levels of decision makers are recounted in the decision of the Supreme Court and may be helpful in appreciating the thinking of the authorities. Including the Supreme Court, there were four levels of decisions. The first was the Patent Examiner, the second was the Patent Office Board of Appeals, the third was the Court of Customs and Patent Appeals, and the fourth was the U.S. Supreme Court. The patent was denied by the first two, granted by the third, and affirmed by the fourth.

According to the Patent Examiner, microorganisms are "products of nature," and as living things, they are not patentable subject matter that could be included under the legislation (35 U.S.C. § 101). At first instance, therefore, the claim was denied. The Patent Office Board of Appeals affirmed the Patent Examiner's decision on the second ground that living things are not patentable under the applicable legislation. In doing so, the second authority relied on the 1930 Plant Patent Act, which it interpreted to mean that Congress did not intend to include laboratory-created microorganisms as patentable subjects.

The third level was the Court of Customs and Patent Appeals, which had two opportunities to consider the case. In the first attempt, the Court relied on its 1977 decision in the case of *In re Bergy (Bergy)*, in which it held that the fact that the microorganism was alive is without legal significance for purposes of patent law. The Supreme Court then granted the petition for that decision to be set aside and returned it for further consideration in light of the case of *Parker v. Flook*. In its second attempt, the Court of Customs and Patent Appeals consolidated the *Bergy* and *Chakrabarty* cases and affirmed its earlier judgments.

The case then returned to the Supreme Court by itself when *Bergy* was dismissed as moot.

Patent law exists in the context of a market economy in which competition is a central feature. To preserve competition, the law prevents monopolies in the market. Patent law provides an exception and permits a monopoly for the precise purpose, in Thomas Jefferson's words, to promote progress of "Science and the useful Arts." Jefferson drafted the law, and the word "Arts" was later substituted with the word "process." The intended promotion is typically achieved through patent law by offering inventors exclusive rights for a limited period, during which they may profit from their invention. This opportunity for profit is provided as an incentive for their inventiveness and to recoup research costs. The intent and hope is to foster the introduction of improvements in society, thereby increasing employment and bettering people's lives.

For the majority in the Supreme Court, the issue was whether the bacterium submitted constitutes a manufacture or composition of matter within the meaning of the statute, 35 U.S.C. § 101, which provides: "Whoever invents or discovers any new and useful process, machine, manufacture, or composition of matter, or any new and useful improvement thereof, may obtain a patent therefor, subject to the conditions and requirements of this title."

By comparison, for the four dissenting judges, "The only question we need decide is whether Congress, exercising its authority under Art. I, § 8, of the Constitution, intended that he (Chakrabarty) be able to secure a monopoly on the living organism itself, no matter how produced or how used."

In the majority line of thinking, in U.S. patent law there is no consideration given to whether the subject of the patent is living or not. Since the purpose of the law here is to address that which is not foreseen, the word "any" in the statute should not be limited or qualified unless otherwise constrained by the legislature. Accordingly, the majority declined to consider the extensive and passionate submissions entered into the record by Sidney A. Diamond, the Commissioner of Patents and Trademarks, or the government. The majority recognized that

To buttress his argument, the petitioner, with the support of *amicus*, points to grave risks that may be generated by research endeavors such as the respondent's. The briefs present a gruesome parade of horribles. Scientists, among them Nobel laureates, are quoted suggesting that genetic research may pose a serious threat to the human race, or, at the very least, that the dangers are far too substantial to permit such research to proceed apace at this time. We are told that genetic research and related technological developments may spread pollution and disease, that it may result in a loss of genetic diversity, and that its practice may tend to depreciate the value of human life. These arguments are forcefully, even passionately, presented; they remind us that, at times, human ingenuity seems unable to control fully the forces it creates—that, with Hamlet, it is sometimes better "to bear those ills we have than fly to others that we know not of."

It is argued that this Court should weigh these potential hazards in considering whether respondent's invention is patentable subject matter under § 101. We disagree. (*Diamond v. Chakrabarty*, 447 U.S. 303 [1980])

The majority did also consider the 1948 decision in *Funk Seed Co. v. Kalo Co.*, in which a combination of bacteria was found to be nonpatentable, as the bacteria jointly and severally served the same ends nature originally provided, and there was nothing new. For the majority of the Court in *Chakrabarty*, when a bacterium acts as a host and delivery vehicle for plasmids symbiotically residing in the bacterium with the aid of antibiotics to suppress rejection, and the plasmids are doing what nature originally provided and the bacterium is acting as nature originally provided, limited as it is by the antibiotic, it is patentable.

To the question raised by the dissenting judges, the answer of the majority of the Court appears to be that since there is no prohibition by the legislature, it is intended that Chakrabarty or General Electric Company or anyone else be able to secure a monopoly on the living organism.

See Also: Agribusiness; Crop Genetic Diversity; Genetically Modified Organisms.

Further Readings

Diamond v. Chakrabarty, 447 U.S. 303 (1980). http://supreme.justia.com/us/447/303/case .html (Accessed at February 2008).

Funk Brothers Seed Co. v. Kalo Inoculant Co. 333 U.S. 127, 333 U.S. 130 (1948).

In re Bergy. 563 F.2d 1031, 1038 (1977).

Kevles, Daniel J. "Ananda Chakrabarty Wins a Patent: Biotechnology, Law, and Society, 1972–1980." *HSPS: Historical Studies in the Physical and Biological Sciences*, 25/1: 111–36 (1994).

Parker v. Flook. 437 U.S. 584 (1978).

United States v. Dubilier Condenser Corp, 289 U.S. 178, 289 U.S. 199 (1933).

Lester de Souza
Independent Scholar

Dioxins

There are hundreds of chemical compounds that are included in the catch-all term *dioxin*. These compounds have had widespread uses in industry and agriculture. They are some of the most toxic chemicals known, do not break down readily in the environment, and thus can easily have unintended effects.

Some dioxins are polychlorinated biphenyls (PCBs), which are a broad family of chlorinated hydrocarbons. PCBs were used in a wide range of manufacturing from 1929 until they were banned in 1979. The compound 2,3,7,8-tetrachlorodibenzo-p-dioxin (TCDD) is the most toxic. The toxicity of other dioxins and chemicals similar to PCBs that act like dioxin are measured in relation to TCDD.

Often dioxin is formed unintentionally when chlorine-based chemical compounds react with hydrocarbons. Many industrial processes that use chlorine use processes that create dioxins as a by-product. Waste incineration, chemical manufacturing, pesticide and herbicide production, pulp wood processing, and paper bleaching also produce dioxin.

Several localities and groups of people have been tragically affected by dioxin. It was the primary ingredient in Agent Orange used during the Vietnam War by the United States' military. It had previously been used as an herbicide on crop land, but in Vietnam, it was used to defoliate vast areas of jungle. Two specific chemicals used were 2,4-D dichlorophenoxy ($C_8H_6Cl_2O_3$) and 2,45-T Trichlorophenoxy ($C_8H_5Cl_3O_3$). These chemicals are directly harmful to humans, and together they form dioxin (2,3,7,8-tetrachlorodibenzodioxin).

Dioxin is an infamous carcinogenic agent in humans that once caused numerous health problems at Love Canal, New York. The old chemical waste dump at Love Canal was used by people, with serious health consequences for many.

People were forced to evacuate from Times Beach, Missouri, because of the spraying of dirt roads that were later paved with waste oil containing high levels of dioxins. The city was disincorporated and demolished, forcing the removal of over 2,300 people.

Seveso, Italy, not far from Milan in northern Italy, suffered an industrial accident that exposed its people to dioxin. The town of 17,000 was forced to evacuate, and thousands more in surrounding towns were affected.

A common source for environmental pollution by dioxin is the unintentional production of dioxin in waste-burning incinerators. These can range from backyard barrel-burns to city or industrial waste incinerator burning. Other sources of dioxin pollution include paper mills that use chlorine bleaching. The chlorine reacts with organic compounds, which includes a huge number of carbon compounds. Other sources include the manufacture of polyvinyl chloride plastics and other chemicals including numerous chlorinated pesticides and herbicides.

Cancer and other diseases have been linked with dioxin. The World Health Organization's International Agency for Research on Cancer issued a report (February 14, 1997) stating that TCDD is now considered to be a class 1 carcinogen, which means that it has been scientifically identified as a known cancer-causing agent (carcinogen) in humans. In January 2001, the U.S. National Toxicology Program reclassified dioxin in the form 2,3,7,8-TCDD as a class 1 carcinogen. In 2002, dioxin was shown to be increasingly related to breast cancer.

Other health problems associated with dioxin include disruption of the reproductive system, interference with the immune system, and interference with the hormone system (endocrine system disrupters). Even low levels of dioxin can act to bind the compound to the hormone receptor of a cell. This in effect modifies how the cell functions and in some cases seems to adversely affect the genetic mechanism of cells. When dioxin is present in developing fetuses, it can promote miscarriages or birth defects; in infants, it can cause nervous disorders. Other implicated diseases are spina bifida (split spine), autism, liver disease, endometriosis, chronic fatigue syndrome, and blood disorders.

Most people in North America are exposed to dioxin from their diet. Dioxin's use and its unintentional pollution of the environment, as well as its careless distribution, have poisoned vast areas with these long-lasting chemicals. Meat and dairy products are the main sources of exposure. However, fish is also a major source, as are poultry and eggs. In contrast, a simulated vegan diet was the least exposed to dioxin. This is because vegetables, greens, fruit, and grains have the lowest levels of dioxin in them.

Dioxins are hydrophobic (water-fearing) but are lipophilic (fat-loving). Therefore, if dioxin is spewed from an incinerator and the dust settles on water, it will not be absorbed by the water but will rather quickly be taken in by fish. Other wildlife also absorb dioxin from exposure, even in the wild.

In humans, males are only able to purge dioxin from the body by allowing its half-life breakdown to proceed. Women have two other ways by which it is expelled from their bodies: It crosses the placenta to be absorbed by their fetus and is fed as mother's milk to nursing infants. Obviously, neither of these is desirable because the infants can be radically affected.

The high level of dioxin in the food chain, which is concentrated in meat and dairy products, has an oddity to it: The use of pesticides and herbicides on cotton and on vegetable crops does not seem to have affected these crops in a manner that allows the absorption of dioxins.

The pathway for dioxins into the food chain comes when mammals and fish ingest the chemical in the air or through soil polluted with the chemical. These levels of dioxin are background levels. However, the chemical accumulates as the small fish are eaten by larger fish, which are eaten by larger fish, and so on. Mammals, rabbits, or other herbivores eat the chemical in dust on plants or in the soil in which they burrow. Dioxin is then passed to predators or to humans, who eat the larger stock herbivores (cattle, sheep, goats, and others). An additional pathway is the use of dioxin-polluted ball clay in animal feeds.

Human exposure to dioxin through inhalation or water is not considered to be a significant problem—food animals and fish are the primary pathways for dioxin exposure for humans. Soil ingestion by grazing animals is a primary mechanism for its entry into livestock. In addition, for dairy and meat cattle, the use of pentachlorophenol-treated wood in barns and other livestock shelters is considered to be an ingestion pathway. Sewage sludge applied to pastures and crop land is also a pathway.

To decrease exposure, it is necessary to protect the food supply from pollution by dioxins. At the least, this requires that pastures and fish sources be placed away from sources of dioxin pollution.

See Also: Pesticide.

Further Readings

Allen, Robert. *Dioxin War: Truth and Lies About a Perfect Poison*. London: Pluto, 2004.

Committee on the Implications of Dioxin in the Food Supply, National Research Council, Committee on the Implications of Dioxin. *Dioxins and Dioxin-Like Compounds in the Food Supply: Strategies to Decrease Exposure*. Washington, D.C.: National Academies, 2003.

Crummett, Warren B. *Decades of Dioxin: Limelight on a Molecule*. Bloomington, IN: Xlibris, 2002.

D'Mello, J. P. F., ed. *Food Safety: Contaminants and Toxins*. New York: Oxford University Press, 2003.

National Research Council. *Health Risks From Dioxin and Related Compounds: Evaluation of the EPA Reassessment*. Washington, D.C.: National Academies, 2006.

Schecter, Arnold and Thomas A. Gasiewicz, eds. *Dioxins and Health*. New York: John Wiley & Sons, 2003.

Andrew Jackson Waskey
Dalton State College

DISAPPEARING MIDDLE

The "disappearing middle" is a phenomenon in which the extremes of a spectrum grow at the expense of the middle. Mathematically, the disappearing middle represents a shift from a bell curve—where most of the population is grouped somewhere in the middle range—to a well curve, where the middle is less populated than the extremes. The phrase is used in the discussion of many such phenomena, often in reference to economics and economic behavior, but not always. In American politics, especially in Congress, the disappearing middle between the left and right has been a much commented on trend dating from the reorganization of the right in the aftermath of Watergate. When liberals move to the right or conservatives to the left—both moving toward the moderate middle—they are likely to lose their seat. Democrats running against a moderate Republican can point out that the Republican candidate is "still too conservative"; fellow Republicans running against the moderate Republican in the primary can argue that their moderate opponent has lost sight of party goals by compromising with the Left. Bit by bit, the middle shrinks, strengthening the extremes. This distribution is also sometimes called "bimodal."

The phenomenon is visible in many areas of life, some of them likely interconnected. Perhaps in part because of the Internet, which provides opportunities for small business while globalization encourages a new wave of massive mergers, companies are getting both smaller and larger—the number of small and large companies is growing, whereas the number of midsized companies is falling. We see this in the retail field, as well, as Blockbuster, Wal-Mart, and Barnes and Noble continue to expand at the "big" end of the scale, as do the smaller specialty stores that do not compete with them so much as supplement them: video stores specializing in independent movies (and Greencine, an indie and multiregion DVD mail-in service supplementing Netflix's coverage), boutique shops, and science fiction bookstores. Meanwhile, the midsized stores that would constitute direct competition with the megachains are on the decline. Even the price of the items we buy is experiencing a shift to the well curve, as more and more $100 hamburgers make headlines—even as chain restaurants play up their dollar menus—and as HDTV becomes the new standard, with bigger and bigger televisions becoming common, making the tiny television-watching screens of cell phones and handheld computers seem even smaller in comparison, and the middle drops away as the "starter television" of years past disappears from shelves. In many industries, consumers are shying away from the middle ground while buying up the expensive premium items and the cheap discount items.

Though it is not as pronounced, the effect is noticeable even with nation-like entities, as the years since the end of the Cold War have seen an increase in multinational state-like organizations like the European Union (which are far stronger than the treaty organizations of the Cold War) and in small independent states, whereas midsized European nations are actually declining in population.

Much of the focus on the disappearing middle is in reference to incomes. Although American incomes had, at the time the global financial crisis began, been steadily rising across the board, the top and bottom tiers were growing fastest, with fewer and fewer individuals and families in the middle of what was once a well-defined bell curve. The adage "the rich get richer and the poor get poorer" is an old one—President Andrew Jackson referred to it in 1832—and in the current generation in the United States it is proving true, as the middle class loses members to the growing numbers of working poor, and the upper-upper class continues to accumulate more and more wealth. The trend of extraordinary corporate welfare in the 21st century, however necessary it may have been for the common good, certainly contributes to this, as profits remain privatized but losses are socialized through bailouts and the use of the taxpayer as the lender of last resort.

The accelerating spread of economic inequality, and the resulting disappearing middle, is often blamed on the Washington Consensus: the "market fundamentalist" approach to economic policy promoted by both the U.S. federal government and D.C.-based international bodies like the International Monetary Fund and the World Bank, especially as a tonic for developing nations in economic crisis. Since the Reagan years, the Washington Consensus has won more and more battles, both minor and major, in issues of national and international policy, and although it generally loses on the protectionist front, as the General Agreement on Tariffs and Trade led to the much more staunchly anti-protectionist World Trade Organization, Washington Consensus adherents have been especially successful in the reactionary deregulation of the banking industry and financial sector. Though market fundamentalism generally accepts the necessity of government intervention to some degree, the term was coined to reflect the fact that its proponents tend to argue as though a free—that is, unrestricted and unregulated—market is a priori a good (and in fact unimprovable) thing, treating any such restrictions as, at best, necessary evils. This position makes deregulation a constant goal even when all the evidence—such as the savings and loan crisis of the 1980s and the resulting government bailout—suggests that deregulation leads to irresponsible banking behavior.

See Also: Export Dependency; Fair Trade; Trade Liberalization.

Further Readings

Bradley, Stephen. *In Greed We Trust: Capitalism Gone Astray.* Victoria, British Columbia, Canada: Trafford Publishing, 2006.

Kaynak, Erdener. *Cross-National and Cross-Cultural Issues in Food Marketing.* New York: Routledge, 2000.

Stanton, John L. "The Disappearing Middle Class: You May Need to Learn How to Market Food to the "Have-Nots." *Food Processing,* 66/5:28 (May 1, 2005).

Bill Kte'pi
Independent Scholar

Doha Round, World Trade Organization

The Doha development round is the current trade-negotiation round of the World Trade Organization (WTO); it has proven to be a contentious round marked by stalled negotiations and disputes between the developed nations (the United States, Japan, and the European Union) and the BRIC-caliber developing nations like Brazil, India, and China.

The World Trade Organization commenced in 1995, succeeding the General Agreement on Tariffs and Trade (GATT) as the primary international trade organization. Nearly all world trade—about 95 percent of it—occurs between member nations of the WTO, of which there are 153 full members and 30 observers (many of whom seek future membership). Similar to the GATT, the WTO conducts its negotiations in rounds, moving trade liberalization forward from the previous round (at least in theory) while addressing concerns that have come up and adapting policies in response to new issues and the expanding membership. The GATT was primarily concerned with tariffs—the taxes levied by countries on products imported from foreign nations—which were the main barrier to international trade and which had often been used to promote weaker (or more expensive) domestic products at the expense of superior (or cheaper) foreign products. Beginning with the Tokyo Round of GATT (1973–79), nontariff trade barriers and trade concerns became topics of discussion, which was the first move toward the more generalized nature of the WTO. The next GATT round was the Uruguay Round (1986–93), which was the most ambitious round of negotiations, embracing a broad range of issues and resulting in the deepest reforms to international agricultural trade in history—as well as in the creation of the WTO, membership in which would require obligations above and beyond those covered by the GATT. GATT members ended the binding nature of the GATT as of December 31, 1995, having joined the WTO over the course of the year.

The GATT was a treaty, whereas the WTO is an international institution, headquartered in Geneva, Switzerland. However, it should not be confused with a policymaking organization or international legislature: It provides a forum for delegates of member-nations to discuss the trade agreements they will all agree to, but nothing is decided unilaterally, and the staff is primarily administrative and secretarial.

The Uruguay Round, the last round of GATT, essentially constituted the first round of the WTO, in that the WTO was created directly out of those discussions. The Doha Round is the first WTO round since its inception; it began in November 2001. The round was supposed to begin at the 1999 Ministerial Conference (a regular meeting of WTO delegates called ministers), held in Seattle, and would have been called the Millennium Round, but negotiations were postponed until the next Ministerial Conference in 2001 because of protests in the city. The WTO protests of late 1999 were the largest antiglobalism demonstrations in American history, with the "Battle of Seattle" involving tens of thousands of activists, ranging from those with labor concerns to conservative religious groups to college students to radical anarchists—not everyone opposed the WTO for the same reason. Though the numbers of protesters were inherently disruptive, they were primarily peaceful; media reports of violent demonstrations and Molotov cocktails thrown at police were erroneous, though the corrections were not reported as frequently as the errors. In any event, the two-year delay of the WTO's negotiations is considered a tangible victory by many of the protesters and their sympathizers.

The Doha Round Begins

The 2001 Ministerial Conference was held in Doha, Qatar, November 9–13, 2001, just two months after the September 11, 2001, attacks in the United States; it thus took place amid the hope that new international trade negotiations could mitigate some of the terrorism-related economic slump the world faced. The Doha conference accomplished three things: China's membership in the WTO was approved, a declaration was made incorporating flexibility into the WTO's intellectual property rights agreement to allow member nations to work around it when patented medicines were necessary for national health emergencies, and the formal beginning of the Doha development round was begun. A March 31, 2003, deadline was set for determining the modalities (specifics) of the round's objectives, and the three pillars of the round's agricultural negotiations were put in place:

reduction of export subsidies and domestic support and improvements in market access. The Doha Round was set to conclude in December 2005, after the second subsequent Ministerial Conference.

Both of those deadlines were missed. Though the modalities were hoped to be determined before the 2003 Ministerial Conference in Cancun, Mexico, the Cancun talks collapsed almost immediately, after only four days. The main bone of contention was the G20's demands that agricultural subsidies in the European Union and the United States end, because of their poor effect on international free trade. The G20 is a group of developing nations who recognized their common concerns at the Cancun conference, where 20 of them signed a document stating their position on subsidies and other issues of trade liberalization. Membership in the G20 has fluctuated since Cancun; there are currently 23 members: Argentina, Bolivia, Brazil, Chile, China, Cuba, Ecuador, Egypt, Guatemala, India, Indonesia, Mexico, Nigeria, Pakistan, Paraguay, Peru, the Philippines, South Africa, Tanzania, Thailand, Uruguay, Venezuela, and Zimbabwe. Brazil, China, India, and South Africa (collectively, the G4) have been the most vocal.

Another area of disagreement was in these developing nations' refusal to come to an agreement with the developed nations on the "Singapore issues," the four topic groups established at the 1996 Ministerial Conference in Singapore: transparency in government procurement, customs, trade and investment, and trade and competition. Though the EU expressed a willingness to give ground, the G20 bloc seemed unwilling to negotiate or to compromise, and the talks collapsed in part because of the perception that not all the countries participating had arrived with the intent of open discussion. Instead, the G20 bloc used the spotlight of the conference to publicize their document, which had been written the previous month. Developed and developing countries, sometimes called the "North" and the "South," disagreed on almost every Doha issue. Rather than focus on those issues that seemed to offer the most chance of healthy discussion, the chair of the conference, Mexico's representative Luis Ernesto Derbez, chose to end the conference.

Progress Is Slow

Negotiations resumed in 2004 in Geneva, with some of the Singapore issues removed from the Doha development round agenda and the European Union agreeing to eliminate agricultural export subsidies at some point; the United States, where agricultural subsidies are already a highly controversial topic with stubborn adherents on both sides of the debate, made no such promises. India and Brazil were heavily involved with the Geneva negotiations and were more active than developed nations are used to seeing developing nations being during multilateral talks. Finally, two and a half years after the Doha Round began, member nations agreed to the "July package"—a framework agreement providing guidelines for the round's objectives, consisting of a four-page general declaration and specific sections on agriculture, nonagricultural market access, services, and customs. The original 2005 deadline was tossed out, and the next Ministerial Conference was scheduled for December 2005 in Hong Kong. Modalities had still not been determined and agreed on.

Negotiators met in Paris in May 2005 in the hope of ironing out some issues before the conference, with little luck. The Hong Kong conference took place December 13–18, after an official announcement that no agreement was expected to be reached on modalities and that the main purpose of the conference was to get all the issues on the table and take stock of the ongoing discussions. Nevertheless, the talks did result in a declaration that developed nations would be required to open their markets to imports from developing nations, finally making real a United Nations goal that had been discussed for decades. Furthermore, many nations agreed to end agricultural export subsidies by 2013.

However, subsequent talks in Geneva in the summer of 2006 made it clear how slow the progress was in reaching agreement on all the issues, and tensions rose because of a ticking clock: the trade authority enjoyed by U.S. President George W. Bush under the 2002 Trade Act would expire in 2007, after which point the United States' agreement to any trade pact would have to go through Congress—a much more difficult task. Indeed, the Doha Round was nowhere near complete when that deadline came and went, and in the summer of 2008, negotiations collapsed completely over a dispute among the United States, India, and China. The most recent farm bill had been passed in the United States in the interim and was widely criticized by WTO proponents, especially the G20 nations, because of its five-year program of agricultural subsidies. India and the United States were completely unwilling to budge on their positions on the "special safeguard mechanism," under which a tariff could be imposed on the import of specific goods if the price of those goods dropped precipitously or saw a surge of imports; the point of disagreement was the exact threshold at which the mechanism would be allowed. India, widely criticized by the European Union for its unwillingness to compromise on the matter, claimed that its was the majority opinion, though fellow G4-member Brazil withdrew its support of India's position in the interest of trying to keep the talks from stalling. When no movement was made, negotiations ended and, a year later, had not yet resumed.

See Also: Farm Bill; Trade Liberalization.

Further Readings

Djurfeldt, Goran, et al. *The African Food Crisis: Lessons From the Asian Green Revolution.* New York: CABI, 2005.

Ingco, Merlinda. *Liberalizing Agricultural Trade: Issues and Options for the Middle East and North Africa in the Doha Development Round.* Washington, D.C.: World Bank Publications, 2005.

McCalla, Alex F. and John Nash. *Reforming Agricultural Trade for Developing Countries: Key Issues for a Pro-Development Outcome of the Doha Round Negotiations (World Bank Trade and Development Series).* Washington, D.C.: World Bank Publications, 2006.

Bill Kte'pi
Independent Scholar

Eco-Labeling

Eco-labeling is a voluntary strategy that involves placing identifiers on goods and services to represent ecological and/or social criteria. It is a market-based approach to achieve sustainability goals by providing consumers with information that would otherwise be difficult to ascertain. This allows interested consumers to support the practices embodied in these criteria through their purchases. It also assists producers to support their sustainability efforts, if providing such information leads to increased sales and/or the ability to charge higher prices. Information on eco-labels is typically represented in the form of a symbol, but may include descriptions of the criteria, or even specific figures. An eco-label for greenhouse gas reduction, for example, might include a symbol that represents reduced emissions, a brief definition of carbon footprints, or the number of grams of carbon dioxide reduced.

Growth of Organic Foods and Eco-Labels

Eco-labels have been developed for goods and services in industries including forestry, energy, tourism, fisheries, and agriculture. The food and agriculture sector has initiated some of the most successful eco-labels, such as organic and fair trade. Organic standards prohibit the use of synthetic pesticides and fertilizers, as well as antibiotics and synthetic growth hormones for animals. Organic sales in the United States have increased at rates approaching 20 percent annually in the last two decades, and currently comprise more than 3.5 percent of all food sales. Fair trade standards apply to products from the global south, and include higher levels of compensation for farmers and farmworkers, as well as other social and environmental criteria. Fair trade was introduced later in the United States, and until recently was only available for coffee, tea, and chocolate. It currently comprises less than 1 percent of total food sales, but has experienced annual growth rates as high as 100 percent. Products with organic and fair trade eco-labels can now be found in retail outlets as mainstream as Wal-Mart, McDonald's, and Dunkin' Donuts.

There are currently several dozen food eco-labels available on a global basis. They address a number of political and ethical issues, including synthetic pesticide use; use of genetically engineered organisms; farmer and farmworker compensation and working conditions; animal welfare; reduction of greenhouse gases; soil and water conservation;

and wildlife or habitat protection. Most eco-labels have standards that focus on just one of these major areas, although a few have been developed that incorporate multiple issues. The nonprofit Food Alliance eco-label is one example; it represents standards for protecting wildlife, soil, and water quality; prohibiting certain materials (but not synthetic pesticides); animal welfare; and working conditions.

Eco-Label Credibility

To be successful in the marketplace, eco-label claims must be trusted by consumers. Organic and fair trade are examples of labels that rely on third-party certification systems to verify compliance with specific criteria. This involves hiring an independent auditor, with no direct financial stake in the outcome, to certify that a standard is achieved. Some eco-labels rely on first-party certification, without independent auditing of the claims made by the firm selling a product or service. Second-party certification refers to verification of guidelines by an organization with a direct financial stake in the outcome, such as retailer verification of producer standards. A less common second-party model is employed by the Certified Naturally Grown eco-label—a peer review process that engages the farmers participating in the program in the certification of other farms. This strategy reduces costs and paperwork requirements in comparison to organic certification, even though Certified Naturally Grown adopts organic standards as a baseline.

The development of eco-labels may include participation from multiple types of organizations, including those that represent producers, distributors, retailers, consumers and scientists, as well as government bodies. The organic eco-label, for instance, was initiated by organic farmers to distinguish their products in the marketplace. Farmers formed state and regional organizations to certify the standards they developed, beginning in the 1970s. In the United States, the U.S. Department of Agriculture (USDA) gained legal oversight of these certifiers in 2002, and now enforces one uniform standard. Organic certification in other parts of the world, as with most other types of eco-labels, tends to involve much less government supervision, and instead relies on nongovernmental organizations to regulate standards. Dissatisfaction with federal control of the organic eco-label in the United States has encouraged efforts to create new alternatives that embody some of the organic movement's original ideals. These initiatives are referred to as *beyond organic* or *post-organic*. One example is the Mendocino Renegade eco-label in Northern California, which incorporates additional criteria, such as reducing fossil fuel consumption, in its standards.

Market Criticism

The proliferation of eco-labels has received some criticism for the burden it places on consumers to distinguish among numerous claims. *Label fatigue* is a term used to describe these concerns, although increasing interest in political and ethical consumption indicates that more consumers are willing to take a reflective approach to their purchasing choices, if given the opportunity. For producers, there are additional costs to participation in eco-labeling schemes, such as certification fees and paperwork requirements, and potentially the expense of segregating products to be sold under differing standards. Businesses that engage in international trade of organic products, for example, must meet differing criteria for markets in the United States, Europe, and Japan. These costs must be offset by increased sales volume or higher prices to be financially viable, unless there is a perceived public relations value in supporting these efforts. Label fatigue and the transaction costs of multiple, differing standards have been cited as arguments for harmonizing eco-labels, or making the standards more uniform. These include proposals to develop a global "super-label" that harmonizes the most successful eco-labels. One concern with harmonization efforts is that they may weaken existing standards to the level of the lowest common denominator.

Market-based strategies to achieve ecological and social goals, including eco-labeling, have limitations. One limitation is that they tend to merely shift purchasing patterns, rather than reducing consumption itself. Another limitation is that they are dependent upon economic purchasing power, yet income is not evenly distributed, resulting in a disproportionate ability for consumers to support their political and ethical principles. In addition, eco-labels encourage individual-focused, rather than group-focused collective action to address ecological and social goals. Individualistic approaches may be more vulnerable to the free rider problem—some may choose to receive the benefits from the actions of others, without paying the costs.

Eco-labeling is also susceptible to the practice of "greenwashing." This involves businesses either making inflated claims, or engaging in eco-label initiatives at a very minimal level, in order to deflect attention from the social and ecological impacts of their conventional practices. Fair trade initiatives, for instance, have recently been joined by large, multinational corporations, but typically for only a tiny fraction of their product lines. Co-optation is another potential hazard, even for the most established eco-labels. U.S. organic standards, for example, have been challenged by businesses that seek to weaken existing criteria and allow lower production costs, yet maintain the higher prices that consumers are willing to pay for foods labeled organic.

See Also: Beyond Organic; Certified Organic; Fair Trade.

Further Readings

Bostrom, Magnus and Mikael Klintman. *Eco-Standards, Product Labelling and Green Consumerism*. New York: Palgrave Macmillan, 2008.
Consumer Reports Greener Choices Eco-Labels Center. http://www.greenerchoices.org/eco-labels/eco-home.cfm (Accessed September 2009).
Rubik, Frieder and Paolo Frankl, eds. *The Future of Eco-Labelling*. Sheffield, UK: Greenleaf Publishing Limited, 2005.

Philip H. Howard
Michigan State University

EXPORT DEPENDENCY

Export dependency is the condition of being dependent on exports as a source of national revenue; it is especially common among developing nations that can produce goods at a cheaper price than the developed nations that constitute a significant part—and for some companies or industries, even the whole—of their market. When too much of a country's gross domestic product (GDP) comes from exports, it finds itself at the mercy of the countries to which it exports—of their economic health and possibly their demands. Much attention is paid to the fact, for instance, that half of Asia's collective GDP comes from exports—twice what it was in 1980, which is a significant factor in understanding events like the Asian financial crisis of 1997.

Asian countries have been exporting electronics products, in particular, at a significant profit for decades. The 1997–98 crisis was a perfect illustration of the connectedness of the world's economies: foreign debt and the decision to float Thai currency (previously pegged to the U.S. dollar) precipitated the crisis, which spread throughout southeast Asia, devaluing currencies and assets, increasing private debt, and tromping through stock exchanges. The International Monetary Fund helped to stabilize the currencies of Indonesia, South Korea, and Thailand, but much of the recovery—which came by the end of the decade, sooner than would have been expected from similar calamities in the West—was a result of the export-based economy. Countries without such heavy reliance on exports would have found themselves short of customers, who would be affected by the inflation and rampant unemployment, but so many customers of Asian manufacturing were far outside the range of the crisis, and their spending habits continued as normal. At the same time, this recovery was short-lived: by 2002, Asian exports were slumping, especially in electronics, as the Western economic slowdown and the cautious spending that followed the September 11, 2001, attacks cut into the demand for high-end electronics and consumer luxuries. The United States, which is the biggest market for many Asian exporters, saw the biggest dip in demand, but it was pronounced across the West, and economic growth in many Asian countries slowed to the lowest it had been, apart from the crisis years, in decades. Taiwan, which was barely affected by the financial crisis despite analysts' fears, was the worst hit by the export slump and experienced its worst quarter-to-quarter decline in GDP in over 20 years. Japan saw its unemployment reach a record high of 5.4 percent in October 2001, as companies downsized and closed plants in response to the slowdown.

Many have argued for decades, raising their voices in recent years, that East Asian nations in particular need to switch to domestic-demand-based economic growth, away from the export-dependent growth model currently holding sway, to reduce their vulnerability to the slings and arrows of overseas fortunes. Saying this, however, is perhaps akin to saying that farmers need to decouple themselves from the random chance of the weather and adopt a model in which only their own efforts affect the quality of their crops: Saying it does not make it so, and the path from one model to another is not necessarily obvious. However, a more realistic or attainable option might be to better diversify export markets while strengthening domestic institutions and the domestic financial sector to better weather storms—growing more crops, figuratively speaking, and being prepared to run for cover if the rivers rise.

One of the real concerns with export dependency is the possibility of *unequal exchange*. The term is associated with Marxian economics but refers to a situation more and more in evidence as the economies of the developed and developing nations intermingle. Marxians tend to look at unequal exchange in terms of labor—consider, for instance, what to think about a situation in which more labor is required to create a product than is required to earn the money to buy the product. When products are made or farmed in a part of the world where labor is very cheap, and sold where labor is very expensive, this can certainly be the case. In particular, we talk about unequal exchange when the producers of a good receive an unfairly small percentage of its price. For instance, though specifics will of course vary, the wages of plantation workers, who contribute the bulk of the labor—the bulk of the effort—responsible for the product, account for about 5 percent of the price of coffee, which is less (and usually significantly less) than the amount of the price created by tax. In the case of coffee, that money is perhaps being nickeled and dimed away—20 percent for various tariffs and sales taxes, 10 percent for the plantation owner, 5 percent for the roaster, 20 percent for the various stages and costs of transportation and middle-management, and so on. The figure seems shocking, but after some noodling around, it can start to seem reasonable, if unfortunate. There is, however, the more extreme case of the banana. A 1990 study of the commodity chain of bananas produced in Central America and sold in Germany found that the plantation workers' wages accounted for .01 percent—one-hundredth of one-hundredth—of the price, whereas the plantation owner received 22.2 percent of the price, the United Fruit Company 37.6 percent, and the retailer 11.2 percent. Those numbers reveal several inequalities—not only the great difference between the plantation owner and the plantation worker, or the corporation taking more money than either of them, but the large share going to the retailer: 11.2 percent of the price of a banana going to a retailer may not sound like much, but "over 1,000 times the wages of the workers who grew the banana" is another matter. It takes more work to grow and harvest a banana than it does to sell one—but the banana is sold in a place where costs are higher, where rents are high, and where there are expenses that must be attended to, both banana-related and banana-independent. The retailer is not, in other words, necessarily doing anything wrong, nor is he or she exploiting the plantation worker. And yet an unfair condition clearly exists—one for which there is no obvious remedy.

The condition of export dependency is, at its worst, that of being the banana plantation worker and facing no other choice but unemployment. One's fortunes are tied not to one's neighbors, domestic institutions, or representative government, but to some far-off German greengrocer and the economic and dietary trends of that country.

See Also: Commodity Chain; Doha Round, World Trade Organization; Food Security.

Further Readings

Allen, Gary J., et al., eds. *The Business of Food: Encyclopedia of the Food and Drink Industries*. Westport, CT: Greenwood, 2007.

Djurfeldt, Goran, et al. *The African Food Crisis: Lessons From the Asian Green Revolution*. Wallingford, Oxfordshire, UK: CABI, 2005.

Jackson, Peter, et al. "Mobilizing the Commodity Chain Concept in the Politics of Food and Farming." *Journal of Rural Studies*, 22/2 (April 2006).

Pollan, Michael. *In Defense of Food*. New York: Penguin, 2009.

Pollan, Michael. *The Omnivore's Dilemma: A Natural History of Four Meals*. New York: Penguin, 2007.

Bill Kte'pi
Independent Scholar

FACTORY FARM

Depending on the economies of scale, the term *factory farm* refers to industrialized livestock rearing in which animals are confined at high densities and brought to production or slaughter as rapidly as possible. Also called intensive, industrialized, or confinement agriculture, factory farming is dominated by agribusinesses, which have invested a great deal of capital into standardizing and mechanizing the "growing" and processing of animals to produce meat, eggs, and milk at the lowest possible costs. Chickens, turkeys, cattle, and swine/pigs are the most common factory farm animals, an estimated 10 billion of which are slaughtered each year in the United States alone. Given predictions that the global demand for livestock foods will more than double over the next 20 years, factory farming will likely continue to expand around the world. Critics argue that the environmental, social, health, and animal welfare costs of expanding factory farming practices are too high.

Large turkey farms like this one in Benton, Arkansas, have multiple buildings that can house as many as 10,000 turkeys each.

Source: U.S. Department of Agriculture, Natural Resources Conservation Service/Jeff Vanuga

History of Factory Farming

For at least 10,000 years, humans have intentionally and unintentionally reconfigured their environments as they have devised ways to raise, store, harvest, and selectively breed their domesticated plants and animals. This article highlights significant reconfigurations in livestock rearing that have taken place since the 19th century.

In North America, the development of mechanical refrigeration enabled the dramatic expansion of the meatpacking industry and thus led to further intensification in livestock production. By the late 19th century, stockyards in Chicago, Illinois, served as a striking example of the mass-production trend. Meat had become yet another of the wide variety of factory-produced consumer items epitomizing American capitalism and its growing dependence on economies of scale. Although the working conditions in the U.S. meatpacking industry led Congress to enact the Pure Food and Drug Act and the Federal Meat Inspection Act shortly after the publication of Upton Sinclair's *The Jungle* (1906), legislation on behalf of farm animals would not exist for several more decades.

In the 1960s, the U.S. company Iowa Beef Packers triggered a major revolution in meatpacking by redesigning the slaughter and packing process to fragment tasks and deskill work. This allowed for the reduction of wages while making it easier to replace laborers; sped up the chain of production; and relocated plants to rural areas, which were less expensive operation locales and usually did not have organized labor. The Iowa Beef Packers model dominated beef processing from that point onward and was adopted in poultry and swine operations, all of which depended increasingly on factory farming.

The post–World War II boom in U.S. and European industries and populations contributed to this expansion of factory farms, as urban and suburban regions overflowed into surrounding farmland. Attendant shifts in production occurred as large-scale commercial agribusiness replaced subsistence and small-scale farming, so that by the 1990s, less than 2 percent of Americans were even involved in production agriculture. Just before World War II, one-fourth of the U.S. population worked in agriculture.

Agricultural industrialization has emphasized fossil fuel–derived forms of energy to power its machines, while human and animal power have become less important and work is fragmented and deskilled. Shifts in discourse about the inputs themselves also index the trend from agriculture to agribusiness, and many agriculture-oriented universities have altered the names of their departments, changing their departments of "animal husbandry" to those of "animal science."

Many steps in the production of milk, meat, and eggs have been consolidated by large, even transnational agribusinesses. For example, particular breeds of swine and poultry may now be raised entirely indoors to regulate their intake and growth. Concentrated (or confined) animal feeding operations restrict beef cattle to small areas and feed them large amounts of grain to fatten them up before slaughter—factory farming practices counter the difficulties the cattle have in digesting the grain by dosing the animals with antibiotics. In fact, antibiotics, artificial insemination, synthetic hormones, genetic engineering, and/or other modifications are commonly used in industrial livestock to control for breed quality and production.

Case Study: Factory Turkey in the United States

Factory farming has made chicken a common, inexpensive meat, although it was once considered a delicacy. Several agribusinesses (e.g., Butterball, Cargill, Jennie-O/Hormel, Perdue) have done the same to transform turkey from a holiday meal to an everyday form of protein.

Each agribusiness typically has its own breeder facilities and may own and operate some turkey farms. More often, however, individual farmers sign exclusive contracts in which they grow turkeys provided to them by the particular agribusiness that holds the contract. The farmers must construct and maintain turkey houses that meet the specifications of their contracts; this allows the agribusiness to outsource the costs of land and infrastructure. The turkeys remain inside these houses from the time they arrive as chicks until they are trucked out to slaughter. Each turkey house can accommodate 6,000–12,000 turkeys at a time, and a turkey farmer may undertake three to four turkey-growing cycles per year.

Hens are raised separately from toms (males), and turkeys receive feed produced or at least supplied by the agribusiness partner. Farmers pay for this feed—although usually the cost of the feed is simply deducted from what the farmers would be paid for the ready-for-slaughter turkeys. The farmers must use agribusiness-approved antibiotics and other injections. This ensures quality control and also allows turkey production to be vertically integrated so that the agribusiness can maximize profits.

It takes three to four months to "grow" the turkeys to typical slaughter age. Toms are usually ready to slaughter between 100 and 120 days of age, or older if they are "roasters," hens around 90 to 95 days of age. When the turkeys are of slaughter age, they are transferred into cages and transported by truck to the processing facility. A single truck can haul 900–1,200 toms or 1,600–2,400 hens. An individual live bird can weigh 30-40 pounds, although an older breeding tom can be 80+ pounds. "Processing" refers to transforming live turkeys into packaged commodities, such as whole birds, cutlets, ground turkey, turkey sausage, and so on. Turkeys can go from live in the truck to packaged meat ready to ship in as little as three and a half to four and a half hours. The farmers get paid according to market prices, which vary depending on grade of meat and intake at a given time. Given the costs of inputs and managing wastes/by-products, profit per animal is relatively small for turkeys, and for factory farm animals in general.

Negative Effects of Factory Farming

The European Union currently has higher minimum standards than the United States for the legal protection of farm animals, but myriad problems persist. Critics note the negative repercussions of factory farming include compromised livestock health and welfare; threats to the health of people and other animals in the ecosystem; high outputs of methane and high concentrations of manure, leading to air, ground, and water pollution; heavy reliance on water and fossil fuels, as well as increasing amounts of land, for the production of feed; and fuel and pollution issues linked to the processing and transport of the animals.

Factory farming has led to the rise of a class of illnesses among livestock now called "production diseases" that result from diets designed to maximize growth instead of maintain long-term health, as well as injuries and infections resulting from the conditions in which they are kept (e.g., wire cages, concrete floors with slats, etc.). Bovine spongiform encephalopathy, or mad cow disease, is one striking example of a "production disease." Factory farms often have inadequate staff (or incentive) to respond to sick or injured animals.

With regard to air and water pollution, the United Nations Food and Agriculture Organization reports that the livestock sector (which includes factory farm and nonfactory farm livestock) accounts for about two-thirds of human-related nitrous oxide production, mostly from manure. Nitrous oxide is estimated to have 296 times the global warming potential of CO_2. Further, the digestive systems of ruminants produce nearly 37 percent of all human-induced methane (23 times as warming as CO_2), and livestock also produce some 64 percent of ammonia, which significantly contributes to acid rain. The United Nations Food and Agriculture Organization estimates that one-third of global arable land is used to produce feed for livestock and describes the livestock business as "among the most damaging sectors to the Earth's increasingly scarce water resources" because of the amount of water used in feed production, as well as the extent of water pollution resulting from animal wastes (runoff, leakage from waste lagoons, etc.), fertilizers and pesticides used during the production of feed crops, chemicals used in growth and processing of animals, and more. These costs to the environment are not figured into the prices of livestock products.

See Also: Agribusiness; Animal Welfare; Confined Animal Feeding Operation; Contract Farming; Mad Cow Disease.

Further Readings

Food and Agriculture Organization of the United Nations. "Livestock's Long Shadow." 2006. ftp://ftp.fao.org/docrep/fao/010/A0701E/A0701E00.pdf (Accessed February 2009).
Pollan, Michael. *The Omnivore's Dilemma: A Natural History of Four Meals.* New York: Penguin, 2006.
Stull, Donald D. and Michael J. Broadway. *Slaughterhouse Blues: The Meat and Poultry Industry in North America.* Belmont, CA: Wadsworth/Thomson Learning, 2004.
Sunstein, Cass R. and Martha C. Nussbaum, eds. *Animal Rights: Current Debates and New Directions.* New York: Oxford University Press, 2004.

Jennifer Ellen Coffman
James Madison University

FAIR LABOR ASSOCIATION

Sustainability is often defined using the "Three E's:" environmentally sound, economically viable, and socially equitable. However, "socially equitable" is typically the least-developed piece of the equation, particularly in terms of food. The conventional food system does not ensure fair prices to farmers or fair wages and working conditions for farmworkers. Several international labels, such as TransFair USA and FLO International, exist to ensure social justice for agricultural producers and workers. The creation of these labels parallels the anti-sweatshop movement in many ways, with many of the same goals, benefits, and criticisms.

In the 1990s, labor and human rights activists launched organized campaigns to expose the growing problem of worldwide labor injustices in the apparel industry, which led to a demand for international fair labor standards. The Fair Labor Association (FLA) is a non-profit organization formed in 1999 to develop and monitor a voluntary labor code of conduct that addresses sweatshop labor, child labor, and other international labor issues in the apparel and footwear industry. The FLA's standards, monitoring practices, and governance have stirred a great deal of controversy and criticism, and the FLA is one of several prominent organizations involved in an ongoing debate over international labor standards and monitoring.

The FLA developed out of the Apparel Industry Partnership (AIP), a task force that was convened by the Clinton administration in 1996 in reaction to growing public outrage about sweatshop labor and the lack of international labor standards in the apparel industry. Clinton's creation of the AIP was prompted by several successful campaigns in the 1990s that drew widespread attention to sweatshop labor abuses, sparking criticism of the garment industry. To address the highly adversarial relationship between industry on one side and labor and human rights organizations on the other side, the AIP began a dialogue between apparel industry leaders and human rights and labor activists. The goal was to create international industry-wide labor standards that would be acceptable to both sides. After the AIP produced a "Workplace Code of Conduct," the FLA was created to enforce and to monitor the code.

The FLA includes many prominent apparel brands such as Nike, Eddie Bauer, Patagonia, and Adidas, among others. Governed by a board of companies, nongovernmental organizations, and colleges and universities, the FLA primarily serves to monitor the apparel and footwear industries. As of 2008, 28 firms and over 200 colleges and universities had signed on to the voluntary code of conduct, which addresses the areas of forced labor, child labor, harassment, discrimination, health and safety, freedom of association, wages and benefits, working hours, and overtime compensation. Companies participating in the FLA conduct internal monitoring and audits and are also subject to independent external monitoring. The FLA also allows for third-party complaints from anyone believing there has been a violation of the code of conduct.

The demand for such voluntary codes of conduct stems from the rapid growth of global industrial production and the differences in labor standards between different countries. Some countries have stricter labor (and environmental) standards and enforcement than others, leading companies to move their production to the locations with fewer regulations to take advantage of cheaper labor costs. In response, human rights activists and labor organizations, particularly student groups, have organized campaigns and protests exposing abuses and calling for stronger international labor standards and monitoring to end the exploitation of workers on a global scale.

The FLA has been criticized by many other organizations, including student groups, nongovernmental organizations, and unions, for having weak standards and for being excessively controlled by industry. The United Students Against Sweatshops (USAS) has been a prominent critic of the FLA. Specifically, the USAS and others have criticized the FLA because its standards do not require that workers be paid a living wage. In addition, the USAS has charged the FLA with failing to provide a meaningful role for students, who have been critical drivers of change in labor standards, as well as for workers themselves. Unions have critiqued the FLA for shifting attention away from union organizing. Perhaps most significantly, critics have raised concerns around transparency and public disclosure, arguing that the FLA is a mouthpiece for the industry and that it is covering up poor labor practices and corporate interests, rather than providing meaningful reform.

Colleges and universities have played a major role in both the campaigns demanding fairly produced apparel and in the criticism of the FLA. To address what they perceived as critical flaws in the FLA and to provide an alternative, the USAS teamed up with the American Federation of Labor and Congress of Industrial Organizations union and other nongovernmental organizations critical of the FLA to start the Worker Rights Consortium (WRC). The WRC is another organization formed in 1999 to monitor labor in international apparel production. Whereas FLA-certified firms do much of their own monitoring, the WRC conducts monitoring of production plants independently of companies and includes a requirement that firms pay workers a living wage. Unlike the FLA, WRC membership is limited to universities and colleges, which restricts the scope of its influence.

The FLA has responded to criticism by encouraging its members to make their audits public. Some companies did so by posting their audits on the FLA website in 2003, which led to increased calls for other firms that had not done so to follow suit. In 2005, Nike responded to concerns about transparency by making some factory locations public, which allowed independent organizations to inspect the factories for themselves. However, despite these actions, the USAS and other opponents continue to maintain their criticisms of the FLA.

Other voluntary labor standards organizations exist in the United States, most notably Social Accountability International (SAI), a third major organization, along with the FLA and the WRC. SAI has developed SA8000, another voluntary code of conduct. SAI was started in 1996 by a group of businesses and organizations including the Body Shop, Reebok, and Toys "R" Us. The SA8000 code includes provisions for a living wage, as defined in the code, and is based on standards defined by the International Labour Organization—an agency of the United Nations.

The FLA and other voluntary codes of conduct represent a new approach to international regulation in that they are not sponsored by any state or governmental body but, rather, are based on boycotts and "buycotts" and consumer pressure. There is some indication that the growing trend of voluntary corporate social responsibility efforts is causing industry change; firms that used to be under heavy criticism for poor labor practices are now signed onto voluntary codes of conduct. However, only a small percentage of global corporations report on their social and environmental practices. There is also concern that these voluntary codes of conduct are not democratic enough and that global governmental bodies such as the United Nations International Labour Organization should be more active rather than relying on industry to voluntarily monitor itself.

See Also: Fair Trade; Labor; Trade Liberalization.

Further Readings

Broad, Robin, ed. *Global Backlash: Citizen Initiatives for a Just World Economy*. Lanham, MD: Rowman & Littlefield, 2002.

Fair Labor Association, "Nonprofit Organization Dedicated to Ending Sweatshop Conditions." http://www.fairlabor.org (Accessed January 2008).

Lappé, Frances Moore. *Democracy's Edge: Choosing to Save Our Country by Bringing Democracy to Life*. San Francisco: Jossey-Bass, 2006.

Seidman, Gay W. *Beyond the Boycott: Labor Rights, Human Rights, and Transnational Activism*. New York: Russell Sage Foundation, 2007.

United Students Against Sweatshops. "FLA Watch: Monitoring the Fair Labor Association." http://www.flawatch.org (Accessed January 2008).

Vogel, David. *The Market for Virtue: The Potential and Limits of Corporate Social Responsibility*. Washington, D.C.: Brookings Institution, 2005.

Alida Cantor
California Institute for Rural Studies

Fair Trade

Fair trade seeks to enhance producer-consumer relationships in ways that reestablish trust, more equitably distribute the benefits of trade, and promote empowerment among impoverished smallholder farmers, artisans, and workers. The initial intentions were to create a real-life retort by developing alternative trade production, trade, and distribution channels that contrasted sharply with the profit-focused "free" trade system. Fair trades pioneers claimed that the mainstream fair trade system had contributed to chronic poverty and environmental degradation in the global South and thus needed to be reorganized and "transformed." The fair trade approach poses an alternative trade system characterized by transparent trading partnerships connecting consumers and retailers with producer and/or worker organizations in a way that ensures human rights, livelihoods, and sustainable development. The current certified fair trade market finds its roots in the international solidarity work that aimed to support empowerment and provide postdisaster relief in ways that more highly valued the labor and artistry of impoverished peoples. The fair trade of agricultural products also started with an intercultural partnership connecting an indigenous coffee cooperative through a religious leader and into a solidarity-oriented market. A core element of this system is to build a more direct relationship that advances empowerment among producer organizations. The most important fair trade standard is related to the guarantee that disadvantaged small-scale farmers and artisans will receive minimum prices sufficient to cover the costs of sustainable production and premiums that enable social development.

The Fair Trade–certified coffee, chocolate, flowers, and crafts found on today's supermarket shelves and in hip cafes made their way to these places through an unlikely history that begins in the late 1940s with international people-to-people solidarity. The primary partnerships consisted of impoverished small-scale producer cooperatives and artisans selling food, beverages, and crafts to Northern volunteers from churches and nongovernmental organizations (NGOs). A small, yet thriving, alternative trade system began to emerge. Producers and buyers shared the risk: on the one hand, farmers and artisans sometimes provided their products months or even years before receiving full payment after volunteers and alternative trade organizations sold their goods into distant and uncertain markets. On the other hand, fair trade organizers from the North provided producers with loans that would otherwise be unavailable and bought crafts and coffee before they had established demand in their home markets.

Leading the Way

Early alternative trade organizations emerged around handicrafts, often connecting religious and politically motivated Northern groups with small groups of female artisans. Several examples of these pioneering organizations include the Mennonite Central Committee in Pennsylvania, which started buying quilts directly from seamstresses in Puerto Rico in the late 1940s and a decade later created Ten Thousand Villages, an alternative trade organization that as of 2006 connected to some 100 artisan groups and had annual sales in excess of $20 million; SERVE International (Sales Exchange for Refugee Rehabilitation and Vocation); and a campaign by Oxfam UK called the Helping-by-Selling Project. The first Worldshop opened in the 1950s, and by 2005 there were more than 2,800 Worldshops throughout Western Europe, selling mostly fair trade products with annual sales of about $151.8 million. These alternative trade organizations and their producer partners are a core part of the solidarity-based root of this movement and marketplace.

After more than 40 years of solidarity-based alternative trade, Northern alternative trade organizations shared a series of reflections with several producer partners and realized that although these relationships were rewarding and contributed to the education of thousands of new citizen consumers while maintaining practices that were different from the mainstream commercial trade, the volumes actually traded and the total revenues generated for Southern producers were still far too small to begin fulfilling their ambitious development goals. This prompted a bit of a split in the movement, with several fair trade organizations deciding that they needed to engage mainstream corporations and develop a certification scheme and others opting for a strategy focused on scaling up the alternative trade organizations. Many involved in these more grassroots efforts to expand the alternative trade channels continued their involvement with an international association of alternative trade organizations. Those seeking to expand the market more quickly created Trade Labeling Organizations International to develop a product label and certification processes that would allow participation by mainstream corporations. Despite their differences in strategy, fair trade advocates continued to agree on the basic goals of partnering with producers and workers to secure livelihoods and human rights and to advance their empowerment.

A Dutch Jesuit Priest; Mayan coffee producers from Oaxaca, Mexico; and several European NGOs led the early efforts to create a certified fair trade system. They partnered with several other advocates to create a product certification and labeling system called Max Havelaar. In 1988, Max Havelaar united with European and North American NGOs to create Fairtrade Labeling Organizations International in 1997. Fairtrade Labeling Organizations International is an international nonprofit multistakeholder association that seeks to establish fair trade standards; support, inspect, and certify disadvantaged producers; and harmonize the fair trade message across the movement. Although fair trade coffee finds began in these same solidarity-based relationships, this bean and beverage is also the pioneer agricultural commodity in the certified fair trade system. The emergence of a product certification system allowed the participation of more conventional companies, expanded fair trade markets, and shifted the ratio of global fair trade goods from crafts to crops.

Although fair trade generally accounts for 1–5 percent of the global trade in the specific agricultural commodities and handicrafts, market growth continues at rates of 10–25 percent in many countries. As of October 2006, the global fair trade–certified network included 586 producer organizations and 1.4 million farmers, artisans, and workers in 58 developing countries from Latin America, Asia, and Africa. Consumers worldwide spent $2.024 billion on fair trade–certified products. The expanding list of fair trade products includes coffee, cocoa, tea, fruits, wine, sugar, honey, bananas, rice, crafts, and some textiles.

The rapidly expanding retail sales figures tell us little about the ability of fair trade to deliver on its stated empowerment and sustainable development goals. An important consideration in assessing fair trade impacts is the fact that as of 2006, only 20 percent of the agricultural goods produced by fair trade–certified organizations are sold according to generally accepted fair trade terms. The remaining 80 percent of products are generally sold into domestic and international markets under less favorable terms. An important percentage of these products is often consumed within the household or traded locally. Most scholars agree, however, that small-scale producers linked to fair trade are better off than producers who lack these connections.

Many producers have advanced their sense of collective empowerment by building stronger organizations, and they have also conserved biological and cultural diversity through their farming practices. However, the combination of fair trade sales and additional support from allied international development NGOs is not a panacea for eliminating poverty or stopping outmigration even within fair trade organizations, although the minimum coffee prices are especially important when conventional market prices fall. Artisans have generally been able to sustain their crafts and cultures and partially support their livelihoods even as they have improved their organizations' business capacity through direct connections to better markets for their products.

Just Useful or More?

In the early 20th century, producer cooperatives organizing for survival and budding Northern social movements working for fairer food both found fair trade useful. For example, student social justice leaders with the United Students for Fair Trade used this concept to link university-based campus groups together in campaigns to push universities toward sustainability. Southern producer organizations used fair trade to gain visibility, and within the context of neoliberal markets, this also provided one of the few places they could have a small degree of leverage in the markets for their goods.

However, fair trade advocates have yet to make significant contributions to reforming the international free trade and development policy agendas. In fact, some certification agencies seem to have dropped this agenda altogether. Small-scale producer organizations have also used their participation in fair trade to strengthen their alliances, increase their visibility, and expand their negotiating power. For example, the Latin American and Caribbean Network of Small-Scale Fair Trade Producers, which represents more than 200,000 producer families, has used its participation in fair trade to win a seat on the Fairtrade Labeling Organizations International board of directors, to gain partial owner-ship of the certified fair trade system, and to advocate for minimum prices that keep up with inflation and cover the costs of sustainable production.

The central paradox within the certified fair trade system is that it sets out to achieve social justice and environmental sustainability within the same market system that it seeks to transform. In fact, the founders of this movement clearly identified with that free trade system and the impoverished small producers in the first place. However, the rapid growth of the certified fair trade system has started to expand its reach beyond the traditional social justice audience and has enrolled new consumers contributing to simultaneous expansion of many alternative trade organizations such as Equal Exchange in the United States. The fair trade system is also experiencing increased competition from an exploding array of sustainable product certification programs, such as the Rainforest Alliance and Utz Certified. Many of these programs have lower social and environmental standards. As a sustainably certified market develops and market-based competition accelerates, there is a growing risk that real minimum prices could drop even while standards and costs to pro-ducers increase. These risks suggest an increasingly important role for civil society organiza-tions and perhaps national governments in the governance of an evolving fair trade system.

See Also: Cash Crop; Family Farm; Retail Sector.

Further Readings

Bacon, C. M., et al., eds. *Confronting the Coffee Crisis: Fair Trade, Sustainable Livelihoods, and Ecosystems in Mexico and Central America.* Cambridge, MA: MIT Press, 2008.

Barratt-Brown, Michael. *Fair Trade: Reform and Realities in the International Trading System.* London: Zed, 1993.

DeCarlo, Jacqueline. *Fair Trade: A Beginner's Guide*. Oxford: One World, 2007.

Fairtrade Labeling Organizations International. *Shaping Global Partnerships: FLO Annual Report 2006/2007*. Bonn, Germany: FLO International, 2007.

Hernández Navarro, L. "To Die a Little: Migration and Coffee in Mexico and Central America." Special Report. Silver City, NM: International Relations Center, Americas Program, December 2004.

Jaffee, Daniel. *Brewing Justice: Fair Trade Coffee, Sustainability and Survival*. Berkeley: University of California Press, 2007.

Krier, J.-M. *Fair Trade in Europe 2005: Facts and Figures on Fair Trade in 25 European Countries*. Brussels: Fairtrade Labeling Organizations International, International Fair Trade Association, Network of European World Shops, and European Fair Trade Association Fair Trade Advocacy Office, 2005.

Latin American and Caribbean Network of Small-Scale Fair Trade Producers. *Estudio de Costos y Propuesta de Precios para Sostener el Café, las Familias de Productores y Organizaciones Certificadas por Comercio Justo en América Latina y el Caribe*. Dominican Republic: Assemblea de Coordinadora Latinoamericana y del Caribe de Pequeños Productores de Comercio Justo, 2006.

Leclair, M. S. "Fighting the Tide: Alternative Trade Organizations in the Era of Global Free Trade." *World Development*, 30/6 (2002).

Oxfam UK. "Documents Oxfam's Pioneering Work in Humanitarian Assistance and Fair Trade (2007)." http://www.oxfam.org.uk/oxfam_in_action/index.html (Accessed January 2009).

Raynolds, L. T., et al. "Regulating Sustainability in the Coffee Sector: A Comparative Analysis of Third-party Environmental and Social Certification Initiatives." *Agriculture and Human Values*, 24/2 (2007).

Renard, Marie-Christine. "Quality Certification, Regulation, and Power in Fair Trade." *Journal of Rural Studies*, 21 (2005).

Vanderhoff Boersma, Franz and Nico Roozen. *La Aventuta del Comercio Justo*. Mexico City: El Atajo, 2003.

Christopher M. Bacon
University of California, Berkeley

FAMILY FARM

As the production of agricultural products and control of the world's food systems becomes increasingly concentrated in the hands of a few global agricultural processing corporations, the family farm faces increased threat of extinction while simultaneously becoming increasingly necessary as a complementary, sustainable local source of healthy foods. As the necessity for a vitalized system of healthy local foods increases and becomes more broadly recognized, the primary value of a family farm may shift from being primarily a sentimental lifestyle to being a vital source of basic nourishment that sustains a community. As the emphasis on family farms increases, the task of defining them grows increasingly more complex.

The Struggle for Definition

Classifying farms is challenging. As a result, the most common definitions come from governing agencies that must define them for the purpose of census, taxation, research, and subsidies. Currently, the U.S. Bureau of the Census in cooperation with the U.S. Department of Agriculture (USDA) and the Office of Management and Budget defines "farm" as any place that sells $1,000 or more of agricultural products or would normally do so during a given year. The USDA Economic Research Service provides the most familiar taxonomy of farms, broadly classifying farms by amount of their gross sales. The USDA collapsed typology includes the following:

- *Rural residence farms:* those with gross sales of less than $100,000 where the operator is living below the poverty level, pursues a significant off-farm vocation, or is retired.
- *Intermediate farms:* those with gross sales of less than $100,000 where the operator lives above the poverty level and pursues farming as a major vocation, or farms with gross sales between $100,000 and $250,000.
- *Commercial farms:* those with gross sales greater than $250,000, or farms organized as nonfamily corporations or cooperatives or that operate with hired managers.

Within this taxonomy, a family farm is classified not by size of sales or acreage but by ownership organization. If a farm is a closely held family proprietorship, partnership, or corporation and not operated by a hired manager, it is a "family farm." Therefore, by USDA standards, of the 30,000 largest farms in the United States (farms with gross sales greater than $1 million), over 80 percent are considered family farms.

Because by USDA definitions over 95 percent of American farms can justify identifying themselves as "family farms," people from a variety of backgrounds have struggled to refine the definition of "family farm" to include only those farms with smaller acreages, smaller gross sales, or more localized markets. Alternatively, some small family agriculturalists struggle to meet the USDA definition of "farm" at all. A small family farmer whose farm produce feeds himself and others but who does not meet the $1,000 gross sales threshold might muse, "I have 20 acres. I feed four households. I own one tractor, two horses, a small herd of cattle, 60 chickens, and cultivate a large garden. Talk to the people at my local farmers' co-op. They love my feed, equipment, and supply purchases. Technically, however, I am not a farmer. What am I?"

The perspective of some larger farmers further clouds the issue of definition. Some who pursue agriculture as their sole vocation insist that an operator should derive 60 percent or more of his/her income from farming to qualify as a "farm." Anything else, in their opinion, is a "hobby farm"—what the USDA taxonomy likely would describe as a "residential/lifestyle farm" in its most detailed typology. Though they yield the word "farm," these vocational farmers prefer to stratify farming among those whose agricultural endeavors produce enough revenue to provide significant support to a family and those whose endeavors do not.

A number of nongovernmental organizations have championed family farms and offer their own definitions. Organizations like the National Family Farm Coalition (NFFC) suggest that family farms should be defined by labor supply and sovereignty. The NFFC contends that family farms are those where the family supplies the majority of the labor and exercises autonomy in the management decisions of the farm. For organizations like the NFFC, the alternative to "family farm" is the vertically integrated "factory farm," whose management decisions are primarily controlled by the large agricultural processing corporations that they supply. Classifying farms is a challenge.

Rural Development and Healthy Food Systems

Among the champions of family farms, there are some for whom the interest in family farms is mostly sentimental. Visions of pastoral scenes, simpler times, and meaningful labor seem worthy of preservation, much like motivations that spur appreciation for and safeguarding of antiques. In addition, there are a growing number of people who sense that the preservation of sustainable, autonomous family farms is essential to the social and economic development of rural America. In addition, they contend that locally oriented, diversified family farms are crucial to a food system that can sustain the health of all Americans.

Advocates of family farms insist that locally marketed foods are an obvious solution to the questions of sustainability associated with the global distribution of foods that consumes tremendous quantities of fossil fuel as they travel from field to table. Since the late 1960s, research has consistently demonstrated that the average meal that is not produced locally travels 1,000–1,500 miles from farm to plate.

Environmentally sensitive family farm advocates contend that the diversified nature of family farm production is better for the Earth's ecosystem than are conventional commercial farms. Diversified farms recycle nitrogen-rich animal waste to improve pasture, which subsequently produces healthier animals. Those animals consume fewer petroleum-based, chemically treated, and genetically altered foods and pharmaceuticals. Advocates of this system of farming contend that it creates a sustainable bioconsistent cycle that is not only sustainable but also beneficial for the ecosystem and ultimately for the planet.

Proponents of autonomous family farms that are not dependent on the modern vertically integrated food system insist that autonomous farms are more secure against widespread intentional and accidental food hazards. They contend that family farms vitalize rural economies by the local recycling of wealth, by significantly contributing to stable social systems in rural areas, and by protecting against not only the monopolization of food processing but also potentially exploitative land ownership by vertically integrated food processing corporations that might acquire lands abandoned by family farmers.

Finally, as Americans become increasingly health conscious and sensitive to the consequences of the processed, chemically tainted, genetically modified products of the current food system, proponents of family farms insist that without these vital farms, there will be no alterative to the modern diet–induced, medically dependent epidemic of disease among American adults and children.

Trends Among Family Farms

The number of U.S. farms is believed to have peaked in the 1930s at approximately 7 million farms. The current number of U.S. farms is estimated to be near 2 million. Less than 1 percent of the U.S. population claims farming as any portion of their vocation. Two percent of the population reports to live on a farm. Of the approximately 2 million U.S. farms, the vast majority would be considered family farms by USDA typology. Less than

25 percent of American farms produce gross revenues of more than $50,000. Even at that level of gross revenue, the net income from the farming enterprise alone unlikely would sustain the farm family's basic needs. As a result, the vast majority of America's family farmers require off-farm income, live in poverty, accumulate substantial debt, or ultimately leave farming altogether.

In the early 1990s, 50 percent of America's agricultural sales were produced by 3 percent of America's farms. In the late 1990s, that number had diminished to approximately 2 percent of America's farms accounting for 50 percent of sales. The concentration of production in the hands of fewer farms has continued into the new century. As the quantity of farmers has declined, the world's population and related demand for agricultural products has increased. Integrated mechanized agricultural processing has more than kept up with demand by replacing people and land with mechanization, chemicals, and genetic modification. In the century between 1890 and 1990, labor efficiency in farming increased from one worker per 28 acres to one worker per 750 acres. As farm populations dwindle, the average age of the American farmer is rapidly approaching 60 years, with very few optimistic young farmers coming along to replace them.

For American family farms, the real opportunity may be for incoming young (25–35 years old) and increasingly female farmers to create local markets of naturally grown foods on small diversified farms. However, although requiring less-expensive inputs, this type of farming is less efficient than mechanized commodity farming on a commercial scale. Therefore, its products are ultimately somewhat more expensive to consumers than those produced in the conventional food system. The fate of these agricultural entrepreneurs is uncertain. Much will depend on whether American consumers respond to the proponents of family farms who contend that for a variety of reasons, both sentimental and practical, the family farm is worth not only preserving but preferring.

See Also: Community-Supported Agriculture; Concentration; Mechanization; Sustainable Agriculture; Vertical Integration.

Further Readings

Hoppe, Robert, et al. *Structure and Finances of U.S. Farm: Family Farm Report (2007 ed.)* U.S. Department of Agriculture Economic Information Bulletin No. 24, June 2007. http://www.ers.usda.gov/publications/eib24/eib24.pdf (Accessed January 2009).

National Family Farms Coalition. "Food From Family Farms Act: A Proposal for the 2007 U.S. Farm Bill." http://www.nffc.net/Learn/Fact%20Sheets/FFFA2007.pdf (Accessed January 2009).

Stofferahn, Curtis. "Industrial Farming and Its Relationship to Community Well-Being: An Update of a 2000 Report by Linda Lobao." *State of North Dakota v. Crosslands.* (September 2006). http://www.nffc.net/Learn/Reports/IndustrializedFarmingonCommunity.pdf (Accessed January 2009).

Strange, Marty. *Family Farming: A New Economic Vision*, 2nd ed. Lincoln: University of Nebraska Press, 2008.

Richard B. Gifford
University of Arkansas Community College at Morrilton

FAMINE

Famine is the extreme scarcity of food, to such a degree as to result in widespread starvation, swallowing entire segments of the impoverished masses. Famine results from the immediate consequences of the lack of sustenance on a population, whereas hunger is a persistent, chronic, long-term, and slowly debilitating problem associated with insufficient food. Hunger is more widespread and problematic than famine, yet it receives considerably less public attention.

The popular image of famine portrayed by media and pushed by political elites and agribusiness is a person of color, wide-eyed, with a bloated belly, surrounded by swarming flies. This image implies the innocence of such victims, whose misfortune is being born on a continent—usually Africa, Asia, or South America—where people are not educated enough to produce food for themselves, where population growth surpasses ecological limits, and where political conflicts and/or environmental disasters tax the available food supply. The prevailing view depicts famine as a problem of agricultural production and overpopulation. This picture perpetuates the assumption that famine cannot be redressed and will only worsen with global population growth.

The basis for the dominant paradigm of famine is the thinking of Thomas Malthus, who famously argued in the early 19th century that population growth and ecological limits in combination produced famine, as population growth exerts constant pressure on food supply. Without empirical evidence, Malthus contended that population increased geometrically (2, 4, 8, …), whereas agricultural production increased only arithmetically (1, 2, 3, …). Although history has proven Malthus wrong time and time again, his thinking resurfaced in the 1970s, wrapped in a green facade, and has haunted environmental academics, activists, and policymakers ever since. Malthusian thinking perpetuates the myth that famines are natural and inevitable, allowing food aid to be used as a political weapon throughout the global South.

Despite the persistence of Malthus's problematic assertion, there are, in fact, several opposing perspectives that explain famine. Central to the differing views are opinions as to the relative weights of two factors relating to the causes of and solutions to famine: agricultural production and population growth. On the question of agricultural production, productionist viewpoints situate agricultural expansion as necessary to reduce incidents of famine, and nonproductionists argue that social factors such as inequality in food access and distribution are to blame. Population growth is thought to be the underlying problem by neo-Malthusians, whereas adherents to non-Malthusian views argue that population pressures are not causal factors of famine. On the basis of these two dichotomies, four general typologies of famine emerge:

1. Agricultural Expansionist (neo-Malthusian; productionist)

2. Free-Marketer (non-Malthusian; productionist)

3. Ecological Malthusian (neo-Malthusian; nonproductionist)

4. Political Economy/Political Ecology (non-Malthusian; nonproductionist)

Agricultural Expansionists view population control coupled with production increases as the path to addressing famine. Free-Marketers argue that free market expansion will increase agricultural production and reduce the incidence of famine. Ecological Malthusians link population growth with ecological degradation in general as the cause of famine. Finally, Political Economy and Political Ecology approaches understand famine as being rooted in social systems and do not view population pressures and agricultural production levels as causal factors.

Despite the variety of explanations for famine and the nagging persistence of Malthusian thinking, most greens (both inside and outside the academy) widely agree that famines are problems of food access and distribution. Widespread starvation results from inequality, not from ecological limits to food production—an argument that has been historically and empirically substantiated.

Historical records indicate that famines in colonial (and postcolonial) Africa, the great Victorian famines that swept the British colonies during the 19th century, the infamous Irish potato famine, and the famine that accompanied China's Great Leap Forward are all directly attributable to social causes. In each case, the underlying forces driving famine were complex, owing to their historically and geographically specific contexts. Consistent evidence demonstrates that changing political economic conditions undermined traditional coping mechanisms and that foodstuff surpluses were available elsewhere that could have thwarted famine (often within the same state). For example, it is widely known that Ireland remained a net exporter of food to Britain during the potato famine. We can conclude that famines today are not a problem of production, as ample global food supplies exist.

Michael Watts's 1983 tome *Silent Violence: Food, Famine and Peasantry in Northern Nigeria* forever changed academic discourse and understandings of famine. In his comprehensive study, Watts correlated colonial and postcolonial political economic transformations with Nigerian famine. This "political ecology of famine" situates assumed ecological limits to food production not in deterministic natural relations but as reflecting the failures of political economic institutions. Empirical studies such as Watts's demonstrate that famines are the result of structural vulnerability and that a full understanding of their causes is contingent on uncovering the social, political, and economic structures of the affected society within their historical and geographical context.

Although famines have been commonplace since the advent of agriculture, they now persist in the context of greater quantities of food production. Contemporary famine is attributable to commodified agriculture that binds poor households to global markets through new relations of production in the most violent of ways. Commodified agriculture does not produce food for human need; rather, it produces goods for the market. More precisely, commodified agriculture produces food (a use-value in political economic terms) for the purposes of producing a commodity that yields a market price, covering the cost of production plus a profit (exchange-value). As a commodity, agricultural products must be kept in high demand to be profitable; food must be scarce if it is to be used as a tool for capital accumulation—profit maximization trumps famine prevention.

As record levels of agricultural production have been achieved in both absolute and per capita terms, famines can only be explained as the result of inequality in social institutions. Within the current political and economic context, the paradox of want amid plenty is much more pronounced and systemic than in previous eras. Furthermore, current levels of food production are, for now, entirely dependent on petroleum. In the face of the materializing realities of global climate change and peak oil, the popular view of famine may, in fact, be darkly prophetic.

Potential solutions to famine correlate with the aforementioned assumptions underlying explanations of famine. The standard litany of suggestions includes international food aid, green revolution prescriptions, structural adjustment programs, genetically modified organisms, and market expansion. In fact, famine is often used as a cover to support these so-called remedies, even serving as ideological cover for colonialism and imperialism. Critics of these approaches charge that famine will only disappear as a daily threat facing the world's poor following expanded "capabilities," democracy, food sovereignty, and food justice.

See Also: Agrarian Question; Agribusiness; Food Security; Food Sovereignty; Green Revolution; Malthusianism.

Further Readings

Brown, Lester. *Tough Choices: Facing the Challenge of Food Scarcity.* New York: W. W. Norton, 1996.
Davis, Mike. *Late Victorian Holocausts: El Niño Famines and the Making of the Third World.* London: Verso, 2001.
Ehrlich, Paul. *The Population Bomb.* New York: Ballantine, 1968.
Lappé, Frances Moore, et al. *World Hunger: Twelve Myths*, 2nd ed. New York: Grove, 1998.
Magdoff, Fred, et al., eds. *Hungry for Profit: The Agribusiness Threat to Farmers, Food, and the Environment.* New York: Monthly Review, 2000.
Malthus, Thomas. *An Essay on the Principle of Population and a Summary View of the Principle of Population.* New York: Penguin, 1970.
Sen, Amartya. *Poverty and Famines: An Essay on Entitlement and Deprivation.* Oxford: Oxford University Press, 1983.
Watts, Michael. *Silent Violence: Food, Famine and Peasantry in Northern Nigeria.* Berkeley: University of California Press, 1983.

Evan Weissman
Syracuse University

Farm Bill

Farm bills are the primary legislative instruments of agricultural and food policy in the United States and typically have far-reaching effects across multiple areas of U.S. life and policy. When reference is made to "the farm bill," what is referred to is the most recently passed farm bill—the one that is currently applicable. A new farm bill is passed every few years, generally in expectation of the preset expiration of the previous one; in the past, farm bills have held for as little as one year and as long as 10 years.

The first farm bill was the Agricultural Act of 1933, a key piece of later New Deal legislation introducing several concepts important to U.S. agricultural policy ever since. The full list of farm bills follows:

1933: Agricultural Adjustment Act of 1933

1938: Agricultural Adjustment Act of 1938

1948: Agricultural Act of 1948

1949: Agricultural Act of 1949

1954: Agricultural Act of 1954

1956: Agricultural Act of 1956

1965: Food and Agricultural Act of 1965

1970: Agricultural Act of 1970

1973: Agricultural and Consumer Protection Act of 1973

1977: Food and Agriculture Act of 1977

1981: Agriculture and Food Act of 1981

1985: Food Security Act of 1985

1990: Food, Agriculture, Conservation, and Trade Act of 1990

1996: Federal Agriculture Improvement and Reform Act of 1996

2002: Farm Security and Rural Investment Act of 2002

2008: Food, Conservation, and Energy Act of 2008

The 1933 farm bill created the Agricultural Adjustment Administration, a New Deal agency that oversaw agricultural subsidies. Controversial, misunderstood, and perhaps misused since its introduction, the 1933 bill introduced the idea of paying farmers not to grow crops, leading to urban legends then and now about farmers growing rich on acres of empty land. The country at the time faced a crop surplus that kept prices dangerously low. The 1933 act raised the value of crops by enforcing scarcity, meaning farmers could afford to stay in business, which was not only of macroeconomic benefit but also intercepted the possibility of real scarcities down the line, should enough farmers find other means of providing for their families. Millions of head of livestock and acres of cotton were destroyed to create this scarcity.

Consumers were more concerned with the microeconomic effects: the sudden increase in prices for food and cotton clothes. Most Americans opposed agricultural subsidies on these grounds and because of the counterintuitive head-scratching notion of "paying someone not to work." In actuality, food prices did not go up very much—the principal effect was to prevent the decreases that had been forecast. Foreshadowing much of the agricultural legislation to come, large farms were the main beneficiaries of crop reduction subsidies because they could make substantial reductions while still having a significant amount of crop remaining. Smaller farms were more likely to go out of business, with their owners selling land to and taking jobs with these increasingly large farms, though this was certainly not solely caused by subsidies but rather was part of a general trend, exacerbated by the toll the Depression and the Dust Bowl took on small farms.

The original act was funded by a tax on agricultural goods, but it was ruled unconstitutional by the Supreme Court in the 1936 case *United States v. Butler*. The next farm bill, in 1938, funded subsidies from the general fund instead, and it went unchallenged. The 1938 act was essentially a refinement of the 1933 act, establishing ongoing research facilities to study farm needs, empowering the U.S. Department of Agriculture to fix market quotas and make available price supports for certain cash crops like tobacco and cotton, and integrating and renewing previous legislation. The 1948 farm bill modified these price supports somewhat.

Except in certain cases (such as the agricultural goods tax in the 1933 farm bill that was ruled unconstitutional), farm bills constitute permanent legislation. Subsequent farm bills supersede them in whole or in part, but they are binding in whatever cases in which they are not so superseded, or when more recent legislation reaches a set expiration date. The 1938 farm bill and the 1949 farm bill still constitute the bulk of the farm bill legislation in effect today in the 21st century. Although the 1930s farm bills were clearly legislation passed by a nation in financial crisis, the 1949 farm bill is an act passed by a nation coming into prosperity. It not only maintains the price controls and subsidies of previous legislation but provides a mechanism by which to donate excess food to overseas nations as development aid. The Commodity Credit Corporation, part of the U.S. Department of Agriculture, handles such donations. The subsequent 1954 farm bill modified and expanded the Commodity Credit Corporation's role in development aid, whereas the 1956 farm bill was concerned with acreage reserve for wheat, corn, rice, peanuts, cotton, and tobacco.

The 1962 and 1964 farm bills were both attempts to draft an emergency wheat program, authorizing such a program contingent on the approval of wheat producers; both times, producers rejected it. The 1965 farm bill focused on commodity programs for cotton, feed grains, and wheat, whereas the 1970 farm bill again paid producers to "set aside" crop acreage to avoid surpluses. The 1973 farm bill adjusted price supports and authorized a disaster relief program for dealing with natural disasters. The 1977 farm bill readjusted these adjustments and set ceilings on the payments for wheat, cotton, rice, and feed grain and also adjusted the food stamp program, bundled in with the 1977 Food Stamp Act. The 1981 bill lowered dairy price supports and eliminated marketing quotas for rice, setting new target prices for various commodities to try to make U.S. goods competitive in the foreign market.

The 1985 farm bill provided a five-year framework for its provisions, which included a conservation reserve program targeted at erosive croplands, as well as various price supports. Upon its expiration, the 1990 farm bill instituted another five-year framework, with various modifications to previous legislation. The 1996 farm bill simplified price supports and eliminated them for dairy, with a six-year framework.

The six-year framework of the subsequent 2002 farm bill included 10 titles (sections) and directed $16.5 billion toward agricultural subsidies. The bill, controversial as farm bills making changes to subsidies always are, was debated in the aftermath of the September 11, 2001, terrorist attacks, in an emotionally charged atmosphere, leading some—especially in areas where agriculture accounts for much of the economy—to worry that the more headline-friendly national security concerns would overshadow the arcane but critical ritual of constructing the farm bill. Furthermore, the subsidies called for by the farm bill could eat up budget surplus money that was needed to fund the invasion of Afghanistan.

The Senate version of the bill included subsidy caps—a maximum amount of subsidy that any farm could receive—to try to curtail the tendency of the bulk of subsidy money going to large agricultural corporations that did not need it as much the smaller farms, which in some years depended on subsidies to prevent layoffs. Other than Senator Blanche Lincoln (D-AR), the Democratic Senators supported caps; Republicans largely opposed them, though one Republican—Senator Charles Grassley of Iowa, himself a former farmworker—was among their most outspoken advocates. In the end, the final version of the bill as signed into law by President George W. Bush included a $360,000 subsidy cap ($110,000 more than Grassley called for). In addition to adjusting and reaffirming previous legislation, it greatly increased the amount of money spent on land conservation programs and, under Title IX, created new programs and new funding for the development of bio-energy sources.

When the 2002 farm bill expired, the 2008 bill was passed, a five-year, $288 billion program of agricultural subsidies and assorted other programs. The 2008 bill greatly increased the focus on the commercial development of biofuels, including grants, guaranteed loans, and tax credits in addition to annual discretionary funds specifically for biofuel-related expenditures. The bill also authorizes the Commodity Credit Corporation to purchase sugar from U.S. producers to sell it to bio-energy producers and makes more money available to rural renewable energy programs and a vague "Rural Energy Self-Sufficiency Initiative" with $5 million a year in discretionary funds.

Agricultural subsidies were largely unchanged, with minor modifications to peanut subsidies; the addition of dry peas, lentils, and chickpeas to those crops eligible for countercyclical payments (made whenever the effective price is less than the target price, according to the same calculations used in previous legislation); opt-in alternatives to the countercyclical payment program; minor changes to the calculation of base acreage of farms; a suspension of most subsidy payments to farms with a base acreage of 10 acres or less unless the farm is owned by a socially disadvantaged farmer; specific pilot programs in Illinois, Indiana, Iowa, Michigan, Minnesota, Ohio, and Wisconsin (for cucumbers, green peas, lima beans, pumpkins, snap beans, corn, and tomatoes for processing).

The 2008 farm bill also added provisions authorizing the use of funds to provide commodities in response to food crises or natural disasters, increased food stamp funding (and eliminated from consideration of eligibility any income received by a member of the Armed Forces deployed in combat, helping soldiers' families), gave priority for farm loans to those most in need (beginning farmers, socially disadvantaged farmers) and those who will use the loans to convert their farms to sustainable or organic farms, made national a 2002 pilot program making additional loan money available to beginning farmers and extended it to cover socially disadvantaged farmers, increased the borrowing limit of farm loans, made emergency loans available for horse ranchers, and created a New Farmer Individual Development Account pilot program to provide matching-funds savings accounts for beginning farmers in at least 15 states, operating through 2012.

Although all farm bills see significant debate within the United States, much of the criticism of the 2008 farm bill came internationally. Globalist organizations like the World Trade Organization criticized the U.S. farm subsidies, along with those of other prosperous nations, for their anticompetitive effect on international agricultural trade.

See Also: Agribusiness; Agricultural Commodity Programs; Department of Agriculture, U.S.; Family Farm; Farm Crisis; Food Quality Protection Act.

Further Readings

Gardner, Bruce. *American Agriculture in the Twentieth Century*. Cambridge, MA: Harvard University Press, 2002.

Halcrow, Harold, ed. *Food and Agricultural Policy*. New York: McGraw-Hill, 1994.

National Agricultural Law Center. "Index of United States Farm Bills." http://www
.nationalaglawcenter.org/farmbills (Accessed June 2009).
Pollan, Michael. *In Defense of Food*. New York: Penguin, 2009.
Pollan, Michael. *The Omnivore's Dilemma: A Natural History of Four Meals*. New York:
Penguin, 2007.

Bill Kte'pi
Independent Scholar

FARM CRISIS

A farm crisis is an economic event that affects farms and may be caused by various factors. There have been a number of farm crises in U.S. and world history. In fact, the observed relationship between droughts and economic calamities was one early explanation for the business cycle in the then-young science of economics, as various economists struggled to find predictable periodic phenomena such as weather patterns or sunspots that could be incorporated into models demonstrating an orderly system of agricultural and economic fluctuations. No such model was ever successfully constructed—there are too many interlocking factors, and although a low crop yield one year can affect sectors of the economy beyond the agricultural sector, a new tax law the following year can affect farmers' ability to bring crop to market, without needing the involvement of sunspots or any such external phenomena.

Though there have been economic calamities affecting farmers in all eras of U.S. history, and a growing divide between farmers and bankers in the 19th century that echoed the agrarian/industrialist, Republican/Federalist, Jefferson/Hamilton divide of the nation's infancy, the first such crisis to be called a farm crisis was that faced by farmers in the 1920s. Tellingly, it was as of the 1920 census that the United States officially became an urban nation, with more Americans living in cities than in rural areas. A generation earlier, the Census Bureau had declared an end of the frontier, since there was no area of the United States left unpopulated, outside of a few nature reserves or uninhabitable patches of desert. Farmers in 1920 represented 27 percent of the workforce; this would decline further to 21 percent in 1930. The world suffered from a global recession in the immediate aftermath of World War I, and although much of the United States recovered quickly and enjoyed the prosperity of what is remembered as the Roaring Twenties, U.S. farmers were barely able to hang on. It would be sorely tempting to blame Prohibition and the resulting decreased demand for grain—the corn, wheat, and rye used to make American whiskeys, the barley used for beer—but the abundant surpluses of crops affected far more than just grains, and the crisis began in 1919, before Prohibition went into effect.

Crop prices around the world fell in response to increases in supply and decreases in demand, and even as these surpluses became more and more problematic, improvements to farm technology made farms more and more efficient and more and more productive. In part, this was simply the result of the ongoing technological revolution, particularly as mechanical equipment became cheaper to manufacture and vehicles of all kinds became cheaper and better, which not only improved farm equipment like tractors but also improved and made cheaper the modes of transportation used to bring crops and livestock to market. In part it was the result of farmers attempting to make their operations more efficient—aiming for their crops cost them less per bushel, since they were selling for less per bushel—but this efficiency made farms even more productive and only worsened the surpluses, keeping prices down. When the stock market crash of 1929 brought about the Great Depression, farmers were already suffering, and only when the rest of the country joined them in that misery were serious remedies discussed. Much of President Franklin D. Roosevelt's New Deal legislation concerned U.S. farms, addressing issues that had been problematic for years before the Depression began and that had already driven many farmers out of business. The federal school lunch program, food stamp plan, and Rural Electrification Act improved rural communities and the lives of the working poor; other programs targeted farm-specific problems. The first two Farm Bills were passed and instituted agricultural subsidies, which protected farmers from the reduced prices caused by surpluses—an institution still in place three-quarters of a century later. Farmers recovered, alongside their fellow Americans, in the 1940s, when the combination of the New Deal reforms and the industrial buildup of World War II rejuvenated the U.S. economy.

The next farm crisis began in the 1970s, and many would argue that it never went away. It 1980, farmers constituted only 3.4 percent of the workforce—a figure that would continue to slowly decline. Although the 1970s were a generally positive time for rural areas, in part thanks to President Lyndon B. Johnson's war on rural poverty in the 1960s, farmers ended the decade with a growing amount of debt. Farm incomes and crop prices had been rising as a result of the General Agreement on Tariffs and Trade's reduced trade barriers and the record demand for U.S. grain exports to Russia, and a loosening of banking restrictions encouraged farmers to take advantage of low interest rates to buy new equipment or expand their land. Although some blamed the crisis on farmers who overborrowed, history textbooks sometimes neglect to mention the extent to which banks overlent—there are numerous cases in which farmers who applied for a loan of a certain amount were approved for a higher, unrequested amount and encouraged to expand further than they had planned. Although the choice was still the farmer's to make, turning down a loan when it is offered is a difficult thing for anyone to do, and most took the banks up on their offers. The banks, of course, expected to benefit from these higher loans, seeing them as investments in prosperous U.S. farms.

The boom could not last forever, and when it subsided, American farmers were in debt to an extent never before experienced. Farmland values dropped by more than half in much of the Midwest, America's heartland, during President Ronald Reagan's first term in office (1981–85); interest rates soared in that same period, and banks became less free with their farm loans. Farming household incomes, which had briefly risen above the national average in the 1970s, dropped to below the national average in the 1980s, where they had been for much of the century and where they remain today. Agricultural exports declined by a fifth, with crop prices dropping by about the same amount. The midsized farmers were the hardest hit—those who earned between $40,000 and $500,000 a year, bigger than a small family farm but smaller than the huge agribusinesses that began to buy up smaller farms as they went bankrupt. Rural communities nearly became ghost towns as the farms that had supported them were foreclosed and the banks that had held those farms' loans went into receivership. The Reagan administration's position was that it was the job of government to protect farms from natural disasters and disastrous price fluctuations, but not from changes in the marketplace or technology or the results of those changes. Most federal money distributed to agricultural enterprises went to large corporations, which collected billions of dollars in subsidy money while smaller farms went out of business. An attempt to privatize the Farm Credit System, which provided special low-interest loans to U.S. farms, died in committee. Protesters in the mid-1980s demanded legal aid and debt relief for farms and a moratorium on foreclosures until the crisis had passed; referring to the ongoing Iran-Contra Affair, protesters who surrounded the Federal Housing Administration office in Chillicothe, Missouri, with tractors chanted, "Farm aid, not Contra aid."

One of the eventual remedies for this crisis was the diversification of crops, which was to some extent reflected in subsequent farm bills (which provide crop-specific subsidies, thus demotivating to some degree any reliance on crops not listed in the bill). Throughout the 1990s, farm incomes and the number of farms decreased globally, but the growing reliance on huge agribusinesses helped to cloak public awareness of what constituted either a new farm crisis or a continuation of the crisis that had begun in the 1970s. Those trends continue today and inform the ongoing debates about agricultural subsidies—most of which still go to the largest agricultural corporations, representing the smallest number of farms—and the current negotiating round of the World Trade Organization.

See Also: Agribusiness; Doha Round, World Trade Organization; Family Farm; Farm Bill.

Further Readings

Davidson, Osha Gray. *Broken Heartland: The Rise of America's Rural Ghetto*. Iowa City: University of Iowa Press, 1996.

Greider, William. "The Last Farm Crisis." *The Nation* (2 November 2000).

Huntley, Steve. "Winter of Despair Hits the Farm Belt." *U.S. News & World Report* (20 January 1986).

McBride, Bob. "Broken Heartland: Farm Crisis in the Midwest." *The Nation* (8 February 1986).

Bill Kte'pi
Independent Scholar

FARMERS MARKET

Farmers markets have existed in the United States since the earliest period of urban development. Farmers markets were already well established in Boston, New York, and Baltimore by the 1650s. Neighborhood markets continued to flourish through World War II but declined precipitously in the late 20th century in the face of the growth of national chain supermarkets.

Farmers markets are seeing a resurgence in the early 21st century. The number of markets recognized by the U.S. Department of Agriculture increased from 3,100 in 2002 to 4,685 in 2006. Over three million Americans spend about $1 billion at farmers markets annually. These markets can range from small seasonal roadside stands to large year-round central market areas that often serve as major tourist destinations. Examples of these include Pike Place in Seattle or the Reading Terminal Market in Philadelphia. Others, such as Cleveland's Westside Market and Detroit's Eastern Market (the largest historic public market in the United States), continue to primarily serve their historic function as local food suppliers—although they also draw customers from a broader region.

Farmers markets address what the Farmer's Market Coalition calls the "triple bottom line." They ensure a fair market price for small growers; improve access to fresh, healthy produce for consumers; and provide a community-building function by providing a modern-day "town square." The renewed support for farmers markets in the United States is driven by several trends including growing concern for health and the environment, decreasing confidence in commercial food production and delivery systems, and economic considerations.

For those moved primarily by environmental concerns, farmers markets offer fresh, locally grown produce, reducing the environmental impact of transportation and packaging, while simultaneously contributing substantially to local economies. Commercially produced foods travel an average of 1,500 miles from producer to plate. Farmers markets, in contrast, typically serve consumers within 100 miles of the producers they support. They are also often seen as good sources for organic or naturally produced vegetables and fruits, in part because it can be easier for local organic producers to gain access to farmers markets than to major distribution networks. As a result, farmers markets serve as outlets for a variety of non-

Food at farmers markets is often produced within 100 miles of the market, resulting in much lower fuel use for transportation in comparison with commercial foods, which travel an average of 1,500 miles. This market is in Montreal, Canada.

Source: Photos.com

conventional local producers such as urban gardeners/farmers, home bakers, and cooperatives. These producers are presumed to have a greater sense of responsibility to the local environment because they live in the area and depend directly on the health of the local ecosystem for their livelihood. Reduced exposure to pesticides and other agricultural inputs is also a motivating factor for increasingly health-conscious consumers. Farmers markets may also provide greater diversity in produce and other goods than conventional grocery stores, which can appeal to consumers driven by both environmental and health considerations.

Finally, some believe that the smaller producers who frequently distribute their products at farmers markets are less apt to rely on genetically modified plant species. However, given the rising costs of agricultural chemicals, species genetically modified for pesticide and herbicide resistance may be more attractive to small producers than was previously common. Some buyers are reassured by the ability to have face-to-face discussions with producers at the farmers market about their products and production methods. The sense of assurance resulting from this direct contact with producers is shared by those who have growing concerns about the commercial food distribution system. In the wake of several national outbreaks of food-borne disease and incidents of contaminated food products, knowing where one's food comes from and who produced it is seen as one defense against the anonymity and lack of direct accountability fostered by industrial agricultural production.

Although they may greatly reduce overhead and increase direct contact between farm/farmer and consumer, their association with health and environmentally conscious consumers can, ironically, create a sense of elitism around farmers markets. They are sometimes seen as inaccessible or unaffordable to lower-income consumers, who are also more apt to live in "food deserts" and other areas in which fresh, wholesome foods are less accessible overall. As a result, these consumers may rely more heavily on processed foods and have limited intake of fresh fruits and vegetables. The U.S. Department of Agriculture has developed two programs specifically to promote accessibility and use of farmers markets among groups especially likely to be affected by these concerns.

Participants in the Supplemental Nutrition Program for Women, Infants and Children (WIC) are eligible for the WIC Farmers Market Nutrition Program (FMNP). WIC FMNP was designed to provide food vouchers to increase intake of fresh, nonprocessed fruits and vegetables among lower-income families while improving accessibility and viability of local farmers markets across the country. The program receives $19.86 million annually, and over 2.3 million individuals were served by the WIC FMNP in 2007, for an average of $8.60 per person per year. Perhaps in part as a result of the financial support provided by WIC FMNP, 75 percent of consumers spending $2,550 or more at Detroit's Eastern Market in 2007 reported annual household incomes below $60,000. One recent study demonstrated that although WIC FMNP participants did not enjoy greater food security (defined by the study as the "ready availability of nutritionally adequate and safe foods for all people, at all times, for an active, healthful life") than other WIC participants, they did report both more healthy, balanced diets and greater awareness of the health benefits of consuming fresh produce.

Similarly, the Seniors Farmers Market Nutrition Program provides vouchers to lower-income seniors to increase access to and consumption of fresh, unprocessed fruits and vegetables to improve health and nutrition among older Americans. Although this program receives $15 million annually, in 2007 it served only 825,691 individuals, for an average of $18.67 per individual.

Although these programs are not sufficient to address the dietary gap among U.S. consumers, they reflect a federal recognition of the relationship between local production, consumption, and health represented by farmers markets.

See Also: Agribusiness; Community Gardens; Department of Agriculture, U.S.; Food Safety; Food Security; Locavore; Slow Food Movement; Urban Agriculture.

Further Readings

Farmers Market Coalition. "Purpose." http://www.farmersmarketcoalition.org/joinus/purpose (Accessed March 2009).

Kropf, Mary L., et al. "Food Security Status and Produce Intake and Behaviors of Special Supplemental Nutrition Program for Women, Infants, and Children and Farmers' Market Nutrition Program Participants." *Journal of the American Dietetic Association*, 107/11 (2007).

National Farmers Market Coalition. http://www.farmersmarketcoalition.org (Accessed March 2009).

Seniors Farmers Market Nutrition Program. http://www.fns.usda.gov/wic/SeniorFMNP/SFMNPmenu.htm (Accessed March 2009).

U.S. Department of Agriculture. "Wholesale and Farmers Markets." http://www.ams.usda.gov (Accessed March 2009).

WIC Farmers Market Nutrition Program. http://www.fns.usda.gov/wic/FMNP/FMNPfaqs.htm (Accessed March 2009).

Kerry E. Vachta
Wayne State University

FAST FOOD

Fast food is a meal that is prepared and served quickly. Fast food restaurants typically have a limited menu, items prepared in advance or heated rapidly, no table orders, and food served in disposable wrapping or containers. Although many cultures in highly populated areas have developed some form of fast food, the U.S. model has had the most influence worldwide. The rapid growth of the fast food industry since the 1960s has contributed to important changes in food production and consumption, such as more intensive animal agriculture and diets that are high in fat and sugar. Criticism of these trends has increased in recent years, as movements have coalesced to challenge fast food corporations politically or challenge the cultural values they promote.

The history of fast food restaurants in the United States in the first half of the 20th century included New York automats, which served take-out food in vending machines, and the hamburger restaurant chain White Castle, established in Wichita, Kansas. Hamburger chains that were founded later in Southern California, however, such as McDonald's, Jack in the Box, and Carl's Jr., have had a much greater influence. These evolved from drive-in restaurants that were popular in the early 1940s. In 1948, Richard and Maurice McDonald applied the principles of a factory assembly line to their restaurant and eliminated drive-in service to dramatically reduce labor costs. Their techniques spawned numerous imitators extending far beyond California, including the Burger King chain in Florida. In 1961, the McDonald brothers sold their business and name to Ray Kroc, who refined their methods of breaking down every task to make them more efficient and achieve consistent quality. Kroc successfully expanded the business to more than 7,500 restaurants worldwide by the time he died in 1984.

Although the top three U.S. fast food or "quick service" restaurants by sales are hamburger chains, other popular formats include fried chicken, pizza, sandwiches, and Mexican food. The majority are operated as franchises. Under this arrangement, a businessperson is primarily responsible for financing and operating a restaurant, but an initial fee and a continuing percentage of the sales must be paid to the parent company. McDonald's Corporation, the largest franchise in the world, makes most of its money not through these fees but through leasing land that it owns to its franchisees. Other fast food chains, such as Subway, rely on requiring the payment of a much higher percentage of sales.

The fast food industry in the United States has benefited from the development of an interstate highway system and the rise of automobile culture. Many outlets are located near freeway off-ramps, and the introduction of the drive-up window by Wendy's in 1972 made access even more convenient for drivers. In 1970, spending on fast food in the United States was $6 billion, but by 2000 that amount rose to $110 billion. On a given day, one in four Americans will consume fast food.

Fast food can be viewed as a symbol of globalization, as U.S.-based chains and their imitators are rapidly expanding in other parts of the world. There are currently more than 30,000 McDonald's outlets in over 100 countries, making it the largest global food retailer. Menus may be customized for different cultures, such as McDonald's Teriyaki Macs in Japan and Pizza Hut's Masala Pizzas in India.

More Fast Food, Fewer Owners

The fast food industry has experienced numerous mergers and acquisitions, resulting in a smaller number of corporations dominating the market. Wendy's and Arby's merged in 2008, for example, and the company currently holds a majority stake in Tim Hortons. Yum! Brands, which was once owned by PepsiCo, controls Taco Bell, KFC, Pizza Hut, Long John Silver's, and A&W. McDonald's once held a majority stake in Chipotle, Boston Market, and Donato's Pizza but had divested its ownership in these firms by 2007.

The growth of the fast food industry could be viewed as having a number of positive effects. Fast food is convenient and easily accessible, particularly for residents of urban areas or those with access to automobiles. Fast food also tends to be inexpensive. Prices continue to decline as more chains are offering selected menu items for approximately $1 or providing a discount for ordering a combination of items called a "value meal."

Critics have argued that the low price of fast food does not reflect its true cost. The sociologist George Ritzer sees the fast food chain McDonald's as embodying societal trends toward rationalization, including efficiency, calculability, predictability, and control. He suggests that these processes, which are replacing more traditional ways of thinking, have many negative consequences for society, such as denying our basic humanity and homogenizing cultures.

Rising obesity rates have led to criticism of the nutritional composition of fast food. These meals tend to be high in saturated fats, trans fats, and high-fructose corn syrup, all of which have been associated with negative health outcomes such as diabetes and heart disease. In addition, portion size and the average number of calories in a meal have increased over time. Fast food companies have found "supersizing," or suggesting a larger amount of food for a small additional cost, to be a profitable strategy.

The fast food industry has been accused of exploiting vulnerable populations. Their marketing efforts make use of toy giveaways, mascots, and television advertising aimed at young children. Several studies have suggested that fast food outlets are more likely to be located in low-income and minority neighborhoods and near schools. A survey of California school districts in 1999 found that more than half sold Taco Bell, Subway, Domino's, Pizza Hut, or other brands of fast food within their schools.

More Fast Food, More Meat

Fast food's popularity has helped to increase the global demand for meat, which has a number of ecological and health impacts. The number of meat suppliers has decreased, even as the scale of their operations has increased. Animals are raised more intensively and confined indoors for much or all of their lives. These conditions have been described as inhumane by animal advocacy organizations. More energy, land, and water resources are required to produce the same number of calories in animal-based foods when compared with plant-based foods. Pollution from the highly concentrated production of animal waste is another concern. In addition, the use of antibiotics has increased to keep animals healthy under such crowded conditions, contributing to antibiotic-resistant diseases. Large-scale epidemics of food-borne disease have been associated with fast food consumption, such as hamburgers contaminated with *Escherichia coli* O157:H7.

Fast food restaurants and their suppliers both have been accused of exploiting workers. Jobs in fast food and meatpacking are among the lowest paid of any occupation, with extremely high turnover rates. Companies in these industries are also notoriously anti-union, sometimes shutting down operations that have experienced successful union organizing efforts.

Many of these criticisms were summarized by activists in London who distributed a pamphlet titled "What's Wrong with McDonald's." Two of these individuals, Helen Steel and David Morris, refused to settle a libel accusation from the corporation in 1990. Their trial began in 1994 and lasted three years, making it the longest-running case in the United Kingdom. Although Steel and Morris eventually lost, it was a public relations disaster for McDonald's. In his ruling, the judge agreed with a number of points in the pamphlet, and the corporation was exposed as having employed private detectives to spy on activists.

Another well-publicized action to oppose fast food was the dismantling of a partially built McDonald's restaurant in Millau, France, by farmers, including prominent antiglobalization activist José Bové, in 1999. A less-confrontational approach has been taken by Slow Food, which was founded to oppose a McDonald's in Rome, Italy. The organization now has chapters all over the world and works to preserve ecological and cultural diversity in food. Critiques of fast food in the United States have included Eric Schlosser's best-selling book *Fast Food Nation* and the documentary *Super Size Me.*

Some efforts to change practices at fast food restaurants have been successful. McDonald's removed Styrofoam packaging that was produced using toxic and ozone-depleting chemicals in 1990, although only in the United States. Targeted boycotts led to agreements with Taco Bell (Yum! Brands), McDonald's, and Burger King to increase the prices they pay for tomatoes to improve the wages and working conditions of farmworkers. Other campaigns resulted in chains requiring suppliers to treat animals more humanely. Some localities have introduced stricter regulation of fast food. This has included requirements to label nutritional content and to remove trans fats. Moratoriums on new fast food restaurants have even been enacted in Los Angeles and Berkeley, California.

See Also: Food Safety; Retail Sector; Slow Food Movement.

Further Readings

Collins, Tracey Brown, ed. *Fast Food*. San Diego, CA: Greenhaven, 2005.
Quick Service Restaurant Magazine. http://www.qsrmagazine.com (Accessed January 2009).

Schlosser, Eric. *Fast Food Nation: The Dark Side of the All-American Meal.* New York: Houghton Mifflin, 2001.

Philip H. Howard
Michigan State University

FEDERAL INSECTICIDE, FUNGICIDE, AND RODENTICIDE ACT

The Federal Insecticide Fungicide and Rodenticide Act (FIFRA) allows the U.S. Environmental Protection Agency (EPA) to regulate the manufacture, sale, and application of pesticides to protect human health and the environment. FIFRA requires registration of pesticides, specifies the label to be placed on pesticides, provides for postregistration review of pesticide registrations and possible suspension or cancellation of those registrations, imposes restrictions on pesticide containers and specifies containment structures for stored pesticides, establishes standards to protect workers, and provides for enforcement of regulatory requirements through civil and criminal penalties.

FIFRA does not generally preempt state pesticide laws: States may place more restrictive requirements on pesticides than the EPA, although states may not require different labels.

Registration and Labeling Requirements

FIFRA requires pesticide manufacturers to register a pesticide with the EPA before selling that pesticide in the United States. Pesticide registrations specify the crops, sites, timing, amount, and conditions of application for the pesticide. Manufacturers must submit extensive environmental, health, and safety testing data showing whether the pesticide has potential adverse effects on humans, wildlife, fish, or plants and whether the pesticide may contaminate surface or ground water through leaching, runoff, and spray drift. Pesticides must be registered both by the EPA and the state before distribution.

The EPA also must approve the language that appears on each pesticide label. A pesticide product must be used according to the directions on the label. Because the EPA allows the registration of pesticides based on the assumption that those who use pesticides follow the label instructions, EPA carefully specifies the label language, and "the label is the law" for those who store, apply, or dispose of pesticides.

To register a pesticide, EPA must find that:

- the pesticide's composition warrants the proposed claims for it regarding efficacy;
- the pesticide's label complies with statutory labeling requirements;
- the pesticide will perform its intended function without unreasonable adverse effects on the environment; and
- the pesticide will not generally cause unreasonable adverse effects on the environment when used in accordance with widespread and commonly recognized practice.

FIFRA defines an "unreasonable adverse effect on the environment" as any unreasonable risk to man or the environment, considering the economic, social, and environmental costs and benefits of the use of the pesticide. This amounts to a cost-benefit analysis of whether the product's benefits outweigh its costs, even if those costs include severe impacts on human health or the environment. For example, if a pesticide has severe neurological or reproductive effects on agricultural workers applying the pesticide, the EPA may nonetheless register it if these costs are outweighed by the value of the pesticide in protecting crops.

FIFRA also considers human dietary risk from pesticide residues in or on any food to be an "unreasonable adverse effect on the environment" if the residue exceeds tolerances set under the Federal Food, Drug, and Cosmetic Act. Before the EPA registers a pesticide used on raw agricultural products, it establishes a tolerance, which is the maximum amount of pesticide residue that can remain on food. The EPA must find that the pesticide residue poses a "reasonable certainty of no harm." The EPA considers (1) aggregate, non-occupational exposure (from diet, drinking water, and home use); (2) cumulative effects from exposure to different pesticides that produce similar effects in the human body; (3) whether there is increased susceptibility to infants and children or other sensitive sub-populations from exposure; and (4) whether the pesticide produces effects similar to the effects of naturally occurring estrogen or produces other endocrine-disruption effects.

Postregistration Risk Reviews

In 2009, the EPA was completing a one-time program to review older pesticides initially registered before November 1984 to determine whether the pesticides meet current scientific and regulatory standards. This re-registration process considers the human health and ecological effects of pesticides and may result in labeling and other changes to reduce risks from older pesticides.

The EPA has also created an ongoing registration review program, which reevaluates the active ingredients in registered pesticides on a regular basis. The goal is to review each active ingredient every 15 years to ensure that the agency's judgment about risks to human health and the environment from the pesticide reflect evolving science, policies, and practices.

When the EPA discovers that use of a previously registered pesticide may result in unreasonable adverse effects on human health or the environment, it may initiate the pesticide special review process. The special review process generally involves an intensive review of one or just a few potential risks from a pesticide. The review includes evaluating existing data, gathering new data, reassessing the risk, and deciding whether additional measures are necessary to reduce the risk.

Suspension and Cancellation of Registrations

The EPA may suspend or cancel the registration of a pesticide that has been registered. Cancellation is a lengthy proceeding, so the EPA typically seeks to negotiate the terms of a voluntary cancellation with the pesticide manufacturer. In the event that a pesticide poses an imminent hazard, the EPA may immediately suspend a registration pending cancellation. However, FIFRA provides for indemnification of end users, distributors, and dealers of pesticides against losses occurring as a result of suspension or cancellation. As a result, the EPA typically allows existing stocks of pesticides to be used under special conditions, rather than immediately precluding the use of a pesticide altogether.

Pesticide Container and Containment Rule

Under this rule, the EPA regulates both refillable and nonrefillable pesticide containers, as well as the repackaging of pesticides into refillable containers by pesticide manufacturers, formulators, distributors, and dealers who refill pesticide containers. The EPA also prescribes standards for secondary containment structures for pesticide dealers, commercial applicators, and custom blenders operating bulk storage sites or pesticide dispensing operations. Various requirements of the 2008 rule become effective between 2009 and 2011.

The EPA also utilizes FIFRA to regulate some products of genetic engineering. For a class of genetically engineered crops that produce their own pesticide, called plant pesticides or plant incorporated protectants, the EPA requires the same registration process as for pesticides. Commonly these are crops like those that incorporate the genes from *Bacillus thuringiensis*, or Bt. Also, since pesticide use can affect its own long term efficacy, as overuse may lead to pesticide resistance in the target organism, the EPA has required Bt crops to follow a strict resistance management plans (in this case a high-dose/refuge model) to ensure that the Lepidopteran insects that are targeted by Bt do not evolve to become resistant to the plant pesticide.

Worker Protection Standard

The EPA has also issued a Worker Protection Standard for Agricultural Pesticides, a regulation aimed at reducing the risk of pesticide poisonings and injuries among agricultural workers and pesticide handlers. The standard offers protections to approximately 2.5 million agricultural workers and pesticide handlers who work at over 600,000 farms, forests, nurseries, and greenhouses from occupational exposure to agricultural pesticides. The regulation covers two types of employees: (1) pesticide handlers—those who mix, load, or apply agricultural pesticides; clean or repair pesticide application equipment; or assist with the application of pesticides in any way; and (2) agricultural workers—those who perform tasks related to the cultivation and harvesting of plants on farms or in greenhouses, nurseries, or forests.

The Worker Protection Standard for Agricultural Pesticides contains requirements to improve worker safety, including pesticide safety training and provision of pesticide label requirements to workers and handlers; prohibitions against applying pesticides in a manner that will expose workers, exclusion of workers from areas while pesticides are being applied or during the specified period of restricted entry, and notice to workers about treated areas to avoid inadvertent exposures; and requirements concerning personal protective equipment, decontamination, and emergency procedures.

Enforcement

Individuals storing, applying, and disposing of pesticides must do so in accordance with both federal and state law. In general, states have primary authority for monitoring compliance with pesticide regulations and for taking enforcement action against those using pesticides in violation of the labeling requirements.

The EPA and authorized state officials may inspect pesticide products where they are stored for distribution or sale. The EPA may issue orders to stop sales and to seize supplies of pesticide products. Persons violating FIFRA are subject to substantial civil penalties, as well as criminal prosecution.

See Also: DDT; Food Quality Protection Act; Food Safety; Pesticide.

Further Readings

Container and Containment Rule. 40 CFR Part 165.
Federal Insecticide Fungicide and Rodenticide Act. 7 U.S.C. 136 et seq.
U.S. Environmental Protection Agency. "Pesticide." http://www.epa.gov/pesticides/ (Accessed January 2009).
Worker Protection Standard. 40 CFR Part 156.

Susan L. Smith
Willamette University College of Law

FERTILIZER

Since medieval times, fertilizers have been used to help increase crop production. The earlier fertilizers were organic and consisted of decayed plant or animal matter, as well as various other items, whereas inorganic fertilizers consist largely of chemicals. Essentially, when crops are produced through the use of organic fertilizers, they are deemed to be "green," whereas when inorganic fertilizers are used, there are certainly queries about whether the crops grown through their use can be regarded as "green."

In ancient and early medieval times, the system of crop rotation was found to be very good in helping with crop yields. Part of this involved leaving the fields to lie fallow for a year in the process, and often being used for grazing animals. The resulting animal dung contributed to helping fields recover from intensive cultivation, and the Koreans introduced similar systems that involved fields being fallow for many years after the growing of ginseng.

Technically, the use of "slash and burn" (or swidden) agricultural techniques, still practiced by some people such as jungle tribes, used carbon to fertilize the soil, but as they found out, the nutrients thus provided were often rapidly used up, causing them to move to a new location. Gradually, during the late medieval and early modern period in Europe, it was determined that some forms of natural fertilizers

Modern methods of fertilizing sometimes make use of computer technology and other instruments. This sprayer features crop canopy sensors on either side to monitor the greenness of plants and an onboard computer that adjusts the application of fertilizer accordingly.

Source: U.S. Department of Agriculture, Agricultural Research Service/James Schepers

could be used. The Englishman Gabriel Potts, in the 17th century, worked out that "colored water" from flooded land, and soil from streams, as well as malt dust, blood offal, and the entrails of animals helped plant nutrition. This was followed up by Walter Blith in the 1640s, who started using farmyard manure, human urine, bones, horn, fish, seaweed, and later even coarse wool clippings and soot—the latter essentially replicating some aspects of "slash and burn" agriculture. Blith also experimented with the use of chalk and lime, depending on the soil where the crops were being grown. By experimentation, agriculturalists learned that particular natural fertilizers were more suitable to specific crops and, indeed, specific species of vegetables.

During the 1730s, Viscount Charles Townshend (1674–1738) studied methods of improving crop production by having a four-year rotation system instead of the three-year system that had been used in Britain for hundreds of years. More than 100 years later, the German chemist Justus von Liebig (1803–83) recognized the importance of ammonia in helping plants grow, and he started promoting the manufacture and use of inorganic minerals involving ammonium salts to help plant nutrition and increase crop yields. He also came up with a method of using bone meal after it was treated with sulfuric acid, but it was not successful. Von Liebig's work went on to inspire the French scientist Jean Baptiste Boussingault (1802–87), who, when he returned to France from Latin America in 1832, worked on the introduction of nitrogen fixation to help in the production of legumes.

The importance of phosphates in increasing crop production was shown by the work of the English agriculturalist Sir John Bennet Lawes (1814–1900), who experimented with their use on his farm at Harpenden, north of London. This led to the British metallurgists Percy Gilchrist (1851–1935) and Sidney Gilchrist Thomas (1850–85) developing a process of using phosphates. However, the major advances were made by the Germans Fritz Haber and Carl Bosch, who won the Nobel Prize in Chemistry in 1918 and 1931, respectively, by which time phosphates and guano were being sourced from various Pacific Islands such as Nauru.

It can be argued that the initial use of chemical phosphates from Nauru on crops elsewhere in the world could still lead to the increased crop production being "green," as it involved scattering powdered phosphorus-rich soil from one part of the world on another, in much the same manner that ground bones had been used since at least early modern times. However, it was not long before the British company Imperial Chemical Industries started producing synthetic sulfates, and they and other companies started working on the development of inorganic fertilizers.

Essentially, the main objection to the use of synthetic fertilizers is that they interfere with the soil. Although the fertilizers are developed to enrich the soil, heavy overuse can lead to an exhaustion of soil, as has happened in some parts of the world such as in North Korea, which made very heavy use of artificial fertilizers in the 1960s, 1970s, and early 1980s. The other problem is that the crops grown as a result of this use of fertilizer can be seen as having been contaminated with potentially disastrous effects in the years to come. Gradually, over many years, there can be an increased concentration of trace elements in the fertilizers that can lead to a buildup of metals or salts, which can, in turn, affect the crops being grown in fields treated with fertilizer.

Just as important, the use of chemical fertilizers has often led to a buildup of nitrates that can leach into groundwater and cause other environmental problems. Some of these, involving the gradual buildup of trace elements, can escape scientific notice for many years, but the problem is still all too real. In addition, there have been serious queries about the environmental problems in Nauru and in northern Chile—the source of the phosphates.

For a food to be deemed "green," it is possible to use fertilizers, but they must be natural, and naturally occurring. This has been done for centuries. In addition, the fertilizers must not have any detrimental effects on the rest of the environment, both where the fertilizer is used and also from where it has been sourced.

See Also: Composting; Crop Rotation; Food Safety; Nitrogen Fixation; Roundup Ready Crops; Swidden Agriculture.

Further Readings

Addy, John. *The Agrarian Revolution*. London: Longmans, 1972.

Bhattacharya, P. *Dictionary of Biofertilisers and Organic Fertilisers*. New Delhi: FDCO, 2002.

The Complete Technology Book on Bio-Fertilizer and Organic Farming. New Delhi: Natural Institute for Industrial Research, 2007.

Cooper, Mark and John Davis. *The Irish Fertiliser Industry: A History*. Dublin: Irish Academic, 2004.

Justin Corfield
Geelong Grammar School, Australia

FISHERIES

The term *fisheries* refers to the sum total of activities involved in the capture, harvesting, processing, trade, and consumption of freshwater and saltwater fish and fish products. These activities contribute to food security and poverty alleviation by providing animal protein, fatty acids, and other micronutrients for billions of humans and nonhumans; giving support and employment to 250 million people, including many small-scale fishing communities; and acting as a source of export earnings for many countries, particularly in the developing world. The greatest dependence on fish for dietary purposes is found in the least developed countries, and in 40 such countries, fish is the third-largest export industry. Fish are also consumed for subsistence purposes and traded locally. Over three billion people, mainly in developing countries, obtain 20 percent or more of their animal protein intake from fish and fish products, and over 400 million people in parts of Africa and South Asia derived more than 50 percent of their protein intake from fish. Despite the increase in global production of the past 50 years, many developing countries have experienced a decline in per capita availability and consumption as a result of population growth, reduction in access to common pool resources, growing domestic income inequality, and the increased cost of fish, making it less affordable for the poor. Efforts are being made to prevent further stock declines, promote more sustainable fisheries policies, and improve the access of poor people to fish.

Fishing activity takes many forms, and fishing communities are highly varied. Conventional classifications recognize three main fisher categories: recreational, large-scale/industrial, and

Small-scale fishers in Asia like this man fishing alone in a small boat in Indonesia land close to half of the world's total open-capture marine and inland fish catch.

Source: World Bank

small-scale/artisanal fishers. Recreational fishers are noncommercial fishers who do not fish for a livelihood and who account for 12 percent of total world catches. They are found throughout the world, with the largest numbers found in developed countries. Large-scale and small-scale fishers generally fish for economic gain, to meet household dietary needs, or some combination of the two. The distinction between large-scale and small-scale fishers is best seen as a continuum ranging from single-line individual fishers operating in limited ranges along coasts and riverbanks and fishing largely for local consumption to highly capital-intensive commercial factory ships, which sweep large areas of the ocean using advanced sonar technology and industrial nets and process and freeze catch onboard. Over 90 percent of the estimated 29 million capture fish workers worldwide are officially classified as small-scale fishers, of which more than 80 percent are in Asia, accounting for 45 percent of total open capture marine and inland fish catch. Both large-scale and small-scale fishers include boat and net owners and investors, owner-operators, seasonal workers, share-catch workers, and wage workers. Although large-scale fisheries directly employ fewer people and are concentrated in fewer countries, their catches per unit of effort and their aquatic environmental impacts are larger than those of small-scale fishers, although the latter do cause environmental damage.

Developing countries contain the largest concentrations of small-scale fisher communities and account for over one-fifth of global fish exports destined mainly for developed countries. Developed country markets are also supplied by foreign large-scale industrial fisheries who fish legally and illegally within developing-country exclusive economic zones. Over the past 50 years, there has been a major expansion of global fisheries and aquaculture production. In 2006 production reached 143,647,650 metric tons made up of 91,994,321 metric tons (64 percent) from capture fisheries and 51,653,329 metric tons (36 percent) from aquaculture. Official sources estimate that there are over 40 million full- or part-time fishers and fish farmers in the world and several hundred million others working in the wider industry. These are underestimates of the true total, as a high proportion of the global fish catch goes unrecorded because of illegal, unreported, and unregulated fishing. The expansion of global production has been at the expense of many of the world's freshwater and saltwater fishing grounds, which were once abundant but are now increasingly overfished and depleted. Thus, in 2005, the United Nations Food and Agriculture Organization estimated that 77 percent of global fish stocks were fully exploited, overexploited, depleted, or recovering from depletion. Overfishing, depletion, and destruction of fish stocks is not a new phenomenon, but over the past 150 years, there has been an extensive—if uneven—global decline in fish stocks caused by a combination of the expansion of commercial fishing effort, the growth of fishing communities dependent on fishing for a living, terrestrial and aquatic habitat degradation, and an increasing global demand for fish and fish products. This decline, considered by some as giving rise to a world fisheries crisis, threatens food security, employment, community survival, and the wider environment. There have been a variety of responses to the crisis from the fisheries industry, national and international policymakers, nongovernmental organizations, local fishing communities, and consumers; these include improved management of existing fish stocks, restoration of aquatic ecologies, promotion of fisheries comanagement schemes, boycotts of the commercial sale of threatened species, creation of marine fisheries councils, strategies to enhance people's capacity to access fish for dietary protein, and support for sustainable livelihood programs for fishing peoples.

Declining Fish Stocks

Much of the debate in academic and policy circles has focused on declining fish stocks—how extensive such declines are and how to deal with them. Some marine scientists argue that fishing down the food chain is undermining ocean ecologies, which will result in a collapse in fish catches by the mid-21st century. Others question this view, arguing that declining abundance of fish does not mean declining maximum sustainable yield if proper management policies are in place. However, there is agreement that many fishing grounds are "overfished," by which is meant a situation in which fishing results in a decline in fish stocks below acceptable levels, even to the point of threatening their existence. There are several types of overfishing, including demographic overfishing, growth overfishing, recruitment overfishing, and ecosystem overfishing. In addition, habitat destruction, pollution, and introduced species—all of which overlap with overfishing—contribute to declining fish stocks.

The result of overfishing and related causes is that global capture fisheries production has at best leveled off, and a range of measures are required to halt stock declines. Ecosystem management has become increasingly popular as a means of protecting and renewing fish stocks, as it focuses on the complex relationships among sustainable yields, reduction in environmental degradation, and the protection of biodiversity, rather than simply on the protection of a single species. Ecosystem overfishing is identified by reductions in ichthyodiversity, declines in aggregate production of exploitable resources, decline in mean trophic level (feeding position), increased bycatch, and human-induced habitat modification. Other measures to protect fish stocks include a reduction in the industrial trawling fleet size, fewer financial subsidies and other incentives to such fleets, the creation of Marine Protected Areas as sanctuaries, better-enforced and more-binding global agreements relating to conservation and sustainability of stocks within and across national boundaries, the establishment of Marine Stewardship Councils, ecological rejuvenation of fishing grounds, controls on illegal fishing, a reduction in the number of fishers, and, with growing concerns about climate change, improved local and national adaptive capacity to deal with its effects. Aquaculture is also promoted as both an alternative and a complement to capture fisheries and is likely to outproduce capture fisheries by midcentury. Aquaculture schemes vary from completely closed culture systems to pro-poor pond fishing, to sea cage harvesting, and to sea ranching. However, many environmentalists and others are critical of "industrial" aquaculture, as its demand for fish meal draws from open capture fishing as well as competing for markets, polluting waterways, spreading pathogens, and releasing farmed fish into wild fish habitats.

It is increasingly recognized that protection of the world's fisheries requires greater involvement of fishers in managing aquatic resources and a more equitable distribution of the benefits and products of fishing among stakeholders, including nonfishers.

A Risky Business

Fishers have historically been exposed to many natural and human-induced hazards such as floods and cyclones, as well as more slow-acting changes in fish availability, coastal erosion, sea level rise, ecosystem changes, and urbanization. However, fishing has become increasingly hazardous in the face of declining availability of and access to fish stocks in many inshore areas of the world as industrial fishing fleets compete with each other and many small-scale fishers extend their operations into deeper and more dangerous waters in search of a dwindling and fragmented supply. The increased risk associated with fishing translates into increased vulnerability of fishers, their families, and communities, especially where communities are highly dependent on fishing for their livelihood. Thus, a central and necessary aim of any fisheries management strategy is to enhance and sustain fish stocks to reduce the risks and costs of fishing and the vulnerability of households, communities, and states to shortages of fish and fish products. However, the physical availability of a sustainable fish supply is insufficient to ensure that fish become socially and economically available to those most in need. This requires a greater understanding of the global and local social, political, and economic contexts within which fishing and fish-dependent communities are embedded.

Globally, large-scale and many small-scale fishers are integrated into global and regional agrofood systems linking producers to consumers through transborder commodity networks controlled by developed-country retailers and supermarkets, who determine price, consumer outlets, and nonprice conditions of purchase through quality control and other regulatory measures. This globalization of fishing has made it possible to transport large quantities of fish across large distances, put pressure on fish stocks in once relatively stable aquatic ecosystems, and threatens the livelihoods of local coastal and inshore fishers. It has been argued that the shift of much global fisheries' production to serving the elite markets in rich countries has resulted in fishing becoming more precarious in the face of unstable global seafood prices and declining access of the poor of developing countries to affordable fish. There have been calls by proponents of fair trade and by many small-scale fisher communities to enhance the local availability of fish and the financial and other returns to fishing through measures such as restrictions on foreign trawler operations, greater fisher control over postharvest processing and trade, and more South-South regional trade. Others have expressed skepticism of such measures, arguing that fundamental changes in national and local governance structures are necessary to ensure that fishery earnings actually flow to poor fishers and nonfishers.

At the local level, fishing communities are often among the most marginalized, as fishing is considered low-paying, dangerous, unpredictable, and low in social status. However, many communities contain wealthy fishers with political influence, and fishers in several countries have organized regional and national fishers associations to press for improved conditions such as greater security of land and water tenure, coastal zoning to restrict outsider entry, comanagement of common pool resources, territorial use rights in fisheries, access to affordable loans, greater control over the postharvest distribution of fish and fish products (processing and trade), and improved working conditions. An improved small-scale fishing sector can have direct and indirect knock-on effects to the wider economy by providing direct primary and secondary employment to local people, provide incomes and employment to local businesses supplying the fishing sector and households, and act as a safety net to protect families from falling into poverty.

The future of global fisheries and their contribution to improving food security depends on adopting long-term integrated fisheries management strategies that seek to combine production, equity, and sustainability objectives adapted to the conditions of local fisheries. Such strategies also require national political support; greater involvement of fishers in the design, planning, and implementation of management strategies; and the linking of fisheries objectives to wider policies on poverty alleviation and food security.

See Also: Aquaculture; Commons; Food Security.

Further Readings

Béné, Christophe, et al. *Increasing the Contribution of Small-scale Fisheries to Poverty Alleviation and Food Security.* FAO Fisheries Technical Paper 481. Rome: FAO, 2007.

Food and Agriculture Organization of the United Nations (FAO). *The State of World Fisheries and Aquaculture, 2008.* FAO Fisheries and Aquaculture Department. Rome: FAO, 2009.

Helfman, Gene S. *Fish Conservation: A Guide to Understanding and Restoring Global Aquatic Biodiversity and Fisheries Resources.* Washington, D.C.: Island, 2007.

Roberts, Callum. *The Unnatural History of the Sea.* Washington, D.C.; Covelo, London: Island/Shearwater Books, 2007.

Bob Pokrant
Curtin University of Technoology

FOOD AND AGRICULTURE ORGANIZATION

The Food and Agriculture Organization (FAO) is a United Nations agency charged with the international fight against famine; the motto appearing on its emblem, *Fiat Panis,* is Latin for "let there be bread." The FAO is made up of 191 member states, both developed and developing nations; the European Union and the Faroe Islands are associate members. Of the major nations of the world, two that have not joined the FAO are Russia and Singapore.

The FAO was established in 1945 among the first wave of United Nations agencies, and after originally working out of Washington, D.C., offices, relocated to its current headquarters in Rome in 1951. Its work is reviewed every two years by the Conference of Member Nations, and it is governed by a director-general and a council of representatives of 49 member states who serve three-year terms. The biannual conference also approves the budget and plan for the next two years and establishes contributions to fund the "regular program"—the core work of the FAO—for that time. The annual budget is in roughly the $750 million range. FAO activity is handled by eight departments: Administration and Finance, Agriculture, Economic and Social, Fisheries, Forestry, General Affairs and Information, Sustainable Development, and Technical Cooperation.

Since 1994, the Director-General of the FAO has been Jacques Diouf, the former Senegalese ambassador to the United Nations and a former secretary-general of the Central Bank for West African States.

Farmer Field Schools

Since the 1990s, the FAO has been active in promoting integrated pest management (IPM), especially in Asia. The Farmer Field School (FFS) was instituted by the FAO in 1989 in Indonesia before being expanded throughout China, Vietnam, and the Philippines. The FFS teaches farmers self-sufficiency, showing them how to improve the yields and sustainability of their crops while reducing their reliance on expensive and harmful pesticides. The focus on small farms was a result of the recognition that members of Asian farming households make up more than a quarter of the world's population. Even small relative gains for this population add up to tremendous total gains, and as pesticides became cheaper and cheaper in the 1970s and 1980s, this population began to rely on them more and more, repeating the same errors made by previous generations in the West. The FFS was the first successful attempt to reform pesticide usage on small farms in Asia, after little success with the FAO's Intercountry Program for the Development and Application of Integrated Pest Control in Rice in South and Southeast Asia, which had started in 1980 and continued until it was discontinued in 2002.

Through experiential learning activities, farmers participating in the FFS become aware of the ecology of their farms and the role of beneficial insects that could be killed by the use of pesticides. The FFS focuses on local issues and is conducted over the course of a full season, typically meeting once a week at a spot near the farmers' homes. A given FFS usually consists of five to six groups of five farmers to maximize participation in activities that focus on group dynamics, ecosystem analysis, and special local/regional topics. Common to every FFS is a comparison of the performance of plots of land handled with IPM with the performance of plots handled with other methods, and explanations of why IPM leads to better yields and long-term performance. The FFS was augmented by the national IPM program of Indonesia, which also began in 1989 and ran through 2000, funded by the Indonesian federal government, a World Bank loan, and a United Nations grant. Originally focusing on rice, the FFS spun off regional IPM education programs for cotton and vegetable farms and helped develop national IPM programs in Bangladesh, Cambodia, China, and Nepal.

In addition to the direct improvements to a farm's production and operating costs, IPM programs like the FFS have a significant value to public health, potentially preventing the health crises that could be precipitated by overuse of pesticides and the entrance of such into the groundwater and food supply. Benefits have varied considerably in accordance to local factors.

World Food Day

Every year since 1979, the FAO promotes World Food Day on October 16, the anniversary of its 1945 founding. World Food Day is celebrated in most of the member states and other countries, raising awareness of poverty, famine, and food-related issues. Various activities are held, including festivals and sporting events, conferences, dances, and media broadcasts; the Pope generally gives an address in Rome, supporting the goals of the FAO.

Recent Initiatives

In recent years, an important role for the FAO has been opposing the rise of biofuels. When oil price shocks led agricultural producers to devote their resources to growing corn and other crops for biofuel, those resources were diverted from their original food use. In developing countries, this leads to critical food commodity crises, necessitating importing staple goods—an untenable economic situation for the country as a whole, and one that puts the poor and impoverished at much greater risk. 40 million people were put at risk of starvation as a result of this and the global financial crisis.

At the beginning of 2009, the FAO called for global food production to be doubled by 2050, to head off what Ireland's Minister for Overseas Development Aid, Peter Power, called "this silent tsunami."

Criticism

Conservative organizations like the Heritage Foundation have criticized the FAO for being inefficient, bureaucratic, and criticized, and for its lack of transparency in budget allocations and operations. Other groups have criticized it for being biased in favor of Western-style agricultural practices and crops, and for giving too little credit to traditional farming methods in the East and for paying too little attention to the input of the people it seeks to help.

Primarily, though, the common feature of criticism has been that the FAO has been around for 60 years and poverty continues to be a pressing concern worldwide. More people die of hunger and malnutrition every year than died in the Holocaust; there are two starving people in the world for every citizen of the European Union. It is easy to see why there is a perception that progress has not been made, and difficult to measure what progress there may have been. This criticism has become especially common in the recent years, as the world food crisis has caused people to question which anti-poverty programs around the world have done the most good, and which were consuming financial resources that might do more good elsewhere.

The Future of the FAO

The FAO's approach to poverty has been principally from the same angle: increasing and improving production in the developing world. While this may be more politically expedient than taking on the agricultural and trade policies of the developed world, there are clearly food surpluses throughout the developed world, and just as clearly people elsewhere in the world that need to be fed. Opponents of agricultural subsidies—which hamper the developing world's ability to export food to the United States, and to other countries that use subsidies to keep the prices of domestic goods artificially cheap— criticize the FAO for not taking their side. Furthermore, some believe that local economies have been destabilized by attempts to westernize their agricultural sector.

When Louise Fresco, the assistant general director of the FAO, resigned in 2006, she did so amid widespread criticism, frustration, and dismay over the passing of the 10th anniversary of the Rome Declaration, a global commitment to reduce the number of people suffering from malnutrition by 50 percent by 2015. At the midway point, not only had the number not declined, it had increased by 7 percent. Even attributing that to simple population growth, the effect of efforts of organizations like the FAO could not be found in the numbers. In her open letter to the Guardian, Fresco lamented the "bureaucratic paralysis" of the FAO and the fact that "the Organization has been unable to adapt to a new era" and "has not proposed bold options to overcome this crisis."

See Also: Agroecology; Food Security; Integrated Pest Management; Millennium Development Goals; Sustainable Agriculture.

Further Readings

Federico, Giovanni. *Feeding the World: An Economic History of Agriculture, 1800-2000 (Princeton Economic History of the Western World)*. Princeton, NJ: Princeton University Press, 2008.

Ikerd, John E. *Crisis and Opportunity: Sustainability in American Agriculture (Our Sustainable Future)*. New York: Bison Books, 2008.

International Business Publications. *Food and Agriculture Organization of the United Nations Handbook*. Washington, D.C.: International Business Publications USA, 2009.

Pasour, E.C., Jr. and Randall R. Rucker. *Plowshares & Pork Barrels: The Political Economy of Agriculture (Independent Studies in Political Economy)*. Oakland, CA: Independent Institute, 2005.

Patel, Raj. *Stuffed and Starved: The Hidden Battle for the World Food System*. Brooklyn, NY: Melville House, 2008.

Peterson, E. Wesley F. *A Billion Dollars a Day: The Economics and Politics of Agricultural Subsidies*. New York: Wiley-Blackwell, 2009.

Shiva, Vandana. *Stolen Harvest: The Hijacking of the Global Food Supply*. Cambridge, MA: South End Press, 2000.

Staples, Amy L. S. *The Birth of Development: How the World Bank, Food and Agriculture Organization, and World Health Organization Have Changed the World 1945–1965 (New Studies in U.S. Foreign Relations)*. Kent, OH: Kent State University Press, 2006.

Weis, Tony. *The Global Food Economy: The Battle for the Future of Farming*. London: Zed Books, 2007.

Winders, Bill, and James C. Scott. *The Politics of Food Supply: U.S. Agricultural Policy in the World Economy (Yale Agrarian Studies Series)*. New Haven, CT: Yale University Press, 2009.

Bill Kte'pi
Independent Scholar

FOOD AND DRUG ADMINISTRATION

The Food and Drug Administration (FDA) is a regulatory agency of the U.S. government. It is housed in the Department of Health and Human Services. Its mission is to protect the health of the public by regulating human and veterinary drugs, biological products, medical devices, cosmetics, products that emit radiation, and the nation's food supply. It also promotes public health by encouraging the development of innovation in medicine and food science and technology.

The FDA was formed in 1906 as what was in essence a consumer protection agency by the Pure Food and Drugs Act. The act outlawed shipping in interstate commerce adulterated or mislabeled drinks, foods, or drugs. In the case of drugs, the claims of many patent medicine sellers that their medicine was a cure-all were false—the "medicines" were worthless or dangerous. Many contained narcotics that simply numbed consumers.

The Meat Inspection Act was signed into law by President Theodore Roosevelt on July 1, 1906. Public support for it came from popular disgust arising from reading muckraking news reports and literature like Upton Sinclair's *The Jungle* (1906). The novel described corruption in the meat packing industry. It also reported the harsh labor conditions of the day. Sinclair was a journalist and a socialist, and he hoped that his muckraking novel would cause the country to adopt socialism. Instead, a bill to regulate business misconduct was adopted to end the used of poisonous preservatives and dyes in foods.

The Food, Drug and Cosmetic Act was adopted in 1938. It increased federal regulatory action in response to a number of scandals involving food and drugs, some of which had resulted in deaths. The law was the foundational core of the FDA. It added regulatory authority and sought to increase consumer safety by mandating a premarketing review of all new drugs. It also added authority for preventing quackery, which profited from the sale of products that made false therapeutic claims. The law also gave the FDA the authority to inspect factories and to establish regulatory standards for foods, cosmetics, and therapeutic devices. The 1938 act has been amended numerous times since its adoption.

After 1938, the FDA developed a system for the approval of certain drugs by having them evaluated by members of the medical profession. The Insulin Amendment (1941) and the Penicillin Amendment (1945) required testing for potency. In 1951, the Durham-Humphrey Amendment applied testing to the category of prescription drugs.

In 1940, the FDA was transferred to the Federal Security Agency, led by Walter G. Campbell, the first commissioner of the FDA. In 1953, the Federal Security Agency was organized as the Department of Health, Education, and Welfare. It was aided in evaluating thousands of new drugs by rapid developments in toxicology, which were needed to meet the requirements of the Miller Pesticide Amendment, requiring testing for the amount of pesticides in agricultural products.

Food testing for radioactive contamination began in 1954. It was followed by testing of food additives. In 1958, it published, through the Federal Register, a list of substances believed to be safe. In 1959, a massive recall of cranberries three weeks before Thanksgiving captured national attention. The recall was instituted to check for aminotriazole, a weed-killer that induced cancer in laboratory animals.

In 1960, the Color Additive Amendment was enacted, requiring manufacturers to certify that the color additives in foods, drugs, and cosmetics were not carcinogens. The Federal Hazardous Substances Labeling Act (1960), requiring warning labels on hazardous household chemicals, was also added to FDA responsibilities.

Thalidomide, a sleeping pill, was sold in Europe, but tragically was later found to cause serious birth defects. When its danger was revealed in 1961, it incited public concern for better drug regulation. Drug abuse from the growing list of stimulants, depressants, and hallucinogens available through synthetic chemistry was only one area of drug safety that concerned the FDA.

Other areas of safety included toys for children and foods. In 1969, sanitation programs for food and food preparation were transferred from the Public Health Service to the FDA. The transfer gave the FDA responsibility for enforcing consumer product labeling under the Fair Packaging and Labeling Act of 1966 for drugs, cosmetics, and other products.

In 1966, the effectiveness of drugs became a concern that moved beyond safety. The FDA undertook studies of the effectiveness of thousands of drugs that had been approved for safe use between 1938 and 1962 in response to investigations by the National Academy of Sciences. Work with animal drugs was also conducted to determine their efficiency.

In the 1970s, over-the-counter drugs were reviewed for safety, adequate labeling, and effectiveness. However, the consumer protection portions of the FDA responsibilities were transferred to the new Consumer Protection Agency in 1973. In 1977, Congress overruled the FDA by overruling its ban on saccharin as a carcinogen, limiting the substance's regulation to a warning label. In 1980, the Infant Formula Act gave authority to the FDA to ensure formula's nutritional content and safety. In 1982, tamper-resistant packaging was mandated to prevent the poisoning of consumers.

In 1988, the Food and Drug Administration Act placed the FDA in the Department of Health and Human Services as an agency led by a commissioner appointed by the president. The newly promoted agency was faced with the problem of pharmaceutical counterfeiting and other trade practices that were inimical to consumer safety. It also had to deal with the development of generics for both humans and animals.

During the 1990s, the FDA was tasked with enforcing the Anabolic Steroid Act to deal with the emerging black market in sports-enhancing drugs. It also continued to advance its regulations on clear labeling for foods, including those claiming to be "low fat" or "light." Nutrition facts were also made clearer by a newly designed food label. Standards for dietary supplement manufacturing practices were promulgated in the Dietary Supplement Health and Education Act of 1994.

At issue in 2009 were the use of medical maggots and leeches. The use of traditional medical remedies relying on herbs or other methods is expected to become an issue of promoting good health while protecting consumers from quackery.

See Also: Food Safety; Meats; Substitutionism.

Further Readings

Daemmrich, Arthur and Joanna Radin, eds. *Perspectives on Risk and Regulation: The FDA at 100.* Philadelphia: Chemical Heritage Foundation, 2006.

Diane Publishing Company, ed. *Requirements of Laws and Regulations Enforced by the U. S. Food and Drug Administration.* Darby, PA: Diane, 1992.

Hawthorne, Fran. *Inside the FDA: The Business and Politics Behind the Drugs We Take and the Food We Eat.* New York: John Wiley & Sons, 2005.

Hickmann, Meredith A. *The Food and Drug Administration.* Hauppauge, NY: Nova Science, 2003.

Parker, Florence R. *FDA Administrative Enforcement Manual.* London: Taylor & Francis, 2005.

Pisano, Douglas J. and David Mantus, eds. *FDA Regulatory Affairs: A Guide for Prescription Drugs, Medical Devices, and Biologics.* London: Taylor & Francis, 2003.

Pray, Leslie and Sally Robinson. *Challenges for the FDA: The Future of Drug Safety, Workshop Summary.* Washington, D.C.: National Academies, 2007.

Whitmore, Elaine. *Development of FDA-Regulated Medical Products: Prescription Drugs, Biologics, and Medical Devices.* Milwaukee, WI: ASQ Quality, 2003.

Andrew Jackson Waskey
Dalton State College

FOOD FIRST

Food First, also known as the Institute for Food and Development Policy, is a nonprofit think tank located in Oakland, California. Its goal is to analyze the causes of global hunger, poverty, and ecological degradation and to work with social movements to develop solutions. Food First carries out research, analysis, advocacy, and education, empowering farmers and communities to achieve three interconnected goals: wrest control of the food system from transnational agrofood industries, claim their right to define their own food and agricultural systems, and ensure access to healthy and culturally appropriate food produced through ecologically sound and sustainable methods.

Food First was founded in 1975 by Frances Moore Lappé and Joseph Collins. Both were groundbreaking researchers and authors. Lappé's *Diet for a Small Planet* was among the first tomes to connect individual eating practices to ecological degradation and world hunger, whereas Collins's *Global Reach* focused on the effect of multinational corporations on poverty in the global South. The two met at the first World Food Day shortly before founding the organization. Food First continues its founders' tradition of publishing. Early highlights include Lappé and Collins's *Food First: Beyond the Myth of Scarcity* and David Weir and Mark Shapiro's *Circle of Poison: Pesticides and People in a Hungry World*. *Circle of Poison* contributed to the formation of the Pesticide Action Network, a leader in the pesticide reform movement.

Food First studies and publicizes the effects of the international political economy and political ecology on food and hunger. Its work uses food as a lens to critique global public policies such as U.S. involvement in Central America and the rise of neoliberal privatization and deregulation. It has inspired thousands of students, educators, and citizens to better understand that hunger is not a consequence of lack of food but of policies designed to maximize wealth at the expense of people's abilities to meet their most fundamental needs.

Food First is aligned with social movements opposing the architects of corporate globalization and takes particular issue with structural adjustment programs requiring poor nations to privatize resources in exchange for development aid. In addition to research and education, it has enacted this opposition through strategies ranging from congressional briefings to nonviolent protest. Notable publications in this area include Walden Bello's *The Future in the Balance: Essays on Globalization and Resistance* and current Executive Director Eric Holt-Gimenez's *Ten Reasons Why the Rockefeller and the Bill and Melinda Gates Foundations' Alliance for Another Green Revolution Will Not Solve the Problems of Poverty and Hunger in Sub-Saharan Africa*.

In addition to its critique of transnational agribusiness, Food First envisions an alternative model that meets the needs of small producers and low-income consumers. To this end, it opposes the development and planting of genetically modified foods, which deepen producers' reliance on agribusiness corporations while threatening both human and environmental health. Instead, they support organic farming and agroecology, as well as local, farmer-led control of food policies. Food First has published many works opposing genetically modified organisms, including Ivette Perfecto and John Vandermeer's *Breakfast of Biodiversity: The Truth About Rain Forest Destruction* and Miguel Altieri's *Genetic Engineering in Agriculture: The Myths, Environmental Risks, and Alternatives*.

Food First works to ensure that knowledge of alternative agriculture methods is shared globally. Because the U.S. blockade prevents the import of agricultural inputs, Cuba has become the world leader in these methods. Food First has sent delegations to Cuba and sponsored trainings and exchanges to make Cuban expertise available to organic farmers worldwide. Food First has published leading research on this topic, such as Laura J. Enríquez's *Cuba's New Agricultural Revolution: The Transformation of Food Crop Production in Contemporary Cuba.*

Food First recognizes that a sustainable agriculture developed according to market-based principles will necessarily privilege the desires of wealthy consumers. For this reason, it attempts to bridge what it calls the "industrial agri-foods divide" that separates sustainable producers from low-income consumers. Food First has allied itself with the growing struggles of low-income communities of color for food security and food justice, documenting local experiences and contributing to relevant policy debates. It is currently engaged in several research projects on this issue. Through its research, Food First helps to build a national coalition of urban communities of color working for food security.

Food First also links local struggles for food justice to similar movements worldwide, enabling grassroots activists to create coalitions beyond national borders. It has worked with Via Campesina, the worldwide peasants movement, to document the struggles and alternatives of farmers and communities. Food First has been involved in the Campesino a Campesino project, helping small producers to share their struggles and successes with one another. This project is documented in a book of the same name by Eric Holt-Gimenez. Through this work, Food First animates global political will for food system transformation.

"Without land," states the Food First website, "there is no food." Food First sees land reform as essential to the creation of food sovereignty and the end of world hunger. To this end, it has partnered with organizations in Thailand, Brazil, and South Africa to convene the Land Research Action Network, which links activists and researchers working to democratize access to land. Food First publications on land reform include Peter Rosset, Raj Patel, and Michael Courville's *Promised Land: Competing Visions of Agrarian Reform* and Angus Wright and Wendy Wolford's *To Inherit the Earth: The Landless Movement and the Struggle for a New Brazil.*

In sum, Food First compiles a rigorous critical analysis of the international political economy of agriculture while highlighting farmer and community-based alternatives. Working against the assumption that hunger is caused by lack of food, Food First maintains that the trouble lies with corporate control of the food system. Through its research and educational work, Food First has allied itself with social movements working to bring about an alternative model that is ecologically healthy and socially just.

See Also: Agroecology; Food Security; Food Sovereignty; Peasant.

Further Readings

Altieri, Miguel. *Genetic Engineering in Agriculture: The Myths, Environmental Risks, and Alternatives.* Oakland, CA: Food First, 2004.

Bello, Walden. *The Future in the Balance: Essays on Globalization and Resistance.* Oakland, CA: Food First, 2001.

Enríquez, Laura J. *Cuba's New Agricultural Revolution: The Transformation of Food Crop Production in Contemporary Cuba.* Oakland, CA: Food First, 2000.

Food First. "Food First History." http://www.foodfirst.org/en/history (Accessed December 2008).

Holt-Gimenez, Eric. *Campesino a Campesino: Voices From Latin America's Farmer to Farmer Movement for Sustainable Agriculture.* Oakland, CA: Food First, 2006.

Holt-Gimenez, Eric, et al. *Ten Reasons Why the Rockefeller and the Bill and Melinda Gates Foundations' Alliance for Another Green Revolution Will Not Solve the Problems of Poverty and Hunger in Sub-Saharan Africa*. Oakland, CA: Food First, 2006.

Lappé, Frances Moore and Joseph Collins with Brian Fowler. *Food First: Beyond the Myth of Scarcity*. Oakland, CA: Food First, 1977.

Perfecto, Ivette and John Vandermeer. *Breakfast of Biodiversity: The Truth About Rain Forest Destruction*. Oakland, CA: Food First, 2005.

Rosset, Peter, et al. *Land: Competing Visions of Agrarian Reform*. Oakland, CA: Food First, 2006.

Weir, David and Mark Schapiro. *Circle of Poison: Pesticides and People in a Hungry World*. Oakland, CA: Food First, 1981.

Wright, Angus and Wendy Wolford. *To Inherit the Earth: The Landless Movement and the Struggle for a New Brazil*. Oakland, CA: Food First, 2003.

Alison Hope Alkon
University of the Pacific

FOOD JUSTICE MOVEMENT

Food justice is a term that refers to social equity problems and solutions in the food system. Food justice is about both distributive justice and productive justice. Distributive justice can be thought of as the process of allocating the surplus of a society in an equitable manner among all people. Productive justice, in contrast, is about the ways in which society is structured and the equal ability of people to participate in the economic and social systems of that society.

Distributive Justice in the Food System

The most telling indicator of the degree of distributive justice is the presence or absence of hunger or food security. The term *food justice* represents a focus on food as a fundamental human right, asserting that no one should go without sufficient healthy food as a result of economic restraints or social inequalities.

Throughout history, at various places and at different times, people have gone hungry. In some cases, this has been a result of crop failures or natural disasters. Today, however, the world produces more than enough food to feed every woman, man, and child, and yet people still go hungry. In 2007, there were 923 million chronically hungry people worldwide, and an estimated 24,000 people die every day as a result of hunger and undernutrition. Despite the development of technologies for increasing food production (many of which have had serious environmental consequences), more people go hungry today than at any point in human history.

Between 2003 and 2005, there was a 75 million person increase in hunger. Those hardest hit include the rural landless, pastoralists, small-scale farmers, and the urban poor in impoverished nations. Women and children living in rural areas of Asia, Africa, and Latin America are especially at risk of hunger. Although there are higher proportions of food-insecure people in some regions, there are greater inequalities in income and food consumption within regions, and even households, than between countries. Even in the United States, the world's largest producer of food, 36 million people—11 percent of the population—were food insecure in 2007, meaning that at times they did not have enough money for food. The prevalence of U.S. hunger is not equally distributed; it is disproportionately experienced in single-mother households and by low-income people, children, and people of color.

In a market economy, people go hungry if they do not have enough money to buy food or the resources to produce food. Poverty, therefore, is the cause of hunger. Almost 50 percent of the world's population—over three billion people—live on less than $2.50 per day. At least 80 percent of humanity lives on less than $10 per day. Poverty levels are highest in impoverished countries in South Asia and sub-Saharan Africa, where, respectively, 40 percent and 50 percent of the population lives on less than $1.25 per day. In the United States, poverty rates are highest in inner cities and rural areas, especially in the southern states. Children under age 18 years and people of color have poverty rates that greatly exceed the average. Poverty among U.S. farmworkers is more than double that of all wage-and-salary employees. In fact, farmworkers are often too poor to purchase the produce they harvest.

Food insecurity and high poverty rates are often the result of historical circumstances that have shaped people's abilities to participate in current economic and social systems. This is at the root of productive justice.

Productive Justice in the Food System

The existing distribution of food, land, income, and wealth is governed by longstanding global and local structural inequalities. Throughout the world, people have been dispossessed of their ability to produce food by colonizers who have taken land and extracted resources and labor from the colonized people. Africa, for example, was self-sufficient in food before colonization.

Thus, one indicator of the degree of productive justice in the food system is the distribution of ownership of land used for food production. In the United States, for example, land ownership is highly concentrated. According to the 2007 U.S. Census of Agriculture, only 3.6 percent of principal operators manage 54 percent of farmland. As well, women own only 5 percent of U.S. farmland, Latinos own 1 percent, and African Americans own less than 0.5 percent. In contrast, nearly all hired farm laborers are people of color. For example, California's farm labor force is composed almost exclusively of ethnic minorities, 95 percent of whom are foreign born. Farmworkers and food processing workers not only earn poverty-level wages but also have little power, and therefore suffer from rights abuses and injury and illness rates far greater than those in other occupations. In several federally prosecuted cases of tomato pickers in the southeastern United States, for example, farmworkers have been enslaved, forced into debt-servitude, beaten, sexually harassed, charged rent for living in trailers where 8–10 workers live together, and have had their families threatened.

Another indicator of productive justice is the distribution and equity of decision making within the food system, which is far from democratic. At a global level, decisions are made in transnational corporations and advanced industrialized countries, with little or no representation from labor, consumer, or environmental groups. Women and people of color are underrepresented as owners and leaders of agrifood-related businesses and organizations.

Even in small rural organizations, men, rather than women, are often the key decision makers, and in U.S. agricultural agencies, the majority of those in decision-making roles are men of European-American descent.

To solve the problem of a lack of social justice in the food system, issues related to both distributive justice and productive justice need to be addressed.

Solving Problems of Food Justice

Throughout the world, people are resisting the forces of injustice and working to create food systems that meet conditions of distributive and productive justice. International organizations and networks such as La Via Campesina and Campesino a Campesino have promoted the concept of "food sovereignty," for example. This concept builds on the Universal Declaration of Human Rights, which was adopted by the United Nations in 1948 and asserted the rights of all people to have a standard of living adequate for health and well-being. The right was strengthened in 1999 with the statement that the right to adequate food is realized "when every man, woman and child, alone or in community with others, has the physical and economic access at all times to adequate food or means for its procurement."

In the United States, the long-standing antihunger and agricultural labor movements have been joined by other movements that also prioritize social justice in the food system, including movements for sustainable agriculture and community food security. In its recognition that the modern food system was built on a history of genocide, slavery, and land appropriation, the food justice movement builds on antioppression and antiexploitation movements such as the antislavery, anticolonialism, civil rights, and farmworker movements.

The first step in solving a problem is articulating a definition of the problem. In 2004, a group of activist-researchers in California defined a socially just food system as one in which "power and material resources are shared equitably so that people and communities can meet their needs, and live with security and dignity, now and into the future." In 2008, a collaborative group of food activists and researchers developed the Food Justice Manifesto, which describes the context of food injustice and states that everyone has "a moral imperative to address the root causes of hunger and starvation." As well, the movement calls for the dismantling of structures that reproduce violence, oppression, and exploitation at all levels and in all components of the food system, from agrifood system laborers, to animals, to nature.

There are several facets to the act of bringing social justice into the food system: ensuring that all have equal access to food, income and wealth, farmland, and fair labor conditions and the ability to be a part of decision-making processes in the food system. Food justice organizations work to increase food justice using varying strategies. One of those strategies is to empower oppressed people to become involved in the leadership and membership of food justice organizations and projects. Mo' Better Food farmers market sells low-priced produce that is desired by the primarily African American community in which it is located. Market coordinators offer free public information on the food system and how community members can get involved. Another strategy is to teach people how to grow their own food and to provide the land and resources needed to do so. For example, City Slicker Farms in Oakland, California, helps low-income residents set up gardens in their yards and provides the continuous mentoring, plant starts, seeds, and compost needed to keep the garden going. Just Food in Brooklyn, New York, teaches community members how to garden and how to teach others how to garden. As well, food justice organizations employ people directly from the communities and demographics that experience food insecurity. For example, the staff and leaders of People's Grocery come directly from the West Oakland community, and the organization hires neighborhood youth to attend to their gardens and office duties.

Organizations that focus on alleviating injustice in the food system exist on both large and local scales. To fight pervasive hunger around the world, they develop solutions to the underlying issues of racism, class disparity, and unequal distribution of resources in the food system, as well as issues in larger economic and political systems that are the causes of hunger and food insecurity.

See Also: Food Security; Food Sovereignty.

Further Readings

Allen, Patricia. *Together at the Table: Sustainability and Sustenance in the American Agrifood System*. University Park: Pennsylvania State University Press, 2004.
Food and Agriculture Organization of the United Nations. "Food Security Statistics." http://www.fao.org/es/ess/faostat/foodsecurity/index_en.htm (Accessed September 2008).
Food Justice Manifesto. "Food Justice: A People's Movement Whose Time Is Now." http://foodjustice.wikispaces.com (Accessed April 2009).
Lappé, Frances Moore. *Diet for a Small Planet*. New York: Random House, 1991.
United Nations. "Universal Declaration of Human Rights." http://www.un.org/Overview/rights.html (Accessed April 2009).
U.S. Department of Agriculture and National Agricultural Statistics Services. "2007 Census of Agriculture." http://www.agcensus.usda.gov (Accessed September 2008).
World Bank. "PovertyNet." http://www.worldbank.org/poverty (Accessed April 2009).

Patricia Allen
Hilary Melcarek
University of California, Santa Cruz

FOOD PROCESSING INDUSTRY

The food processing industry includes a vast variety of businesses that prepare fresh food for sale and manufacture prepared food products. This is a significant sector of the global food industry that includes agricultural, manufacturing, retail and distribution, research and development, and lobbying businesses; often a food processing company will be owned in whole or in part by a larger corporation that has its fingers in the rest of the pie as well.

Essentially, food processing is the preparation of food for sale. This does not include the washing of produce, which is typically done at the production end, but does include the peeling and slicing of carrots sold ready-to-cook—either fresh or frozen—as well as the canning of such carrots, the production of jars of carrot puree for baby food, the production of bottled or boxed carrot juice, the production of freeze-dried carrot snacks, and the production of jarred pasta primavera sauce using tomatoes, carrots, and other vegetables. Food processing includes the production of pet food and animal feed as well as the production of food for human consumption.

One of the oldest food processing businesses is the slaughterhouse, also known as the *abattoir* or (in the case of open-air slaughterhouses) *shambles*. In the slaughterhouse, food

animals like cattle, sheep, pigs, and poultry are received from the farm or feedlot, herded
into holding pens, and then slaughtered and butchered. Typically, the animal is rendered
unconscious—for instance, cattle are knocked out by electric shock before being hung
upside down and exsanguinated (the carotid artery and jugular vein are severed, allowing
the animal to rapidly bleed to death). Various parts are removed—in the case of cattle, the
head, hide, feet, and internal organs—and examined for signs of disease. A government
inspector then confirms that the carcass is safe for sale and some kind of antibacterial
measure is taken, such as exposure to boiling water. These days, beef carcasses are typically "tenderized" after death with electrical shocks. Meat must be chilled quickly to
inhibit bacterial growth and deterioration. The carcass is usually butchered into smaller
pieces—the primals and subprimals, various muscle groups—depending on the needs of
the customer. Slaughterhouses may also perform special procedures—for instance, they
may follow kosher or halal procedures, or they may dry-age the beef, a process that is safer
than the wet-aging used for supermarket beef and that results in stronger-flavored, more
tender beef, but that is resource-intensive and tends to be reserved for special customers,
such as steak houses. (Dry-aging is usually not performed at the slaughterhouse, but beef
can be prepared for dry-aging as part of the slaughter process.)

By-products from slaughterhouses are sent to rendering plants, which may also take in
expired meat from retail stores, dead animals from animal shelters, and restaurant dis-
cards. Rendering plants create a variety of products from animal products like bone, lard,
and tallow. The end product, and whether it will be used in human food, determines the
procedures used. For instance, finely chopped beef fat from a cow carcass can be heated
in a centrifuge, which separates the liquid from the fat; the fat—beef tallow—can then be
used for cooking purposes (it was once common to use it to fry French fries). Fats can also
be processed for use in soap-making or for pet food. The solids left from rendering can be
ground with bone to make animal feed.

Meat packing plants also receive their materials from slaughterhouses and are responsible for canning, freezing, and otherwise packaging meat for sale to the consumer.
Butchers and supermarkets are alternative destinations for a slaughterhouse's wares, with
the products usually passing through a distributor first.

Another major food processing industry is the sugar processing industry, which takes
plant sugar sources (sugarcane and sugar beets are the best known, but sugar is also derived
from the date palm, sorghum, and at times the sugar maple tree, though its syrup is usually
more valuable without being processed into table sugar) and extracts the sugar from them.
Brazil is far and away the largest sugar-producing country in the world; India, China,
Thailand, Mexico, and Australia follow behind. Sugarcane is crushed, juiced, filtered, and
boiled to produce sugar crystals that are removed from the surrounding syrup with a centrifuge. Further processing turns tan-colored sugar into white sugar, though ironically
darker sugars sell for more, because they are a gourmet product and white sugar enjoys
economies of scale. Beet sugar is extracted by using a diffuser and an alkaline solution and
uses a quarter as much water as the processing of cane sugar does, making it attractive in
dry climates. Various grades of sugar are sold on the market, with the options varying in
different regions of the world; sometimes the grades refer to the degree of refining, sometimes to other acts of processing. Brown sugar, for instance, is not less-refined sugar but is
refined white sugar with a portion of the removed syrup (molasses) added back in.

Just as rendering plants make use of the by-products of slaughterhouses, sugar by-
products have a long culinary history. The principal by-product of sugar refining is molasses, which can be sold for baking or table use or fermented and distilled into rum, an

alcohol product that evolved and was popularized in large part to take advantage of the vast molasses resources made available by the rise of the sugar industry.

See Also: Agribusiness; ConAgra; Food Safety; Meats; Sugarcane.

Further Readings

Allen, Gary J., et al, eds. *The Business of Food: Encyclopedia of the Food and Drink Industries.* New York: Greenwood, 2007.

Berry, Wendell. *The Unsettling of America: Culture and Agriculture.* San Francisco: Sierra Club, 1996.

Shapton, David. *Principles and Practice for the Safe Processing of Foods.* Boca Raton, FL: CRC, 1998.

Bill Kte'pi
Independent Scholar

FOOD QUALITY PROTECTION ACT

The 1996 Food Quality Protection Act (FQPA) is a U.S. federal law that reformed the way the Environmental Protection Agency (EPA) regulates pesticides. The FQPA was a minor amendment to the Federal Insecticide, Fungicide, and Rodenticide Act (the last significant revision of which was in 1972; it is the descendent of 1910's Federal Insecticide Act) and a significant amendment to the Federal Food, Drug, and Cosmetic Act (FD&C), which was first passed in 1938.

The EPA was established in 1970 as an independent federal agency. Similar to other such agencies—for instance, the National Aeronautics and Space Administration, the Federal Communications Commission, and the Central Intelligence Agency—it is not under the jurisdiction of the federal executive departments; unlike most of those agencies, its administrator is traditionally given cabinet-level rank. The 12th and current administrator of the EPA, appointed by President Barack Obama, is Lisa Jackson, who had previously worked for the EPA in the 1980s and 1990s before taking a job in New Jersey state government. The EPA is charged with the protection of human health by the preservation of the environmental quality and safety of air, water, and land, and by the regulation of chemicals. Pesticide regulation is one of the duties the EPA inherited from the U.S. Department of Agriculture when the agency was created and is, therefore, one of its oldest tasks.

The FD&C is best known for its appearance on ingredient labels in the form of certified and numbered food colorings, such as "FD&C Blue No. 2," which refers to food-safe natural indigo dye. Though related to the general flood of New Deal legislation that characterized the 1930s, the FD&C Act was also a response to growing public concerns with food safety as a result of new technologies and demands. With more and more food prepared in factories instead of in kitchens, often with additives and ingredients unheard of in those kitchens, before being packaged and shipped around the country, the consumer had become divorced from the producer, and the process of food production had been, if not explicitly hidden, at least removed from public view and common knowledge. Everyone knows what is in a hamburger; few know what is in a hot dog.

The FD&C itself superseded the earlier 1906 Food and Drugs Act; the FD&C was a massive overhaul of previous legislation, which was shown to be insufficient when 107 sick people, many of them vulnerable children, died as the result of a toxic but legal treatment: a raspberry-flavored preparation of sulfanilamide (an antibacterial) with diethylene glycol, a sweet-tasting solvent. The pharmacist who formulated the concoction for Massengill, the manufacturer, was not aware of diethylene glycol's toxicity, and because there were no laws in place requiring strict oversight and testing of pharmaceutical products, Massengill's owner announced that "we could not have foreseen these unlooked-for results. I do not feel that there was any responsibility on our part." The pharmacist himself committed suicide.

Over the years, many amendments have been made to the FD&C, following a general trend of caution and placing the burden of proof of safety (and in the case of medicine, effectiveness) on producers. These amendments have expanded regulations to cover everything from allergen labeling on food products to the expansion of information included in nutrition labeling. The FQPA is somewhat different from many of these amendments in that it is concerned not with increasing the amount of information for consumers, nor even directly with the production or processing of food, but, rather, with the regulation of pesticides, which affect the safety of food in a number of indirect ways (not only by remaining present on raw fruits and vegetables but also by potentially affecting the health of livestock and the quality of meat and milk and by entering the water supply). The FQPA requires that the EPA consider aggregate risk and cumulative exposure when evaluating pesticides, imposes a new safety standard on all pesticides used on food commodities ("reasonable certainty of no harm") and in other respects revises standards of risk assessment, requires that pesticides be evaluated in terms of their effect on the endocrine system, requires the EPA to reassess all existing pesticides under the new guidelines and to periodically review all pesticide registrations, and substantially promotes integrated pest management through research and education programs.

All in all, the FQPA was the most significant and wide-reaching pesticide-related legislation since the inception of the EPA, and in part for that reason, it took a long time to actually be implemented. A goal of 10 years was set in 1996 for the reevaluation of existing pesticides, with periodic review of registrations every 15 years. Although some evaluation could begin immediately, evaluation of pesticides' effects on the endocrine system—the Endocrine Disrupter Screening Program—did not begin until 2009 and is expected to take two to three years to complete.

See Also: Food and Drug Administration; Integrated Pest Management; Pesticide.

Further Readings

Goodlatte, Robert. *Impact of the Food Quality Protection Act Implementation on Public Health: Hearing Before the Committee on Agriculture, U.S. House of Representatives.* Darby, PA: Diane Publishing, 1999.
Pollan, Michael. *In Defense of Food.* New York: Penguin, 2009.
U.S. Food and Drug Administration. "Food Quality Protection Act of 1996." http://www.fda .gov/RegulatoryInformation/Legislation/FederalFoodDrugandCosmeticActFDCAct/ SignificantAmendmentstotheFDCAct/ucm148008.htm (Accessed June 2009).

Bill Kte'pi
Independent Scholar

FOOD SAFETY

Food safety is the management of food in ways that prevent food-borne illness and negative social and environmental consequences. Food can contain or grow bacteria that can cause

food poisoning and transmit diseases. Food therefore requires scientific examination to ensure safety to its consumers. Food safety therefore emerges as a new scientific discipline that examines the production and processing of food to prevent negative health, environmental, and social implications. Green or organic food is often presented as an example of safe food because of the debate on the health and environmental consequences of genetically modified food. Previously, food safety was understood just as freshness and taste free from any hazardous bacteria; at this time, however, different social and environmental attributes are also part of it.

This food technology scientist is comparing the efficacy of several types of washes that may reduce microbes and contaminants on spinach and other greens.

Source: U.S. Department of Agriculture Agricultural Research Service/Peggy Greb

Evolution of Food Safety System

The hazard analysis critical control point (HACCP) is one of the main tools and mechanisms for food safety. The HACCP system has a long history of development and evolution. The current global food safety system, under the auspices of the United Nations, began in 1945 with the organization of the Food and Agriculture Organization (FAO). The General Agreement on Tariffs and Trade, concluded in 1947, included provisions for countries to apply measures necessary to protect human, animal, or plant life or health. Several of its stipulations were that measures adopted by an individual country must not unjustifiably discriminate between countries in which similar conditions prevail and must not act as disguised restrictions on international trade. After that, the HACCP system took a long path of evolution until, in 1997, it reached the "Codex Document on HACCP Principles and Application."

Although the early HACCP system (in place until 1972) was quite simple and consisted of only three principles (conducting hazard analysis, determining critical control points, and establishing monitoring procedures), modern HACCP is built on seven principles (conducting hazard analysis, determining critical control points, establishing critical limits, establishing monitoring procedures, establishing corrective actions, establishing verification procedures, and establishing recordkeeping procedures). The United States, Japan, and several European Union countries largely use modern HACCP systems to ensure food safety. They also require that their food exporters in the developing nations apply HACCP systems. The current food safety system, however, goes beyond the HACCP manuals.

Privatizing Food Safety Governance

Since time immemorial, "food" has been playing a pivotal role not only in human health and pleasure but also in economics, culture, and politics. Whereas early human societies—hunters and gatherers, for instance—were dependent on food sources within their tribal territories, nowadays most people in richer countries consume food from all over the globe. As a consequence, the physical distance between the places of production and places of consumption is growing fast, whereas the time gap between producing and consuming a food is closing rapidly. At this time, because of the globalization process, developing nations are orienting their food products to meet global market demands, and as a result, many local food production systems in developing countries are increasingly linked to global commodity chains or networks, generating a complex governing system between the local and the global.

Traditionally, government agencies were responsible for monitoring food safety standards and other food quality attributes. However, the recent emergence of privately regulated supply chains organized more around principles of safety or quality has precipitated a shift in governance. Previously, the notion of food safety was understood as freshness and taste; however, recent movements have extended this notion to include other social and environmental attributes. Local and global environmental and civil rights movements have launched campaigns to address social justice issues by making sure that agrofood products are environmentally friendly and socially responsible and have meaningful community participation. These are sometimes known as "credence" or nonmaterial characteristics of food safety—characteristics that the consumers cannot detect after purchase in the same way that they detect freshness and taste. These credence qualities include the environmental and ethical conditions of production. For example, is the food commodity produced organically? Do foods contain genetically modified varieties? Are bean or coffee producers being paid a fair price? Are dolphins affected when tuna are caught? Under what conditions were animals raised? As these safety attributes are confined to the production and processing of food commodities and are not readily apparent in the physical products that reach the consumer, consumers can make a choice or understand the credence qualities if the product has trustworthy labels.

This shift toward a broader definition of safety (appearance, size, consistency, taste, freshness, quality, fair trade, environment, etc.) is driven, partly, by the neoliberal turn in agrofood regulation, with states ceding responsibility for regulation just as corporate actors step in to provide a proliferating array of quality assurance schemes and voluntary standards—a trend that is often characterized as "privatizing regulations" or "certification regimes." Among the important mechanisms used in certification schemes are identity preservation, segregation, and traceability systems allowing for "field-to-plate" monitoring of supply chains. The shift to "safety" or "quality" involves new dilemmas for agricultural sectors, regions, and individual producers, as privately regulated supply chains define different sets of winners and losers. For farmers, the exacting contract specifications required for participation in quality chains can increase costs and tend to marginalize smaller, less-sophisticated producers. As large seafood buyers such as Wal-Mart, Darden, and Lyons have already committed to buy only certified seafood, it is now realistic to imagine that significant portions of global industrial aquaculture production could come to be certified within the next few years.

Food Safety Schemes

To address the physical, social, and environmental characteristics of food safety, there are an increasing number of schemes covering eco-labeling, organic certification, and recently, fair trade. The growing number of certification programs and possible competition among certification schemes has the potential to result in confusion among buyers and consumers. Despite this confusion, third-party certification is becoming a powerful means of demonstrating to customers and to government a commitment to environmentally responsible production of safe food products. Wal-Mart, the world's biggest retailer, has set 2012 as a target date for selling only certified fish, and fast food giant McDonald's has switched from threatened species to more plentiful pollack for its fish sandwiches. As examples, some existing certification programs are given below with no endorsement or support for any particular certification.

The Codex Alimentarius Commission was probably the first certification program, created in 1963 by the FAO and the World Health Organization to develop food safety standards, guidelines, and related texts for protecting the health of consumers; ensure fair trade practices in the food trade; and promote coordination of all food standards work undertaken by international governmental and nongovernmental organizations. The Codex alimentarius system presents an opportunity for all countries to join the international community in formulating and harmonizing food standards and ensuring their global implementation. Many other certification schemes follow Codex standards.

The Accredited Fish Farm Scheme is a governmental scheme developed by the Agriculture, Fisheries, and Conservation Department of Hong Kong to brand local products and to increase consumer confidence in fish quality and safety. The Alter-Trade Japan is a Japanese company dedicated to food safety by ensuring fair trade with several commodities, including bananas, coffee, and shrimp. The Aquaculture Certification Council, a certification body of the Global Aquaculture Alliance, has developed "Best Aquaculture Practices" standards that address social, environmental, and food safety of shrimp aquaculture. The Marine Stewardship Council—an independent, global, nonprofit organization—is working toward environmentally responsible fisheries through certification programs. Based on the FAO Code of Conduct for Responsible Fisheries, the Marine Stewardship Council has developed an environmental standard for sustainable and well-managed fisheries and uses a product label to reward environmentally responsible fishery management and practices.

BioGro New Zealand is a not-for-profit organic producer and consumer organization that has been actively working to grow organics in New Zealand since 1983. The Bio-Gro Organic Standards looks at all production and processing segments including farms, processors, exporters, input manufacturers, distributors, and retailers to ensure food safety. Bio Suisse is an umbrella association of organic farming organizations and farms based in Switzerland and working to ensure food safety. Carrefour, the first retailer in Europe and the second-largest in the world after Wal-Mart, has developed the Carrefour Quality Line for food products that are safe from farm to table and that comply with international food safety standards.

The Ethical Trading Initiative is an alliance of companies, nongovernmental organizations, and trade union organizations that aims to promote and improve the implementation of corporate codes of practice that cover supply chain working conditions, particularly to ensure that the working conditions of workers producing for the United Kingdom market meet or exceed international labor standards. EurepGAP is a private-sector body represented mainly by a group of retailers who set voluntary standards for the certification of agricultural products to ensure food safety. Fairtrade Labelling Organizations International is a worldwide fair trade standard-setting and certification organization that permits more than 800,000 producers and their dependants in more than 40 countries to benefit from products labeled fair trade. Fairtrade Labelling Organizations International guarantees that products sold anywhere in the world with a fair trade label marketed by a national initiative conform to fair trade standards and contribute positively to disadvantaged producers.

GLOBALGAP is a private-sector body that sets voluntary standards for the certification of agricultural products around the globe. International Social and Environmental Accreditation and Labelling Alliance (ISEAL) members set voluntary standards in sectors including forestry, agriculture, fisheries, manufacturing, and textiles. They operate programs that reward producers for social and environmental performance and are backed by independent third-party certification, enabling supply chain companies and end consumers to make more sustainable purchasing decisions. The ISEAL Alliance developed and complies with the "Code of Good Practice for Setting Social and Environmental Standards" to strengthen the credibility of standard-setting procedures. The Gold Standard for Sustainable Aquaculture Ecolabel Design, developed by the Environmental Law Institute and the Ocean Foundation, is another certification scheme that seeks to surpass other existing schemes in standards and values to ensure food safety. By complying with the ISEAL Alliance Code of Practice, Gold Standard currently seeks to comprehensively consider effects on the environment, society, human health, and animal welfare; scientific standard-setting; careful controls on certification decisions; transparent review and reporting on performance; and robust objections procedures.

International Federation of Organic Agriculture Movements (IFOAM) is a global umbrella body for organic food and farming. IFOAM's goal is the worldwide adoption of ecologically, socially, and economically sound systems that are based on its "Principles of Organic Agriculture." Naturland is one of the certification bodies for IFOAM organic standards. International Standards Organization (ISO)14001:2004 is the standard for environmental management, not for a product, but for minimizing harmful effects on the environment caused by shrimp farming. ISO 9001:2000 is a quality management systems standard, and ISO 22000:2005 is for a food safety management system for facilities such as processors. ISO technical committee ISO/TC 234, Fisheries and Aquaculture, was set up in February 2007 for standardization in the field of fisheries and aquaculture. Quality Assurance International offers verification services, ensuring that the organic integrity of food and fiber products is preserved from seed to shelf. Headquartered in San Diego, California, Quality Assurance International maintains operations in Japan, Canada, and Latin America, with satellite U.S. offices in Minnesota and Vermont.

At this time, the broader definition of food safety also includes the essential components such as "acceptability" (access to culturally acceptable food, which is produced and obtained in ways that do not compromise people's dignity, self-respect, or human rights), availability (providing a sufficient supply of food for all people at all times), accessibility (the equality of access to food), adequacy (adequate measures are in place at all levels of the food system to guarantee the sustainability of production, distribution, consumption, and waste management), and agency.

See Also: Fair Trade; Food Security; Food Sovereignty; Genetically Modified Organisms; Mad Cow Disease; Organic Farming; Supply Chain.

Further Readings

Busch, L. and C. Bain. "New! Improved? The Transformation of the Global Agrifood System." *Rural Sociology,* 69/3 (2004).
Josling, T., et al. *Food Regulation and Trade: Toward a Safe and Open Global System.* Washington, D.C.: Institute for International Economics, 2004.
Koc, M. et al., eds. *For Hunger Proof Cities: Sustainable Urban Food Systems.* Ottawa: International Development Research Centre, 1999.
McMichael, P. and F. H. Buttel, eds. *New Directions in the Sociology of Development.* Elsevier, 2005.
Wal-Mart. "Wal-Mart Takes Lead on Supporting Sustainable Fisheries" (2006). http:// walmartstores.com/pressroom/news/5638.aspx (Accessed April 2011).

Md. Saidul Islam
Nanyang Technological University

Food Security

Food security is a contentious issue, and attempts to define it align with particular approaches to what should be done. At the most basic level, food security indicates access to sufficient, safe, and nutritious food. The U.S. General Accounting Office indicates that up to 2 billion people lack food security worldwide. According to the U.S. Agency for International Development (USAID), 75 percent of the world's food-insecure people are located in rural areas, predominantly in the global South. In this context, food security is the opposite of starvation or hunger.

Food insecurity also exists in industrialized nations, largely among low-income communities and communities of color. Activists working to increase food security in the United States tend to define it as access to healthy, affordable, culturally appropriate food through nonemergency means. Their strategies often consist of efforts to increase access to locally grown produce through farmers markets and community gardens. Some activists have moved beyond the concept of food security to food justice, which calls for a deeper analysis of structural inequality with regard to the distribution of food.

The Modernizationist Approach in the Global South

One approach to food insecurity, embodied by USAID and others working with them, is aligned with what those studying development call modernization theory. This approach argues that rural residents of underdeveloped countries need to modernize and participate more fully in the global economy to increase their ability to purchase food. USAID, for example, views the key to addressing food insecurity as increasing agricultural productivity. Their plan to increase food security contains six points:

- Improving policy frameworks to catalyze economic growth
- Bolstering agricultural science and technology
- Developing domestic market and international trade opportunities to ensure rural farmers adequate returns
- Securing property rights and access to finance
- Enhancing human capital through education and improved health
- Protecting the vulnerable through conflict resolution and transparency in public institutions

Taken together, these strategies encourage nation-states to create an economic climate in which rural, food-insecure farmers can sell increased quantities of food, thus raising their incomes and achieving food security.

Organizations approaching food security through the lens of modernization theory argue that rural producers need increased access to foreign and domestic markets. To this end, such organizations work to increase product quality standards, develop infrastructure for transport, and increase access to market information. For example, USAID works with small coffee farmers to increase quality, improve business practices, promote value-added approaches, and encourage producers to diversify into niche markets such as gourmet fruits and vegetables or environmental services. In doing so, producers can increase their incomes and thus their food security.

Modernizationists working to address food security are often strong proponents of increasing access to technology among rural farmers in the global South. They tend to support what is commonly called the green revolution, in which agricultural technologies such as pesticides, irrigation projects, synthetic nitrogen fertilizers, and improved crop varieties were made available to farmers in the global South. Modernizationists claim that production increases enabled by the green revolution have helped India to avoid famine. Indeed, renowned economist Jeffrey Sachs, director of Columbia University's Earth Institute and adviser to United Nations Secretary-General Ban Ki-moon, has stated that lack of chemical fertilizers is one reason for massive rates of hunger in Africa. Sachs and other modernizationists also tend to support the introduction of genetically modified crops. This technology, they argue, can provide food-insecure farmers with improved crop and livestock varieties, increasing agricultural productivity and economic growth.

The Dependency Approach

Critics of the modernizationist approach to food security are generally aligned with the dependency (or dependencia) paradigm in development studies. Dependency theorists believe that food security is not the result of lack of food but of unequal relations between the global North and South. The North's imperialism, these theorists believe, destroyed indigenous cultures and economies while exploiting the global South's natural resources for capital gain. These relations of production forced nations in the global South to accept a low position in the global division of labor. Imperialism also created a transnational capital class in which elites within the global South aligned their own interests with those of the global North and perpetuate their nations' dependence. Dependency theorists sometimes refer to the global South not as developing countries, which implies that they are moving toward a modernized state, but as underdeveloped countries. This latter term emphasizes the purposeful creation of dependency.

Dependency approaches argue that food security requires not only increased food production but also increased food access. According to Food First, a nonprofit food policy think tank located in Oakland, California, current rates of food production are sufficient to provide each human with more than 3,500 calories, or 4.3 pounds of food, per day. Rather than production, dependency approaches see food insecurity as rooted in poverty. As evidence, they cite the many nations in the global South that are home to millions of food-insecure people but are net food exporters. Ghana, for example, contains plentiful natural resources and has twice the per capita output of the poorer countries in West Africa but remains heavily dependent on international financial and technical assistance. Tellingly, its chief exports are nonfood products, such as timber and cocoa, which garner resources that might otherwise be used to advance food security. Even the nations most often hailed as green revolution successes, including India, Mexico, and the Philippines, have seen large increases in food production for export while hunger persists. Often, these agricultural exports feed livestock rather than humans. Recently, investment in food crops for ethanol has been linked to the current food crisis, as highlighted by dependency theorists focused on food security.

Those who view food security through the lens of dependency argue that agricultural participation in the global economy can increase hunger and that increased foreign aid cannot create food security. This stands in stark contrast to modernizationist approaches. USAID, for example, often mandates that receiving nations accept free trade and free market policies, thus undercutting local production. These dictates can be imposed through structural adjustment programs similar to those of the International Monetary Fund and World Bank.

Instead of increased foreign aid, dependency approaches promote eliminating obstacles to local production to allow the world's poor to feed themselves. Thus, their vision is not only food security but also food sovereignty, or the right to define one's own food system. Across the global South, a variety of social movements empower food-insecure people to achieve food security through food sovereignty. For example, Food First highlights the Campesino a Campesino (Peasant Farmer to Peasant Farmer) project in Mexico and Central America. Through this program, indigenous small farmers teach one another about soil conservation and water retention techniques including mulching, composting, and terrace planting on hillside slopes. These technologies enable campesinos to farm without dependence on pesticides and other inputs that make them dependent on U.S. agribusiness companies. Moreover, the Campesino a Campesino Project emphasizes a culture of simplicity and mutual respect, in contrast with the hierarchies of knowledge often imposed by agricultural extension agents and researchers. Projects such as these encourage food-insecure people to adopt techniques that eliminate their dependency on foreign aid, thus building their capacity to provide their own food security.

The dependency approach is most often embodied by grassroots projects in the global South and small development organizations in the industrialized North. However, it has also affected larger development projects. For example, the United Nations' 1996 Rome Declaration on World Food Security and World Food Summit Plan of Action recognized poverty, rather than food production, as the leading cause of food insecurity. Conflict, terrorism, corruption, and environmental degradation were also named as contributors. This document does call for increased food production, but it also acknowledges the need for sustainable management of natural resources, as well as the elimination of unsustainable patterns of production and consumption in industrialized countries.

Although many United Nations projects remain reliant on the import of both technology and expertise, some also favor using local production to directly address hunger. For example, Njaa Narufuku Kenya (Ban Hunger in Kenya) provides community nutrition and school meals, as well as food-for-work programs during times of low labor demand, and relies on local food purchasing. Thus it is Kenyan farmers, rather than U.S. commodity producers, who provide for local food needs. This represents an important shift in the approach of one of the largest international organizations working to bring about food security. Civil society continues to apply pressure on international agencies to move in this direction. In their response to the Rome Declaration, the Forum of Nongovernmental Organizations that paralleled the United Nations gathering released a statement titled "Profit for Few or Food For All? Food Sovereignty and Security to Eliminate the Globalization of Hunger." This statement provided a radical critique of the United Nations' approach to food security, emphasizing economic globalization and the global concentration of wealth and power.

Food Security in the Industrialized World: A Community-Based Approach

In the global South, food security stands in stark contrast to hunger. In industrialized nations, however, food security presents a paradox. Those who lack it are often overweight and obese and suffer from high rates of diet-related illnesses, including diabetes and heart disease. This paradox exists because the food most available to low-income people in industrialized countries, although containing plentiful calories, is utterly lacking in basic nutrition. The commodity foods distributed through federal assistance programs also tend to be low in nutritional value.

In the United States, a growing number of activists work to increase community food security. By adding the term *community*, these activists indicate that food security is not merely about an individual's needs but also a product of the built environment. The Community Food Security Coalition, a network of 300 North American organizations, defines food security as "a condition in which all community residents obtain a safe, culturally acceptable, nutritionally adequate diet through a sustainable food system that maximizes community self-reliance and social justice." Food security, according to this organization, is an interdisciplinary problem and can encompass factors such as income, transportation, food prices, nutritious and culturally acceptable food choices, food safety, environmental hazards, and access to adequate, local, nonemergency food sources.

The community food security movement has taken many cues from the dependency approach described above. It works to develop a local community's ability to become self-reliant, often through the implementation of alternative agrifood systems, rather than increasing their abilities to consume industrially produced foods. In this way, the community food security movement strengthens local and regional food systems, as well as the abilities of low-income people to participate in them.

For example, community food security activists often promote farmers markets. Although farmers markets tend to be located in affluent areas, those begun or embraced by the community food security movement can be found in neighborhoods that lack access to fresh, affordable, culturally appropriate produce. Moreover, the community food security movement encourages markets to accept Women, Infants & Children coupons or food stamps, gleaning programs through which farmers donate excess produce to food shelters, and opportunities for consumers and farmers to share ideas concerning market operation. The community food security movement has also embraced community gardens, farm-to-school programs, and economic development initiatives focused on local, sustainable agriculture and has encouraged these programs to locate in low-income areas and to maintain democratic participation. In addition, the movement encourages the development of food policy councils that can advocate for favorable food and nutrition policy.

Beyond Food Security: A Call for Food Justice

Although the community food security movement emphasizes social justice, critics argue that as a framework, community food security does not properly address the role of structural racism within the food system. These critics embrace the term *food justice* instead. The concept of food justice more fully integrates food security with an environmental justice perspective, addressing the racial and economic distribution of environmental benefits. The movement for food justice works to dismantle racism and to empower low-income people of color to promote sustainable food systems for their own communities.

One noteworthy example is Growing Power in Milwaukee, Wisconsin, which works to transform communities through the establishment of community food systems. Growing Power began in 1993 when Will Allen, an African American farmer, designed a program that gave local teenagers the opportunity to grow food for their community. Today, Growing Power operates a variety of urban and periurban farm sites in Wisconsin and Illinois, offers training in agricultural techniques, establishes youth programs, and works on agricultural policy initiatives. They also house an initiative called Growing Food and Justice for All—a network of individuals, organizations, and community-based entities working toward a food-secure and just world. This initiative's annual conference, listserv, and other activities enable those working for food justice to share strategies integrating antiracism with the creation of sustainable food systems.

Although the food justice movement has been largely domestic, its emphasis on antiracism and local empowerment is ideologically aligned with international peasant movements demanding the right to define their own food systems. Such movements are again aligned with the dependency approach described earlier, as they promote food sovereignty and local self-sufficiency, rather than increased participation in international markets. All of these movements share not only a critique of modernizationist approaches to food security but also a determination to build peoples' capacities to feed themselves.

See Also: Department of Agriculture, U.S.; Food First; Food Sovereignty; Green Revolution; Modernization.

Further Readings

Community Food Security Coalition. "What Is Community Food Security?" http://www .foodsecurity.org (Accessed December 2008).

Food and Agriculture Organization of the United Nations. "Improving Kenyan's Access to Food." http://www.fao.org (Accessed December 2008).

Food and Agriculture Organization of the United Nations. "Rome Declaration on World Food Security." http://www.fao.org (Accessed December 2008).

Growing Food and Justice for All Initiative. "About Us." https://www.growingfoodandjustice .org/About_Us.html (Accessed December 2008).

Growing Power. "About Us." http://www.growingpower.org/about_us.htm (Accessed December 2008).

Holt-Gimenez, Eric. *Campesino a Campesino: Voices From Latin America's Farmer to Farmer Movement for Sustainable Agriculture*. Berkeley, CA: Food First, 2006.

Lappé, Frances Moore, et al. *World Hunger: 12 Myths*. New York: Grove/Atlantic and Food First, 1998.

NGO Forum Statement to the World Food Summit. "Profit for Few or Food for All? Food Sovereignty and Security to Eliminate the Globalization of Hunger." http://www.converge .org.nz/pirm/food-sum.htm#ngo (Accessed December 2008).

Sachs, Jeffrey D. *Common Wealth: Economics for a Crowded Planet.* New York: Penguin, 2000.

U.S. Agency for International Development. "Food Security." http://www.usaid.gov/our_ work/agriculture/food_security.htm (Accessed December 2008).

Alison Hope Alkon
University of the Pacific

FOODSHED

The term *foodshed* describes the geographic area from which food flows into a community. It is analogous to a watershed—the catchment area for rainfall. A foodshed includes the rural and urban farmland, processing and distribution facilities, transportation systems, and wholesalers and retailers that make up a region's food system. Natural features like soils, topography, and land cover, as well as constructed infrastructure and public policies, determine the shape, size, and productive capacity of both foodsheds and watersheds. In recent years, the foodshed has been used to describe any localized food system that serves as an alternative to the global food marketplace.

A variety of benefits are attributed to a productive foodshed that can supply a particular community with some or most of its nutritional needs. These include energy and material efficiency, because of the shorter distance between producers and consumers, and decreased need for processing and packaging; improved taste and nutritional quality of foodstuffs that are able to be harvested at peak ripeness with minimal chemical or mechanical processing; regional farmland preservation, and the continued viability of small and medium-size family farms near cities; community food security in both urban centers and rural farming communities as a result of a thriving regional agriculture sector; and closer relationships between producers and consumers that also improves democratic decision making, trust, and accountability among individuals within the food system. To advocates of preserving cultural traditions, unique heritage breeds, heirloom fruits and vegetables, and established culinary techniques, the foodshed represents a place with distinctive physical and cultural characteristics (i.e., *terroir*).

The first use of the term *foodshed* is generally attributed to Walter P. Hedden, who in 1929 wrote about New York City's food supply in *How Great Cities Are Fed*. Hedden described the sources of the foods eaten by New Yorkers, the paths of food from producer to consumer, and the factors that directed the flow of food to the city's consumers. In contrast to New York's "milkshed," a relatively small, well-defined region circumscribed by the short distance that milk could be safely and cost-effectively transported in 1929, Hedden noted that the city's overall foodshed had become quite expansive, with fruits and vegetables traveling an average distance of 1,500 miles from farm to table—a figure still used today to emphasize that food travels great distances from farm to table.

Concerns about food security made the foodshed concept particularly salient in the 1920s. In October 1921, a planned nationwide railroad strike made real the possibility that an urban area as large as New York City, dependent on distant food supplies, would be at risk if those supplies were cut off and revealed that little attention had been paid to understanding and protecting the sources of food for New York and other U.S. cities. Metropolitan areas like New York also began to experience increased displacement of farmers by suburban residential development and, consequently, a sharp decline in the numbers of farmers selling directly to the public at city farmers markets. Expanding agricultural production in California and the South, coupled with improved transportation technologies and the rise of middlemen to broker and distribute shipments from producer to consumer, lengthened the distance from which perishable food, including milk, meat, and fresh produce, could be delivered to urban markets and increased the number of intermediaries in the supply chain. These changes made large cities like New York increasingly dependent on—and vulnerable to—disruptions in the food supply.

Some 60 years after the publication of Hedden's book, the concept of the foodshed was revived, although with a different, normative meaning centered on sustainability. To writers like bioregionalist Arthur Getz and rural sociologist Jack Kloppenburg, foodsheds are the areas that support locally based, alternative food systems. They are quite different from placeless global food supply chains that stretch around the world. In a global economy, with few barriers to food imports, the boundaries of a foodshed are neither precise nor impermeable, yet foodsheds, in contrast, are real places, with unique features such as soil types and plant communities, as well as identifiable cultural traditions and culinary patterns. Eating from within one's foodshed has the potential to minimize the negative social and environmental impacts of a globally based food system by reducing the distance between producers and consumers and by fostering sustainable production and distribution methods appropriate to the conditions and needs of the communities within the region.

Proponents of alternative food systems have described foodsheds as places within which networks of farmers, processors, marketers, and consumers can cooperatively support sustainable food production. This sense of the term *foodshed* includes conventional farmers and distribution networks and may also include cooperative forms of production and consumption, such as community-supported agriculture programs, farmers markets, community gardens, and larger urban and peri-urban farms. Described by Thomas Lyson as "civic agriculture," local foodsheds consist of clusters of smaller-scale businesses using sustainable practices and working together to supply fresher, more nutritious foodstuffs to small-scale processors and consumers, to whom producers are linked by the bonds of community as well as the market. These farmers, distributors, and consumers focus on protecting the foodshed in the same way that networks of citizens and environmental organizations work to protect the source waters of a watershed.

The term *foodshed* has also been used as an organizing principle for opponents of globalization and trade liberalization. The very notion of a foodshed, in contrast to the global food market, causes people to think critically about global food systems. The term *foodshed* helps to frame people's understanding of the food system—it not only encourages thinking about the kind of food system desired but may foster attachment to place and serve as the location for initiating changes to the food system. The foodshed is the region within which advocates can build relationships among consumers and producers.

Communities have begun to define the contours of their foodsheds for planning and policy development through the process of foodshed analysis—the study of a community's sources of food and the systems of food processing, transportation, and sales within the foodshed. Foodshed analyses can be assessments of existing conditions or strategic plans for increased sustainable food production within a foodshed. Foodshed analysis typically includes a community food assessment that identifies the farms, gardens, and other production facilities that provide food for a particular community; uncovers gaps between supply and demand in the local food system; maps available land for food production; tallies the needed processing, transportation, and sales infrastructure; and evaluates the policies that enhance or inhibit food production within a foodshed.

Researchers use a variety of methodologies to assess foodsheds. Some use life cycle analysis to determine the material and energy efficiency of feeding communities from local foodsheds compared with global food systems. Others examine the potential to increase production capacity on urban and periurban land adjacent to population centers. Still others examine the infrastructure needs of a foodshed to determine how to make the food system within a foodshed more efficient. Some foodshed studies examine networks of people and technologies that arise within a foodshed and their interactions.

Foodshed analysis enables activists to unmask the inefficiencies and environmental risks in the global food system and to propose alternatives. These alternatives can include programs to increase institutional purchasing of locally grown food, land use policies to support urban agriculture, economic development support for regional food-processing infrastructure, and a host of other policies to improve the viability of the local foodshed. Urban planners and designers use foodshed analysis to understand the movement of food within the food system of a community and as a framework for land use planning and design to support local foodsheds. They work to carve out land to support new farmers markets, provide technical assistance to community-supported agriculture programs, develop distribution infrastructure such as regional wholesale farmers markets, and design efficient transportation systems that both help farmers within the foodshed process and ship their food to urban markets.

Foodshed analysis also helps to highlight the unique food products, styles, tastes, production methods, and distribution channels that reflect the culture and history of a region. A variety of organizations advocate eating food produced within one's own foodshed to support unique food systems. Slow Food, an international membership organization founded in Italy to fight "fast food" and globalization, supports culinary traditions and the locally grown and raised foods on which they are based and emphasizes the environmental and cultural importance of eating within one's foodshed. Other groups, known as "locavores," have formed to commit to sourcing ingredients primarily from the local foodshed to demonstrate the feasibility of eating locally and seasonally for environmental, social, and gustatory reasons. Farmers, too, with the help of government agencies and nongovernmental organizations, have launched "buy local" marketing campaigns to highlight the qualities of food produced within the local foodshed and to urge consumers to preferentially buy and eat food grown locally.

See Also: Community-Supported Agriculture; Locavore.

Further Readings

Hedden, W. P. *How Great Cities Are Fed*. New York: D.C. Heath, 1929.

Kloppenburg, J., et al. "Coming in to the Foodshed." *Agriculture and Human Values*, 13:3:33–42 (Summer 1996).

Lyson, T. A. *Civic Agriculture: Reconnecting Farm, Food, and Community*. Medford, MA: Tufts University Press, 2004.

Nevin Cohen
Eugene Lang College
The New School for Liberal Art

FOOD SOVEREIGNTY

Food sovereignty was introduced into international policy debates at the 1996 World Food Summit by the international federation of farmers' organizations, La Via Campesina. The goal of food sovereignty unites rural social movements and international nongovernmental organizations working to right what they believe are injustices built into the rules of the World Trade Organization and other "free trade" pacts. These groups contend that the 1994 World Trade Organization Agreement on Agriculture and other trade-liberalization policies subordinate ecologies and human needs to the logic of profit and therefore cause increased environmental degradation and hunger.

A simple definition of food sovereignty is the ability of countries and communities to control their own food supplies—to have a say in what is produced and under what conditions it is produced, and to have a say in what is imported and exported. At the local level, food sovereignty entails the rights of rural communities to remain on the land and to continue producing food for themselves and for domestic and other markets if they so desire. Food sovereignty is an evolving concept; how and at what scales food sovereignty can be attained is a focus of debate and experimentation. After World War II, Japan became one of the first countries to adopt a strict food sovereignty policy to protect its domestic rice production, although the country remained a net food importer.

Advocates counterpose food sovereignty to food security, which, they say, is an inadequate framework for eliminating hunger. In World Trade Organization, United Nations, and other international negotiations, governments of major food-exporting countries—particularly the United States—contend that food security can be achieved by increased food imports by food-deficient countries. Food-sovereignty supporters say that this approach leads to greater vulnerability, which was dramatized when global food-price spikes in 2007–08 left many poor countries unable to purchase needed food.

Food sovereignty proponents stress that food is a universal human right recognized in the 1966 International Covenant on Economic, Social and Cultural Rights and other United Nations agreements. They maintain that human rights must take priority over trade policies that, in effect, protect the putative rights of private investors to pursue profits across borders. Although World Trade Organization rules enforce narrowly economic criteria for trade regulation, a food-sovereignty strategy would enable states to use multiple criteria in development planning and trade decisions. It would permit governments at various levels to pursue greater domestic food self-reliance and apply policies that discriminate in favor of productions and production methods on the basis of ecological sustainability, humane animal treatment, gender equity, fair labor practices, and other social goals.

Proposals to implement food sovereignty and realize the right to food include:

- Eliminate dumping—sales of farm commodities for less than their costs of production—and allow countries to protect themselves from predatory underpricing.
- Ban subsidies for export crops to limit overproduction and food dumping.
- Use domestic reserves and global supply commodity–supply management agreements to ensure adequate but not excessive food production and to increase prices that farmers receive for their products.
- Permit countries to prevent the ruin of domestic food producers by means such as import controls—quotas, tariffs, or price band systems—and preferential agricultural credit for farmers producing staples for local markets.
- Enact land reform to put neglected lands to productive use; recognize the individual or collective rights of food producers and do not saddle them with debt.

- Recognize farmers' rights of access to adequate water; maintain water resources as public goods.
- Accept the authority of municipal, state, and national governments to regulate food supplies in the public interest by: (1) requiring labels stating the origins and production methods of foods and crops; (2) deciding whether to accept genetically modified food imports and whether and on what terms to permit the use of genetically engineered crops; and (3) banning the private patenting of living organisms and genetic information.
- Protect the rights of farmers to save seeds and breed livestock, including patented varieties, for exchange, replanting, and improvement.
- Require living wages and safe working conditions for agricultural and food-sector workers.

Food sovereignty is more than an alternate set of trade rules, it is a different way of understanding agriculture and the role of food, farming, and rural life in national and community development. Its advocates hold that food is first a source of nutrition and cultural meaning and only secondarily an item of commerce. Trade is beneficial as a means to social well-being, they say, but not as an end itself.

Food-sovereignty proponents argue that the maintenance of healthy agrarian communities, backed by policies to support domestic food production, are better guarantors of food security than what they describe as the "global-corporate food regime." They point out that a shrinking small number of transnational firms dominate food production, processing, transport, and retailing. Farm inputs and animal feeds are transported to distant feedlots and fields; food commodities then travel again around the globe to reach consumers.

Supporters of food sovereignty contend that decentralized, diverse, and locally adapted farming systems can be more environmentally sustainable than such a globalized agrofood system. Where livelihoods and family goals are tied to the longer-term health and productivity of the land, they say, farmers have more incentive to conserve and improve soils, landscapes, and water systems. In contrast, in a food system dominated by transnational agribusiness, the competitive imperative to maximize profits compels companies to externalize their environmental costs, shifting them onto the public and future generations.

Food-sovereignty supporters are skeptical about technological solutions to hunger. They point out that sufficient food is produced globally but contend that lack of control over food supplies denies poor communities and countries access to adequate food. They observe that genetic engineering and other technology-centered agricultural research has brought no significant productivity breakthroughs since the green revolution, which itself resulted in environmental damage and displaced millions of farmers. They note that small-scale farms are often more productive than large-scale, high-external-input "industrial" farms and argue that agroecological methods can raise small-farm productivity. Thus, for most proponents of food sovereignty, hunger and inequality are mutually produced and inseparable from ecological sustainability. As food-sovereignty activists express this idea, there can be "no ecology without equity, and no equity without ecology."

See Also: Agribusiness; Agroecology; Agrofood System (Agrifood); Doha Round, World Trade Organization; Export Dependency; Land Reform; Sustainable Agriculture; Trade Liberalization.

Further Readings

Desmarais, Annette. *La Via Campesina: Globalization and the Power of Peasants*. London: Pluto, 2007.

Friedmann, Harriet. "The Political-Economy of Food—A Global Crisis." *New Left Review*, 197 (1993).

McMichael, Philip "Global Development and the Corporate Food Regime." http://www.tradeobservatory.org/library.cfm?refID=37655 (Accessed February 2009).

Smaller, Carin and Sophia Murphy. "Bridging the Divide: A Human Rights Vision for Global Food Trade." http://www.tradeobservatory.org/library.cfm?RefID=104458 (Accessed February 2009).

Via Campesina. "The Doha Round Is Dead! Time for Food Sovereignty." Statement July 2006. http://www.viacampesina.org/main_en/index.php?option=com_content&task=view&id=196&Item (Accessed February 2009).

Windfuhr Michael and Jennie Jonsén. "Food Sovereignty: Towards Democracy in Localized Food Systems." FIAN International/ITDG Publishing. http://www.ukabc.org/foodsovereignty_itdg_fian_print.pdf (Accessed February 2009).

Kathleen McAfee
San Francisco State University

Fruits

Fruit production during much of the 20th century depended heavily on the use of chemical pesticides, fungicides, herbicides, and fertilizers. Public concern emerged over the effects of chemical residues on human health. Environmentalists realized that production methods could also have harmful effects on soil, groundwater, and wildlife. Better methods of fruit production have been developed, based on a combination of preindustrial practices and new techniques. As the agricultural community becomes increasingly aware of these new techniques through extension services, suppliers, and education programs, more farms are expected to shift toward sustainable production.

Bananas are susceptible to parasites and other problems, leading to high levels of chemical use by growers. This banana plant was damaged by an infestation of red palm mites, which caused the discoloration on its leaves.

Source: U.S. Department of Agriculture, Agricultural Research Service/Amy Roda

The Rise of Large-Scale Production

In the 19th century and early 20th century, many rural families grew fruit for their own consumption. Small commercial growers provided fruit for local markets. Once insulated and cooled railroad cars became available around 1870, fruit could be shipped longer distances, although it was not until 1889 that fruit was shipped across the continent from California to New York. With transcontinental shipping, large-scale fruit production was feasible. By 1890, both insecticides and fungicides were being applied to fruit crops, and although there were concerns regarding pesticide residues on fruits, the small number of applications applied over the growing season minimized health risks. As time went on, however, insect pests developed resistance to these chemicals, and application rates were increased. New compounds were developed to control pests, and large-scale spraying techniques were employed. People grew to expect unblemished fruit, and ultimately, consumer demand for unblemished fruit encouraged agricultural practices employing the extensive use of chemicals.

Growing Public Concerns

From the 1960s through the 1980s, public pressures arose over the health risks associated with consuming chemical residues on fruits. Activists for farm laborers publicized concerns over the risks of their pesticide exposure. Environmentalists noted that commercial fruit production often resulted in pesticide-contaminated soils and waterways, increasing pest resistance to chemicals, erosion, excessive water use, habitat destruction, and loss of wildlife, as well as atmospheric emissions from agricultural equipment, fertilizer production, and fruit transport. Because many fruits cannot be grown locally for all markets, the economics, as well as the energy and carbon footprints, of transport and storage were also analyzed.

The combination of health and environmental safety concerns forced a reexamination of the way that fruit is grown. Although the consumer still wanted unblemished fruit, consumer demand also induced changes in fruit production. By 1993, the U.S. Department of Agriculture (USDA), the Environmental Protection Agency (EPA), and the Food and Drug Administration (FDA) announced a commitment to reducing pesticide use and promoting sustainable agriculture. The United States withdrew many compounds formerly used to treat crops and increased restrictions on the use of many others. At the same time, fruit consumption was promoted as an important component of a healthy diet. As nutritionists learned more, USDA recommendations for fruits and vegetables rose from 2 cups per day in 1979 to 4.5 cups per day in 2005.

Although chemicals are still widely used in fruit production, in both Europe and the United States, there is an increasing emphasis on integrated production, which advocates sustainability and environmental health. Integrated Fruit Production (IFP) was first introduced in Europe in the 1950s but was not widespread there until the 1980s and 1990s. More recently, regional IFP guidelines have been drawn up for areas of the United States. IFP is practiced primarily for pome fruits and grapes. IFP-certifying organizations award certification points to farms for following recommended practices but do not require farms to follow all practices. Although these techniques have not spread to all regions, they represent environmentally friendly practices. Organic fruit producers follow similar guidelines but must also be certain to avoid synthetic chemicals, with the exception of approved chemicals that are permitted when biological and cultural controls are insufficient. In the United States, organic fruit must be certified by a USDA-approved certification agency that ensures that the producer meets federal organic standards.

Sustainable Fruit Production Methods

Sustainable fruit production stresses field-level assessment, which means examining all aspects of the growing process to reduce environmental impact and to minimize inputs such as fertilizer, water, and chemicals. If the producer has the luxury of developing a new farm, this begins with examining the local environment and the landscape for planting. Spacing of plants and orientation of crop or tree rows is important in ensuring that each plant has adequate sunlight and ventilation to minimize molds. Planting in relation to the contour of the land can minimize erosion and ensure adequate drainage. Before planting, fields are tilled, sometimes after several repeated plantings of a cover crop that increases organic material in the soil. Organic material helps reduce problems with root fungi. Organic fertilizers such as manure, fish emulsion, and compost may also be added to the soil. After planting, straw, leaves, woodchips, or other organic materials may be placed between plants, or a cover crop may be planted. Mulching can provide habitat for insect predators. Cover crops can be rotated over different years to minimize disease organisms. Weeds can be removed manually, or geese or chickens can help keep them under control. Steam and flame weeding are also effective, nonchemical methods for controlling weeds.

Ideally, the landscape surrounding the orchard or field should not harbor alternate host plants of disease organisms. However, there should be sufficient natural habitat to maintain good populations of favorable organisms such as pollinators; insect predators such as bats, spiders, and birds; and beneficial insects such as parasitic wasps that can control pest outbreaks. Native insects are often good pollinators of fruit crops, and maintaining natural habitat can encourage their presence. Cover crops with abundant flowers, planted between rows, may do the same. Even if the grower begins with an established orchard, he may be able to establish some wild spaces along adjacent access roads, fence lines, and hedgerows or near outbuildings.

Next, choosing varieties of fruits that do well under local growing conditions is important. Factors to consider are temperature regimes, length of growing season, and moisture availability. Choosing varieties that are resistant to locally known disease organisms and buying disease-free stock will reduce the need for chemical pesticides. In the United States, the USDA Cooperative Extension System can often provide information about the best local varieties, as well as information about what pests are likely to be problems in the region. Market values, storage requirements, and salability are also taken into consideration.

Once the crop is planted, monitoring and forecasting help prevent pest outbreaks. Prevention, rather than treatment, is the goal of sustainable fruit production. Monitoring means constantly checking the crop for the presence of pest organisms and noting the levels at which they occur. Sticky traps and other types of insect traps can help the grower detect the presence of unwanted pests. Proper identification of insects or diseases prevents unnecessary spraying. Again, the Cooperative Extension Service can often be of use by providing identification materials. Local Integrated Fruit Production organizations may provide threshold guidelines for different pest organisms, so that growers know when actions should be taken. The use of chemicals is discouraged when preventative measures are available, but they are still an option if pest levels exceed the threshold. Then, the least harmful chemicals are employed. Forecasting uses weather information to predict when particular pest organisms are likely to become problematic, so that they can be recognized, monitored, and dealt with in the least environmentally harmful way possible.

Several other methods for controlling pest insects are available. First, preventing water stress and maintaining healthy soil makes plants more resistant to insects. Biological controls such as predatory mites, ladybird beetles, and parasitoid wasps can be purchased to keep certain problematic insect populations low. Pheromonal control that disrupts the mating of lepidopterans can reduce the numbers of fruit moths and borers without harming other insects or human health. Avoiding high humidity from irrigation and pruning to reduce interior foliage of orchard crops can minimize fungal problems. The removal of dropped fruits and diseased plant material that would harbor pests also is effective.

For larger problem animals, other techniques can be employed. Scare devices (such as mylar flash tape that flutters in the wind) and netting can help discourage fruit-eating birds, and tree guards can protect trunks from rabbits. Keeping mulch several inches away from tree trunks helps prevent rodents from damaging trunks, although additional trapping may be necessary. Sometimes owls and hawks can be encouraged by providing suitable perches or nesting sites. Repellants are available to help keep deer away.

Banana Production—A Special Case

Most of this article has focused on fruit production in Europe and in the United States. However, the world's fifth-largest agricultural commodity is bananas, which are produced in tropical climates. In some parts of Central and South America, banana plantations have been responsible for tropical deforestation. Banana cultivation depletes the soil, resulting in the expansion of plantations into the rainforest. Bananas are susceptible to a number of fungi, insect pests, and viruses that have spread widely because bananas are planted as monocultures, and most of the world's commercial bananas are of one variety. This has resulted in heavy use of fungicides and insecticides that are harmful to laborers and to the environment. Contaminated runoff has been implicated in coral reef destruction and in the death of algae important to marine life. Several major suppliers have recognized these problems and implemented new standards. Smaller-scale organic producers have developed polyculture plantations using crop rotation and have started plantations in drier areas, where fungal diseases are less problematic. The Rainforest Alliance runs a banana certification program, and several of the large international banana suppliers have adopted these standards for the farms that produce their fruit. Only about 15 percent of bananas are produced with this certification, but even this has produced notable changes in the environment. Since the average American eats about 28 pounds of bananas a year, American consumers can have a large effect on market demand.

Pesticide Residues on Fruit

The Food and Drug Administration analyzes both domestic and imported fruit for pesticide residues. In its 2006 report, 0.9 percent of domestic fruit violated EPA-acceptable tolerance levels of pesticide residues, 54.9 percent showed residues that did not violate tolerance levels, and 44.2 percent showed no residues. Of imported fruits, 3.6 percent violated EPA tolerance levels and 22 percent showed residues below tolerance levels; 70.4 percent was pesticide-free. Some of the violations represent pesticides for which no tolerance level has been established for a particular fruit, indicating that it was used on another crop and the fruit was contaminated, or it was used in a way that was not licensed. The presence of pesticide residues is measured on whole, unpeeled fruits—even for bananas, oranges, and other fruits that are peeled before consumption.

Toward a New Future

Growing fruit using cultural, mechanical, and physical techniques, with biological controls, reduces the need for chemical pesticides, fungicides, and synthetic fertilizers. Fruit production tends to be labor-intensive, and these techniques tend to make it even more so.

Thinning of flowers and fruits is done by hand rather than chemically, and weeding and culturing may require additional hands. In contrast, laborers tend to prefer working on organic farms, and there are often waiting lists of people looking for these healthier jobs. Organic certification requires recordkeeping, reporting, and paying certification fees. Thus, the produce from these types of farms may be more expensive. Many consumers are willing to pay the cost, either for their own health or because of their concerns for the environment. As more people become aware of the advantages of sustainable fruit production, commercial growers may find a greater demand for these products.

See Also: Agriculture Extension; Certified Organic; Department of Agriculture, U.S.; Food and Drug Administration.

Further Readings

Ames, Guy K. and George Kuepper. "Tree Fruits: Organic Production Overview" (2004). ATTRA, National Sustainable Agriculture Information Service. http://attar.ncat.org/attar-pub/PDF/fruitover.pdf (Accessed January 2009).
Cornell University Cooperative Extension. "Integrated Crop and Pest Management" (2008). http://ipmguidelines.org/TreeFruits/content/CH01/default.asp (Accessed January 2009).
Granatstein, David. "Tree Fruit Production With Organic Farming Methods" (2003). Wenatchee, WA: Center for Sustaining Agriculture and Natural Resources, Washington State University. http://organic.tfrec.wsu.edu/OrganicIFP/OrganicFruitProduction/OrganicMgt.PDF (Accessed January 2009).

Carol Ann Kearns
Santa Clara University

FUNCTIONAL FOODS

Although there is no universally accepted definition of functional foods, this term essentially describes foods that contain bioactive compounds believed to result in particular health benefits. These health benefits range from the facilitation of physiological functions (e.g., digestion, bone growth) to the prevention of disease (e.g., heart disease, breast cancer). Tomatoes, for example, are considered a whole functional food. They contain significant amounts of lycopene, a compound thought to maintain prostate functions. Functional foods also include processed foods with functional additives such as energy drinks and fiber-enriched cereals. In some ways, "functional food" is just a new term for the age-old belief that the consumption of certain foods will result in particular bodily changes. However, in a world of increased biochemical research and regulatory control, functional foods blur the line between food and drug (evidenced best by their alternative name of "nutraceuticals") and thus warrant careful study. A social examination of functional foods is quite relevant to the larger study of green foods for several reasons. On one hand, linking health benefits with fresh, organic fruits and vegetables may inspire people to support more local, organic, and diversified farms, thereby increasing the viability of sustainable agriculture. On the other hand, increased emphasis on functional foods may also usher in a greater privatization of the agrofood system, especially in terms of food science research.

Functional food's recent growth is indicative of a larger ideological swing wherein people understand a healthy diet as a collection of particular nutrients, as opposed to particular foods or meals. Many Westerners today strive less for a balanced diet of the major food groups and more for consumption of "good" nutrients (e.g., omega-3 fatty acids, fiber, antioxidants) and avoidance of "bad" nutrients (e.g., omega-6, saturated fats, simple sugars). Accordingly, "good" nutrients lead to better physiological functions, therefore driving consumers toward foods associated with particular functional ingredients. Much of the nutritional information that informs consumers about bioactive compounds and the foods that contain them comes from television, magazines, and other media sources. In addition, a great amount of information linking certain nutrients to health effects comes in the form of functional-food packaging.

Allowing health claims on food packaging is very controversial. Food manufacturers argue that there is no harm in placing health claims on food packaging and that instead, such packaging will lead to greater public health. Accordingly, they contend that such labels will better inform consumers about healthy food choices, leading to a more nutritionally conscious and healthy society. They also argue that allowing health claims on food labels gives private companies a vested interest in supporting more food science research that will otherwise go unfunded.

Many public health professionals disagree with the belief that health claims on packaging will lead to improved public health. First, nutritionists question how generalizable functional foods and their attached health benefits are. Second, they argue that the supposed health benefits of functional foods are often overstated. For example, studies that link the health claims to particular nutrients or foods are often inconclusive or limited to certain demographic groups, making their relevance to greater public health questionable. Some nutritionists also doubt the effectiveness of processed functional foods, questioning whether the functionality of certain ingredients is overridden by other non-health-promoting ingredients (e.g., salts, sugars, pesticides). Public health professionals also worry that allowing health claims on packaging will place the responsibility for food research funding in the private sector, resulting in research that will benefit a few private companies and their products more so than the public good. Instead, these professionals argue that public health will be best ensured by environmental protections, equal access to healthy foods, and health education.

Another controversy regarding functional foods is regulatory. At this time, only several countries around the world—the United States, the United Kingdom, and Japan—allow for the marketing of health claims on functional-food packaging, and those claims are carefully regulated. Since the Food and Drug Administration (FDA) in the United States regulates the marketing of foods, dietary supplements, and drugs very differently, functional food's blurry food/drug status leads to new regulatory terrain for the FDA to negotiate. At this time, there is no specific legislation for functional foods, leaving the Dietary Supplement Health and Education Act of 1994 as the closest precedent for control. The Dietary Supplement Health and Education Act allows for supplement packaging to boast structure/ function claims (e.g., lycopene maintains prostate function), but specific disease-prevention health claims (e.g., lycopene prevents prostate cancer) need to be individually approved by the FDA. Food manufacturers argue that these FDA regulations ensure that only well-researched claims make it on their packages, thus protecting and educating consumers. Opponents of health claims on packaging worry that the high public costs of regulation outweigh the benefits to public health available through functional food consumption.

Ultimately, how much functional foods influence our diet choices depends on which public health paradigm dominates our society. If the medical paradigm achieves more dominance with policymakers, we will likely see a marked increase in functional food advertising. Conversely, if the community health paradigm wins out, health education will deemphasize the benefits of particular functional foods and instead promote a diet high in whole grains, fruits, and vegetables. Either way, we will most likely see an increased demand for fruits and vegetables, especially those with known functional ingredients.

See Also: Agrofood System (Agrifood); Food and Drug Administration.

Further Readings

Brogan & Partners. "Functional Food Fights." *Environmental Health Perspectives*, 107/9 (1999).

Crawford, Robert. "Health as a Meaningful Social Practice." *Health: An Interdisciplinary Journal for the Social Study of Health, Illness and Medicine*, 10/4 (2006).

Lawrence, Mark and John Germov. "Future Food: The Politics of Functional Foods and Health Claims." In *A Sociology of Food and Nutrition: The Social Appetite*, edited by J. Germov and Lauren Williams. Oxford: Oxford University Press, 2004.

Milner, John A. "Functional Foods: The U.S. Perspective." *American Journal of Clinical Nutrition*, 71/6 (2000).

Scinis, Gyorgy. "On the Ideology of Nutritionism." *Gastronomica*, 8/1 (2008).

Christie Grace McCullen
University of California, Davis

Genetically Modified Organisms

All cultivated plants have been genetically modified over centuries through traditional processes of selection and breeding. The term *genetically modified organism* (GMO) specifically describes a type of genetic modification in which the DNA (deoxyribonucleic acid) of microbes, plants, and animals is directly altered. GMOs resulting from recombinant DNA technology—moving genes from one species to another—are also called "transgenic." Unlike genetically modified (GM) products in the pharmaceutical sector, such as insulin, GMO food and animal products have engendered public resistance and are highly controversial on account of mutually overlapping concerns of health, environment, economics, and ethics. There are numerous controversies surrounding the use of GMOs, but there is a general consensus about the fact that they are very complex and that there should be more research before taking further decisions about their use.

GM Products

First-generation GM crops enable producers to reduce production costs and more easily control disease, insects, and pests. These crops are more or less similar or "substantial equivalent" to non-GM counterparts when it comes to appearance, taste, and nutrition value. The second-generation GM crops, also called value-enhanced crops, focus on consumer-oriented benefits like enhanced nutritional quality. Third-generation GM crops include those that are altered to produce pharmaceuticals, vaccines, or biologics.

The first GM crop to appear in the market was Flavr Savr™ tomatoes, with slower ripping and longer shelf life. Roundup Ready soybean and corn were introduced in 1996 and 1998, respectively. Designed by Monsanto, Roundup has an herbicide called glyphosate as an active ingredient. Roundup can be sprayed on GM soy and corn cultivation to kill the weeds without damaging the actual crops. In 2005, about 85 percent of the soybean cultivated in the field was glyphosate tolerant.

Bacillus thuringiensis (Bt) plants and crops are the most widely used GM products. Bt is the name of a bacterium that was isolated from soil in 1911. It has been used in pest control since 1930, but in the last few decades, its consumption has increased. Monsanto and other biotech corporations found a way to insert the toxin-producing gene from Bt

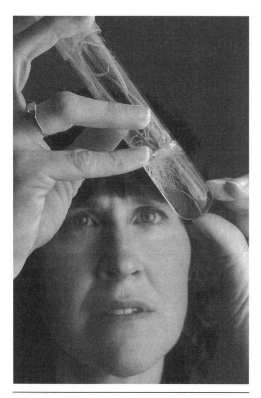

This geneticist is working with genetically modified wheat plants that may be resistant to the fungi *Fusarium* which can lead to costly crop losses.

Source: U.S. Department of Agriculture Agricultural Research Service/Jack Dykinga

bacteria into plants. This enabled the Bt plant or crop to produce its own pesticide, and therefore kill the insects eating it. Statistics indicate that in 2006, approximately 11 percent of corn and 33.5 percent of cotton grown around the world was Bt.

Another GM product on the market is recombinant bovine somatotropin milk. An artificial growth hormone, recombinant bovine growth hormone is synthetically produced to mimic bovine growth hormone protein hormone, which is produced in the pituitary glands of cows and other cattle. When injected in cattle, recombinant bovine somatotropin increases milk production by preventing mammary cell death.

Advantages of GMOs

GMO crops have the potential to improve crop productivity on existing land and water resources, either by increasing the yield potential of plants or by reducing the effect of biotic and abiotic challenges. Several GMO crops are resistant to pests and disease. Crops that are resistant to abiotic challenges, such as drought, soil acidity, and salinity, have the potential to bring marginal land under agriculture while increasing productivity on existing farmland. For example, drought-resistant soybeans and rice, salt-tolerant rice, maize, tomatoes, rapeseed, and so on all have been engineered. Cost-saving measures like improving storage stability, delayed ripening, and other changes that provide flexibility to manage products have also being explored.

By not requiring pesticides, GM crops like Bt corn, soy, and cotton can also remove the negative effects of such chemicals on health of the producers. GM crops prevent soil and water degradation through reduced tillage and reduced application of pesticides and herbicides. There also are improvements in soil organic matter, soil structure, soil water relation, and reduced soil compaction that contribute positively to the agroecosystem.

GM crops can enhance nutritional and health benefits and have the potential to produce pharmaceuticals and vaccines. For example, engineered rice can help target the deficiency of vitamin A, which causes partial or complete blindness, and iron, which causes anemia. Cultivation of engineered crops used to produce pharmaceuticals and vaccines, called pharming, has the potential to serve the poorest population, as these pharmaceuticals and vaccines are significantly cheaper and delivered easily without requiring refrigeration, trained administrators, sterile needles, and so forth.

Environmental Concerns

GM plants that are designed to resist pests and disease can adversely affect nontarget organisms, however. This concern surfaced with the controversy involving the possibility of deleterious effects of Bt corn pollen on the monarch butterfly population in 1999. Since then, researchers have tried to ascertain the direct and indirect effects of GM on nontarget and beneficial organisms. Critics of GM crops also brought attention to the negative effects of GM-fed animals that become prey of beneficial organisms, thereby disturbing the delicate agroecosystem.

The possibilities of gene flow—transfer or exchange of genes between species—also pose severe environmental risks. An herbicide-resistant plant can interbreed with its wild relatives and other plants that are capable of hybridization and make them herbicide resistant. In fact, GM crops are more likely to interbreed with other species because they are engineered to surpass natural barriers of reproduction. Another ecological implication is the development of weeds related to the GM crops into superweeds that become herbicide resistant. This can lead to crop losses and increased use of herbicides. Superweeds can lead to "bioinvasion," where they take over the entire ecosystem. GM crops used to produce pharmaceuticals present the chances of environmental and human disaster and, if not confined, can spread to non-GM crops.

There also can be the development of resistance in target insects by consumption of Bt crops, which excrete Bt even during the growing season. In traditional agriculture, pesticides are sprayed only during short periods, and so the danger of pests developing resistance is small. The U.S. Environmental Protection Agency and biotech companies propose different techniques to tackle this problem. Use of a high-dosage strategy ensures that insects are not exposed to sublethal levels of BT that can help develop resistance. Another technique is to plant non-Bt crops near Bt crops to provide refuge for the susceptible insects and pests. When the resistant insects interbreed with the nonresistant insects from non-Bt crops, their hybrid offspring will be nonresistant to the toxins. However, these insect-resistant management techniques are not easy to enforce, and hence pose environmental risks; namely: (1) insect-resistant management sometimes demands community-level coordination among neighborhood farms, which is often not complied with; (2) farmers with small farms who grow Bt crops are hesitant to grow less-productive non-Bt crops on their land; and (3) if non-Bt crops require herbicides, the whole crop has to be sprayed, which makes the use of Bt crops immaterial.

With the cultivation of GM crops, the land use pattern is changing from polyculture to monoculture. With increasing commercialization of agriculture and the advent of the green revolution, monoculture—planting the same crop in the entire field—has gained prominence. However, with the introduction of GMOs, monoculture has reached a new level, as now all the crops in the field are not only of the same species but are also identical genetically. This ignores the known fact that symbiotic relationships between different plants increase the quality and production of the food grain and require very little fertilizer and other chemicals. Monoculture, hence, can affect biodiversity and increase ecological vulnerability. For example, in the district of Warangal in Andhra Pradesh, India, the mixed farming pattern based on cultivation of millets, pulses, and oilseeds was rapidly converted to a monoculture of hybrid cotton. Hybrid cotton planted year after year led to very high pest buildup, despite the use of high doses of chemicals, and resulted in the failure of crops in 1998.

Economic Concerns

Some farmers and food system activists maintain that cultivating GMOs is more expensive because of the higher costs of patented seeds, technology fees, use of chemicals, and costly related insect-resistance management. Genetic use restriction technology, also known as terminator seeds, can further increase the cost by forcing farmers to buy other products such as gene activator and rule out the possibility of saving seeds for future planting.

Poor farmers are thus subordinated to the neoimperialist control of rich corporations like Cargill, Monsanto, and so on, and have to pay for expensive seeds, insecticides, and other chemicals. Also, reliance on seeds from biotech companies runs counter to the idea of self-sufficiency.

Gene flow from GMOs contaminate natural crops and can affect the business of organic farmers, who rely on the trust of their consumers, and also of traditional producers supplying their crops to the countries that do not allow genetic engineering. Critics claim that big corporations take advantage of gene flow, and when GM crops are discovered where they are not supposed to be, the poor producers are sued for the violation of intellectual property rights.

Vandana Shiva, a world-renowned social activist and a strong critic of GMOs, maintains that the weeds that are killed by herbicides like Roundup are in fact food for the poor farmers. In polyculture farming, about 150 varieties of plants that grow in the field with the main produce are consumed either as food, medicine, or fodder and so the purported benefits of high production of GM crops are very myopic.

Big corporations can slow the research in the public sector, demanding heavy costs related to the intellectual property rights. Some maintain that corporations like Monsanto spend large amounts of money to develop patented GM food where lucrative markets exist. Corporations, however, are unlikely to be interested in developing crops to feed the poor and starving, as these people are without money to create a profitable market.

Health Concerns

Health concerns include

- allergic reactions resulting from consumption of GM food—a major health concern;
- gene transfer from GM food to the cells of bacteria found in the gastrointestinal tracts of humans, which can affect human health. Critics maintain that the antibiotic-resistant marker genes in GMOs can pass into humans and restrict their ability to fight infections;
- according to critics, a large range of potential risks to health resulting from uncertainty about the stability of genes and the effects of cell insertion. It is also maintained that because the whole GM process is so complex, the nutritional effects of the modified genes are incomprehensible.

Cultural Concerns

GM crops are changing the land use pattern from a community-based model of agriculture to monoculture and commercial production of crops. With the privatization of the seed sector, major changes have occurred in the cultural practice of food production. For example, traditional Indian agriculture is marked by the exchange of seeds and a barter system between different producers. The intellectual property rights for GMOs make this culture of seed reciprocity illegal. Treating cows with growth hormones is especially unethical on religious grounds in India, where cows are considered sacred.

Ethical Concerns

On ethical grounds, critics maintain that the development of GMOs is tampering with nature and disturbing the balanced ecosystem for human greed. Also, many cultures around the globe consider life to be sacred and do not believe in patenting or ownership of life forms. Ethical debate also surrounds the controversial issue of whether or not to label GM products, which is critical to the rights of individuals to know and say no to what they consume. Radical greens argue that GM technology follows the capitalist patterns of industrialization and intensification of agriculture, with heavy reliance on a technical fix rather than attending to the underlying problems. For example, Dave Toke maintains that even though biotech companies promise better environmental sustainability, it is only in relation to conventional agriculture and not in comparison to low-input or more sustainable organic farming. These alternative practices deal with agroecological problems in more sustainable ways. Another important issue is when the genetic traits used for GMOs come from local seed banks meant for collective use. Thousands of species of grains that have been selected and bred naturally by our ancestors are stored in such banks. Companies like Monsanto develop GMOs by using genes from these species and then collect royalties even from the communities whose ancestors created it.

GMOs: Controversies and Debate

Resistance to and debate about GMOs surfaces across multiple axes of confrontation between developed/developing, rich/poor, and producer/consumer. Although GMOs have more readily been accepted in the United States, the European public has actively resisted the use of GM food. There are various reasons for this difference. First is the contract between American values of industrial agriculture, grounded in values of competition and accumulation, versus European values of agriculture, associated with the traditional, idyllic way of life in rural areas and principles of solidarity. In many Third World countries, agriculture defines the way of life and influences and organizes the culture, religion, and social life. Second, the unwavering trust in U.S. of government agencies like the Food and Drug Administration (FDA) and the U.S. Department of Agriculture versus European views and skepticism about the political economic nexus of government regulators and biotech corporations like Monsanto, Cargill, and so on. Critics maintain that in the United States, FDA involvement is more for consultation than for approval. The FDA merely does an audit of risk assessment of the products provided by the biotech companies and does not conduct any major scientific review of the data provided by the companies. There is also a general feeling that big companies are willing to play with public health and the environment by downplaying potential risks for profit. In Europe, the European Commission demands verification of the information provided by companies and may conduct further risk assessments if not satisfied. Third, Americans follow the principle of substantial equivalence that first emerged in the 1990s to remove bureaucratic red tape and overregulation of biotech companies, according to which, if a new product is found similar to the naturally existing counterpart in composition and nutritional value, it can be considered safe to consume. This depicts the general behavior of the U.S. government, which ignores the scientific unknown by evoking the term "sound science." The European view is framed by the "precautionary principle"; that is, if potentially dangerous effects of anything are identified, lack of certainty about them should not bar preventive actions. On the basis of this principle of "guilty until proven innocent," the European Union issued a five-year moratorium from 1999 until 2004 on GMOs for environmental and health safety.

Controversies surrounding the principle of "substantial equivalence" also surface in connection with GMO patent rights. Critics maintain that if GMOs are identical to the non-GM products, then the intellectual property rights should not be applied; and if they are not identical, then the principle of substantial equivalence is spurious. Consumer politics also take up the issue of labeling of GM crops as their right to information. On the grounds of "substantial equivalence," major companies in the United States have completely ignored the need to label GM products. Of course, there were practical problems with labeling, such as mixing GM and non-GM produce by farmers themselves, extra conscientiousness during various processing stages, and so on. However, consumer groups in the United States maintain that if labeling can be done in Europe, it can certainly be done in the United States. However, large companies maintain that this can be very costly, and some find it more reasonable for the producers of non-GM products to label their products "GM-free" when more than half of the products in grocery stores contain some amount of GMOs.

Although the critics maintain that there is enough food to feed the world, proponents of GMOs use the backdrop of a starving Third World population to promote their case. They say that the consumer politics of the First World ignore the desperate needs of the Third World population. If GMOs can be readily accepted as life-saving medicines, then they should also be used to feed starving masses of the Third World, who may soon die because of hunger. Some maintain that even though the staple crops in several African countries are cheap and easily available, there is still malnutrition because the nonstaple foods rich in micronutrients, vitamins, and so on are very expensive. In such areas, GMOs such as functional foods can provide a cheaper alternative to the nonstaple foods like fish, fruits, pulses, and so forth. The proponents of GMOs also maintain that there is enough food in the world, but feeding everyone will require an expensive global food distribution system, the possibilities for which appear bleak. The opponents respond to this debate of feeding the Third World population by saying that this is just another way for the First World to dump toxins into the Third World. Some African countries, like Zambia in 2002, have refused to accept GMOs as aid in crises. Critics also say that companies simply use the discourse of "feeding the starving Third World" to capture First World markets.

Finally, both sides maintain that facts about GMOs are often inaccurately represented. For example, the research on the potential decrease in the population of monarch butterflies attributed to Bt-corn pollen was proven false, but this result was never reported by the media. Similarly, the research and controversy of Arpad Pustai, which informed the public of some potential health risks, made headlines in Europe but were hardly seen in U.S. media.

See Also: Bt; Functional Foods; Monsanto; Recombinant Bovine Growth Hormone; Roundup Ready Crops.

Further Readings

Anderen, Per Pinstrup and Ebbe Schioler. *Seeds of Contention: World Hunger and the Global Controversy over GM Crops.* Washington, D.C.: International Food Policy Research Institute, 2001.

Chetty, L. and C. Viljoen. "GM Biotechnology: Friend and Foe?" *South African Journal of Science*, 103 (July/August 2007).

Shiva, Vandana. *Stolen Harvest: The Hijacking of the Global Food Supply*. Cambridge, MA: South End, 2000.

Smith, Jeffrey. *Seeds of Deception: Exposing Industry and Government Lies About the Safety of the Genetically Engineered Foods You're Eating*. Portland, ME: Yes!, 2003.

Thies, Janice and Medha Devare. "An Ecological Assessment of Transgenic Crops." *Journal of Development Studies*, 43/1 (2007).

Toke, Dave. *The Politics of GM Food: A Comparative Study of the UK, USA and EU*. New York: Routledge, 2007.

Priyanka Jain
University of Kentucky

GRAIN-FED BEEF

The term *grain-fed beef* refers to beef that comes from cattle raised on a diet of grains such as corn, rather than a natural diet of grasses and vegetation. Some grain-fed cattle are entirely raised on grain, whereas others are grass-fed for much of their lives, but finish on a diet of grain shortly before being sold for beef. Most grain-fed beef cattle are housed and fattened on large factory farm confined feedlots, whereas grass-fed beef cattle are raised in open free-range pastures. Grain-fed beef became popular in the 20th century as large agri-businesses replaced many small family farms, U.S. animal production became increasingly specialized, and consumer demand for beef on the global market grew along with the world's population. Grain-fed beef cattle housed in feedlots grow fatter faster, allowing more beef to reach the market in a shorter period of time. Lower production costs resulting from lower land and herd number requirements have also aided the rise of grain-fed beef production.

The growth of grain-fed beef, coupled with a renewed interest in grass-fed beef, has led to a debate over which method is better in terms of environmental impact, animal welfare, consumer health, and the marketplace. A key environmental concern with grain-fed beef production is the concentrated amount of manure produced by large numbers of cattle in close confinement. The high cost of manure disposal can inhibit feedlot farmers from using proper disposal methods. If not properly disposed of or recycled, this waste could result in soil and water pollution. The United States passed environmental protection laws in the 1970s in response to this hazard. The Environmental Protection Agency (EPA) requires concentrated (confined) animal feeding operations, also known as intensive livestock operations, to obtain permits and develop management plans to prevent the contamination of surface and groundwater. Concentrated (confined) animal feeding operations are defined as operations that feed large numbers of animals in a confined area for 45 days or longer in a 12-month period. Residents who live near feedlots complain of excessive odors, among other issues.

A second environmental concern centers on possible pollutants related to the production of grain-feed beef, including antibiotics, pesticides, pathogens, and nutrients. Antibiotics and pesticides are used to prevent the spread of diseases, which tend to occur more frequently in overcrowded conditions like those found on feedlots. Pathogens contracted by the cattle, including parasites, bacteria, and viruses, can spread to livestock handlers, manure, or the beef itself. Manure can release nutrients into the soil, which can result in lethal algal blooms in nearby waterways or the contamination of drinking water.

Animal rights activists and environmentalists charge that the confined life and unnatural diet of grain-fed beef cattle represent animal cruelty. Close confinement can lead to psychological stress, resulting in problems such as repetitive or injurious behaviors. Critics also note that cud-chewing animals fed an unnatural grain-based diet—most of which are genetically modified—can have adverse health effects. Some cattle develop subacute acidosis, which causes them to kick their stomachs and eat dirt instead of feed. Critics also oppose the common use of single cattle breeds on feedlots for the consistency of the beef produced—a practice that results in low genetic diversity and animals that are not adapted to their local habitats. Feedlot owners and their supporters counter that the larger feedlots that raise grain-fed beef cattle have the financial resources to hire employees to closely monitor the cattle's health and can afford sometimes expensive veterinary care when problems are discovered.

Some consumers express a preference for the more-fatty, marbled grain-fed beef, which they find more tender and better tasting, whereas other consumers express a preference for leaner, lower-fat grass-fed beef for health reasons. Other consumers find no discernible taste difference between the two types of beef. Health-conscious consumers note that beef cattle's diet affects not only the taste of the beef but also its nutritional content. Grain-fed beef is higher in fat, cholesterol, and calories and lower in vitamins and nutritional value. Grass-fed beef contains more good, omega-3 fatty acids and conjugated linoleic acid.

Price and supply are also concerns among both producers and consumers of grain-fed beef. Feedlots have lower overhead expenses because they require less land and feed and produce greater quantities of beef more efficiently. For these reasons, grain-fed beef can be supplied at cheaper prices than grass-fed beef. Grain-fed beef production is also not as affected by droughts or reliance on seasonal vegetation, allowing year-round operations. It also brings outside money into the local rural economy because it represents one link in the meat industry, selling cattle to distant buyers. Such economic benefits have recently been partly negated, however, as a result of a declining economy, lower consumer demand, and consumer preferences for cheaper cuts of beef. A second concern has been a global rise in the cost of feed such as corn, as demand for feed in countries such as India and China increases and as the use of corn in ethanol production has grown.

Environmental and consumer activists have expressed concerns over the common treatment with antibiotics of the grains served to feedlot cattle to promote growth and to prevent disease. The overuse of antibiotics encourages the development of disease-resistant bacteria such as methicillin-resistant *Staphylococcus aureus*, which can cause illness in both cattle and humans. Proponents of organic foods, chemical-free farming, and grass-fed beef note that cattle raised on natural vegetation are not exposed to the antibiotics, steroids, hormones, and pesticides that can contaminate grains used as feed.

Grain-fed beef cattle are also more prone to acid-resistant strains of *Escherichia coli*, as their diets result in abnormally high acid levels in the digestive tract. Cattle infected with *E. coli* or other bacteria such as *Campylobacter* can later infect humans. One of the greatest disease concerns associated with the beef industry has been mad cow disease (bovine spongiform encephalopathy) and the similar Creutzfeldt-Jakob disease. Animal feed containing animal by-products can lead to the disease's spread. The practice of using such feed was once common in Europe but was never popular in the United States and has been banned in the United States since 1997. In addition, cross-contamination of animal feed remains a concern. Fears of disease outbreaks have led to public avoidance and beef embargoes in many parts of the world. Proponents counter that the concentrated conditions found on feedlots could instead ease the inspection process and improve safety.

See Also: Agribusiness; Confined Animal Feeding Operation; Factory Farm; Mad Cow Disease; Meats.

Further Readings

Fussell, Betty Harper. *Raising Steaks: The Life and Times of American Beef.* Orlando, FL: Harcourt, 2008.
Rifkin, Jeremy. *Beyond Beef: The Rise and Fall of the Cattle Culture.* New York: Dutton, 1992.
Simpson, James R. and Donald E. Farris. *The World's Beef Business.* Ames: Iowa State University Press, 1982.

Marcella Bush Trevino
Barry University

GRAZING

Originating from the Old English for grass, *grazing* refers to the consumption of grasses and other plant materials by herbivores and some omnivores. Wild or domestic grazing animals (e.g., cattle, sheep, deer, horses) consume portions of low-lying plants, often only eating part of the plant and leaving the rest behind (most importantly, the root) to allow for new growth. Sometimes multiple species will graze on the same plants, with smaller animals eating the tender parts exposed after larger animals have consumed the tougher, woodier parts. When grazing focuses on higher vegetation, such as the twigs and leaves of shrubbery and trees, it is called "browsing." For instance, the caprid subfamily (or "goat-antelope family") includes grazers (e.g., sheep, musk-oxen), browsers (e.g., goats), and animals that both graze and browse (e.g., ibexes). Giraffes have evolved to browse on the higher-level foliage that other browsers cannot reach. In the agriculture, domestic grazing animals are traditionally food sources themselves and/or produce other food byproducts (e.g., milk).

Most domestic livestock are, and historically have been, grazers, and grazing has been an opportunity for people since the earliest days of man to make indirect food use of grasses and other vegetative resources that are of limited direct use. At the same time, because cattle and sheep were feeding on grasses, they were not competing for food with their human ranchers. Though we do not think much about the fact that we do not raise carnivorous species for food, one reason early man made that choice was because of the inefficiencies involved. (The principal exception is the pig, which is omnivorous; the pig can eat meat, but does not require it as long as it obtains protein from some source. Omnivorousness has an efficiency appeal all its own, as a pig can be fed unusable or undesirable scraps or leftovers, much as the family dog would be. The trade-off is that, unlike

These Colorado ranchers have brought a herd of cattle to graze on a healthy pasture (at right) while avoiding the overgrazed land on the other side of the fence.

Source: U.S. Department of Agriculture, Natural Resources Conservation Service/Irv Cole

the grazing species, it does not provide milk for its owners.) Furthermore, through a lucky ecological quirk, most land that is suitable for grazing is not suitable for growing crops, and vice versa. It is easy in the 21st century to lose sight of how well-adapted farming and animal husbandry are to the natural dispositions of the environment.

When livestock graze, the grazing can be controlled (with the farmer regulating what is available to the animals, through one means or another) or continuous (where the animals have free access to whatever is available to forage). The growth of the cattle ranching industry and its continuous grazing in the 19th century highlighted the problems of grazing large numbers of livestock without any method or system to prevent overgrazing, in which grasses are eaten too quickly to grow back (which could lead to root systems dying). Overgrazing can lead to soil erosion, as those root systems are no longer there to tether the soil in place, and it can affect local water quality—soil erosion, left unchecked, will reduce the amount of plant life that can be sustained by the land, which in turn will lead to further soil erosion, and so on. The intentional clearing of land to turn it into grazing land usually involves the destruction of woodland or wetland, both of which have their own consequences, both short and—especially—long term. Over the course of the 20th century, particularly as the frontier closed and public lands were no longer available for indiscriminate cowboys to let their cattle graze at will, various grazing systems were developed with an eye toward sustainability.

Seasonal grazing shifts the animals to different pastures at different times of the year, letting new plants grow in one area while the animals feed in another; this is the simplest form of controlled grazing and mimics the natural effect of roaming herds.

Rotational grazing divides the land into multiple pastures that are grazed on in sequence. Deferred rotation grazing, for instance, alternates grazing between different pastures with the goal of allowing new growth to reach maximum growth before it is grazed. Patch burn grazing, in contrast, burns part of the pasture while livestock graze on the other parts; soon enough, the new growth of the burned pasture attracts their attention and they migrate there of their own accord.

Different grazing systems make sense on different ranches, according to the local natural resources, the grasses being used, and the size of the herd. Though overgrazing has dire consequences both to the herd and to the environment, grazing itself offers a host of advantages, many of which are still being studied and explored, having evolved as part of the natural balance of the complicated ecosystem. Though grazing involves the animals eating the plants, for instance, it is actually beneficial for both the plants they feed on and others in the area. The act of grazing itself stimulates plant growth, and the animal's waste returns nitrogen, potassium, and other plant nutrients to the soil, as well as organic feeding material for insects and microorganisms—the presence of which improves both soil and water quality. The trampling of the animals helps to embed recently fallen seeds in the soil to germinate, and keeping grasses from becoming overgrown keeps them from running out of water, whereas preventing the accumulation of dead grasses significantly reduces brushfires—the effect on the latter is profound enough that in some parts of the country, park rangers arrange for their lands to be grazed to control the fire hazard. Even before the arrival of ranchers, much of the ecosystem of the American prairies was governed by a millennia-old interaction between grazing herds and periodic fires, which between them created a complicated system of checks and balances responsible for the evolution of the native flora.

It is difficult to do deliberately that which nature has evolved over thousands or millions of years. Even when overgrazing is avoided, livestock grazing can cause complications. Bringing a herd near a stream can disrupt the stream (as trampling and soil erosion increase sediment), as the animals' waste can pollute it, and thus the local water supply. The elimination of wetlands to create grazing lands has drastic consequences; it is widely agreed that the loss of wetlands on the Gulf Coast of Louisiana and Mississippi was responsible for the severity of the effects of Hurricane Katrina, as wetlands act as a natural buffer against severe weather systems. The effect of grazing on biodiversity is still little understood, as grazing can have both positive and negative effects on an ecosystem's biodiversity; certainly, however, when a grazing herd is introduced to a new ecosystem, it competes, however unknowingly, with local wildlife. The American prairie dog population, for instance, has literally been decimated since the introduction of grazing livestock to the prairie lands.

See Also: Agroecology; Confined Animal Feeding Operation; Grain-Fed Beef.

Further Readings

Berry, Wendell. *The Unsettling of America: Culture and Agriculture*. San Francisco: Sierra Club, 1996.

Pollan, Michael. *In Defense of Food*. New York: Penguin, 2009.

Pollan, Michael. *The Omnivore's Dilemma: A Natural History of Four Meals*. New York: Penguin, 2007.

Samson, Fred and Fritz Knopf. "Prairie Conservation in North America." *BioScience*, 44 (1994).

Bill Kte'pi
Independent Scholar

GREEN REVOLUTION

In the latter half of the 20th century, international agricultural research centers (IARCs), alongside national research programs, contributed to the development and distribution of modern high-yielding varieties (HYVs) for many principal food crops in the developing world. Although these HYVs undoubtedly generated enormous increases in production and substantial reductions in food prices, the process was inherently controversial. Productivity gains have been argued to be inconsistent across crops and regions, and the new farming practices and artificial additives introduced alongside the high-yielding crops have been charged with an array of environmental problems. Today the green revolution continues, albeit in an altered state, and debates persist about the achievements of crop genetic improvement and its effect on agrarian livelihoods and the environment.

Agricultural Innovation

The 1940s heralded the beginning of concerted efforts from the global North toward extensive "development assistance" to the South. In 1943, the Comparative Wheat Research and Production Program in Mexico was created under the powerful Rockefeller Foundation and the Mexican Ministry of Agriculture. Leading this venture's International Maize and Wheat Improvement Center was a pioneering American microbiologist named Norman Borlaug. Borlaug, later to be awarded the Nobel Peace Prize for his work, directed the organization's research toward agricultural innovation through plant breeding, with the view of reducing the risk of unpredictable yields faced by poor farmers. Borlaug was born in 1914 in Cresco, Iowa, and as a young man during the Great Depression of the 1930s, he studied plant pathology at the University of Minnesota. At that time, the U.S. Midwest was experiencing a relentless drought, affecting numerous states in the southwestern plains. Widespread overgrazing and extensive dryland farming, caused by the mounting demand for wheat products, aided the drought in the near obliteration of the region's farmland and natural prairie grasses, causing recurrent crop failures. However, in little over a decade, this American Dust Bowl saw the beginnings of a spectacular turnaround. Through scientific advancements, including technological advances in high-yield agriculture, the Midwest was transformed into what some called "the world's breadbasket," and by the 1980s it had more than doubled its annual agricultural outputs. Conversely, for Borlaug, these early successes represented nothing more than an unsatisfactory application of the new technology, and he stressed that only in those areas where the high-yielding techniques were fully applied were the problems wholly eradicated. The soon-to-be Dr. Borlaug became motivated to spread the benefits he saw in high-yield farming techniques to those constantly facing the threat of starvation and undernourishment.

Drawing on the extensive breeding and agricultural experience of the global North, the Mexican program's work incorporated the development of new "dwarfed" wheat and rice varieties that were bred to be particularly reactive to synthetic nitrogen fertilizer. Through a system of hybridization and back-crossing, in which a hybrid individual is crossed with one of its "parents," or a genetically comparable individual, to produce offspring with hereditarily desirable traits, new genetically engineered crop seeds were fashioned. Borlaug's team started with five established Mexican wheat varieties and meticulously crossed them with 12 imported strains. The investigations delivered promising results, and four hybrids were selected for development. The resulting HYVs were seen to have several distinct advantages over the customary crop varieties: Because the new strains were dwarfed, the plant exerted less energy in growing inedible components like the stalk. Likewise, the head of the plant (that which carries the ears of wheat or grains of rice) could be heavier without the risk of the increased leverage bending the stem and restricting the plant's growth ("lodging"). In addition, the altered photosensitivity of the new strains was premeditated to allow the plants to mature unaltered by annual climate variation or local-ized day-length, thus allowing multicropping.

By the 1960s, the need for innovations in agricultural production in the world's devel-oping nations was becoming increasingly desperate: Global food security became para-mount, as the equilibrium between populace and food production was believed to be under insurmountable pressure from the "population monster," as Borlaug later termed it, of the Third World. Indeed, the 1960s saw the highest rate of population growth in human his-tory. Beginning in the 1950s with the reduction of mortality rates in less-developed coun-tries, population increases peaked at 2.04 percent, with an average of 2 percent lasting until the mid-1970s. Asia presented particularly alarming statistics. In 1960, 70 percent of the total world population lived in developing countries; Asia was home to nearly half of this total population and dominated other regions in terms of absolute numbers of people added annually. Even when excluding mainland China from the data, the continent tow-ered above comparative regions in terms of total population expansion.

Equally, famine and drought were taking their toll on the continent's ability to sustain life. India was still reeling in the wake of the 1943 Bengal famine and a succession of disastrous harvests, causing the subcontinent to fall dangerously behind in food security. Likewise, following several consecutively disappointing growing seasons and Chairman Mao Zedong's catastrophic attempt to develop China's agriculture and industry to rival that of the West (known as the Great Leap Forward), in which private ownership of the means of agricultural production were given over to poorly managed publicly owned "communes," China was in the grip of arguably the largest famine in world history. Although these events can be seen as comparatively localized episodes, policies such as the Great Leap Forward have been identified as direct contributors to the stagnation of global cereal production and the concurrent downturn in the growth rate of worldwide cereal yields.

Observers were turning to the work of early-19th-century Anglican clergyman Thomas Malthus, who forecast that unrestricted population growth would increase geometrically while the resource base (the means of supporting the population) could only increase arith-metically. That is to say, ultimately for Malthus, unconstrained population expansion will exceed the ability to support itself. The recognition that the dramatic population increase being witnessed throughout much of the world, teamed with the global South's inability to produce adequate food levels, equated to the archetypal Malthusian model cemented the need for large-scale action to prevent a seemingly inevitable spread of famine.

A Technological Solution

During the same period, the landscape of agricultural research institutions had also changed. Seeking to replicate Borlaug's successes with wheat, the Rockefeller Foundation teamed with the New York–based Ford Foundation and the Philippine government in set-ting up the International Rice Research Institute and appointed future World Food Prize Laureate Dr. Robert Chandler, Jr., as its director. The Rockefeller's Mexican Wheat Program had been developed into the International Center for the Improvement of Maize and Wheat—frequently referred to by its Spanish acronym CIMMYT (Centro Internacional de Mejoramiento de Maíz y Trigo). The original program had revolutionized Mexico's agricultural landscape, and by 1963 nearly 95 percent of Mexico's wheat crops were prod-ucts of the semidwarf "miracle seeds"—the end result was an annual harvest six times that of Borlaug's arrival in 1943.

Borlaug was eager to transfer the successes of the Mexico program's technological advances to parts of south Asia, where food security appeared ever more tenuous. He concentrated his efforts on the Indian and Pakistani state agriculture companies that, at that time, controlled the countries' seed distribution networks. This undertaking offered substantially larger hurdles than Borlaug had experienced in South America: he was attempting to introduce high-yielding wheat varieties to a region that had previously relied on beans and pulses such as lentils, and he would have to promote a shift in cultural acceptance as well as new agricultural techniques to guarantee success. However, Borlaug's work had noteworthy supporters within the subcontinent. C. Subramaniam, India's agricultural minister, and M. S. Swaminathan, who became director of the Indian Agricultural Research Institute in 1965 and who had been trained by Borlaug in Mexico, were strong backers of the new seed technology. After a visit to India in 1963, Swaminathan shipped 400 kilograms of the semidwarf varieties to be tested in-country. In the following year, rice seeds were dispatched from the International Rice Research Institute in the Philippines. The director of the International Crops Research Institute for the Semi-Arid Tropics, Ralph W. Cummings, declared that enough trials had been carried out in the region to promote and implement the full introduction of what would become known as green revolution technology and approached Swaminathan for support.

Drought in 1966 and the subsequent Bihar famine dramatically reduced India's food output, creating an increased need for food grain from the United States. Then-president Lyndon B. Johnson and his administration were reluctant to commit to any long-term food aid plan beyond a month-to-month basis until an agreement to adopt the green revolution package was signed between India and the United States. Under these circumstances, and following the death of Prime Minister Lal Bahadur Shastri, who was skeptical about the introduction of a new agricultural framework, the sweeping introduction of HYVs in India went largely unchallenged.

The HYVs and related technologies were promoted as a package and aimed at areas of comparatively high potential. Implementation programs designed to promote and apply the package were introduced or developed using ongoing schemes. For instance, the centrally sponsored Farmers' Training and Educational Scheme, introduced in 1966 and designed to cover up to 100 Indian districts cultivating HYV crops, provided for 15 expert-led demonstrations per district in agronomy (the management of crop plants to supply socially useful products), plant protection, and agricultural engineering. Other formats included the Multiple Cropping Scheme in Tamil Nadu, which was introduced as part of the government's New Agricultural Strategy. Because the innovative technology was seen as exhausting local soils through repeated cropping, the scheme facilitated the upgrading of irrigation facilities to meet the HYV demands, continuous soil testing, and the promotion of HYV rice-crop rotation with traditional crops such as jute, beans, or lentils. Importantly, grants and loans were made available to assist those unable to transfer to the new methods.

The uptake of the new high-yielding "package" in Latin America and Asia was by now extensive, and the immediate effects were clear: Where farms replaced their customary farming practices with a combination of synthetic (primarily nitrogen) fertilizers, expanded irrigation systems, and the new HYVs, there were substantial increases in harvests. Mexico had been transformed into a sizable exporter of wheat, and Pakistan and India became self-sufficient. In fact, by 1974, India became self-reliant in all cereals. These feats were recognized in the annual report of the United States Agency for International Development, where then-administrator William Gaud christened the developments as a "green revolution." The programs continued and expanded, seeing Pakistan produce approximately 8.4 million tons in 1970, up from 4.5 million tons in 1965. Similarly, India, commonly recognized as the polestar of the revolution, increased its production by almost 9 million tons. By 1979, Indian farmers, using the HYVs, had increased the country's wheat yields to such an extent that they were enough to provide nearly 70 percent of the nation's calorie requirements.

The early achievements have since been seen to have shifted the organizational driving force behind agricultural research toward multidisciplinary commodity-directed agendas, based chiefly on centralized breeding programs guided by socioeconomic dynamics. Furthermore, the widespread adoption of HYVs and new agricultural technology during the green revolution propelled agricultural research to the forefront of investments in rural development, solidifying itself as an instrument of development policy. Agricultural innovation research thus became a significant component of development studies.

Numerous aid agencies and international organizations were subsequently introduced to advance the successes of the new technologies. In 1971, following a recommendation from the Rockefeller Foundation, executives of the World Bank, led by Robert McNamara, formed the Consultative Group on International Agricultural Research (CGIAR) to mobilize and finance the growing network of IARCs under a permanent administration. Alongside partner institutions such as the United Nations' Food and Agriculture Organization, Development Programme, and International Fund for Agricultural Development, the CGIAR brought together four existing IARCs to support and implement multidisciplinary research to generate global agricultural public goods. It has been suggested, however, that in its early years, the CGIAR operated a predominantly "top-down" approach to agricultural research, with little precedence afforded to collaborative engagements with the research's beneficiaries—namely, the poor rural communities.

The Agronomical, Social, and Environmental Implications

In all, a review of the literature offers one solitary concurrence: green revolution techniques introduced from the mid-1950s to the mid-1970s generated no less than an agricultural explosion in those areas where they were fully applied. This, however, is where agreement ends.

In terms of environmental consequences, the green revolution has garnered severe criticism. Censure has typically come from those who argue that the successes of HYVs are only realizable when all requisites—namely, intensive synthetic fertilizer and pesticide application, optimal irrigation levels, and agricultural monoculture (the practice of growing a single crop variety over a large area)—are met. In the same way, skeptical observers have drawn attention to the newly introduced farming techniques and accompanying environmental risks. Soils were argued to become "addicted" to synthetic fertilizers, requiring higher and increasingly recurrent doses. Such intensive fertilizer use was shown in many cases to result in nitration (the harmful reaction between nitric acid and an organic compound) and the eutrophication of local bodies of freshwater (in which abnormally high weed and algal growth, caused by an enrichment of nutrition levels from heavy chemical use, eventually dissipates the oxygen levels of the affected area) and had predictably harmful effects on natural ecosystems and the wider rural population.

Likewise, the loss of indigenous varieties, particularly of wheat, has provoked condemnation from proponents of biodiversity. Certainly, figures show that in China, for example, between 1964 and 1970, numbers of unique wheat varieties plummeted to approximately

10 percent of the country's original total. Critics of this departure from natural genetic miscellany argued that advocates of the increases in grain production had not taken into account the loss of the vital secondary outputs cultivated in traditional mixed crops, such as husk for fuel and straw for building and livestock feed, all of which are reduced in HYVs.

Environmental activist Dr. Vandana Shiva argues that the precondition for agricultural monoculture is a Western monocultural mindset, in which the destruction of natural diversity is promoted under the dubious justification of progress, and where, in actual fact, the methods promoted are dangerously unsustainable. That is, for Shiva, through the introduction of high-yielding uniformity, the implementation of the green revolution package irrevocably marginalized local agricultural knowledge and practices, replacing them with a homogenous Western "scientific" knowledge. Equally, Shiva has been criticized for romanticizing the "local" and indigenous.

Architects of the green revolution sought to go beyond the expansion of existing cultivatable land and introduced the systematic double-cropping of farmland in current use. Of course, the logic of traditional single-cropping is that there is only one monsoon season per year. As such, to introduce biannual harvests, a second "monsoon" must be established. Water, and therefore irrigation, is central to food security, and provided that water demands are low relative to availability, there is typically little competition. Creating a "second monsoon" required a remarkable transformation in the irrigation practices of those regions adopting the new methods. Some areas suffered from rising water tables, which, if they reached the soil level, often caused the salinization of the soil (in which water-soluble salts ascend to the surface with the rising water and are deposited during evaporation, often reducing the amount of water later available for plant uptake). Other areas suffered from devastating falls in water table levels resulting from excessive water use. This, again, is more likely to affect poorer rural communities, as they are the ones that use traditional open wells, as opposed to the mechanically dug tube and bore wells of their wealthier counterparts.

Socially, the new agricultural methods and practices were also argued to be well-matched to the capacities of larger, more established, "progressive" farms, whereas many small-holders suffered from the reduction in prices caused by the considerable increases in large farm production. Likewise, as word spread and prospective profits increased, share-croppers (occupants authorized to work the property of a landowner in return for a share of the profits) and others of limited tenancy were often dispossessed. Naturally, as the increase in output and resulting drop in food costs benefited consumers, farmers were only able to gain where the reductions in cost surpassed the decrease in price. With such social concerns becoming gradually more widespread, development circles became increasingly aware of the significance and benefits of direct beneficiary involvement. Accordingly, the CGIAR launched progressively more "bottom-up" focused programs. One such example is that of Collective Action and Property Rights, led by development sociologist Dr. Ruth Meinzen-Dick, which sought to use collective action and property rights as modes of empowerment for rural communities through agricultural innovation.

Future Prospects

Although many areas within low-income nations have now experienced the effects of green revolution technology, the countries that inhabit Africa have, in many respects, been left out. Data show that compared with Asia, where investment in agricultural research tripled in the last two decades, Africa witnessed only a 20 percent increase. Arguably, malnourishment and famine in sub-Saharan Africa are at the forefront of people's imaginations when considering world food security, and figures showing that average cereal yields in Africa are less than one-third of those in Asia do nothing to shift that trend. To that end, major international development initiatives are beginning to take form to bring the green revolution to Africa. In fact, the majority of funding being donated to secure food sovereignty for the continent is coming from organizations that, similar to the Ford and Rockefeller Foundations before them, made their fortunes in the Western corporate arena. The Bill and Melinda Gates Foundation, for instance, alongside the Rockefeller Foundation, fund the Alliance for a Green Revolution in Africa (AGRA).

AGRA, headed by former UN Secretary-General Kofi Annan, seeks to fund and develop a program for Africa's seed systems in the hope of creating a 50 percent increase in the land area planted with improved varieties across up to 20 African nations. Through this, AGRA calculates that such a system can plausibly bring almost 20 million people out of poverty in a sustainable way. Such confidence is not universally shared, however. Some observers argue that Africa presents a much larger and immensely more complicated challenge than past green revolution locales. The obstacles presented by the social and physical environments in Africa include developing new varieties of five or six staple crops, unlike the one or two that were developed across Asia. Arguably, Africa's soils are a great deal more prone to erosion in comparison with those areas in which the green revolution processes originally flourished and at are a huge disadvantage with regard to water and irrigation. Opponents have argued that, in addition to this, conditions in which HYVs are tested are far removed from the realities of the small African farmer; HYVs were often assessed with levels of expensive nitrogen fertilizer beyond the means of marginalized family farms and required levels of plowing, animal traction, and cultivation that overworked and undernourished communities are often unable to provide.

Interestingly, four years before the inception of AGRA, a small-yet-noteworthy assemblage of United Nations–commissioned scientists expressed similar serious doubts over the applicability of green revolution technology and methods in Africa because of the diverse farming systems in operation across the continent. Their report suggested that there would be no "magic technological bullet" able to alleviate the problems of African agriculture, and that it was far more likely that a series of "rainbow evolutions" would take place across African nations, rather than a single "revolution" as in Asia and South America. That notwithstanding, on leaving his post at the head of the United Nations, Annan became chairman of AGRA and began work on the "uniquely African Green Revolution." This self-named "African-led partnership" maintains that it has learned from the mistakes of the green revolution's past and is set to build on and develop its strengths. Growing emphasis is being placed on participatory methods designed with small-scale farmer needs and circumstances in mind. This is complemented with extensive community-led trials designed to allow local farmers, guided by researchers, to try out new methods within their particular environment and facing their particular constraints. AGRA's aim is to develop crop varieties that produce higher, more stable yields without complete reliance on vigilant management, sufficient water, and synthetic fertilizers and to garner success with low, potentially organic, additives, poor soils, and in unpredictable climates. Constant farmer feedback and subsequent modification, it is hoped, will circumvent past problems and inform further developments to ensure a revolution that can be applied to every situation. Perhaps the most encouraging advance in African agricultural research is that the needs of individual low-income farming communities are being taken fully into account in combination with inimitable indigenous knowledge of local farmland and have begun to influence the way technology transfers are organized.

Conclusion: Feast or Famine?

In all, the green revolution offered a brief respite for a number of developing countries that were staggering under their recently enlarged populations and provided unparalleled levels of food security for many, while allowing scores of communities to avoid the very real threat of starvation. Nonetheless, as Professor Graham Chapman argues, comparatively little has been done with regard to key subsistence crops like cassava and sorghum in Africa and millet in India, and any successes were without doubt geographically specific; indeed, before recent developments, it can be plausibly argued that the majority of African small-scale agriculture had been entirely overlooked. What is more, benefits have been profoundly dependent on specific socioeconomic circumstances, which appear likely to provide similar problems for the green revolution's future advancements.

Scientists continue to believe that biotechnology and gene manipulation offer the solution to world food problems. Indeed, modern science has taken large steps in understanding plant growth, morphology (referring to the composition and configuration of an organism), and agricultural technologies, and many now see genetically modified organisms (GMOs) overtaking and replacing Borlaug's unhurried system of crossing and selecting. Gordon Conway sees the use of GMOs in agriculture as the "Doubly Green Revolution."

The public faces of this research, however, are changing; the CGIAR remains influential, with 15 centers under its umbrella, including several socioeconomic departments introduced alongside its technical mainstays. Even so, the forerunners of new agricultural technology are no longer the IARCs or National Agricultural Research Centres, but (often Western) multinational corporations. Faced with mounting condemnation of scientifically derived agricultural changes, development bodies have retracted substantial amounts of funding from agricultural research and focused their efforts on different concerns. As a consequence, the budgets of NARSs and IARCs are in decline. Simultaneously, global trends in agricultural research, encouraged by the expansion of intellectual property rights, have contributed to an ever more competitive and proprietary research environment that is increasingly occupied by private sector organizations. Cynics argue that any profits or benefits resulting from the new technologies and research establishments will be reaped by the multinational corporations and bypass the poor, who offer little in potential revenue.

Taken as a whole, agricultural research remains unrivaled in its potential to alleviate food shortages throughout the "developing" world. In addition, as recognized by AGRA, the requirements are greatest in Africa—a region with the unenviable distinction of being the only continent in the world where levels of food security are in decline. Some argue that the green revolution is better seen as an ongoing 50-year process of innovation and development rather than a one-time event, and they no doubt view the new era of corporate-sponsored participatory research as the latest phase in the convalescent advancement of HYV technology. All the same, a balance among stable agricultural gains, environmental responsibility, and sustainable social development has yet to be attained.

See Also: Agroecology; Agrofood System (Agrifood); Beyond Organic; Biodynamic Agriculture; Crop Genetic Diversity; Organic Farming.

Further Readings

Chapman, Graham. "The Green Revolution." *The Companion to Development Studies.* London: Arnold, 2002.

Conway, Gordon, et al. "The Rockefeller Foundation and Social Research in Agriculture." In *Researching the Culture in Agri-Culture: Social Research for Agricultural Development,* edited by Michael Cernea and Amir Kassam. Cambridge, MA: CABI, 2006.

Dyson, Tim. *Population and Food: Global Trends and Future Prospects.* London: Routledge, 1996.

Evenson, Robert and Douglas Gollin. "Assessing the Impact of the Green Revolution, 1960–2000." *Science,* 300/5620 (2003).

Harrison, Paul. *The Greening of Africa: Breaking Through in the Battle for Land and Food.* London: Penguin, 1987.

Paarlberg, Robert. *Starved for Science: How Biotechnology is Being Kept Out of Africa.* London: Harvard University Press, 2008.

Pearse, Andrew. *Seeds of Plenty, Seeds of Want: Social and Economic Implications of the Green Revolution.* Oxford: Clarendon, 1980.

Pinstrup-Andersen, Per and Tewodaj Mengistu. "Implications of Globalization for Agricultural Research." In *Globalization of Food and Agriculture and the Poor,* edited by Joachim von Braun and Eugenio Diaz-Bonilla. Oxford: Oxford University Press, 2008.

Shiva, Vandana. *Monocultures of the Mind: Perspectives on Biodiversity and Biotechnology.* London: Zed, 1993.

Shiva, Vandana. *The Violence of the Green Revolution: Third World Agriculture, Ecology and Politics.* London: Zed, 1991.

Willis, Katie. "Norman Borlaug." *Fifty Key Thinkers on Development.* Oxford: Routledge, 2006.

Ziaur Rahman, Abm. "Correlations Between Green Revolution and Population Growth: Revisited in the Context of Bangladesh and India." *Asian Affairs,* 26/3 (2004).

Paul Henry Johnson
University of Durham

High-Fructose Corn Syrup

High-fructose corn syrup (HFCS), a sweetener derived from the industrial processing of corn, increasingly appears on the ingredient lists of American processed foods. Corn syrup began being used in the 1800s, but HFCS was not developed until the 1950s and 1960s. In 1967, HFCS began appearing in U.S. food, and today the average American consumes 63 pounds of the sweetener each year. The development and rise of HFCS in the United States has occurred within the context of an increasingly industrialized and specialized agricultural system. Today, two companies (Cargill and Archer Daniels Midland) buy one-third of all U.S. corn crops and also control the processing steps needed to make HFCS. Corn enters our food system in many forms beyond corn on the cob—as cornmeal, corn syrup, cornstarch, and other substances. In addition to contributing to the American obesity epidemic, HFCS production carries adverse environmental effects at each stage of processing.

Political economic processes have largely worked to produce a contemporary American food system reliant on processed corn. In 1850s Chicago, corn began to be traded as a commodity, encouraging farmers to adjust their practices to focus primarily on corn production. New Deal programs eased the economic uncertainty of farmers' producing storable commodities such as corn by creating government-established target prices. When the market price of corn dropped below the target price, the farmer could take a government loan to store grain until prices recovered and then sell the corn and pay back the government. Farmers were also given the option of keeping their loan and giving the government their corn if prices remained low. By the 1970s, this program was no longer in place. In 1972, the United States sold 30 million tons of grain to Russia as part of a plan to get farmers to vote for Nixon. The sale led to huge price inflation in grocery stores and then to the removal of the New Deal farm programs. The 1973 farm bill included provisions for government institutional corn purchasing and payments to farmers for idling their land. This marked a shift from loans to direct payments to farmers and contributed to the rise of industrial agriculture. Now, the farm bill allows farmers to be paid per bushel of potential corn production. These changing farming policies and practices have created an agricultural system in which the government subsidizes overproduction of corn, creating a need for new uses of surplus corn.

HFCS emerged as a new food product to capitalize on surplus corn production. The technological foundations for producing HFCS were laid in 1866, when the enzyme hydrolysis process was developed to create corn syrup. Then, in the 1960s, Japanese chemists discovered the enzyme glucose isomerase, which produces fructose from glucose and creates HFCS, a blend of 55 percent fructose and 45 percent glucose. Since the 1970s, complex industrial processes have been breaking down different parts of corn kernels into various corn products, including HFCS. Processing plants transform the skin of the corn into vitamins and supplements, the center of the kernel into oil, and the starchy meat into HFCS and other products. In the United States, about 530 million bushels of corn are converted to HFCS through this process annually.

The rise of HFCS has been facilitated by agribusiness intervention in the U.S. agricultural political economy. In the 1970s, Archer Daniels Midland, a company that controlled much of the HFCS, was looking for ways to increase sales opportunities. To accomplish this, it lobbied to restrict sugar imports, which would increase demand for domestic sources of sweeteners. They were successful: By 1982, federal limits on sugar imports spurred increased demand for HFCS. Complex food systems facilitate substitutionalism such as this. As a result, since the 1980s, most soda sold in the United States has been sweetened with HFCS, and the substance also has found its way into many packaged foods on grocery store shelves.

The process of creating HFCS carries considerable environmental impacts, beginning with corn production itself. Industrial agriculture has been fueled by what Vandana Shiva calls the "leftovers of World War II"—industries that shifted to fertilizer and pesticide production after World War II ended the demand for munitions and war materials. Most corn in the United States today is grown either as a farm's sole crop or in rotation with soybeans. This specialization exceeds soils' natural nitrogen availability and requires additional nutrient inputs in the form of fertilizer. Corn crops in the United States receive 162 million pounds of pesticides and 17.8 billion pounds of synthetic fertilizer each year. These two applications contribute nearly 37 billion pounds of greenhouse gas emissions annually. In addition, excess fertilizer washes off the fields and flows into waterways. Runoff fertilizer from the U.S. Corn Belt flows from the Mississippi River to the Gulf of Mexico, where it has contributed to a massive "dead zone," where sea life cannot survive. Industrial corn production is a tremendously energy-intensive process: today, one bushel of industrial corn requires between a quarter and third of a gallon of oil to grow, which amounts to about 50 gallons per of oil acre. This means that we are putting more energy into corn than we extract as calories. The industrial transformation of corn from a recognizable vegetable to HFCS is also energy and water intensive. Throughout the corn milling process, one bushel of corn requires five gallons of water. Also, HFCS contains tremendous embodied energy—10 calories of energy are needed to create one calorie of HFCS.

In addition to its effects on the environment, HFCS may be having adverse effects on public health. Some consumers and scientists suggest that the rise of obesity and diabetes in the United States is partly attributable to the introduction of HFCS into Americans' diets. Increases in HFCS consumption have not been offset by a decrease in consumption of other sweeteners, meaning that Americans are eating more sweets than ever before. The low cost of HFCS may be one reason for this. For example, by 1984 Coca-Cola and Pepsi had switched to HFCS because it was cheaper than sugar. This drove down the price of soft drinks, so the companies began "supersizing" drinks to maximize profits. Also, because HFCS is inexpensive, poor people can buy more calories of HFCS-filled food than of healthy food. Also, although the corn industry asserts that the body breaks down HFCS the same way as it does other refined sugars, other studies suggest that the fructose in HFCS contributes to the body's resistance to leptin, a substance that signals "fullness." In response to growing public concern about the relationship between HFCS consumption and obesity, in 2008 the Corn Refiners Association launched a public relations campaign aimed at convincing consumers that HFCS is healthy "in moderation." In early 2009, news reports cited another alarming potential human health concern: recent scientific studies have found a high incidence of mercury contamination in consumer products in which HFCS is either the first or second ingredient.

See Also: Archer Daniels Midland; Corn; Substitutionism.

Further Readings

Pollan, M. *The Omnivore's Dilemma: A Natural History of Four Meals.* New York: Penguin, 2006.
White, John S. "The Past, Present and Future of High-Fructose Corn Syrup." *Frozen Food Digest,* 23/3 (February 1, 2007).
Woolf, A. "King Corn." Mosaic Films, Inc., and Independent Television Service, 2008.

Kate Darby
Arizona State University

HOLISTIC MANAGEMENT

Successful production on sustainable and organic farms relies heavily on working in harmony with natural ecological cycles. The term *Holistic Management,* coined by Allan Savory, describes a decision-making process for aligning with those cycles. Holistic Management incorporates systems thinking into the task of applied ecology—especially to the sustainable, profitable management of farms and public lands. The Holistic Management approach, which has been applied in organizations beyond agriculture, begins with defining the whole and establishing the boundaries of the enterprise. This is then followed by establishing a metagoal for the enterprise, called a *holistic goal.* The remainder of the Holistic Management process includes harmonizing goals with four natural ecological cycles, using nine strategic tools, testing decisions against the holistic goal, and monitoring results.

The Holistic Decision-Making Process

As the term suggests, a "holistic" farmer's resources are managed as a whole unit rather than as isolated fractions. The farm operator begins by defining the whole. Resources such as vested human decision makers, land, livestock, equipment, money, and community are considered contributors to the whole. From those, a clear, detailed, values-based holistic goal is developed, which then becomes the centerpiece that guides and sustains production and profits.

Holistic financial planning empowers farmers to make decisions that are consistent with their personal values while simultaneously benefiting the environment, their community, and their bottom line. One of the distinctions of holistic financial planning is that a reasonable profit is planned on the front end. This creates a sense of accountability that mitigates the deficit spending in the operations that accompany much of modern agriculture. Expenses are then allocated into one of three categories: wealth-generating expenses that contribute to the current year's profit, inescapable expenses such as taxes and debt payments, and maintenance expenses that, however meaningful, do not contribute to the current year's profit.

One of the principle tenets of Holistic Management is that ultimately it is more expensive to work against nature than to work in harmony with nature. To work in harmony with the natural processes of the Earth's ecosystem, holistic managers are challenged to understand and conform to four fundamental natural cycles.

The first is the *water cycle*—water precipitates to the Earth, filters through the soil, is taken up by plants, or contributes to the water table reserve. When farmers value the water cycle, they contribute to the natural resistance to flooding, release water to flow freely through streams and springs, and minimize soil erosion.

The second cycle, the *mineral cycle*, describes the transition of minerals from the soil through plants and into animals, then back to the soil through composting or in the form of animal waste. Farmers who work in harmony with the mineral cycle mitigate the expense and undesirable effect of chemical fertilizers and amendments.

The third cycle is the *dynamic community cycle*, which produces biodiversity among plant and animal communities in ecosystems. Biodiversity contributes significantly to stability and pest control within the system. Holistic managers who embrace this cycle of community are especially averse to monoculture among crops and livestock.

The final cycle is *energy flow*. Holistic managers embrace the reality that the sun is ultimately the fuel that drives a healthy farm. Sunlight travels through the leaves of green plants that pass their accumulated energy along as they are eaten by other organisms, which are in turn consumed by other organisms. Uneaten plants that decompose pass their energy along to organisms in the soil, which are then consumed by secondary decomposers. During each step of the consumption or decomposition process, the sun's original energy is transferred or lost through heat.

Because these four cycles function as a whole, changes to one often affect one or more of the others—for better or for worse.

To achieve the holistic goal and to profit from producing in harmony with nature's cycles, Holistic Management identifies nine tools at the manager's disposal to maximize production and profit: money, labor, human creativity, fire, grazing, animal impact (by walking on soil), living organisms, technology, and rest (idle land). Each one of these tools produces positive or negative impacts on the ecosystem, depending on when and how it is applied. Many agricultural challenges are solved without adverse effects to other parts of the system by creatively using nontraditional tools. For example, one holistic manager addressed a gopher problem naturally by erecting hawk perches over his pastures. The gophers left for safer feeding areas.

The final steps of the Holistic Management decision-making process include testing decisions against the holistic goal before implementing them, then monitoring the results against the Holistic Management plan. To test decisions, Holistic Management proposes seven questions related to economic, environmental, and social impacts. These questions challenge the holistic manager to consider broader implications beyond mere cost or gut feelings.

Nature is complex. Decisions related to the management of an ecosystem are always made in a dynamic environment. The motivation to measure results in the Holistic Management process emerges from a humble respect for nature. The holistic manager always assumes that a wrong choice was made, given nature's complexity. This compels the manager to consistently monitor progress and results toward attaining the holistic goal. Those who assume correct decisions are at risk of not monitoring and of being unknowingly or adversely affected as a result.

Holistic Management results in a higher quality of life, financial stability, consistent profitability, and a lifestyle consistent with the deeply held values of the holistic manager. Ultimately, Holistic Management offers a process by which to make decisions that harmonize with the way natural systems work (in wholes).

Holistic Management in Practice

Because most practitioners of Holistic Management have not identified "maximizing profit" as their holistic goal, there are presently very few high-profile examples of large-scale Holistic Management. One of the more recently visible practitioners of Holistic Management is Polyface Farms' founder Joe Salatin, a Virginia resident whose adherence to the principles of Holistic Management was highlighted in Michael Pollan's 2006 bestseller, *The Omnivore's Dilemma*. The majority of holistic agricultural managers farm on a small to intermediate scale and typically market their products locally, regionally, and/or directly to consumers. Additional growth among practitioners of Holistic Management comes from new-generation heirs to rural lands who implement the model's decision-making principles on formerly idle lands for personal food sources, recreation, residence, or values-based land stewardship.

The Holistic Management decision-making process has also been adopted into a variety of nonagricultural settings such as education, community development, and the emerging business models of natural capitalism.

History of Holistic Management

Holistic Management practices had been implemented successfully by farmers and ranchers internationally and in the United States for centuries, perhaps millennia, before Savory coined the term. However, it was Savory's model that provided the vehicle for widespread multiplication and application.

Allan Savory originated the Holistic Management model in the 1960s to provide land stewards with a way to improve the health and productivity of the resources under their care. As a young farmer/biologist in his native southern African nation of Zimbabwe (formerly Rhodesia), Savory intensely considered the role that human decisions played in the desertification of lands and the destruction of human health and social systems. Savory's role in the Rhodesian civil war ultimately led to the necessity of his immigrating to the United States. Together with native Zimbabwean Stan Parsons, Savory consulted internationally on the establishment of sustainable grazing systems for improved productivity and land health. In 1983, Savory and Parsons parted, with Parsons establishing Ranch Management Consultants. The Savory Center for Holistic Management in Albuquerque, New Mexico, was established in 1984 and later renamed Holistic Management International.

See Also: Agrodiversity; Agroecology; Grazing; Low-Input Agriculture; Soil Nutrient Cycling.

Further Readings

Dagget, Dan. *Gardeners of Eden*. Santa Barbara, CA: Thatcher Charitable Trust, 2005.

Lovins, Amory, et al. "A Road Map for Natural Capitalism." *Harvard Business Review*, 85/7 (2007).

Malmberg, Tony. "Beyond the Row Crop: Ultra High Density Grazing." *Land & Livestock* (March/April 2008). http://www.holisticmanagement.org/n7/Info_07/InPractice_Archives/Ultra.pdf (Accessed January 2009).

Savory, Allan and Jody Butterfield. *Holistic Management: A New Framework for Decision Making*, 2nd ed. Washington, D.C.: Island, 1998.

Richard B. Gifford
University of Arkansas Community College at Morrilton

HOMEGARDENS

The term *homegardens* refers, generally speaking, to small-scale, domestic agricultural plots. They consist of combinations of vegetable, herb, and flower production and can include small fruit orchards, as well as various decorative features. Though seemingly peripheral, homegardens have historically played a central role in food production and availability in regions across the world. At this time, semisubsistence domestic garden plots serve as key components to burgeoning slow, local, urban, and sustainable agriculture movements, as well to food security and food sovereignty initiatives. Homegardens encompass a range of horticultural and agricultural endeavors, ranging from organic, subsistence vegetable plots to biocide-saturated floral borders, from native plant reservoirs to manicured lawns in the desert. They have come to represent the model of sustainability and the crux of ecological awareness and activism; they also continue to represent an artificial, constrained, constraining, and even colonizing relationship with the natural world.

Epitomizing both simple living and consumerism, homegardens embody a host of compelling environmental and social paradoxes. They span the gamut of being innocuous and revolutionary, the most conspicuous site of bourgeois consumption and leisure, and the most fervent site of subsistence-as-resistance production and labor. This tension permeates the history and current environmental interest in homegardens, which are at times considered harbingers of human manipulation, modification, and control of their environment, and at times, spaces of interaction, harmony, and reciprocity between the human and nonhuman worlds. Michael Pollan's *Second Nature* reflects on this paradox and the intricate interdependence at work in a garden between plants and their growers.

Among its various political ecologies and economies, the household garden also brings forth issues of gender. Similar to domesticity in general, the homegarden has been feminized—and delegitimatized—as the place for quaint, decorative projects and the realm of social reproduction in general. For instance, early archaeological and anthropological accounts of Native American history did not consider their extensive horticulture systems as true agriculture because they were matrilineal "women's work," and so undervalued the ecological, economic, and cultural importance of this small-scale, agribiodiversity-rich mode of growing: "practices associated with modest-scale production were used in the native Southeast for expansive large-scale 'garden-agriculture' in systems for which we lack contemporary parallels" (Scarry and Scarry). Agrarian writers and environmentalists have since worked to counter the secondary status of gardens. (For example, see Gene Logsdon's essay "A Farm is a Large Garden (or a Garden is a Small Farm)".)

Feminists—and now environmentalists—have nevertheless noted and lauded the treasury of local ecological and social knowledge at work in subsistence and semisubsistence gardening traditions. In her collection of essays *In Search Of Our Mothers' Gardens*, Alice Walker offers the metaphor of the household garden to describe all the skills, knowledge, and wisdom she and other women have gained from their mothers and grandmothers. This revaluation of the previously marginalized yet highly valuable knowledge embedded within gardening heritages overcomes and helps uproot long-held biases of the "poverty" of subsistence production. Successful gardening is skilled, highly diversified craft work, incorporating scientific knowledge of the local ecosystem as well as artistic expression.

Household gardens serve as extensions of the house itself, both in terms of domestic processes and social reproduction cycles—through kitchen gardens, compost piles, and laundry lines—and in spatial terms—as places of socializing, dining, rejuvenation, or retreat. Usually noncommercial and voluntary, gardening is often valued as a beneficial

Although homegardens like this one have played a crucial role in world food production and may contribute to soil conservation, biodiversity, and habitat creation, their importance has often been overlooked.

Source: iStockphoto.com

experience and process, as much as it is a series of beneficial products. Within the contemporary sustainability movements, the household garden sheds its peripheral status to become a key arena, locale, and site of ecological and political agency. Through such practices as semisubsistence food production, seed-saving, and home processing, the home garden serves as a means of personal environmental accountability, as well as of economic self-sufficiency, the basis for food security and sovereignty.

Within U.S. history, "victory gardens" offer a historical precedent for attempts at national-scale self-sufficiency. Planted in surplus yardage and other unused spaces, these vegetable gardens aimed to replace the foodstuffs shipped to Europe during World War II. At their peak, "war gardens" provided 40 percent of America's fresh produce. Though they have become models for widespread food security and sovereignty, victory gardens also ushered in a postwar paradigm of consumption governance. Suddenly, home production and consumption garnered mainstream respect as the cornerstone of liberty and independence. One of the most powerful metaphors in early American history—that of self-sufficiency as freedom—returned, though now to support the state and its nationalistic arm: that of the military. Generations later, consumption has come to comprise a key meaning of citizenship—though victory gardens were a momentary transition from celebrating "brought on" to home-grown products.

In contemporary sustainability discourses and practices, homegardens have risen to prominence as effective means of environmental stewardship and physical health. In regions with a long history of sustainable subsistence agriculture, home gardens have been, and often continue to be, models of ecological health. Ecological studies around the world have shown that traditional home gardens perform critical ecological functions such as soil conservation. They serve as invaluable repositories of in situ agricultural biodiversity and are sites of intensive, low external input, culturally appropriate agricultural production and reproduction. They supply fresh produce and herbs at staggered intervals throughout the growing season(s) and, through composting, intercropping, and crop rotation, maintain soil fertility. Low-input, biodiverse homegardens also provide fodder and habitat for bird and butterfly species. Finally, within the modern, suburban setting, homegardens offer a rare space for people to interact with, learn about, and benefit directly from ecological processes.

Homegardens have also been championed as being integral to long-term health. Recent scholarship has shown that the increase of home gardening ensures a more nutritionally diverse and robust diet. Though they do not provide complete levels of protein, gardens supply exceptionally high levels of vitamins and micronutrients. Moreover, in economically vulnerable areas, and with food prices rising globally, homegardens preclude food shortages, with the proven potential to supply large quantities of food. In terms of physical exercise, sustainable, low-input gardening requires demanding manual labor, from cardiovascular exertion in fresh air and sunshine to regular limbering and strengthening activities. Health advocates claim that, collectively, domestic gardens mitigate obesity, disease, and even illnesses such as attention deficit/hyperactivity disorder, anxiety, or depression.

Planting a garden has been celebrated as one of the most powerful endeavors an individual, family, or community can embark on. Advocates note that it reduces overall carbon footprints, enhances health, and strengthens communities. Seed-swapping and bountiful harvests ensure neighborhood networks and interaction. Locavores have popularized the endeavor of semisubsistence gardening, lifting the "drudgery" of growing and processing one's own food from a marker of underdevelopment to an admired site of agency, social justice, and ecological justice. The Food, Not Lawn movement has also worked to re-envision the humble household garden as powerfully, and even radically, green. Homegardens are now widely recognized as pivotal to the transition to sustainable agricultural systems.

See Also: Agrarianism; Food Sovereignty; Locavore; Vegetables.

Further Readings

Hoyles, M. *The Story of Gardening: A Social History.* London: Pluto Journeyman, 1991.

Saguaro, S. *Garden Plots: The Politics and Poetics of Gardens.* Aldershot, UK: Hants; and Burlington, VT: Ashgate, 2006.

Scarry, M. and J. Scarry. "Native American 'Garden Agriculture' in Southeastern North America." *World Archeology,* 37/2:259–74 (2007).

Garrett Graddy
University of Kentucky

HORIZONTAL INTEGRATION

Horizontal integration, an industrial business tactic utilized by agribusiness corporations, involves the consolidation of smaller companies that operated within the same sector or production level or that produced variations within the same product line in order to increase market share and decrease competition. It is also known as horizontal expansion, lateral expansion, or horizontal combination. Vertical integration is a related tactic in which a corporation seeks to dominate all aspects of production, from supply through distribution and retail. Horizontal integration can occur at any level of production or distribution. A narrow definition of horizontal integration would include those corporations that focus on a specific product, such as wheat, corn, chickens, or beef, while a broad definition would include those corporations that focus on a broader single sector, such as grains or meat. The end result of horizontal integration has been the reduction of the number of corporations operating within each sector of agricultural production and distribution.

Traditionally, U.S. agriculture was based on the small, self-sufficient family farm model in which individual farms supplied their own inputs and grew enough diversified agricultural products to support the family. Any surpluses were small and were sold or traded within the local community. This pattern began to change to a commercially based agricultural model with the advent of the Industrial Revolution in England and the United States in the mid-19th century. Growing urban, industrial populations increasingly relied on farmers to produce surplus food for the market. Consequently, the farms themselves became larger and developed into more mechanically based factory-style farms.

The Growth of Agribusinesses

As agriculture became increasingly commercialized, many industrial business practices such as horizontal integration were adopted. The consolidation of various agricultural sectors and aspects of production into the hands of fewer and fewer large corporations slowly evolved through the process of mergers and acquisitions. This process came to dominate the agriculture industry later than it came to other industrial sectors, but was largely in place by the mid-20th century. Examples of large agribusiness corporations include ConAgra, Tyson Foods, Phillip Morris, Altria, Monsanto, Cargill, Maple Leaf Foods, Smithfield, Philip Morris, British Nutrition, Chiquita, and Imperial Foods.

The adoption of horizontal integration among other industrial business tactics helped foster the increasing specialization of agriculture. Rather than producing a variety of crops and animals, most farms now specialize in one particular crop or animal, often grown for a specific purpose. For example, some chickens are destined to become broilers or nuggets, and some potatoes are destined to become a particular brand of fries or chips even before their life cycle has begun. The development of an agribusiness model also led to the growth of corporations that now served as middlemen between the farmers and growers who produced agricultural products and the consumers that purchased them. The growing number of stages within the production of the food supply led to increased opportunities for horizontal integration within these stages. Initially, many firms competed within each agricultural sector. As horizontal integration proceeded, however, fewer and fewer firms operated within each sector.

Impact of Horizontal Integration

Horizontal integration aids in the production of a large affordable food supply for the urban marketplace, but has also negatively impacted the local agricultural industry. The agricultural specialization that resulted in part from horizontal integration meant that farmers or growers often had to acquire large amounts of capital to invest in the necessary inputs of production, such as seed, feed, fertilizer, buildings, machinery, and labor, leaving many farmers in a constant state of indebtedness to agricultural banks or large corporations. Many smaller local companies also lack sufficient capital to compete with larger corporations, leaving them vulnerable to elimination or acquisition. Specialization also left farmers and growers more economically vulnerable to hazards such as crop blights, diseases, weather phenomena, or economic declines in their particular agricultural commodity.

New agribusiness arrangements such as contract farming, where a corporation pays a farmer or grower on a piece rate basis and controls all production decisions, have left farmers with little independent control over their increasingly complex and specialized operations. Farmers and growers also have less choice of corporations to enter into contracts with as horizontal integration increasingly limits the number of competing firms in any given agricultural sector or aspect of production. Large corporations determine and usually supply the type of product grown and the techniques, tools, and other necessary inputs. This also lessens consumers' choices in the retail food market, which is also increasingly horizontally integrated, as large corporations determine what types of products are available and at what price.

Horizontal integration and other monopolistic business practices have also led to an increased need for agricultural inputs, such as fertilizers and pesticides that can harm the environment, and machinery and transportation networks that often run on fossil fuels. Large-scale commercial farming resulted in the clearance of large areas of forest and land for growing larger crops and for grazing pastures. Increased agricultural specialization has lessened biodiversity and made sustainability more difficult to achieve. The continual growth of a single crop can exhaust the land, requiring the use of larger land areas and can lead to dependence on inputs such as fertilizers.

By the late 20th century, horizontal integration began to occur on a global scale with the development of multinational corporations that sought to control a sector or aspect of production within the rapidly developing global food supply chain. Many multinationals began as local corporations that used tactics such as horizontal integration to expand first nationally and then globally. Reduced competition allowed these corporations to dominate the global food supply chain from seed to table, permanently altering the agricultural industry.

These large multinational corporations could afford to overproduce and absorb the economic losses brought by a subsequent drop in prices, effectively eliminating or absorbing smaller competitors who could not afford to sustain such losses over an extended period. Economic power, rather than efficiency of production, became the determining factor of survival in the marketplace. These corporations receive most of the profits from agricultural product sales and either pass those profits on to shareholders, or reinvest them within the corporation or in another geographic area, depriving local rural economies of the benefits once derived from family farming and subsistence agriculture. Food security is also compromised as individual countries have less ability to regulate an increasingly global food supply in which it is much harder to determine an individual product's country of origin.

See Also: Agribusiness; ConAgra; Factory Farm; Monsanto; Vertical Integration.

Further Readings

Blank, Steven C. *The Economics of American Agriculture: Evolution and Global Development.* Armonk, NY: M.E. Sharpe, 2008.

Clay, Jason. *World Agriculture and the Environment: A Commodity-by-Commodity Guide to Impacts and Practices.* Washington, D.C.: Island Press, 2004.

Conkin, Paul K. *A Revolution Down on the Farm: The Transformation of American Agriculture Since 1929.* Lexington: University Press of Kentucky, 2008.

Heffernan, William D. "Agriculture and Monopoly Capital." *Monthly Review* (July–August 1998).

Ikerd, John E. *Crisis and Opportunity: Sustainability in American Agriculture (Our Sustainable Future).* London: Bison Books, 2008.

Mazoyer, Marcel and Laurence Roudart. *A History of World Agriculture: From the Neolithic Age to the Current Crisis.* New York: Monthly Review Press, 2006.

McCullough, Ellen B., Prabhu L. Pingali, and Kostas G. Stamoulis. *The Transformation of Agri-Food Systems: Globalization, Supply Chains, and Smallholder Farms.* Sterling, VA: Earthscan, 2008.

Vogeler, Ingolf. *The Myth of the Family Farm: Agribusiness Dominance of U.S. Agriculture.* Boulder, CO: Westview Press, 1981.

Marcella Bush Trevino
Barry University

HUNTING

Hunting is the practice of pursuing or trapping and killing animals for food, sport, or trade (usually fur). Some people consider trapping to be separate from hunting, but the activities are both governed by the same laws. Illegal hunting (such as of species protected by law or not in season, or on grounds where you do not have the legal right to hunt) is called poaching, whether it involves pursuit or trapping. Mammals or birds that are the object of the hunt are called game. Fishing is generally considered a separate activity, though it is also bound by laws and regulated by the Fish & Wildlife or Fish & Game agencies of the local jurisdiction.

Hunting is a controversial activity, whether or not it is done for food. Hunting advocates claim that hunting keeps wildlife populations in check healthily, the way grazing benefits plant life. Opponents claim otherwise or protest regulations that permit what they consider overhunting or the disruption of the natural ecosystem. Vegetarian animal rights groups like People for the Ethical Treatment of Animals oppose hunting—as they oppose all harm done to animals—to an extreme extent; however, hunting opponents include many unrepentant meat-eaters as well, who argue that hunting is cruel and unusual (because of the fear felt by the animal, or because the weaponry used may be overkill and can cause the animal more suffering than death in a slaughterhouse would), that it is poorly regulated with little oversight, and that it is too often wasteful.

There are few requirements as to what hunters must do with their kills, for instance; although there are regulations about what they may sell, and which animals are too young to be killed, hunters are not required to do anything at all with their kills, and a pursuit that originated to defend the village and provide meat for families may legally result in nothing more than a heap of dead animals. It sometimes seems there is little middle ground in the hunting debates, with the extreme positions co-opted by animal rights organizations on the one side and gun rights organizations on the other, neither of which pursue agendas to which hunting is central, and who therefore are not motivated to compromise or to admit the validity of constructive criticism.

Hunting is one of the oldest human activities, and in fact it almost certainly predates *Homo sapiens* and probably the *Homo* genus altogether: Australopithecines, the early hominids who appeared about 4 million years ago (1.6 million years before genus *Homo*), very likely hunted larger mammals. Hunting provided the motivation and "laboratory" for the creation of stone tools and fire and was probably a key reason for the evolution of bipedalism. Agriculture was invented only about 11,000 years ago. If the history of mankind were a map of the United States, agriculture would take up about as much space as Southern California; the rest of the country would be hunter-gatherers. Prehistoric society was strongly defined by those two occupations: the hunters, who provided meat and guided the invention of technology, and the gatherers, who foraged wild fruits and vegetables and tended to the home. Language, culture, religion, the family—all the things that make humans human—somehow developed out of this hunter-gatherer culture, in ways we will likely never piece together because of the lack of a written record. Hunting may also be responsible for the extinction of the giant animals of prehistoric times—the woolly mammoths, terror birds, and other megafauna.

Hunting is not the same as predation, which is the stalking and killing of one animal by another for food, though the line is sometimes fine; chimpanzees, our fellow omnivores in the family of modern hominids, hunt in troops of males led by an alpha male, and other animals hunt in packs or stalk herd animals, waiting for the right moment to strike, exhibiting behavior similar to that used by hunters today. One difference, of course, is that few humans need to hunt in the 21st century—there are fewer and fewer places where hunting provides the main source of food (or of animal protein), and even fewer where hunting is the only option. (It is a necessary part of life among the Inuits in the Arctic and in other climates where agriculture is impractical or impossible.) In addition, humans are the only animal to use technology in the hunt—not just our modern firearms, but spears, arrows, and even a sling all provide an advantage far beyond what is available when an unarmed man faces a wild animal.

Furthermore, hunting and technology had a symbiotic relationship. As technology enhanced hunters' prowess, it also provided raw materials—skins, furs, and feathers for clothes; sinew for cords; and bones for fashioning tools. Many early arrowheads and spearheads were made of bone. When animals were domesticated, as many of them were put to use as hunting aids (dogs, horses, ferrets, and birds of prey) as were raised for food. In other words, the advent of new technologies and new food technologies augmented rather than displaced hunting.

Because hunting traditions in North America predate the arrival of Europeans and the creation of the United States, various exemptions and special laws are made to try to protect the traditions of Native Americans (most famously, Inuits are permitted to hunt marine mammals—whales, dolphins, seals, etc.—and Native Americans are allowed to possess eagle feathers from protected species). Hunting is strictly regulated throughout the country at the federal level (which protects certain species), and especially the state and sometimes county levels. Most species commonly hunted are protected, which means that hunting them requires a permit and abiding by certain regulations and restrictions; in jurisdictions where there are unprotected species, those species are usually rodents, small predators, and pests. This is one reason there is a tradition of squirrel- and rabbit-hunting in many rural areas of the country—those species are unprotected in many areas, which makes them cheaper to hunt, and unlike many other unprotected species, they provide a sufficient amount of good, healthy meat.

Protected species are regulated according to category: big game (a term associated with tourist safaris but that includes deer, moose, elk, bear, and wild boar), small game (rabbits, raccoons, squirrels), furbearers (beavers, foxes, bobcats), predators (mountain lions, coyotes), upland game birds (turkeys, pheasants, quail), and waterfowl (ducks and geese). These are the typical divisions, but they vary from region to region, according to the local wildlife; for instance, alligators are regulated very specifically in Louisiana, where hunting them is a part of traditional life in Acadiana. Hunting a protected species generally obligates the hunter to abide by a restricted number of animals that they can kill/capture in a day or throughout the season. Specific types of firearms and other methods are regulated; rifles are often disallowed in flatlands or near high-population areas, for instance, whereas a certain minimum muzzle energy is required for hunting big game to assure a clean kill (a deer that is shot once and survives will flee and will be unlikely to be shot again, possibly suffering a slow death; a bear who survives a shot is a danger to the hunters and anyone else nearby).

Other changes in technology have necessitated other regulations: When "Internet hunting" was introduced in 2005, a process that allowed computer users to "hunt" animals by clicking a button to fire a real trigger on a real gun pointed at an animal in a reserve, both animal rights advocates and hunting advocates condemned the activity. Thirty-eight states have since outlawed it, and the industry was never able to become profitable.

See Also: Animal Welfare; Grazing.

Further Readings

Berry, Wendell. *The Unsettling of America: Culture and Agriculture*. San Francisco: Sierra Club, 1996.

Pollan, Michael. *In Defense of Food*. New York: Penguin, 2009.

Pollan, Michael. *The Omnivore's Dilemma: A Natural History of Four Meals*. New York: Penguin, 2007.

Bill Kte'pi
Independent Scholar

INSTITUTE FOR AGRICULTURE AND TRADE POLICY

The Institute for Agriculture and Trade Policy (IATP) is a nonprofit research and education organization that seeks to build fair and sustainable food, farm, and trade systems. IATP research addresses local and global issues and works at the intersection of agricultural policy and practice. Based in Minneapolis, Minnesota, and employing over 40 staff members, the IATP is one of the most influential voices in debates around sustainable food systems and agricultural policy. Through regular news bulletins and an extensive library of research publications, the organization provides easy access to a large amount of information and analysis on trade and agriculture issues.

The IATP aims to help public interest organizations, both in the United States and internationally, to effectively influence agricultural policymaking by monitoring key events related to agriculture, trade, and food systems. Research publications analyze the economic and environmental implications of changes in agricultural and trade policy and present policy options in response to specific issues. The IATP also plays a leading role in the provision of education and training materials to policymakers and the public, using a wide range of media including printed publications, websites, podcasts, e-mail newsletters, fact sheets, blogs, and Web feeds, delivering information and analysis in a variety of formats. Four Resource Centers are accessible through the IATP website, covering the following policy areas:

- trade policy (http://www.tradeobservatory.org)
- agricultural policy (http://www.agobservatory.org)
- community forestry (http://www.forestrycenter.org)
- public health policy (http://www.healthobservatory.org)

Research and Advocacy

Research and advocacy are organized under six programs: Trade and Global Governance, Rural Communities, Food and Health, Environment and Agriculture, Forestry, and Local Foods. Analyzing and responding to global trade agreements is central to the institute's work, which seeks to clarify how World Trade Organization trade negotiations affect U.S. farm and food policies. From an office in Geneva, the IATP's Trade Information Project follows United Nations and World Trade Organization negotiations. Global governance research monitors the standards applied to genetically engineered foods and their presence in the food system and reports on agricultural policy reform in Europe and around the world. The IATP has also played a key role in developing the concept of access to food and water as a basic human right.

In the United States, IATP programs focus on sustainable rural development, aiming to revitalize rural communities by developing alternative economic models and supporting progressive rural leadership and policies. By encouraging agricultural diversification and regional food systems, IATP research seeks to demonstrate how rural economies can be regenerated through community-based development strategies and, in particular, through the development of local food systems. In addition to supporting community-based sustainable development in rural communities, the IATP has identified African refugee populations that have settled in the U.S. midwest as a group in need of further support and is working to ensure their successful integration into the rural communities where they now live and work. The Rural Young Adults program has been initiated to help local communities take control of their future and to train young adults as they become the next generation of community leaders.

The IATP also monitors food safety and the environmental damage caused by industrial agriculture. The overuse of antibiotics in industrial animal and fish farming is of particular concern, and tests of poultry carried out by the institute's Antibiotic Resistance Project have routinely found at least one strain of antibiotic-resistant bacteria. The institute also monitors the release of toxic chemicals and heavy metals into the environment and the food supply, and having identified food as the primary source of exposure to toxic pollutants, it provides information to help consumers choose safer and more sustainable food. The *Eat Well Guide* (http://www.eatwellguide.org) is a searchable online directory of sustainable and antibiotic-free meat, poultry, dairy, and eggs produced by the IATP Food and Health program.

The Forestry program addresses the need of family and community forest owners for advice and guidance about sustainable forest management. Through collaboration with landowners, conservation groups, and forestry cooperatives, the IATP promotes responsible forest management and has developed a certification scheme that gives private landowners access to forest management and certification to Forest Stewardship Council standards.

History

The IATP was founded in 1986, following a meeting of rural and farm organization leaders in Geneva to discuss the deepening farm crisis. During the 1980s, rural communities around the world were affected by international trade agreements that were pushing commodity prices below the cost of production. This group of leaders identified the need for an organization to examine the links between global trade agreements and agricultural policies and changes in farming practice in local communities. In 1987, the newly incorporated IATP reported on the General Agreement on Tariffs and Trade negotiations that led to the formation of the World Trade Organization. During the 1990s, the IATP moved beyond its initial activities of reporting and analyzing the local effects of global trade agreements to develop and advocate for positive alternatives to unsustainable agricultural and trade practices. This broadened focus has allowed the IATP to develop links with a wide range of organizations both in the United States and around the world and has led to campaigns to promote sustainable fisheries, farming and forestry certifications, and fair trade schemes. The IATP has also founded Transfair, an international fair trade certification organization, and Peace Coffee, an award-winning organic and fair trade coffee company. The current board of directors is drawn from nongovernmental and research organizations from around the world, including individuals working in the Netherlands, Brazil, France, Canada, Mexico, and Japan in addition to the United States.

See Also: Doha Round, World Trade Organization; Family Farm; Farm Bill; North American Free Trade Agreement; Trade Liberalization.

Further Readings

Ellis, Sarah. *The Changing Climate for Food and Agriculture: A Literature Review.* Institute for Agriculture and Trade Policy, 2008. http://www.iatp.org/iatp/publications.cfm (Accessed January 2009).

Harkness, Jim and Alexandra Spieldoch. "Rejoining the World." In *Thinking Big: Progressive Ideas for a New Era,* edited by the Progressive Ideas Network. San Francisco: Berett-Koehler, 2009.

Oosterveer, Peter. *Global Governance of Food Production and Consumption: Issues and Challenges.* Northampton, MA: Edward Elgar, 2007.

Edmund M. Harris
Clark University

Integrated Pest Management

Integrated pest management, or IPM, is an ecologically based, multipronged approach to managing damage from crop pests below a level that will cause economic loss if no remedial action is taken. The term *pest* refers to any organism that is detrimental to humans. However, the majority of IPM programs in agriculture have been designed around the control of arthropods, weeds, and plant pathogens. The foundation of decision making in IPM is an understanding of the biology of the pest organism and the ecology of the cropping system. Ideally, IPM relies on cultural, biological, and mechanical practices to prevent competitive species from becoming pests. When crop monitoring indicates that pest populations exceed the economic threshold, a mechanical, chemical, or biological method can be employed to reduce the pest population. By using pest management practices that are effective, economically sound, and ecologically harmonious, IPM is a fundamental component of sustainable agriculture. IPM was developed as an alternative approach to pesticide-centered management programs that were the predominant model for pest management post–World War II, when synthetic pesticides became widely available. Chemical pesticides have been central to industrial agriculture production because they are easy and relatively inexpensive to produce and use, can be persistent, and are effective against a broad array of insects. Undoubtedly, chemical pesticides have been important in reducing economic damage from insect pests and the diseases they can carry.

Integrated pest management requires in-depth knowledge of pest organisms such as this western corn rootworm, photographed in its adult stage while it sought pollen in strands of corn silk.

Source: U.S. Department of Agriculture, Agricultural Research Service/Tom Hlavaty

However, the dependence on chemical pesticides as the principal defense against pest damage has led to many deleterious agricultural, social, and ecological effects including pesticide resistance, pest resurgence, mortality of natural biological control organisms leading to secondary pest outbreaks, health hazards for people in contact with the pesticides (e.g., applicators, farm workers, and consumers), and the contamination of soil, water, and air, which can threaten biodiversity. Rachel Carson's highly influential book, *Silent Spring*, published in 1962, brought international attention to the ecological dangers of widespread pesticide use. In 1972, the term *integrated pest management* was formalized in a letter to Congress by President Richard Nixon requesting further development of IPM concepts and application.

There are six essential components to an IPM program: (1) the people who make decisions about production and manage the crops in the field, (2) the knowledge about a crop production system and a pest that informs management decisions, (3) the cultural practices that give crops the competitive advantage over pests, (4) a system of monitoring the crop for the presence of pests and beneficial organisms and for crop damage by the pest, (5) an economic threshold to aid decisions about treatments, and (6) reduced risk control options including biological control, mechanical control, and pesticides of low toxicity.

How IPM Works

The fundamental guideline of an IPM program is to apply knowledge of the population biology and ecology of the pest organism and its natural enemies to the design of agroecosystems that are resistant and resilient to pest outbreaks. Pest populations are naturally controlled by a number of factors including weather, disease, predators, and parasites. For example, the seeds of many weeds cannot germinate unless the right temperature, moisture, and light conditions are present. Similarly, the growth rate of insect pest populations is directly related to the amount of time accumulated within an optimal temperature range, known as degree-days. IPM practitioners can take advantage of phenological restrictions by cultivating host crops when environmental conditions are less favorable for the development of the pest organism, but sufficient for the crop. Often just small amounts of time without initial pest competition can help crops develop to the point where they can tolerate pest presence without loss in productivity or marketability.

There are numerous other cultural practices that make the crop environment less favorable for pest survival, reproduction, and dispersal. Choosing crop varieties that are resistant or tolerant to pest attack (because of chemical or mechanical defenses) or planting varieties that have nonsynchronous growth cycles is a first line of defense against pest establishment. Planting a small area of preferred host crop, known as a trap crop, can lure insect pests away from the commercial crop. If necessary, the pest can be destroyed in the trap crop. This method reduces the area in which control tactics are employed and can reduce nontarget effects associated with the control measure in the commercial crop. Diversifying the cropping system with nonhost crops in both space and time through intercropping and crop rotation limits the availability of resources for the pest and, consequently, its ability to multiply and persist. Maintaining good sanitation practices—including the removal of pest breeding, refuge, and overwintering sites; the use of pest-free seed and seedlings; and the disinfection of materials and equipment used in cultivation—can be effective at limiting the establishment, growth, and dispersal of pest populations. Finally, conserving habitat that provides food resources, shelter, and overwintering sites for natural enemy organisms can enhance and maintain biological control services.

Monitoring Pests

Even when multiple cultural practices are employed, the dynamic nature of an agroecosystem can lead to the emergence of new and recurring pests. Hence it is crucial that IPM practitioners have a system of monitoring to determine the presence and abundance of pest organisms to detect and manage potential problems early. Monitoring is the process by which the abundance and life stages of pest and beneficial organisms are recorded. The first component of a monitoring system is the method used to sample the pest species, which depends on the biology and ecology of the pest organism. The presence and stage of development of insect pest and natural enemy species can be monitored with traps that sample insects from a general area. An assortment of traps is available to monitor insect pests: sticky traps for flying insects, blacklight traps for nocturnal flying insects, and pitfall traps for insects that travel on the surface of the soil. Pheromone traps, which use olfactory chemicals used in insect communication, have the advantage of being species specific and thus more efficient for monitoring a single pest of economic importance. Pheromone traps can be used in conjunction with mating disruption, which reduces the population by interfering with the male insect's ability to locate and mate with a female insect.

To estimate the density of a pest organism in the field (number of individuals/area), devices that sample pests from a certain volume or area are used. Flying insect pests can be sampled with sweep nets, where the sampled area is a standardized number of sweeps. Pests that live in the soil can be estimated with soil sampling devices such as a shovel or soil corer, or by baiting. Pest populations, such as aphids or plant pathogens, that are relatively immobile and live on the crop plant, are best estimated through visual inspection of a defined area of the crop stand. Depending on the distribution and abundance of the pest and the nature of the crop damage, the sampling area may be as large as multiple plants or as small as a section of the crop plant, such as some leaves or roots. Weed densities are measured within crop fields by counting the number of weeds within a specific area or quadrant.

The second component of a monitoring system is the sampling procedure. Several factors need to be considered in designing a sampling procedure: pest density, pest mobility, required number of samples to accurately estimate numbers or potential crop damage, acreage and value of the crop, time of sampling, and frequency of sampling. A main goal of monitoring is to get an accurate estimate of the density of a pest through random or systematic sampling. Sampling intensity, however, depends on the density and distribution of the species such that abundant and evenly distributed pests require fewer samples than pest populations that are low and unevenly distributed in the field. Larger fields will also require a greater number of samples to get an estimate that reflects the pest density of the entire field. Thus, reliability and cost effectiveness need to be balanced when determining an appropriate sample size used in monitoring. Sampling early in the season and in fields that have a history of pest infestation can increase the chance of early pest detection and prevent pest damage or dispersal into other fields. The frequency of sampling depends on the growth rate of the pest organism and the rate at which injury to the crop is increasing, such that pests with high growth rates and high capability to injure crops require more frequent sampling than pests with low growth rates or damage potential.

Taking Action

Sequential sampling is an efficient and reliable procedure for estimating the density of a pest in relationship to the need to introduce a treatment or control. Sequential sampling incorporates the economic threshold of a pest into a decision-making table or chart, which is then used to gauge the cumulative number of pests found with each additional sample. If pest populations are low or high, then a field manager can quickly make a decision whether to treat the sampled field. Medium pest populations require more sampling effort. Sequential sampling depends on having in-depth knowledge of the population biology of the pest and the associated crop injury levels.

Economic thresholds are essential for gauging the relevance of the collected abundance data. An economic threshold is the density of a pest at which the cost to control the pest is less than the financial loss in crop yield or marketability if no action is taken. In other words, a producer will have a greater economic return if a control treatment is applied. Economic thresholds differ across pest species, crops, and time of year. Aesthetic values, consumer preferences, market channel, and productivity of the crop are all figured into an economic threshold.

When a pest population exceeds an economic threshold, control measures can be used that ideally are nondisruptive to natural control factors. Pests can be physically destroyed via hand-picking, hoeing, trapping, manipulating temperature (used in greenhouses), flaming, and tilling the soil. Biological control uses predator, parasitoid, pathogen, or competitor populations to suppress pest populations. Populations of biological control agents that are already present at the site, but in insufficient numbers to effect control, can be boosted through periodic releases (augmentative biological control) or applied in large numbers, much like a pesticide (inundative biological control).

In cases in which pest populations have reached economic injury levels despite cultural and mechanical practices and biological control, chemical controls can effectively reduce pest populations. It is important that whenever pesticides are used, managers consider: (1) the toxicity of the pesticide and its potential hazard to people and the environment; (2) the specificity of the pesticide to the target pest organism; and (3) the potential effects of the pesticide on nontarget organisms like natural enemies and pollinators. Several classes of pesticides including insect growth regulators, contact and stomach poisons, microorganisms, botanicals, and oils are suitable for IPM programs because of their selectivity, efficacy, and low environmental toxicity. These pesticides work by directly killing the pest, causing disease, or disrupting developmental processes.

See Also: Biological Control; Low-Input Agriculture; Organic Farming; Weed Management.

Further Readings

Benbrook, Charles M. *Pest Management at the Crossroads*. Yonkers, NY: Consumers Union, 1996.

Carson, Rachel. *Silent Spring*. Boston: Houghton Mifflin, 1962.

Flint, Mary Louise and Robert van den Bosch. *Introduction to Integrated Pest Management*. New York: Plenum, 1981.

Kogan, Marcos. "Integrated Pest Management: Historical Perspectives and Contemporary Developments." *Annual Review of Entomology*, 43 (1998).

Norris, Robert F., et al. *Concepts in Integrated Pest Management*. Upper Saddle River, NJ: Prentice Hall, 2003.

Tara Pisani Gareau
University of California, Santa Cruz

INTERCROPPING

Intercropping is an agricultural practice in which two or more crops are planted in a spatial arrangement that facilitates direct interactions between them. In recent years, the study of intercropping has revealed the complex ecological relationships that often make intercropping superior to single-species (i.e., monoculture) cultivation in terms of productivity and environmental impact. A subset of multiple cropping, intercropping is distinct from sequential cropping practices such as crop rotations and cover cropping, in which crops are combined in time rather than in space (see Figure 1).

Intercropping takes four principal forms: relay, mixed, row, and strip intercropping, described here. All forms of multiple cropping systems increase on-farm diversity compared with crop monocultures and also facilitate agroecosystem functioning, which can reduce or remove the need for synthetic or chemical inputs into the farm and contribute to sustainability.

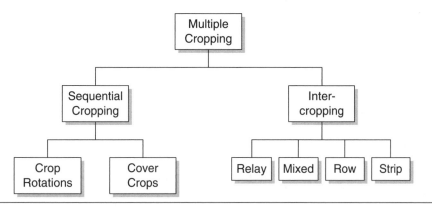

Figure 1   Multiple Cropping Systems Are Based on Temporal Combinations of Crops (Sequential Cropping) or Spatial Combinations (Intercropping)

In the Field

Intercropping is an agricultural management practice with roots in traditional farming worldwide. Examples of intercrops range from simple soybean-corn systems, in which the plants coexist in neatly planted rows, to tropical homegardens with tens of different crop species intermingled in no row pattern whatsoever. When planning an intercrop system, a farmer must consider the different agroecological roles of plants. For example, intercropping similar crops such as broccoli and kale, which are both in the *Brassicaceae* family, could result in nitrogen depletion from the soil because the brassicas are heavy nitrogen feeders. It might be more beneficial to intercrop the kale with a bush bean, which can fix nitrogen and return it to the soil.

Intercropping Ecology

Intercropping can benefit pest and weed management; alter soil, water, and light dynamics; and increase farm productivity. Describing and understanding the ecological mechanisms behind these benefits is a central focus of the field of agroecology. What follows is a brief review of ecological theory relevant to intercropping.

In an intercrop, pest damage may be reduced through a variety of means such as food source interruption, natural enemies, or trap crops, to name a few. Food source interruption occurs when a pest encounters a plant that is not food. For example, an insect that specializes in cucumber foliage will flourish in a monoculture of cucumbers but may leave the system if it encounters a tomato in a cucumber-tomato intercrop. Incorporating plants that attract the natural enemies of a particular pest into an intercrop may help control pests through predation and parasitism. Trap crops are usually noncrop plants that are attractive to crop pests and draw the pests out of the productive portion of the agroecosystem.

Intercropping can reduce weeds by having mixed crops occupying open ground or through allelopathic inhibition. In an intercrop, there is often less bare ground than in a monoculture, which means there is less space for weeds to colonize. For example, in a sugarcane monoculture, weeds could easily fill in bare space between cane plants. If a ground-covering vegetable such as squash were intercropped into the system, that space would be unavailable for colonization by weeds. Allelopathic control of weeds is also possible in an intercrop when a plant with known allelopathic attributes, such as wild mustard, is planted. The chemicals released by the mustard plant through its roots and leaves can inhibit the growth of weeds in the system.

Intercropping can also lead to habitat modification as plants physically alter the abiotic components of the agroecosystem such as soil, water, and light. As has been discussed, some plants fix nitrogen and make it available in the soil. Plants also differ in the amount of organic material left behind in the agroecosystem after harvest. For example, crops whose harvestable portion is leaves, such as lettuce, leave little organic material in the system. Beans leave more biomass in the soil because only the fruits are harvested. By intercropping beans with lettuce, more soil organic material is left in the system than in a lettuce monoculture.

Water can also be affected by intercropping. Certain deep-rooted crops, such as trees, are able to access water from deep in the soil profile and bring it to the surface through a process known as hydraulic lift. This water is then available to crops with shallower roots. An olive tree-grain intercrop has the potential to take advantage of hydraulic lift. Modification of the light environment can affect water evaporation, as well as temperature and shade quality. Planting broad-leaved crops in the canopy of an intercrop creates a suitable environment for shade-loving plants and reduces evaporation from the soil, thereby saving water for use by all crops in the system.

One of the most important outcomes of intercropping is increased productivity. Often, intercropped systems produce more food on less land than equivalent monocultures. This is possible because of facilitation and species complementarity. Facilitation occurs when different species in an agroecosystem modify the environment in a way that is beneficial to other species and is the outcome of all the above-described processes. Less pest and weed damage, combined with optimal abiotic environment modifications, facilitate agroecosystem health and allow crops to produce large harvests. Species complementarity describes the ability of two plants to coexist because they use resources differently. For example, a shallow-rooted crop can be intercropped with a deep-rooted crop because they will access nutrients from different parts of the soil horizon without competing, allowing a farmer to plant crops in a higher density than in a monoculture. By increasing density, more crops can be planted in less space, resulting in overall higher productivity.

Classification

Relay, mixed, row, and strip intercropping combine crops in ways that result in trade-offs between degree of ecological interaction and management type. In "relay intercropping," a second crop is planted into a standing crop that was previously planted. In "mixed intercropping," two or more crops are planted at the same time with no row arrangement, maximizing ecological interaction between plants. "Row intercropping" is when multiple crops are planted together, with at least one of the crops planted in a row. This design facilitates harvesting of a target crop from the row while still promoting ecological interactions through proximity of crop species. Finally, in "strip intercropping," two or more plants are grown together in a field, but each crop type is planted in a strip with like crops. In this way, mechanized management and harvesting is made possible: Diversity on a field scale is maintained, but potential for interaction between crop species is reduced.

Corn, bean, and squash intercrops can be used to further explain these intercropping designs. Traditionally, in the Americas these three crops are planted in a mixed intercrop, where the corn stands as the tallest plant in the system. Bean vines then wind up the stalk of the corn plant, and the squash crop grows low to the ground between corn plants, providing weed control in the system. These same three crops can also be planted in a row intercrop, where the corn is planted in rows and the beans and squash are distributed randomly between the crop rows. In strip intercropping, three different rows are planted side by side. With a strip of corn followed by a strip of beans and then one of squash, a tractor can easily pass through the field for planting and harvest. The trade-off in this example is that in the strip intercrop the beans no longer have the structural stability of the corn stalks, and the corn and beans no longer benefit directly from the weed suppression by the squash plant.

Conclusion

Intercropping is a key agricultural practice that can reduce the need for chemical inputs, prevent soil erosion, conserve water, and increase land use efficiency. In an era of increasing awareness about the need to practice agriculture that is environmentally sound, intercrops represent an important solution.

See Also: Agroecology; Cover Cropping; Crop Genetic Diversity; Crop Rotation.

Further Readings

Altieri, M. A. "The Ecological Role of Biodiversity in Agroecosystems." *Agriculture Ecosystems & Environment*, 74 (1999).

Chou, C. H. "Roles of Allelopathy in Plant Biodiversity and Sustainable Agriculture." *Critical Reviews in Plant Sciences*, 18 (1999).

Gliessman, S. *Agroecology: The Ecology of Sustainable Food Systems*, 2nd ed. Boca Raton, FL: CRC Press, 2007.

Sullivan, P. "Intercropping Principles and Production Practices." *Appropriate Technology and Transfer to Rural Areas Agronomy Systems Guide*, 2003. http://www.attra.ncat.org (Accessed June 2009).

Vandermeer, J. *The Ecology of Intercropping*. New York: Cambridge University Press, 1989.

Kathleen Elizabeth Hilimire
Devon Sampson
University of California, Santa Cruz

INTERNATIONAL COFFEE AGREEMENT

Coffee, a stimulant because of the caffeine it contains, is a very popular beverage around the world. Tea and soft drink beverages are its nearest challengers for popular consumption. Coffee is brewed as a beverage from the roasted seeds of the coffee plant. The seeds are picked as berries from several varieties of the small evergreen bush of the genus *Coffea*. After harvesting, the berries are dried and hulled, and the inner portion is roasted as a coffee bean. Roasting is an important part of the production of coffee because the darker the roast, usually the stronger the coffee will be when brewed.

The genus *Coffea* has over 90 species that are flowering members of the family Rubiaceae. All are either shrubs or small trees that are native to subtropical Africa or to southern Asia. Coffee is made from the roasted berries of several of the species. The two most popular species of the evergreen bush are *Coffea robusta* (*Coffea canephora*) and *Coffea arabica*. There are also less popular species: Liberica, Excelsa, Stenophylla, Mauritiana, and Racemosa.

Liberica (*Coffea liberica*) is an Arabica coffee species first discovered in Liberia, West Africa. Its berries are picked from a tree that grows over 20 feet (9 meters) tall. It was first grown in Indonesia (Java) in the 19th century and is still grown in areas of the Philippines.

Excelsa (*Coffea excelsa*) was first discovered in 1904. It is similar to Liberica and is disease and drought resistant. It is a mild-tasting coffee often used as filler in blends.

Coffea stenophylla was discovered and cultivated in West Africa. It is more resistant to leaf rust than Arabica varieties and is more flavorful, but it lost out to Arabica for economic reasons. It is still grown in limited quantities in Guinea and Côte d'Ivoire.

Mascarene (*Coffea mauritiana*) grows wild on several of the Mascarene Islands in the Indian Ocean. It grows as a shrub or a small tree that produces berries with low caffeine content and a sharp to bitter taste.

Coffea racemosa is found in southern Africa. It was first cultivated in Mozambique. It is now grown in small quantities in a few African countries. Like all coffees, it is grown between 3,500 and 8,000 feet in tropical regions and is especially productive in volcanic soils.

The origin of coffee drinking seems to have begun when a goatherd of the Galla tribe in Abyssinia (now Kaffa province in Ethiopia) noticed that his goats were stimulated from eating the berries of a local plant. He tried some and found the effect enjoyable. The other possible source of coffee is Yemen. In either case, the use of coffee spread to Arabia, where it was named *qahweh* in Arabic and coffee in the world beyond.

Historically, the United States was the largest consumer of coffee, a drink Americans adopted at the end of the Colonial Era after their refusal to pay the tea tax to Great Britain, dumping chests of East India Company tea into Boston Harbor during the Boston Tea Party (1773). However, in recent decades, the Scandinavian countries and some other European countries have surpassed the United States in coffee consumption. In the United States, coffee consumption has remained steady in the face of competition from many new drinks, from herbal teas to new forms of bottled drinks. In addition, tea, long drunk iced in the summer, has increased in consumption, as many people turn to it as a hot beverage.

Coffee is grown in large quantities on coffee plantations. It is also produced on hundreds of thousands of small farms in many parts of the world. In Mexico alone there are over 100,000 small coffee farms. Kona coffee, the Hawaiian specialty coffee, is grown on over 800 small family farms only on the Kona Coast of the Big Island of Hawaii. Maui also has some small coffee farms.

Coffee consumption has increased as world population and trade have grown in the years since 1945. Coffee is very important in the economies of many countries; however, increased demand has not always favored coffee growers. Coffee as a commodity since the 19th century experienced fluctuations in demand that have ranged over long periods of time, from oversupply with low prices to times in which demand rose faster than supply. Shortages in supply have often been the result of weather conditions in Brazil, where drought or unusual frosts have damaged or killed the coffee plants. Both situations—oversupply and lower prices, or drops in supply, usually resulting from shocks of weather or disease—have affected growers to the point where they have not received sufficient returns on their production costs to make any or even a reasonable profit.

Following the 1953 frost that destroyed much of Brazil's coffee production, increased demand led to global increases in production, which led to an oversupply in the mid-1960s. With prices dropping, governments began initiatives to stabilize the price of coffee in the markets. As the economic mainstay of many Latin American countries, coffee price stability was important in maintaining economic and political stability.

In 1962, the International Coffee Agreement (ICA) was adopted by a number of countries. A provision of the ICA was the establishment in 1963 of the International Coffee Organization (ICO). The ICA has continued to operate under successive agreements adopted in 1968, 1976, 1983, 1994, 2001, and 2007.

The ICO is an international organization for coffee that provides a forum for producing and consuming countries. Issues of concern are numerous. They include international cooperation to achieve price stability, environmental integrity in production, maintaining quality, improving living standards in producing countries, which are usually in the poorer areas of the world, and other issues.

The ICA of 2007 was agreed to by 77 member countries of the International Coffee Council. The agreement added strength to the ICO's role as a forum for dealing with coffee issues, including food safety, aid for niche growers, and financial support for small growers. Among the provisions was an adoption of the Millennium Development Goals of the United Nations.

An important goal since 2001 has been development of sustainable coffee production because coffee plants, being evergreens, are important to carbon sequestration. In addition, attending to production methods that eliminate soil erosion is vital to preventing water pollution and to long-term success in rural areas. Crop diversification is also an important goal of the ICO, because fluctuations in the commodity price of coffee adversely affect small rural producers.

See Also: Millennium Development Goals; Soil Erosion.

Further Readings

Arnold, Edwin Lester. *Coffee: Its Cultivation and Profit.* Charleston, SC: BiblioBazaar, 2008.
Clarke, R. J. and R. MacRae, eds. *Coffee: Commercial and Technico-Legal Aspects*, Vol. 6. New York: Springer, 2007.
Raffaelli, Marcelo. *Rise and Demise of Commodity Agreements: An Investigation into the Breakdown of International Commodity Agreements.* Cambridge, UK: Woodhead, 1995.
Talbot, John M. *Grounds for Agreement: The Political Economy of the Coffee Commodity Chain.* Lanham, MD: Rowman & Littlefield, 2004.
Wild, Anthony. *Coffee: A Dark History.* New York: W. W. Norton, 2005.
Wintgens, Jean Nicolas, ed. *Coffee: Growing, Processing, Sustainable Production: A Guidebook for Growers, Processors, Traders, and Researchers.* New York: John Wiley & Sons, 2009.

Andrew Jackson Waskey
Dalton State College

IRRADIATION

Food irradiation involves exposing food to shorter-wavelength ionizing radiation in the form of gamma rays, X-rays, or electron beams as an alternative means of preservation to more traditional methods such as canning. Small doses of irradiation can inhibit food properties such as sprouting or ripening, and larger doses can genetically alter or kill insects and microorganisms that lead to spoilage or human diseases. Irradiated food does not change in terms of appearance, smell, taste, or texture. Nutritional losses during the irradiation process are minimal and are either lower than or equal to nutritional losses from methods such as cooking or freezing. Food irradiation products or facilities can be found in over 30 nations worldwide, including the United States, Canada, Japan, China, Russia, the Netherlands, Belgium, France, and South Africa. Irradiated food does not play a significant role in the U.S. food industry, although research and development are expanding.

The discovery of radioactivity in the late 19th century led to scientific research into its effects and potential uses. French scientists determined in the 1920s that irradiation could be applied to food preservation. The United States did not pursue food irradiation studies until World War II, when the U.S. Army began experimenting with it to meet the challenges of feeding a large military force. Food irradiation is particularly useful in preservation, but it is also used to control sprouting, ripening, mold growth, insect damage, and microorganism control. Food irradiation at higher dosages is used for the sterilization of food for space travel or for hospital use in patients with compromised immune systems. Irradiated foods are still considered perishable, even though their shelf life is extended. Food irradiation at dosages used for consumer products does not sterilize food, so refrigeration and proper food handling and preparation guidelines must still be followed.

Food irradiation facilities are expensive to build but are generally considered safe. They house electron-accelerator machines that create beams of electrons. The most common source of radiation used in the process is cobalt-60, which is

These biologists with the U.S. Department of Agriculture's Agricultural Research Service are preparing to irradiate hotdogs packaged in plastic as part of an experiment.

Source: U.S. Department of Agriculture, Agricultural Research Service/Stephen Ausmus

inside a steel casing in a lead-lined chamber. Conveyor belts carry the food through this chamber to complete the irradiation process. The amount of radiation absorbed by food during the irradiation process is known as a dose, which is measured either in the older form of rads or the newer form of grays. The U.S. Food and Drug Administration (FDA) measures doses in kilograys. Irradiation is a cold process, meaning that there is no significant temperature increase in the food during the irradiation process. U.S. food irradiation facilities must meet plant and worker safety guidelines established by the Nuclear Regulatory Commission and the Occupational Safety and Health Administration.

Within the United States, food irradiation is regulated by the FDA and monitored by the U.S. Department of Agriculture. Regulations specify the types of foods that may be irradiated, the radiation dosages that may be used, and labeling requirements for irradiated foods sold in the marketplace. The first FDA-approved use of food irradiation came in 1963 to control insect infestations of wheat and wheat flour. Since that time, approved uses have expanded to include inhibiting sprouting in white potatoes; controlling trichinosis in pork; controlling insects and microorganisms in certain herbs, spices, vegetable seasonings, dry enzyme preparations, packaged fresh or frozen uncooked poultry, and red meat; and inhibiting growth and ripening in a variety of fruits, vegetables, and grains.

By the 21st century, over 40 different irradiated food products were approved for sale in close to 40 countries around the world. France, Belgium, and the Netherlands lead the irradiated food market. The FDA established labeling requirements for irradiated food in 1986, with revisions in 1997. Irradiated food sold in the retail marketplace must be labeled with an international symbol known as a radura and contain the phrase "treated with radiation" or "treated by irradiation." Companies may also add a phrase denoting the purpose of irradiation if they so choose. Wholesale products must also add the phrase "do not irradiate again." Unpackaged irradiated fruits and vegetables can be labeled individually on a shipping container easily viewed by the consumer or on a nearby sign.

Only limited quantities of irradiated foods reach the marketplace because of the high costs of production and concerns over its safety. Many public consumers avoid irradiated foods because of fears that it may contain traces of radiation and cause cancer or other unknown long-term health risks. Opponents have picketed facilities and stores selling irradiated products to make their views known. Proponents counter with scientific experiments on human volunteers and animal feeding studies in the United States and China that found irradiated foods to be safe for human consumption. Irradiated food does not have direct contact with a radioactive substance, nor does it retain radiation. The chemical changes that occur in irradiated food, known as "radiolytic products," are not harmful and are similar to those caused by regular cooking methods. The United Nations Food and Agricultural Organization, the International Atomic Energy Agency, the World Health Organization, and the FDA have all declared food irradiated at or below a certain dosage to be safe.

Although the food industry has been slow to adopt irradiation because of its expense and fears of consumer reluctance, its potential advantages continue to draw interest. Food irradiation could allow lengthier storage of food, preventing waste and the large percentage of the world food supply lost annually because of insect infestation and spoilage. Future developments in sterilization with irradiation could lead to the expansion of the types of products that can be stored without refrigeration. Modern outbreaks of food-borne illnesses such as *Salmonella, Shigella, Campylobacter, Yersinia,* and *Escherichia coli* have strengthened interest in irradiation as a method of helping prevent food-borne illnesses. A growing market for organic products not treated with chemicals or pesticides could also fuel interest in irradiation if fears over its safety can be overcome.

See Also: Department of Agriculture, U.S.; Food and Drug Administration; Food Safety; Pesticide.

Further Readings

Blumenthal, D. "Food Irradiation: Toxic to Bacteria, Safe for Humans." *FDA Consumer* (November 1990).
"Food Irradiation." In *Science Under Siege: Balancing Technology and the Environment,* edited by Michael Fumento. New York: W. Morrow, 1993.
Jay, James. *Modern Food Microbiology.* New York: Van Nostrand Reinhold Company, 1986.
Redman, Nina. *Food Safety: A Reference Handbook,* 2nd ed. Santa Barbara, CA: ABC-CLIO, 2007.

Marcella Bush Trevino
Barry University

Irrigation

Throughout human history, agricultural societies in semiarid and arid regions have relied on irrigation to sustain food cultivation. Widespread application of irrigation technologies, particularly after the green revolution, resulted in environmental problems including waterlogging, salinization, and land subsidence. Research and development efforts aim at increasing irrigation efficiency while promoting greater socioeconomic equity and ecological sustainability.

In his highly controversial 1957 volume *Oriental Despotism*, historian Karl Wittfogel argued that rulers who controlled expansive hydraulic networks were uniquely able to wield political power. Wittfogel theorized that large-scale irrigation required a form of highly organized and centralized control that led to an absolutist managerial state.

Although many have noted the shortcomings of Wittfogel's "hydraulic society," his writings have spurred many ecological anthropologists and historians to more closely examine the relationship between irrigation control and sociopolitical development. Revising Wittfogel's original thesis, environmental historian Donald Worster has argued that the modes of water control that gave rise to settlement of the American West were based less on self-governed agrarian democracy than on the federal government's vast appetite for public subsidization of large-scale irrigation.

Irrigation is no less important today than it was at the turn of the 19th century. Irrigation can be sourced from surface water, groundwater, or wells that tap deep aquifers. Surface water for irrigation is usually withdrawn from rivers, lakes, or reservoirs created by dams. Large-scale canal systems carry water hundreds of miles from reservoirs to fields through a combination of water-pumping and gravity-fed systems.

This fertilizer and irrigation, or fertigation, system mixes water with chemical fertilizers that are injected into the pipe at several locations. It then delivers the mixture to individual furrows of this romaine lettuce crop in California's Coachella Valley.

Source: U.S. Department of Agriculture Agricultural Research Service/Floyd Adamsen

Modern irrigation systems rely on several different kinds of technologies to deliver water to fields. The main systems for delivering irrigation are:

- *Flood:* a system of applying water in which the entire surface of the soil is covered by ponded water
- *Furrow:* a system of partial surface flooding in which water is applied in furrows or rows
- *Sprinkler:* an automated system in which water is applied at a uniform rate and fixed pattern through small droplets emanating from pressurized pipes; sprinkler irrigation can include center pivots or traveling guns
- *Drip* or *trickle:* a system in which water is applied directly to the root zone of plants through low-pressure hoses that are placed either on or right below the surface of the ground

Impact of Irrigation

Widespread use of irrigation, particularly associated with the green revolution in Asia, has led to a range of deleterious environmental impacts. The main impacts have been waterlogging, salinization, land subsidence, and agricultural runoff of pesticides and fertilizers into downstream surface flows and/or groundwater.

There have also been serious human impacts associated with irrigation from large dams. The World Commission on Dams noted in its 2001 report that large dams have resulted in the forced displacement of approximately 40–80 million people worldwide. This report also claimed that over half of the world's large dams were built for the single purpose of providing irrigation.

With many parts of the world facing extreme water shortage, it is now crucial that irrigation systems become more efficient. This can be achieved through engineering, economic, and biological solutions. For example, agronomists are developing efficient drip irrigation technologies that make the most of each drop of water by limiting evaporation losses. Economists believe that water conservation and increased efficiency will result from water-pricing schemes that take into account the environmental costs of irrigation. Last, plant geneticists are experimenting with improved plant breeds that can thrive in highly saline environments.

The environmental costs of modern irrigation development have also led to an interest in reviving indigenous irrigation technologies from the past. The Centre for Science and the Environment in New Delhi published *Dying Wisdom* to celebrate indigenous water-harvesting techniques from 15 different ecological zones in India.

A number of international organizations support research and development of irrigation data and techniques. These include the United Nations Food and Agriculture Organization, the International Water Management Institute, and the International Food Policy Research Institute. The International Water Management Institute records global irrigation patterns through its Global Irrigated Mapping Area project. Several journals, including *Irrigation Science*, are devoted to presenting new research in the field.

Irrigation research from these institutions has helped document the importance of devolving irrigation management away from the public sector and into the hands of local users. The International Network on Participatory Irrigation Management is a global network that promotes participatory approaches to irrigation and water resource management through the exchange of best practices. The World Bank's 2000 *Case Studies in Participatory Irrigation Management* describes the institution's interests in promoting equity and sustainability through irrigation investments.

See Also: Fertilizer; Green Revolution; Pesticide.

Further Readings

Agarwal, A. and S. Narain, eds. *Dying Wisdom: The Rise, Fall and Potential of India's Traditional Water Harvesting Systems*. New Delhi: Center for Science and the Environment, 1997.

Groenfeldt, D. and M. Svendsen. *Case Studies in Participatory Irrigation Management*. Washington, D.C.: World Bank Publications, 2000.

International Food Policy Research Institute. http://www.ifpri.org (Accessed February 2009).

International Network of Participatory Irrigation Management. http://www.inpim.org (Accessed February 2009).

International Water Management Institute. http://www.cgiar.org/iwmi (Accessed February 2009).

Postel, S. *Pillar of Sand: Can the Irrigation Miracle Last?* New York: W. W. Norton, 1999.

Wittfogel, K. A. *Oriental Despotism: A Comparative Study of Total Power.* New Haven, CT: Yale University Press, 1957.

World Commission on Dams. *Dams and Development: A New Framework for Decision-Making.* London: Earthscan, 2001.

Worster, D. *Rivers of Empire: Water, Aridity, and the Growth of the American West.* New York: Oxford University Press, 1985.

Roopali Phadke
Macalester College

J

JUST-IN-TIME

Just-in-time is a philosophy of manufacturing organization focusing on inventory management, pioneered by Japanese automotive industry worker Taiichi Ohno of Toyota in the early 1970s. The Japanese automotive industry as a whole widely adopted the just-in-time system to make their product more competitive by improving their product quality and corporate image. The just-in-time system also has precedents in the manufacturing process pioneered by Henry Ford at the Ford Motor Company in the first decades of the 20th century. It is also sometimes known by the terms *lean manufacturing* or *lean production*. The just-in-time system spread from the automotive industry to a wide variety of private, government, and nonprofit organizations of all sizes around the world by the 1990s, including the retail food industry. The just-in-time system also spread to include all points of a product supply chain.

In earlier times, societies would store surplus crops to survive times of bad harvests or food shortages, but in modern society, storing large amounts of food results in waste and inefficient food distribution. Inventory minimization is also critical to the food service industry because of the perishable nature of many foodstuffs, which limits the amount of time they can sit in storage before spoilage and loss of quality become issues. The added costs of inventory maintenance led food producers, distributors, retail sellers, and trade associations such as the Food Marketing Institute and the Grocery Manufacturers of America to search for newer management strategies such as just-in-time. Using the just-in-time system, food could move more quickly closer to the point of sale. The just-in-time system also appealed to large food service organizations such as restaurants that needed to receive large quantities of food that was consistent in quality to ensure a uniform retail product.

The costs associated with inventory maintenance, such as transportation and storage fees, do not add value to the company. The just-in-time management system eliminates many of these costs by reducing the amount of inventory maintained at any one time and in any one location. The just-in-time system also eliminates wasted inventory through tight control of what is needed at which location and the delivery of only those supplies or parts needed at the right moment. The just-in-time system results in smaller production lots and batch sizes, improved efficiency, better quality control, and better customer response times. Beginning with the development of the Universal Product Code in the 1970s, distribution centers and grocery stores began using food scanning technology to better forecast

269

sales and control inventory. Scanning also allows the product to be traced from its origin all the way through the supply chain, improving quality control and allowing poor-quality or contaminated foods to be traced back to their source.

The food distribution system involves a supply chain that links together all steps from the farm to the table, including growers, manufacturers, distributors, retailers, and consumers. Farm products make their way to food product manufacturers, who then deliver pallets of food to wholesalers or distribution centers, where they are repackaged for delivery to grocery chains and commercial and noncommercial food service operations. Wholesale operations can be either specialized or broad line, meaning they carry whole grocery lines rather than specific categories of foods. Large retailers such as Wal-Mart, large grocery store chains such as Publix or Kroger, and large restaurant chains such as McDonald's have their own company-owned distribution centers. Some manufacturers also deliver their products directly to individual stores rather than relying on a wholesaler or distributor. Food banks that collect donated foods from multiple sources, maintain warehouses, and distribute collected foods to various soup kitchens, food pantries, shelters, and other organizations that aid the needy also use the just-in-time system. The U.S. military has adopted the system periodically to avoid wartime delays in high-priority items, including food.

Recent problems and criticisms have arisen regarding use of the just-in-time management system in general, and its use in the food industry in particular. One problem involves the food distribution system's heavy reliance on the trucking industry to transport its products along the various points on the food supply chain. Many of the trucks require expensive refrigeration systems to protect perishable foods from spoilage. Rising energy costs and the omnipresent threat of fossil fuel shortages thus add to distribution costs. Consumers also use gasoline to drive to their local supermarkets or restaurants to purchase food, sometimes several times a week if they buy supplies for only that day or the next few days in a process of just-in-time eating.

The system's emphasis on inventory reduction relies on accurate knowledge of inventory needs that is often based on analysis of past needs. This means that the just-in-time system is not as responsive to changes in product demand as are other systems, so food production companies could suffer economic losses by either not having enough product to meet a significant rise in consumer demand or vice versa. Low inventory maintenance also means that adequate supplies may not be available in times of crisis or shortage. For example, grocery stores that maintain a limited supply of food products may run out in the aftermath of a war, terrorist attack, natural disaster, or other event that temporarily closes the roadways, preventing trucks from resupplying empty storerooms. As most people do not grow their own food in modern urban society, mass hunger, hoarding, and price gouging could follow temporary shortages.

See Also: Agribusiness; Commodity Chain; Food Processing Industry; Retail Sector; Supermarket Chains; Supply Chain; Wal-Mart.

Further Readings

Cheng, T. C. E. and S. Podolsky. *Just-in-Time Manufacturing: An Introduction*. New York: Chapman and Hall, 1993.

Godman, David and Michael Watts. *Globalizing Food: Agrarian Questions and Global Restructuring*. New York: Routledge, 1997.

Hay, Edward J. *The Just-in-Time Breakthrough: Implementing the New Manufacturing Basics*. New York: Wiley, 1988.

Moskowitz, Howard. *Food Concepts and Products: Just-in-Time Development*. Trumball, CT: Food and Nutrition, 1994.

Walsh, John P. *Supermarkets Transformed: Understanding Organizational Technological Innovations*. Brunswick, NJ: Rutgers University Press, 1993.

Marcella Bush Trevino
Barry University

LABOR

The term *labor* refers to human beings' physical and mental services or work of any kind. Unlike other animals, human beings cannot get food, shelter, and other necessities directly from nature, and therefore they need to modify nature. The effort of modifying nature is known as labor. Although labor has been an important issue in human history since time immemorial, it has become more important today in the era of globalization, especially in the context of green food. There are different kinds of labor, such as wage labor (in which people sell their labor to their employers in exchange for wage or money), child labor (when children under a certain age determined by a country's law or custom sell their labor), bonded labor (in which a person must work for somebody to pay debt), indentured labor (when people work for a limited term specified in a signed contract), manual labor (physical work), and unfree labor (slave labor, for example), and so forth. In the production and processing of green food, labor has become a very crucial issue.

Labor in Human History

In earlier human societies such as bands or hunting and gathering, people had limited or no permanent settlements. Hunting and gathering were the main means of livelihood. It is believed by most anthropologists that men used to hunt, and women used to gather fruits. Though the societies were highly traditional and simple, they contained the principles of egalitarianism. The form of labor was therefore largely voluntary. In ancient Greek and Roman civilization, slavery was the main mode of production. Slaves—collected through war—were unfree labor and did not have any rights. They were also known as "speaking tools" as opposed to "unspeaking tools" such as domestic cattle. During the time of European feudalism, peasants or serfs were bonded labor—they were bonded to lands and bonded to pay rents to the landlords. In feudalism, there emerged a new social group—the money lenders, who also formed small guilds of business enterprises, used to lend money to the peasants, as well as to feudal lords, and charged high interest. Having failed to pay back money with interest, many landlords surrendered their estates to the money lenders. Money lenders emerged as a new "capitalist class" who now "freed" the peasants from their lands and made them "free labor." This "freedom," however, was double-sided: freedom to work in the factories, and freedom from their means of production (i.e., lands).

As a consequence, the peasants had nothing but to sell their labor in the factory or farms. Though this freedom did liberate them from landlords' exploitation, it subjugated them in a deeper manner. In the previous system, they at least had lands to cultivate in return for paying rent to the landlords. In the new capitalist system, they lost their control over land, and when they started working in the factories, they lost control of their own labor as well. The emergence of wage-labor is nothing but, to quote Karl Marx, "the historical process of divorcing the producer from the means of production."

Classical Marxists think that the process of capitalist accumulation is unleashed by the logic of commoditization, which eventually generates a polarized agrarian class structure, with a capitalist landowning class known as the "bourgeoisie" occupying or owning the large-scale, wage-labor farms, and a "proletariat" comprising the marginalized and ultimately landless peasant class that supplies the wage-labor for capitalists' farms. This notion of an agrarian class dichotomy has, however, been contested by the existence of medium- or small-sized family farms based exclusively on family labor, and with no pressures for the enlargement of the business (i.e., expanded reproduction) beyond demographic and cultural factors. Although some scholars have contested the inevitability of the agrarian class dichotomy predicted by neo-Marxists, they cannot ignore the increasing power of capital and the increasing vulnerability of family farms.

Labor in the Globalization of Production

Over the past two decades, research on the globalization of agrofood production, as well as other export-oriented industries, has shown a rapid process of transformation that has affected labor relations in various ways. It has opened opportunities for women to enter new areas of paid employment, earn an income, gain independence, and participate more actively in social life. However, the transformation has also created new challenges because much of this employment is informal, with poor working conditions and a lack of labor rights, and has to be carried out in addition to household and family responsibilities. The general patterns in the workforce include feminization alongside masculinity, characterized by an increasing female workforce in factories that are still dominated by males; flexibility and informality with rigidity, in which an increasingly precarious and vulnerable workforce, being casual, temporary, and part-time, finds itself in an inflexible and formal workplace with various codes of conduct; and human mobility, with movement from rural to urban areas as well as from one form of subsistence to another. According to Philip McMichael, wage labor is undergoing a profound transformation, signaled by the increasingly unstable terms on which people are hired around the world and the growing range of forms of labor in industry and agriculture—from stable cores of wage work through contract and piecework to new forms of indentured, slave, and child labor—incorporated into global commodity chains under the restructuring of the global economy.

In the globalization of production, good jobs are getting harder and harder to find and to keep, and bad jobs seem to be getting worse, even as they become more plentiful than ever. Corporations are moving to the global South to exploit the cheap labor. Labor in the global South is always gendered, and its endless supply depends on complex patriarchal and subcontracting hierarchies. For corporations, subcontracting is a method of lowering labor costs and controlling labor power while evading responsibilities for exploitation. Global agrofood and other export-oriented industries prefer young, unmarried, and relatively educated women. The precarious nature of labor patterns in the production and processing segments of food and other commodities generates a considerable debate and an emergence of privatizing labor regulations.

Privatization of Labor Regulations

In the global agrofood system, consumer concern not only has centered on long-established issues like food safety and environmental degradation but has also included questions of, among other issues, labor exploitation in food-producing countries. Different private certification agencies have therefore started incorporating labor codes in their certification schemes. Different private certification agencies have incorporated labor issues differently. Some have clear and separate codes on labor, and some have merged it within social principles, whereas others have mentioned social principles without indicating labor issues. The latter case can be found in the Marine Stewardship Council (MSC). The World Wildlife Fund, with the help of Unilever, formed the MSC in the late 1990s, ushering in the modern era of eco-labeling for wild fish. On the basis of the United Nations Food and Agriculture Organization Code of Conduct for Responsible Fisheries, it claims to reward environmentally responsible fisheries management and practices with a distinctive blue product label. Although the MSC is now recognized as the undisputed leader in seafood eco-labeling, focusing more on environmental management for local ecology and ecological communities, its social principles specify only compliance with relevant local and national laws and standards and international understandings and agreements. Though MSC principles do not include any statement on "labor," its commitment to manage and operate in a responsible manner, in conformity with local, national, and international laws and regulations, can broadly include issues of labor.

Another pioneer and leader of private seafood certification is the Aquaculture Certification Council (ACC), a certification body of the Global Aquaculture Alliance. The recent commitment of Wal-Mart, Darden, and Lyons to buy only ACC-certified seafood made ACC a major global regulator. The Global Aquaculture Alliance sets the Best Aquaculture Practices—standards that address social, environmental, and food safety of shrimp aquaculture. The ACC certifies the shrimp hatchery, the farm, and the processor on the basis of Global Aquaculture Alliance standards. Unlike MSC, the ACC has more clear principles on labor standards. Its second standard is "Community," and under community there is a subheading called "Worker Safety and Employee Relations" that clearly says, "Processing plants shall comply with local and national labor laws to [ensure] worker safety and adequate compensation." The ACC recognizes that processing work is potentially dangerous because of the types of machinery needed and the use of potentially hazardous materials, especially coolants. Workers are usually not highly educated, and safety instruction may not be adequate. An uncaring employer may not provide safe and healthy working conditions. The ACC has also accepted the fact that in tropical nations, processing workers' pay scales are low and wage or other labor laws may not be consistently enforced, and therefore processing factories should maintain a good working relationship with not only employees but also the communities in which they operate. To receive ACC certification, processing plant management shall show both compliance with labor laws and a commitment to worker safety. Certified processing plants are required to provide legal wages and a safe working environment, and efforts should be made to exceed these minimum requirements. In addition, the ACC requires the following:

- Workers should be given adequate initial training, as well as regular refresher training, on safety in all areas of plant operation. Workers should also be trained in the first aid of electrical shock, profuse bleeding, and other possible medical emergencies.
- In some locations it is necessary for plants to provide meals for workers. In such cases, food services should provide wholesome meals for workers, with food storage and preparation done in a responsible manner. Safe drinking water shall be available at all times to employees working at the facility.

- During facility inspection, the ACC auditor will evaluate whether conditions comply with labor laws. The auditor will also interview a random sample of workers to obtain their opinions about wages and safety conditions.

Although still a niche market, "organic" certification works as a kind of "gold standard." The International Federation of Organic Agriculture Movements (IFOAM) is a global umbrella body for organic food and farming. IFOAM's goal is the worldwide adoption of ecologically, socially, and economically sound systems based on the Principle of Organic Agriculture. The four principles of organic agriculture are health, ecology, fairness, and care. The principle of fairness is defined in this way: "Organic Agriculture should build on relationships that ensure fairness with regard to the common environment and life opportunities." Though there are no separate labor codes, the principle of fairness is broadly applied to issues of labor as well. IFOAM's Organic Guarantee System is designed to facilitate the development of organic standards and third-party certification. IFOAM Certification bodies are accredited by the International Organic Accreditation Services Inc. on a contract basis.

Naturland is probably the most active company in certifying shrimp among organic certifiers. Naturland has developed standards on several aquaculture commodities and issued its standards on organic shrimp production at the end of 1999. Along with its own standards, Naturland is accredited by IFOAM. Its 12 broad principles mainly focus on farm management to ensure freshness and quality, and there are no separate principles on labor. However, in its last principle, "Social Aspects," there is something that can apply to labor: "The operator of the farm has responsibility as well for the housing and living conditions of employees living permanently or temporarily on the farm area. The respective regulations concerning industrial law shall be adhered to."

Other certification organizations/schemes that either set standards, regulations, or codes or certify food farming and processors include the Accredited Fish Farm Scheme of Hong Kong, Japan's Alter-Trade Japan, New Zealand's BioGro New Zealand, Bio Suisse in Switzerland, Europe's Carrefour, the Food and Agriculture Organization's Codex Alimentarius Commission, Canada's Conseil des appellations agroalimentaires du Québec, Fairtrade Labeling Organizations International, International Standards Organization (ISO 14001; 9001; 22000; and recently, ISO/TC 234), the International Social and Environmental Accreditation and Labelling Alliance, KRAV (Sweden), Marine Aquarium Council, Office international des epizooties, Swiss Import Promotion Programme, Safe Quality Food, Seafood Watch, Soil Association (U.K.), Thai Quality Shrimp, and Quality Approved Scottish Salmon.

Although some of these agencies, such as Thai Quality Shrimp, are operated by state agencies, they include various private standards required by the buyers. A review of all these certification schemes and their codes adumbrates that codes of conduct vary from agency to agency and from product to product. In comparison with other certification agencies, Fairtrade Labeling Organizations International has placed greater emphasis on issues of labor. It certifies agrofood products such as coffee, cocoa, vegetables, fruits, honey, herbs, and spices. In most other food certification agencies, farm management and natural quality are highly emphasized and there is very little or no specific guideline or regulation on labor, though their social principles can be broadly extended to labor issues.

As local labor practices are largely influenced by local ecological and social dynamics (e.g., weather, abundance of cheap labor), fundamental questions arise as to whether and how the private certification schemes/regimes can address labor issues. Who is setting standards for whom, and with what effect? Do they consult with the local stakeholders regarding labor standards that would be consistent with local ecological and social conditions? How will the private certification schemes that aim to certify food production and processing pinpoint these problems and broadly address labor issues for complete certification?

See Also: Eco-Labeling; Fair Trade; Family Farm; Food Safety; Peasant; Supply Chain; United Farm Workers.

Further Readings

Atkins, P. and I. Bowler. *Food in Society: Economy, Culture, Geography*. London: Arnold, 2001.

Islam, M. S. "From Sea to Shrimp Processing Factories in Bangladesh: Gender and Employment at the Bottom of a Global Commodity Chain." *Journal of South Asian Development*, 3/2 (2008).

Jaffee, D. *Brewing Justice: Fair Trade Coffee, Sustainability and Survival*. Berkeley: University of California Press, 2007.

McMichael, P. "Globalisation: Myths and Realities." *Rural Sociology*, 61/1 (1996).

NACA. "Certification Systems." http://www.enaca.org/modules/tinyd11/index.php?id=3 (Accessed August 2007).

Md. Saidul Islam
Nanyang Technological University

LAND GRANT UNIVERSITY

Land grant universities (LGUs) exist in every state in the United States and form the backbone of public research and education in agriculture and food. Today, many LGUs are world-renowned centers of learning, including: Cornell University, Louisiana State University, Pennsylvania State University, Rutgers University, University of California (Berkeley, Davis, and Riverside campuses), and University of Wisconsin–Madison, to name a few.

Origins and Components

LGUs began with the ideas of Jonathan Baldwin Turner, an Illinois College professor who campaigned throughout the 1850s to create a public higher-education system for the working class. Postsecondary education to this point had been largely an elite endeavor. Turner saw a need for an educational system that would provide broader access and enable subjects such as agriculture and engineering to be more relevant to the lives of most Americans.

It was Justin Smith Morrill, a Senator from Vermont, who introduced the bill in Congress. President Abraham Lincoln signed the Morrill Act in 1862, interestingly after the South, the predominant agricultural region of the United States, had succeeded from the Union. LGUs were expanded with the Second Morrill Act in 1890 to include 17 historically black colleges such as Tuskegee University. Grants were of money rather than of land, but they gave the colleges LGU status similar to that of the 1862 colleges. LGUs were expanded again in 1994, when 29 tribal colleges and universities joined the complex.

The first LGUs were established soon after the Morrill Act. Under the act, the federal government granted each state 30,000 acres of land for each member of Congress from the state, thus granting more land to those states with larger populations. The states had the responsibility of using that land, and the income gained from its sale, to establish a university to educate its students in the agricultural and mechanical arts, but not to the exclusion of classical, scientific, and military studies. LGU research activities were bolstered by the Hatch Act of 1887, which created the Agricultural Experiment Stations as research centers to support agriculture and the well-being of the states' citizens. The Smith-Lever Act of 1914 established Cooperative Extension to convey the results of the universities' research to the general population. By 1914, then, all of the major components of LGUs—education, research, and extension—had been created. Although LGUs are public universities serving the constituencies of their home states, the U.S. Department of Agriculture plays an important role in the LGU complex by providing funding and directing research priorities.

Land Grant Universities Today

LGUs were originally created to provide education useful to farmers and tradespeople and to democratize education by expanding it from an elite experience available to very few people to a popular endeavor that would be relevant to practical professions. Today, LGUs still serve as major public academic educational institutions. Although they still teach agriculture and engineering, LGUs also continue to cover a very wide range of other disciplines.

The second purpose of LGUs is to conduct research oriented toward solving problems of interest to the public. The Agricultural Experiment Stations were created to facilitate this type of research, and LGUs today continue to conduct a great deal of basic and applied research. Research at land grant institutions is funded through a variety of sources, including a significant but declining amount of public funding and increasing funding from private sources. The annual U.S. Department of Agriculture budget includes over $1 billion for LGU research and Cooperative Extension activities.

The third goal of LGUs—outreach to the public—was originally meant as a method to disseminate the results of the agricultural research conducted at the universities. Today there are still Cooperative Extension advisers around the country that are focused on agriculture, but there are also specialists in many other areas, including youth development and home economics. Cooperative Extension is responsible for running programs such as 4-H and Master Gardener across the United States.

Accomplishments and Critiques

Land grant institutions also have played an important role in shaping the structure of U.S. agriculture along the lines of productionism. LGUs have created many of the technologies that have industrialized U.S. agriculture, increasing productivity relative to labor and land. One example is seeds bred for specific qualities, such as high yields through responsiveness to nitrogen fertilizer or varieties that ripen all at once for easy harvesting. Mechanical harvesting equipment that reduces the need for manual labor is another example. LGUs have contributed to the creation of genetically engineered crop varieties and animal breeds. Although initially treated with skepticism and ridicule, work on integrated pest management, as pioneered by professors such as Robert van den Bosch, occurred at LGUs. There have also been contributions by social scientists at LGUs, including work in rural sociology, community development, and agricultural economics.

One of the most thorough critiques of LGUs has come from Jim Hightower, who criticizes the land grant complex for becoming "the sidekick and frequent servant of agriculture's industrialized elite." Hightower argues that by working with agribusiness and agricultural commodity groups, the needs of smaller farmers, alternative farmers, farm workers, and rural and urban communities have been neglected relative to the needs of corporations and large-scale farmers. Hightower recognizes that LGUs have had many great successes and achievements but argues that these have been outweighed by the failures and costs of not serving the public interest equally. The original rationale for providing public support to agriculture was that most of the public was farming, but by 1990, less than 2 percent of Americans were actually farmers. Hightower contends that by pushing for mechanization and technologies that help large farms, LGUs have actually harmed the majority of farmers and rural communities by putting workers and small farmers out of jobs; thus, he argues that LGUs are not following their mission of serving the public interest.

This critique of LGUs has been echoed by many other scholars, who argue that the universities have, as a whole, drifted too far from their civic roots and are losing their sense of public purpose. Those critiquing LGUs for straying from their public mission generally call for more public accountability, transparency, and "engaged" scholarship that better identifies and serves the public good. Over time, scholars have documented a loss of public faith in the kind of expert scientific knowledge produced by LGUs; to be a legitimate source of knowledge, critics argue that LGUs must solicit more public participation and recognize types of knowledge beyond positivist scientific knowledge as valid. Along with a need for increased public engagement and participation comes a need for increased public funding to make LGUs less dependent on private funding, and thus less tied to private interests.

Sustainability and LGUs

As the idea of sustainability and concerns about environment crises continue to gain traction, sustainability also is making inroads at many LGUs. Sustainable agriculture activities such as organic farming programs form a small but growing segment of LGU activities. Despite the growing presence of sustainable agriculture and food systems education and research programs, sustainable agriculture still faces a lack of significant financial or institutional support at most LGUs, where commitments to industrial agriculture remain dominant. A few notable programs in sustainable agriculture at LGUs include the Agricultural Sustainability Institute in California, the Leopold Center in Iowa, the Minnesota Institute for Sustainable Agriculture, and the Center for Sustaining Agriculture and Natural Resources in Washington. There are also charges that sustainability is being co-opted by mainstream interests as a form of greenwashing.

Increasingly, LGUs are also home to other sustainability-oriented programs, such as energy efficiency, sustainable transportation, and green engineering. As LGUs strive to combat the criticism that their knowledge is not addressing the issues of concern to the public, focusing on green technology can be one way to gain relevance in a society in which a smaller population is actively farming. LGUs have the potential to serve a major role in addressing important issues such as global climate change and the need to move toward more sustainable sources of energy, transportation, and food production.

See Also: Agribusiness; Agricultural Extension; Department of Agriculture, U.S.; Integrated Pest Management; Labor; Mechanization; Productionism.

Further Readings

Hightower, J. *Hard Tomatoes, Hard Times: The Failure of the Land Grant College Complex.* Washington, D.C.: Agribusiness Accountability Project, 1972.

Kloppenburg, J. R. *First the Seed: The Political Economy of Plant Biotechnology, 1492–2000.* New York: Cambridge University Press, 1988.

National Research Council. *Colleges of Agriculture at the Land Grant Universities: Public Service and Public Policy.* Washington, D.C.: National Academies, 1996.

Parr, D. M., et al. "Designing Sustainable Agriculture Education: Academics' Suggestions for an Undergraduate Curriculum at a Land-Grant University." *Agriculture and Human Values,* 24/4:523–33 (2007).

Peters, S. "Changing the Story About Higher Education's Public Purposes and Work: Land-grants, Liberty, and the Little Country Theater." Foreseeable Futures #6. Ann Arbor, MI: Imagining America, 2007.

Van den Bosch, Robert. *The Pesticide Conspiracy.* Garden City, NY: Anchor, 1980.

Alida Cantor
California Institute for Rural Studies

Ryan E. Galt
University of California, Davis

Land Reform

Land reform is usually understood as a narrow section of more comprehensive agrarian reform that involves not only granting land titles and changing the land property structure but also credit, research, and training. Land concentration and lack of investment are the main issues usually tackled by land reform policies. However, the opposite aim, of promoting land concentration and attracting investment, is also an excuse for what some prefer to call *counter-reform*. However, land reform has not always been successful in achieving these broad goals. From Ancient Egypt and Rome through medieval Europe and Asia to current Latin America and Africa, land reform has always been at the heart of the development policy debate. This debate has been intensified by the contradictions between different development views of urban and rural populations and between those who profit with these reforms and those who afford their costs.

Agrarian reform in early-20th-century Mexico not only transformed its national land property structures but also influenced politics, law, and economics worldwide. The emergence of land issues in the Mexican Revolution contributed to developing the idea of a social function for property—an issue that was later developed by the Chilean experience under Salvador Allende and other nationalist regimes. The 1917 Queretaro Mexican constitution introduced a number of social rights, derived from the agrarian reform carried out, that also influenced law and politics on a broader perspective. The constitution effectively fractured land holdings into small land property, abolished monopoly, and established labor rights.

In 1960s Latin America, agrarian reform sponsored by the U.S. Agency for International Development and the Kennedy administration aimed to reform the structure of land property to avoid the spread of the Cuban revolution in the region. In some Latin American countries, left-wing nationalist de facto governments aimed to steer the development reform through land reform. However, the reforms were resisted and boycotted and only reached a small percentage of agricultural land.

Since the 1990s, following a different approach, land reform has been proposed by World Bank policies that promote land reform, usually fostering modernization, mechanization, agribusiness, and individual property as adequate policies for the vast areas of communal land still held by indigenous peoples and peasants in Latin America, for example.

"A key precondition for land reform to be feasible and effective in improving beneficiaries' livelihoods is that such programs fit into a broader policy aimed at reducing poverty and establishing a favourable environment for the development of productive smallholder agriculture by beneficiaries."

In this case, land reform is part of a more general drive for modernization and development through an export-led development strategy that also aims to develop a land market and a labor force market. The mindset of these reforms is that individual property is a more efficient use of land than common property, that private investors are more efficient than small landholders, and that the world market can only be fed by modern agribusiness with high technology.

This is not necessarily true in multicultural societies such as Latin America. For instance, in the Upper Tambopata River, near the border between Peru and Bolivia, organic coffee is produced in an isolated rural area of money-lacking peasants who, for that very same reason, do not use expensive chemical nutrients and pesticides, thus maintaining organic production of coffee. This coffee is now sold to world-market coffee traders such as Starbucks, with exports doubling many times since 2003. Thus, not only is modernization possible for large investment, but there is also space for small land proprietors—whether indigenous peoples or not—and also for other types of landholders, such as communal landholders. The world market demands agricultural production from these areas but also nature's products, gathered from the rivers, lakes, forests, grasslands, glaciers, and other areas where people in subsistence economies supply the flow line to the cities and ports.

Modernization and agribusiness are both feasible in areas where capital can find comparative advantages; that is, near a port or a road network. However, many products from remote rural areas are subsidized by nature or by small producers in subsistence economies. When such products are available in these subsistence economies, traders from the modern economic sector trade with local producers and channel this production into the national and international market. Some products such as leathers, rubber, organic coffee, and others have been flowing from subsistence economies to the world market. This has been happening, particularly, in the last five centuries since America was connected to the world market, but it is a process that has been affecting Europe, Asia, and Africa since much earlier, and Oceania more recently.

Current land reform proposed by President Alvaro Uribe in Colombia and President Alan Garcia in Peru, but also sponsored by President Luiz Inacio Lula's Brazilian National Development Bank under the South American Initiative for Vial Integration, aim to modernize the rural areas of the continent in the hands of agribusiness and export-oriented crops. This attempt at modernization lacks both strategic environmental assessments and environmental impact assessments. Moreover, these attempts to develop infrastructure also aim to grant land to investors. These administrations are granting investors access to lands already occupied by indigenous peoples in isolation and initial contact, settled indigenous communities, and other rural populations such as riparian populations and colonists.

In these cases, the attempts to modify the land property structure are naturally resisted by these populations. In 2008, Amazonian indigenous peoples' strikes in both Colombia and Peru gave the Uribe and Garcia administrations an idea of the strength of these local peoples' mobilization. In Peru, this resulted in the nullification of two legislative decrees authorizing third parties to request that communal land be parceled. However, another 90 legislative decrees also changed Peruvian legislation to allow private property over natural forests, which are usually occupied by other indigenous peoples and other rural populations, whose possession is not being formalized into property, as the constitution states in support of indigenous peoples' right to land.

These neoliberal attempts at land reform in the Amazon and the South American continent are being questioned in the light of environmental and indigenous peoples' rights. However, the mixture of Brazil's National Development Bank's $530 billion investment funds plus private funds, including investments from a $20 million (Pucallpa) port to $900 million (Callao) port, show that the magnitude of these projects can barely be assessed by the indigenous peoples and the environmental movement.

At the same time, similar attempts are being pursued by China to modernize its rural areas. China has in its far west at least five different economic sectors that are mainly characterized by extensive rural areas where local production and the subsidy of nature and local organization have kept and sustained the commons through millennia.

These publicly funded infrastructure development projects are seen as a stimulus for economic growth and have come at a time when other economies see this drive for public investment as needed in the national and global economies. The question remains of how suited these developments are for the rural populations and the environment.

See Also: Agrarianism; Agribusiness; Land Tenure; Peasant.

Further Readings

Cousins, Ben. "Agrarian Reform and the 'Two Economies': Transforming South Africa's Countryside." In *The Land Question in South Africa: The Challenge of Transformation and Redistribution*, edited by Ruth Hall and Lungisile Ntsebeza. Cape Town, South Africa: HSRC, 2007.

World Bank. *Land Policies for Growth and Poverty Reduction.* New York: Oxford University Press, 2003.

Carlos Antonio Martin Soria
Universidad Nacional Agraria La Molina
Instituto del Bien Común

LAND TENURE

Historically, the concept of land tenure derives from English common law, in which feudalism installed a system where the lord was a tenant-in-chief who "held" the land and granted rights over it to "lesser tenants" in exchange for services or duties. In this system, the individual holds the land of someone else.

In a broader sense, particularly in legal systems such as civil law, the land tenure concept refers to the tenancy of the land; that is, the right to possess or own the land. The Food and Agriculture Organization of the United Nations defines the term *land tenure* as "the relationship, whether legally or customarily defined, among people, as individuals or groups, with respect to land." The Food and Agriculture Organization, in simple terms, proposes that "land tenure systems determine who can use what resources for how long, and under what conditions."

The world economy has three segments: subsistence, mercantilist, and free market. These economies are widespread over the globe. The free market is hegemonic but coexists and links with other economic segments such as subsistence and mercantilist economies to access resources. As trade occurs through the trails between these economic subsystems, it helps fuel the conditions by which the systems survive and coexist with the free market. These economies also use the landscape differently. Subsistence economies are characterized by low population density and large landscapes. Mercantilist economies occupy the networks of towns and cities that channel natural resources into the modern economy through capital cities, main economic cities, and ports to feed the world market. Thus, land tenure as an individual land title registered in a public access record is a reality that barely exists in areas of subsistence economies. Some rural areas in the eastern slopes of the Andes show that less than 10 percent of the land is held with official land titles, but there is a large amount of more local arrangements.

Land tenure is a contested space. Different societies find different meanings for the same concept. For example, common property can mean unlimited access to the resource, restricted collective access, or property rights that are assigned to a group or community of people. Many states have followed the single-oriented policy of promoting individual private property under the idea that private ownership would ensure adequate resource use; however, this is not necessarily the best recipe. On one hand, research has registered many cases of overuse on private lands, and on the other hand, most of the successful collective land management cases have been researched in recent decades. Thus, although in many cases governments aim to spread individual property rights over the landscape, in practice, social and cultural values enforce other land rights arrangements, such as various degrees of common property and open access. An adequate property regime for a certain area has to take into account historical, economic, and cultural issues, but every time it also has become more evident that there is a need to consider ecological interactions to ensure sustainability. Moreover, there is a need to consider the interactions of any property regime with population increase, higher demand for resources, and technological advances that can vary the property arrangements at a local level.

Land tenure is usually a cultural product that emerges from the hinges of culture and history. The resource user who is able to grasp the utility and service of a given property arrangement does so because he or she also shares the cultural values behind these arrangements. On the contrary, those aiming to vary these arrangements might most likely belong to a different cultural group, deny the existence of these other arrangements, or, despite knowing about them, be willing to take the resources regardless of any consequences.

Different land rights have flourished in different societies. The land can be held in public property, communal property, and private property. These lands can be accessed individually or cooperatively. A particular landscape or basin can be occupied by a mixture of land tenure rights polygons, and it can require an understanding of local history and social and economic processes to properly understand the arrangement by which the land is held or the resources are used.

Non-Western societies from Southeast Asia to South America have also possessed land and allocated rights. The study of these arrangements constitutes a central issue for land tenure research, the study of customary law in non-Western societies, and public policy specialists. In a more particular sense, it is necessary to explore tenancy arrangements around different natural resources. In many cases, the rights assigned on land influence the allocation of rights in the stream, which might reflect those rights allocated on the land adjacent to it.

The Pichis sub-basin is a mosaic of public property, private property, and communal property, with indigenous peoples using the last two types of tenancy. The use of fishing areas with communal, private, and open access gives us an idea of a certain complexity in the management of fishing in the Neguache River. This system is flexible enough to adapt to the abundance or scarcity of the resource, which is the final crucial criterion in deciding to set fishing regulations in those gorges. In our view, the criteria for setting fishing regulations are influenced to a certain extent by current land use and the introduction of individual property rights in the landscape. This influence is a composite of state discourse and action, of religious beliefs, and also of the interculturality prevailing in the study area.

The participatory mapping of fishing areas in the four settlements—two of indigenous peoples and two of migrants—in a tributary of the Pichis River in the Peruvian Amazon showed that, despite apparent chaotic, open fishing access, there is a more complex and flexible order in place. Fishing is regulated in those areas with high importance for fishing, whereas there are no regulations in the open river or in areas of fishing of less value. This suggests a process of adaptation of fishing regulations to current ecological and governance conditions on the Pichis River.

The customary legal system and the formal legal system are not adjacent spaces, but overlap each other. There are interconnections, interpenetration, and intersection of legal orders between the two systems. Therefore, interlegality is a useful concept in observing the connections between norms and their symbolic expression at different levels, from the local to the national and international. This interconnectedness becomes apparent when we see that the restriction to use a toxic plant such as barbasco (Rotenona Spp) in a stream bed is not only a result of the need to resolve or avoid conflicts between neighbors (preventing the cattle from imbibing barbasco) but also of the need to recognize that the land where the gorge is located is held by an owner different than the fisherman.

See Also: Land Grant University; Land Reform.

Further Readings

Freyfogle, E. T. *The Land We Share: Private Property and the Common Good.* Washington, D.C.: Island, 2003.

Gray, K. and Gray, S. F. *Elements of Land Law.* New York: Oxford University Press, 2009.

Mackin, A. *Americans and Their Land: The House Built on Abundance.* Ann Arbor, MI: University of Michigan Press, 2006.

Carlos Antonio Martin Soria
Universidad Nacional Agraria La Molina
Instituto del Bien Común

LEGUME CROPS

Legumes are plants in the botanical family Fabaceae, which is also called Leguminosae, or the pea family. Legumes are the second-largest family of flowering plants, with between 14,000 and 17,000 species and as many as 18,000 species that belong to the pea family, with the Composite family being the largest. The Latin origin of the term *legume* (*legumen*) is the verb for "to gather," reflecting the fact that the seed pods of the plants are gathered for food.

The "fruit" of legumes is a seed pod. The seeds (pulses), such as a pea, a bean, or the peanut, split into two valves. The seeds are attached to one edge of the valves. The pod or the seeds are used as food. They include a variety of common and broad beans, as well as acacia, alfalfa, carob, clover, cowpeas, lentils, lupines, mesquites, mimosa, peas, soybeans, tamarind, and peanuts.

Legumes can be found in much of the world and vary so widely that it is often only the trained botanist who recognizes them as legumes. There are three subfamilies of Fabaceae or legumes (Caesalpinioideae, Mimosoideae, and Papilionoideae), all of which produce root nodules. Some are trees, and others are shrubs, herbs, or climbing plants. Papilionoideae are legumes that have flowers that look like butterflies. The sweet pea is a member of this group. The flowers vary from small to large and from regular to irregular.

The ability of legumes to fix nitrogen in the soil makes them an important natural fertilizer. Nitrogen is a gas that makes up much of the Earth's atmosphere. However, it is relatively inert, which creates a problem for soil fertility. Without nitrogen, the fertility of soil is low. Violent lightning can create nitrogen compounds in the atmosphere that can wash into the soil to fertilize it. Much more effective are the legumes that can fertilize

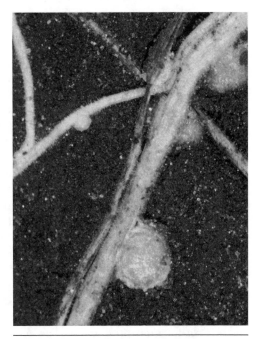

The photo shows the small (2 mm) nodules on the roots of a legume plant where the beneficial bacteria rhizobia establishes itself and helps the plant procure nitrogen from air.

Source: U.S. Department of Agriculture Agricultural Research Service/Markus Dubach

the soil by fixing atmospheric nitrogen. The roots of these legumes take in nitrogen from the air with the aid of a beneficial bacterium.

Legumes have a unique relationship with the bacteria rhizobia, which are soil bacteria. They fix nitrogen (diazotrophy) after the bacteria becomes established in the root nodules of legumes. It is a symbiotic relationship: The rhizobia cannot independently fix nitrogen; they can only do so through a host plant. The rhizobia are usually gram-negative, motile, and nonsporulating rods.

Humans have been growing and gathering legumes for around 8,000 years. Where meat is in short supply, legumes make an excellent source of protein. Legumes such as beans are usually eaten by cultures that depend on rice or corn as their main food; combined, legumes and grains provide a nearly complete diet.

Globally, legumes such as peas, beans, and peanuts are very important economically as foods. The legumes alfalfa, clover, and vetch are grown for animal feed in forage or pasture. Many other legumes are valuable in making edible oils, flavors, fibers, gums, fibers, oils, medicines, dyes, poisons, and timber. Some are used naturally, and others are chemically processed and synthesized into plastics or other products. Some are used as ornamentals.

Farmers have long used pasture legumes as green manure, enriching the soil when the green crop is plowed under. They can remain in the soil as a cover crop that is valuable to agriculture because they build nitrogen in poor soils and restore depleted soils. This is because of their symbiotic relationship with rhizoidal bacteria.

Crop rotation has long been practiced. A depleting crop like corn can be followed by a legume crop, which restores the fertility of the soil. Traditional farming in the U.S. Midwest used crop rotation, with legumes as the crop to follow corn. The development of chemical fertilizers has allowed constant monocrop cultivation, removing the need to practice crop rotation. However, organic farmers still use rotation with legumes as a form of natural fertilization.

The agricultural classes for farmed legumes include forage, grain, bloom, pharmaceutical, industrial, fallow, green, and timber species. Fallow legumes are planted to be turned into green manure. Often, commercial legumes will provide a crop and act as a green manure, enriching the soil with nitrogen after the crop is harvested. Forage legumes are used as cover crops, as pasture, and as green manure. Some are woody shrubs or trees that can also be foraged directly or cut by humans.

Legumes grown as grains are eaten directly by humans or animals. They are also used for industrial production of products such as peanut oil. Bloom legumes include the lupine, which has escaped gardens to grow wild, just as mimosas have. The *Indigofera* and *Acacia* species of legumes are raised for industrial purposes. Indigo has been widely used as a dye, and *Acacia* for gum production. The *Acacia* species and *Castanospermum australe* are grown widely to produce timber.

In a green society, legumes are nutritious foods that are better for the environment and for humans. Legumes are an important dietary factor for people of all ages, although they have been found to be especially important for the elderly, no matter their ethnicity or geographic location. Increasing the intake of legumes for the elderly enhances their health; the elderly in Japan attribute their age to increased intakes of legumes such as tofu, soy, natto, and miso, all found in Japanese cuisine. Korean, Chinese, and Japanese cuisines also include edamame, which is made by boiling baby soybeans and serving them with sea salt. Swedes attributed their longevity to eating brown beans and peas, and the Mediterranean diets comprise lentils such as those used in Greek soups, chickpeas (garbanzo beans) such as those used in Lebanese hummus, and white beans such as those used in various Italian soups.

In terms of green foods that have health benefits, soybeans are important. In Japan, natto has been made from soybeans since the Edo period (1600–1868). Traditionally, it was made by packing cooked soybeans in a bed of rice straw. Left in a warm place, the bacterium *Bacillus subtilis natto* will turn the beans into a slimy substance that smells like an Asian version of limburger cheese. Today, natto is made in sanitary conditions in factories, but its smell and goopy characteristics remain the same. What is important is that this is not simply an ethnic food—even in Japan it is a regional food that is not eaten east of Mount Fuji. Notably, instances of osteoporosis are much higher east of Mount Fuji than they are west of Mount Fuji, where natto is eaten. Natto's osteoporosis-resisting power is the result of vitamin K-2, which is the product of bacterial fermentation and can be obtained as a vitamin supplement made from natto.

Research into the health benefits of legumes for all ages is important. Nutrition science is investigating diets for the elderly as the world's population ages. Green diets would include more legumes than many people now eat or have eaten in earlier decades of their lives.

See Also: Corn; Cover Cropping; Crop Rotation; Nitrogen Fixation; Soybeans.

Further Readings

Bennett, Sarita J. *New Perennial Legumes for Sustainable Agriculture*. Perth: University of Western Australia Press, 2003.

Frame, J., et al. *Temperate Forage Legumes*. Oxford: CABI, 1998.

Kirti, P. B. *Handbook of New Technologies for Genetic Improvement of Legumes*. Boca Raton, FL: CRC Press, 2008.

Plancquaert, P. and P. Haggar, eds. *Legumes in Farming Systems*. New York: Springer, 1989.

Smartt, J. *Grain Legumes: Evolution and Genetic Resources*. Cambridge, UK: Cambridge University Press, 1990.

Sprent, Jane. *Legume Nodulation*. New York: John Wiley & Sons, 2009.

Andrew Jackson Waskey
Dalton State College

LOCAVORE

The term *locavore* describes a person who makes a deliberate choice to primarily or exclusively eat food grown or raised locally, within his or her own foodshed. The term is analogous to carnivore, or meat eater, and herbivore, or plant eater. Its genesis has been attributed to Jessica Prentice, a San Francisco Bay Area food activist, who in 2005 coined the word as a way to define friends and colleagues who accepted her challenge to eat foods grown within 100 miles of San Francisco for an entire month. The popularity of eating locally has grown, and in 2007, the Oxford American Dictionary named *locavore* its word of the year.

Although the term *locavore* is new, throughout much of human history people have primarily subsisted on food produced by themselves or by people living fairly close. As transportation, storage, and processing technology improved in the 19th century, food production and consumption were increasingly separated, leading periodically to concerns about the reliability and safety of distant supplies. Local food production and consumption have been advocated at various points throughout the 20th century in response to concerns about food availability and quality. During the world wars and the Great Depression, people planted backyard gardens and urban farms to supplement their diets. Beginning in the 1930s, the organic agriculture movement raised questions about the health and environmental safety of large-scale conventional agriculture and promoted small-scale, local alternatives. In the 1960s, the environmental, labor, and counterculture movements criticized large-scale agribusiness and called for more ecologically sound, fair, and smaller-scale alternatives, including local food production using few or no synthetic chemicals. Beginning in the 1980s, chef Alice Waters, owner of the Berkeley, California, restaurant Chez Panisse, gained a reputation for her use of local, seasonal foods, helping to increase the popularity of eating locally. The slow food movement, founded in Italy in 1986 to counter the trend toward fast food, supports traditional foods and production practices that risk disappearing as a result of the preponderance of mass-produced food and the homogenization of tastes. The growing numbers of slow food members, although not necessarily locavores, work to preserve and celebrate traditionally prepared food made with ingredients that are unique because they reflect local growing conditions, cultural practices, or processing methods.

All Locavores Are Not the Same

Locavores vary in how strictly they adhere to a local diet and in how they define "local." Some avoid all ingredients that are not locally produced, whereas others include in their diets a limited range of foods that have been traded for centuries, such as chocolate, coffee, oils, and spices. Some locavores define local food as that which is grown or raised within as close as a 50-mile radius or as far as 250 miles, as food produced within a day's drive from the point of consumption, or food produced within geographical boundaries such as the adjacent metropolitan area, nearby counties, the state, or even multistate regions. Locavores also make the distinction between locally sourced food that comes from farms primarily geared to the local market versus food produced nearby by agribusinesses that distribute primarily through global supply chains.

Locavores are motivated by a number of factors, but chief among them is a concern for the environment. A principal environmental benefit attributed to eating locally is the minimization of food miles, or the weighted average distance from farm to table, which in many cases reduces the energy required to process, package, refrigerate, and transport food, and the need to limit production to varieties best suited for long-distance transportation. Another environmental benefit attributed to eating locally produced food is the maintenance of a market for urban and periurban farms, which helps to maintain a working agricultural landscape in and around cities, conserving prime farmland, preventing sprawling residential and commercial land uses, and maintaining open space for humans and nonhuman species. Some locavores eat locally to reduce the perceived environmental and health risks associated with long supply chains that are insufficiently regulated and monitored. In the wake of outbreaks of food supply contamination, from melamine to *Salmonella*, eating locally, locavores argue, ensures a greater degree of producer responsibility and traceability.

Locavores also emphasize the economic development benefits of eating locally. In theory, by keeping dollars circulating in the local economy, spending money on food produced within one's region provides a much greater economic benefit for the local community than purchasing food produced in other parts of the country or the world. Furthermore, supporting the local agricultural economy enables nearby farmers to stay in business, and as farmland contributes more to a municipality in property taxes than the cost of the public services used by farmers (compared with residential taxpayers, who use more dollars of public services than they pay in property taxes), avoiding the conversion of farmland to residential or commercial prevents the ballooning of local government costs.

Locavores and Local Culture

There is an important cultural dimension to eating locally, too. By limiting their diets to food produced nearby, locavores must by definition eat more thoughtfully, paying attention to seasonality and the types of foods suited to the soils, climate, and growing season of their communities. One result is a diet that may be less varied, but that includes seasonally available ingredients that are fresher, with unique flavors and textures not found in foods produced for distant markets. Locavores emphasize the unique attributes of traditional local foods, including heirloom cultivars, and traditional harvesting, processing, and culinary methods.

Some locavores are motivated to eat locally by a desire to support institutions that are embedded in their communities and that constitute the local food system, from family farmers to farmers markets and community-supported agriculture programs. These entities form a social network that involves mutual trust and responsibility, which are not often found in the purely market-based economic relations that occur in conventional business transactions. The embedded economy of a local food system ensures that food producers take responsibility for consumers, in addition to earning profits, and that consumers share responsibility with producers. In transactions that occur at the farmers market, for example, the face-to-face relations between consumer and farmer enable quality control issues to be negotiated directly, giving the farmer an incentive to produce good products. Likewise, community-supported agriculture programs require members to pay for farming costs up-front, sharing the production risks and benefits with farmers, and thus forging a much closer bond than a traditional buyer–seller relationship.

Finally, locavores choose to eat from the local foodshed as a form of protest against the forces, and effects, of globalization. Eating only that which can be produced in one's own community demonstrates the possibility of withdrawing from the global food economy and stimulates thinking about the negative effects of globalization and trade liberalization.

None of these claims about the benefits of eating locally is uncontested. Life cycle analyses have demonstrated that the energy consumption of a food system is quite complex and that more proximate production may be less energy efficient than food transported from farther away, depending on the methods of production, processing, and shipping. Global trade can benefit farmers in developing countries, whereas the practices of farmers within one's region may be just as exploitative, unfair, or unhealthy as those of farmers in distant communities. In addition, the locavore diet can be out of the reach of poor people, who must buy lower-priced food no matter how high the external environmental and social costs.

See Also: Foodshed; Slow Food Movement.

Further Readings

Born, Branden and Mark Purcell. "Avoiding the Local Trap: Scale and Food Systems in Planning Research." *Journal of Planning Education and Research*, 26:195–207 (2006).

Nabham, Gary Paul. *Coming Home to Eat: The Pleasures and Politics of Local Foods*. New York: W. W. Norton, 2001.

Petrini, Carlo. *Slow Food: The Case for Taste*. New York: Columbia University Press, 2004.

Nevin Cohen
Eugene Lang College
The New School for Liberal Art

LOW-INPUT AGRICULTURE

Low-input agriculture describes farming systems that require and use less purchased, synthetic products to effectively grow a crop. Agricultural inputs include fertilizers, fungicides, pesticides, and herbicides, as well as seeds, irrigation water, and fuel. Though necessary in all agriculture, within the context of the terms *low-* and *high-input agriculture*, the additive products are chemical-based or artificial, the irrigation is more intensive and disruptive to underground water tables, and the seeds are hybrid or, more recently, genetically engineered. According to agroecologists, low-input agriculture has numerous ecological benefits in comparison with its high-input counterpart; namely, it conserves and even enhances soil fertility, agricultural biodiversity, water quality, and habitat integrity. As a result of the decreased use of pesticides and herbicides, low-input agriculture protects farmer health and produces less-toxic foodstuffs. Moreover, such farming requires less tillage, thus allowing soil to sequester more carbon, and it consumes fewer inputs manufactured with greenhouse gas emissions. Finally, advocates contend that low-input agriculture relieves the grower of the burden of debt associated with the "treadmill" of annually purchased, technological inputs, and so makes up a central component of food security, food sovereignty, and environmental justice.

Some growers prefer the term *low-external-input agriculture*, as effective, sustainable farming requires a great deal of inputs—just not synthetic or purchased ones. Low-input farmers rely chiefly on managing on-farm resources to plant, cultivate, and harvest their crops. For instance, farmers may save their seed from a previous harvest rather than buy hybrid seeds—that do not save effectively—or genetically engineered seeds—which are patented such that their saving would be a breach of intellectual property rights, punishable by fine or lawsuit. Farmers increase the fertility of their soils through the application of manures, crop residues, compost, and the cultivation of nitrogen-fixing legumes, planted off-season to replenish the nitrogen lost during the regular growing season. Growers may also extend their growing season through simple-technology cold-storage frames, greenhouses, and hoop houses. Although dramatically lessening the amount of purchased, synthetic additions, low-input agriculture does, however, require a significantly high-input of labor, skill, and local ecological knowledge.

Technically, most indigenous agrarian systems around the world and throughout history have been low-input, flourishing with masterful agribiodiversity results. From the Quechuan region of the Southern Andes to the Luzon Cordilla of the Philippines, locally adapted, low-input agrarian traditions now serve as models for long-term ecological sustainability. The legendary tradition of the ancient *milpa* system of North and Central America interplants squash, beans, and corn so as to enhance soil fertility (legumes replenish nitrogen used by corn), prevent weeds (squash leaves shadow the ground), foster timely growth (beans wrap up corn stalk), and make up a complete protein.

Though ancient and universal, low-input agricultural practices have enjoyed a revival within the recent environmental movement. Posited as an alternative to the industrialization of agriculture, the green revolution—and its offspring the gene revolution—uses *low-input agriculture* as an umbrella term for a number of specific agroecological movements. It often encompasses organic food production, following the tenets of British botanist Sir Albert Howard, who demonstrated the correlation of soil health to human health through the detoxifying properties of the composting process. It also partakes of Austrian Rudolf Steiner's theories of biodynamics, wherein the entire farm is understood as a comprehensive, unique organism. The Kyusei Nature farming movement of 1930s Japan introduced the use of effective microorganisms—a low-input strategy—to improve soil, and thus food quality. The Australian permaculture, or "permanent agriculture," movement popularized the goal of designing a farm or garden that mimicked—instead of combated—natural ecological processes. Each of these endeavors emphasizes the process as much as the products of farming, advocating that the long-term health of the soil, and thus of humans, entails careful attention to and nurturance of the ecological balances and systems already at work in any field.

Silent Spring, Rachel Carson's pivotal expose of pesticides' ubiquity in the postwar United States, served as a catalyst for the sustainable agricultural movement, which defines itself as low chemical input. The book brought to popular consciousness the recognition that such inputs do not disappear on harvest but bio-accumulate in watersheds, birds, animals, and eventually humans. To reduce unwanted insects and diseases, low-input growers turned to integrated pest and weed management—multiprong and locally specific series of preventative and proactive strategies. One such tactic involves fostering a wide array of agricultural biodiversity comprising locally adapted, disease- and insect-resistant, open-pollinated seed varieties. Biodiverse fields ensure resiliency, as diseases and insects thrive and multiply on vast genetically uniform fields of fodder.

Farmers also intercrop and plant perennial forage crops to stifle pathogens and unwanted insects and weeds (though often such "weeds" serve as attractions for insect-eating birds, or even as valuable herbs). Agricultural management systems respond, but they also have short- and long-term effects on the weed and insect communities of the soil—density, abundance, and diversity—as well as on surrounding bird species, which can in turn facilitate insect control. Integrated management entails a comprehensive knowledge of the existing local weed and insect dynamics. Transitions to such integrated, diverse, and intensive agriculture have been ecologically, economically, and socially successful in case studies throughout Asia, Africa, and the Americas, showing, for instance, increased yields amid significant reductions in pesticide use.

To decrease the need for artificial fertilizer, low-input growers concentrate on soil—and particularly tilth—conservation. This entails crop rotation, terraces, conservation buffer strips, windbreaks, and regular fallow seasons. Otherwise, intensive production leaves the soil deficient in nutrients, prone to droughts and floods, impacted, dry, and acidic, without the necessary beneficial insects to replenish it. Such methods have been researched extensively by Wes Jackson at the Land Institute, who—with others—points to agriculture's central role both in the global crisis of soil erosion and in any possible amelioration of it. Moreover, peak oil promises to limit nitrate fertilizer, a petroleum by-product.

Agriculture also carries the chief responsibility—and opportunity—with regard to the quantity and quality of global water resources. Low-input agriculture mitigates non–point source pollution and protects groundwater through decreased biocide use and riparian forest buffers, whose root systems provide reservoirs for subterranean water retention. This serves as a buffer for droughts and offers extra absorption during floods—benefits in the increasingly erratic weather patterns accompanying global climate change.

Despite government subsidies and credits for agribusiness—disincentives to farmers considering the transition—substantial numbers of growers and consumers are supporting low-input agriculture. The revival is as social, political, and economic as it is environmental, because of its potential to relieve indebted farmers, bolster food security, honor the social and cultural integrity of small-scale farmers, and bring social justice—as such agriculture does not concentrate wealth the way high-input agrisystems do. Via Campesina and others propose investments in low external input agriculture as critical components of poverty alleviation, food security, food sovereignty, and ecological justice. More than merely a method, low-input farming has become an agrarian ideal and a key to sustainability in general.

See Also: Agrarianism; Agrodiversity; Crop Rotation; Food Sovereignty; Sustainable Agriculture.

Further Readings

Reijntjes, C., et al. *Farming for the Future: An Introduction to Low-External-Input and Sustainable Agriculture*. London: MacMillan Education, 1992.

Schaller, N. "Mainstreaming Low-Input Agriculture." *Journal of Soil and Water Conservation*, 45/1:9–12 (January 1990).

Garrett Graddy
University of Kentucky

M

MAD COW DISEASE

Mad cow disease, also known as bovine spongiform encephalopathy is a brain wasting disease affecting cattle and humans. Once contracted, it is invariably fatal. The disease primarily affects cattle and is perpetuated by industrial animal husbandry practices; namely, feeding cattle by-products to other cattle. It is not thought to spread among livestock any other way. The disease has been observed in cattle in Britain since 1985, and as early as 1996 there were reported cases of a human disease, variant Creutzfeldt-Jakob disease (vCJD), which had many of the same symptoms but an unknown cause. The British government reassured the public that the two diseases were not related until 2000, when they reported a link between them: The newly described human disease was indeed likely caused by mad cow disease. That later announcement caused public panic. The human outbreak sparked conversations throughout Europe and North America about animal husbandry and food safety, elevating concern about food systems and ultimately changing international policies and trade practices. Because scientists and regulators had repeatedly assured the public that humans were not susceptible to the disease before announcing the opposite, the episode changed cultural ideas about the role and validity of scientific knowledge. The epidemic among both cattle and humans has been most sizable in the United Kingdom. As of December 2008, 210 people had been diagnosed with vCJD, the human form of the disease, and affected cattle have been found in most countries that have instituted monitoring programs.

Mad cow is in a class of diseases known as transmissible spongiform encephalopathies (TSEs). Other TSEs include scrapie in sheep and goats, kuru in humans, and chronic wasting disease in deer and elk. These diseases are communicable to varying extents, and some are able to affect other species. Unlike most communicable diseases, which are spread by viruses and bacteria, TSEs spread via exposure to a misshapen protein, called a *prion*. The way TSEs spread was initially disputed because it differs from standard explanations of disease, but scientists have gradually come to accept prion theory. Prions are found in all mammals and do not cause problems in most situations. When exposed to misshapen prions, however, healthy prions can change their folding pattern and themselves become misshapen. When this occurs, plaques of misshapen prions form in the central nervous system, creating holes in the brain material (thus the "spongiform" in the name).

These U.S. Department of Agriculture technicians helped track the first case of mad cow disease in the United States. U.S. efforts to control the spread of the disease have been considered inadequate, but no infected U.S. cows have been identified since 2006.

Source: U.S. Department of Agriculture Agricultural Research Service/Peggy Greb

Mad Cow in Humans

Prion-based brain wasting diseases can occur sporadically in mammal populations independent of transmission from animal to animal. In humans, this condition is called classic Creutzfeldt-Jakob disease. It occurs at a low rate and primarily affects people between the ages of 55 and 75. Creutzfeldt-Jakob does not spread from person to person outside of cases caused by transplants or cannibalism. The human form of mad cow is vCJD, and it can affect people as young as teenagers. People can contract the disease by consuming affected beef, though not all who eat contaminated meat contract the disease. Sheep herds have maintained some level of scrapie for hundreds of years, and no human cases of that disease have been identified. The extent of the human outbreak of vCJD will not be known for some time, in part because most people who consume tainted beef do not contract the disease, and the time between exposure and onset of symptoms is highly variable and can be several decades in length.

The ultimate source of the mad cow outbreak is unknown, but proposed explanations include the idea that the original case was a sporadic incidence of a brain wasting disease, or alternatively, that it may have come about when cattle were fed by-products from sheep affected by scrapie. Using animal products to supplement cattle feed became common practice in the 1970s. Although cattle are herbivores, they are often fed animal by-products (ground meat, bone, brain, and other organs) to induce them to grow more quickly and/or produce greater quantities of milk. It is this practice that perpetuates and spreads the disease; scientists do not believe that the disease efficiently spreads among cattle otherwise.

John Fox and Hikaru Peterson offer a comprehensive history of the mad cow outbreak. The first cases of the disease in cattle were identified in Sussex, United Kingdom, in 1984. By 1987, scientists had established a link between the disease and the practice of feeding animal by-products to cattle, and in 1988 the U.K. government banned that practice. With a sudden surplus of animal by-products, the United Kingdom increasingly exported them to other European Union countries without adequate warnings about the possible consequences of using the products. As it was discovered that mad cow could spread to other livestock, restrictions about feeding cattle by-products to other mammals were gradually put in place and strengthened. To limit the spread of the disease, the United Kingdom now bans the use of all mammalian by-products in farm animal feed and does not allow cattle older than 30 months to enter the human food chain.

Throughout the latter half of the 1980s and the beginning of the 1990s, various consumer groups and scientists raised concerns about the safety of eating affected cattle. Beef consumption dropped significantly, but the official stance of U.K. government scientists and regulators was that the disease posed no human health risk. Government officials felt that this disease among cattle was unlikely to be transmissible to humans because, in part, they thought it was simply the sheep disease scrapie in cows. Scrapie had never been known to affect people, so the logic was that people were similarly invulnerable to mad cow disease. In an effort to assure the public that beef was safe to consume after it became known that humans could contract mad cow, the agriculture minister in the United Kingdom at the time, John Gummer, tried to feed a hamburger to his daughter at a press conference. In an awkward moment, she refused the burger, so he ate it himself. Once they reversed their official stance about the safety of beef, the U.K. government faced great scrutiny, and the public outcry gained momentum.

Eradicating Mad Cow

Once it was discovered that humans could contract mad cow, Europeans began a massive program to eradicate the disease in their herds, slaughtering millions of cattle between 1996 and 2001. Postmortem testing—still the only kind available to identify infected animals—confirmed that a few hundred thousand of the slaughtered cattle were infected. Scientists believe that testing only identifies animals in the latter stages of the disease, however, and therefore fails to detect all affected animals. Estimates that take this into account suggest that the number of infected cattle may be have been as high as 2 million.

The United States and a suite of other countries banned the import of cattle from the United Kingdom in 1989 out of fear of the disease. Cases of mad cow disease have sprung up in a handful of additional countries in the European Union and beyond. Canada announced its first case in 2003, which was a development with profound implications for the beef market there. To date, the United States has discovered two cases in cattle born within its borders and another in an animal that was born in Canada. To prevent further cases, regulations in the European and Japan now prohibit use of mammalian protein in all animal food; the U.S. regulation is weaker, in that it allows cattle to be fed blood from other cattle. Japan tests all cattle bound for the human food chain at slaughter, but until 2004, the United States tested only a fraction of the cattle that showed obvious signs of illness (known as downer cattle). The sufficiency of testing efforts and safety precautions in the United States remains contentious, and the General Accounting Office in 2002 determined that U.S. policies at that time were not sufficient to prevent occurrence of the disease. When a cow with bovine spongiform encephalopathy (later found to have been born in Canada) was discovered in Washington State, testing and other regulations were strengthened. They are, however, still weaker than most of those in the European Union and in Japan.

See Also: Factory Farm; Food Safety; Meats; Origin Labeling.

Further Readings

APHIS. *USDA's BSE Surveillance Efforts.* (July 2006). http://www.aphis.usda.gov/newsroom/hot_issues/bse/surveillance/bse_disease_surv.shtml (Accessed January 2009).

Buzby, J. C. "Agricultural Economic Report No 828: International Trade and Food Safety: Economic Theory and Case Studies." Washington, D.C.: U.S. Department of Agriculture, 2003.

Fox, J. A. and H. H. Peterson. "Risks and Implications of Bovine Spongiform
    Encephalopathy for the United States: Insights From Other Countries." *Food Policy,*
    29:45–60 (2004).
National Creutzfeldt-Jakob Disease Surveillance Unit. "Variant Creutzfeldt-Jakob Disease
    Current Data." http://www.cjd.ed.ac.uk/vcjdworld.htm (Accessed January 2009).
Rowell, A. *Don't Worry, It's Safe to Eat: The True Story of GM Food, BSE, and Foot and
    Mouth.* London: Earthscan, 2003.

Mark Neff
Arizona State University

MALTHUSIANISM

Malthusianism is the belief that a rapidly growing human population faces imminent
starvation and misery as it exceeds the Earth's capacity to feed it. The concepts of
Malthusianism are attributable to the Reverend Thomas Malthus, who first proposed his
ideas—during the middle of the Industrial Revolution—in the 1798 volume *An Essay on
the Principle of Population.* In this book, Malthus contends that humankind is quickly
putting into production all arable land from which to feed its population. Because human
population growth increases exponentially, whereas food production can only increase
linearly, the size of the population will be checked by misery (disease, famine, or war) and
vice (moral abominations) as it exceeds the planet's ability to support it.

Malthus specifically based his arguments on the fact that both food and "passion
between the sexes" are necessary for human survival. However, when human population
levels exceed the planet's carrying capacity to support them, a host of preventative and
positive checks take effect to bring them back to a sustainable level. Preventative checks
encompass a variety of actions taken by humans that both directly and indirectly lead to
reduced fertility. For example, when the population grows larger, the overall economic pie
is sliced more thinly. Therefore, more well-to-do men find it difficult to quickly amass sufficient wealth to support a wife, thus prolonging bachelorhood. Preventative checks also
include vice—activities that satiate the "passions between the sexes" but minimize population growth. Examples of what is termed *vice* are prostitution, contraception, abortion,
and infanticide. Malthus saw a lack of preventative checks leading to positive population
checks. A positive check in the form of food shortages and starvation subjects the poorer
among us to high infant mortality rates, undernourishment, and a general state of misery.

Malthus strongly argued against England's "poor laws" as being counterproductive to
the needs of the indigent population. He contended that reducing the immediate misery of
a few poor by providing them with food and clothing instills in these recipients a false
sense of economic security in their ability to afford more children. With more mouths to
feed, aided families return to their original states of destitution. Thus, governmental welfare programs promote procreation, which ultimately subjugates a larger segment of the
population to misery.

Another concern that Malthus voiced—similar to his ideas on welfare—and that was
shared by the economist David Ricardo was the stimulative effect that the Industrial
Revolution would have on population growth. They argued that industrial workers could

demand higher wages than agriculturalists. With higher wages, these workers would feel a sense of economic security, leading to a desire for larger families than their agrarian counterparts. Therefore, a transition from an agricultural to an industrial economy would further compound stated predictions for an impending population calamity.

Although Malthus accurately characterized the growth of the human population and agricultural production before the Industrial Revolution, his prognostications for explosive population growth following industrialization, and food scarcity leading to mass famine and misery for the poor, did not come to fruition. Other than the fact that his economic ideas regarding the effect of industrialization on human population growth were false, Malthus also failed to consider that per area food production is not necessarily a static process—unaffected by the advent of agricultural technology and changing methods of cropping.

Ester Boserup best articulated several ideas that partially refute Malthus's overpopulation hypotheses—namely, that as human populations become more dense, instead of finding more land to put into cultivation, they will take measures to intensify agricultural production and increase food output on existing agricultural lands. In her study of low-technology countries, Boserup found that communities transitioned from less-intensive hunting and gathering subsistence activities to more-intensive slash and burn, bush-fallow, short-fallow, and annual cropping, eventually leading to multicropping as population density increased. In addition to more frequent crop rotations, agricultural intensification techniques also include frequent weeding, watering, and fertilizer and pesticide application. The advent of these measures ultimately increases food production without subjecting more land to agriculture.

With the rapid acceleration of human population growth both during and continuing after the Baby Boom, self-described Neomalthusianists took up the mantle of humankind's imminent doom. The best known among the neomalthusianists was Paul Erhlich. In his book *The Population Bomb*, Ehrlich writes about impending global famine resulting in the death of millions as an ever-increasing human population overtaxes the Earth's ability to regenerate natural resources. Fortunately, his concerns were not immediately realized. The advent of the green revolution and the discovery of green technology, including chemical fertilizers and pesticides, foiled Ehrlich's predictions. Such technologies were able to dramatically increase food production on good agricultural substrates and to bring poorer soils into agricultural production when soil amendments were applied. These technologies were exported across the planet to lessen food scarcity problems in Africa, Asia, and Latin America. With all the good that green technology brought us in the way of increasing food production, however, it has its share of drawbacks. The continued use of petrochemical fertilizers and pesticides is exhausting the nutrient base of many agricultural soils and leading to the buildup of poisons within—both side effects that are rendering many lands infertile.

With the world population projected to increase from 6.5 billion to 7.4–10.6 billion people by the year 2050, and with the exhaustion of many once-productive agricultural soils, we are certain to enter another era of Malthusianism, with another call for population reduction measures. At this future moment, will technologists find additional methods for averting mass famine, will policymakers institute programs to effectively reduce global population growth, or will Malthus's original prognostications finally be realized?

See Also: Agribusiness; Famine; Green Revolution.

Further Readings

Avery, John. *Progress, Poverty and Population: Re-Reading Condorcet, Godwin and Malthus*. New York: Routledge, 1997.

Boserup, Ester. *Population and Technology Change: A Study of Long-Term Trends*. Chicago: University of Chicago Press, 1981.

Malthus, Thomas Robert. *An Essay on the Principle of Population* [1st ed. of 1798]. London: Pickering, 1986.

Jason Davis
University of California, Santa Barbara

MEATS

Meat production has increased consistently in the last decades through organizational innovations and better technology, despite sporadic high-profile outbreaks of diseases such as bovine spongiform encephalopathy, avian flu, and foot and mouth disease. The growing purchasing power of consumers from developing regions has also boosted the demand for meat. Though the livestock industry is not of the largest economic sectors globally, its social, political, and economic significance cannot be underestimated. The livestock industry accounts for 40 percent of agricultural gross domestic product, creates livelihoods for 1 billion people worldwide, and provides one-third of our protein intake. The sector is touted as a remedy for global undernourishment and also is paradoxically blamed for the rise in obesity rates as well. These realities make the meat sector a rich terrain for analysis—one that a cross-disciplinary group of researchers, hailing from sociology, agricultural economic, and developmental studies, have long undertaken. The question of meats is often drawn into broader debates over culture, economy, society, health, and environmental safety, with politics playing a particularly important role. Pertinent issues in the discussion of meats include the following:

- the cultural and religious dimensions of meat consumption,
- the changing systems of production of meat,
- the politics of meat production and consumption,
- the environmental impacts of the production of meat, and
- nonconventional meat production.

The Cultural and Religious Dimensions of Meat Consumption

Cultural preferences and religious restrictions have always been important considerations in the consumption of meats. For example, the most systematic and codified of all food taboos are the Islamic *halal* and the Jewish kosher dietary laws. Of all the restrictions in their diets, the avoidance of pork is perhaps the most well-known. Hindus are also noted for their general avoidance of meat consumption and their specific avoidance of beef. The latter is directly related to their belief that the cow is a sacred animal. Much research has been attempted to explain the roots of such religious prohibition of meats.

Cultural or regional variations in the consumption of meats are also prevalent in many parts of the world. For example, in the American South, niche meat types like alligator are more commonly eaten than in other parts of the country. At the extreme, horse, whale, and dog meat (considered taboo by many cultures) are consumed as delicacies by groups of people around the world. The intense affinity toward certain types of animals has compelled many to launch protests and campaigns against the consumption of meats like dogs and cats by other cultures. Nonetheless, even at such aggregate societal levels, tastes and perception toward meat do change. For example, as noted earlier, India, a country that is historically averse to meat consumption, has seen steady growth in the demand for meat. Religious and cultural norms have also shaped the production of meat in very specific ways. The slaughtering of animals in Islamic and Jewish traditions has to adhere strictly to a prescribed set of rituals and regulations.

Pigs near the finishing stage on an Arkansas farm in 1983. Contract farming has meant that the percentage of pigs that are raised in a single location dropped to just over 20 percent in 2004.

Source: U.S. Department of Agriculture Natural Resources Conservation Service/Tim McCabe

The Changing Systems of Production of Meat

The defining feature of the contemporary meat industry is that production methods have become increasingly intensified. For example, in the United States alone, the number of pig farms decreased from 2 million in 1950 to 73,600 in 2005, although the production of pigs in the same period rose from 80 million to 100 million. Intensification processes in animal rearing has led to the emergence of megafarms (variously known as factory farms, megabarns, intensive livestock operations, or confined animal feeding operations). In fact, the meat industry has not only intensified (in that more meat is produced by each farm) but has intensified in a very extreme manner, where more meat is being produced by fewer farms in fewer places. The reasons for such increasingly focused geographical concentration in the spatial distribution of farms are largely the result of local economic policies, strong agricultural interest group lobbying, and the changing corporate landscape of meat producers. The former includes favorable tax policies that encouraged meat producers to shift or expand their operations to specific places.

The intensification process runs parallel to two related developments in the meat industry. The first is the growing market consolidation of the top meat producers, and the second is the increasing popularity of contract farming as the main mode of meat production. Major meat producers have consolidated the market through expansion, mergers, and acquisitions.

In the United States, several leading companies control most of the supply of meat in the country. For example, in 2005, the top three beef packers in the United States controlled more than 80 percent of the market, and the pork packing industry saw the top four companies controlling 64 percent of the market in 2005 (up from 40 percent in 1990). The global meat industry has also witnessed a string of transnational acquisitions. For example, Goldman Sachs acquired the biggest pig producer in China for $252 million in 2007. The immediate implication of such transnational investment from developed countries to less-developed countries is the likely transfer of new technologies of meat production. Such a transfer will have significant consequences on the structures of the meat industry in these less-developed countries. For instance, mirroring the trend in the developed countries, contract farming as an organizational feature of meat production may become more prevalent.

Contract farming essentially breaks up the production of meat into different stages in which contractual arrangements (both formally and informally) are made between farmers and meat companies that specify one or more conditions of production and/or marketing of an agricultural product. Clearly, this involves the transportation of animals from farm to farm at different stages of their growth, with the large meat-packing companies as the final point in the production process. The trend toward contract farming as a system of production is noteworthy. For example, as recently as 1992, about 65 percent of all pigs in the United States were produced in farrow-to-finish operations (i.e., the animals were raised from birth to slaughter in the same farm). In 2004, this figure had dropped to slightly more than 20 percent. Specialized farms under the contract farming system themselves have grown larger and intensified through the years.

The Politics of Meat Production and Consumption

The organizational and production changes seen in the meat market are ultimately the result of complex political and cultural forces (sprung from global food companies and national/local governments, as well as the citizenry) that are often cognizant of the implications of their business decisions, policies, and actions. There has been increasing opposition toward the development of large meat producers in terms of their environmental impacts and their treatment of both farm labor and livestock.

The concentration of meat producers in local communities has often been resisted by local residents wary of the environmental risks such megaoperations may bring. Megafarms are also sometimes accused of compromising on workers' welfare. They have also been under persistent criticism for their lack of concern toward the deteriorating welfare of livestock raised under such intensified production systems. Yet others have pointed out the economic gains that megafarms bring to local areas and stressed the importance and effectiveness of regulation in ensuring optimal environmental standards on these farms. Nonetheless, according to many animal welfare proponents, the transportation of animals in cramped conditions is said to result in animal fatigue and the unnecessary deaths of animals. For example, animal activists have criticized the export of Australian sheep to the Middle East by sea, arguing that the animals are under constant duress when traveling such great distances under less-than-ideal conditions.

Different kinds of meat companies present different kinds of politics. For example, poultry companies (i.e., the packers, integrators, or processors) have very prescriptive contracts with the growers (i.e., contract farmers). The former have control of the production process to achieve a "uniform" product. Being the earliest among all meat companies to have modernized and industrialized, broiler production has evolved into a state in which producers/growers have weak bargaining power relative to the processor/integrator companies. This situation presents significant political and legislative challenges for concerned parties to push for more sustainable production practices in the poultry sector.

The debate over animal welfare has grown increasingly vocal. The key contention here is the alleged appalling conditions in which animals are born, raised, and slaughtered for consumption in intensive megafarm systems. In response to such criticisms, many countries have enacted laws and proposed guidelines aimed at safeguarding the interest of livestock that is raised for human consumption. For example, Britain has adopted five key principles of "freedom" that are considered to be a global standard in ensuring the welfare of livestock. They are

- freedom from malnutrition,
- freedom from thermal and physical discomfort,
- freedom from injury and disease,
- freedom to express most normal patterns of behavior, and
- freedom from fear and stress.

Animal rights groups have repeatedly publicized the inhumane treatment of factory animals, and vegetarian advocates have highlighted similar mistreatment, as well as the health problems brought about by eating meat. People for the Ethical Treatment of Animals has specifically singled out the extreme suffering of pigs when being transported to the slaughterhouse (calling the process "Hell on Wheels"), as well as the way in which the animals are slaughtered. Yet scholarly works on the alleged abuse and ill-treatment of livestock remain scarce, largely because of the difficulty of penetrating the inner workings of such factory farms.

The Environmental Impacts of the Production of Meat

Livestock farming has significant impacts on the environment. It has been estimated that meat production contributes about 18 percent of total anthropogenic greenhouse gas emissions and nearly 80 percent of all agriculture-related emissions. This exceeds total emissions from the entire transportation sector. Specifically, the livestock sector accounts for 9 percent of anthropocentric carbon dioxide, 37 percent of methane, and 65 percent of anthropogenic nitrous oxide (296 times the global warming potential of carbon dioxide). Nearly three-quarters of all agricultural land on the planet is used for livestock production, as well as 8 percent of total water use.

Different types of meat production are prone to somewhat different environmental impacts. For example, cattle ranching has now become the major cause of deforestation in the Amazonian forests. Other general environmental impacts include water pollution, olfactory nuisance, and health hazards to farmworkers. The chief sources of environmental pollution come from excretion, belching, and the growing of animal feed (comprising mainly corn and soy) for the animals. For example, nitrates and phosphates from the manure and urine of pigs that are improperly disposed of are the main pollutants. Efforts to solve this problem include, inter alia, better waste management, improved feeding regime, and increased legislation and enforcement.

The growth of megafarms has resulted in particular environmental management challenges. For example, on mixed-use farms, pig manure is used to fertilize the soil for crops to supplement or even supplant inorganic fertilizers. However, this course of action is untenable in megafarms because larger operations are likelier to have an insufficient land base to take in such manure nutrients. This means that the management of manure will be driven by lowering disposal costs rather than optimizing the nutrient needs of crops and pasture. Manure then becomes a form of waste rather than a potential resource. Livestock waste disposal remains a major threat to the environment and aggravates the greenhouse effect.

For the pig and poultry industries, the strong odors that they emit can degrade the quality of life of residents nearby. As smells are more tangible and more immediately apparent than other forms of pollution to local residents, odor complaint is one of the significant public issues faced by the pig and poultry industry. Despite this array of environmental problems, the Organisation for Economic Co-operation and Development reports that there is evidence that many meat producers are improving their environmental performance through better technologies and husbandry practices.

Nonconventional Meat Production

Against this broader climate of larger and larger livestock farms, the continued existence of small-scale livestock production is a contrary trend. Yet nonintensive, small-scale commercial livestock production has always had an important role in the rural developmental process in less-developed regions. This is unsurprising, given that the intensification of meat relies on a relatively high level of technical and infrastructural competence that may not be readily available to all farmers. In recent years, not least because of high-profile health scares concerning factory farming (mad cow disease and avian flu), increasing attention has been turned to organic meat production, which places more emphasis on safety by eschewing the use of hormones and antibiotics.

Production costs for such organic meats are significantly higher than for conventionally produced meats, with price differences ranging from 15 percent to 200 percent. Although demand in the United States has been rising steadily in the past decade, organic meat is still estimated to make up not more than 3 percent of the total meat supply. The cost of organic meat and the related issue of adequate and steady supply are the two main reasons for the limited size of the organic market. The demand for organic meat is also in part driven by consumers who are concerned about where and how livestock are raised. Animals that are reared in nonintensive, organic systems of production are thought to have better welfare than their counterparts on factory farms. Interestingly, the research is still somewhat divided on whether organic meat production would necessarily be more environmentally friendly than conventional intensified modes of production.

The genetic manipulation of meat has been an ongoing process concomitant to the intensification of meat production. Hormone injections, among other measures, have also been routinely administered to livestock to produce "better" meats. Genetic modifications of pigs, for example, have made their excretions less polluting to land and water. The cloning of meat has become a scientific possibility. However, although researchers have been able to clone a small sample of meat, this technical breakthrough is continually implicated in a complex religious-ethical debate that questions such modernized methods of food production. Moreover, cloned meat is at the moment unable to replicate in full the exact taste and texture of conventional meat. It is also an extremely expensive process and hence commercially unviable in the near future, except possibly for selective breeding purposes. Above all, such meats will continue to be met with considerable resistance from consumers because of health and food safety concerns, as well as a prevailing aversion toward cloning technologies. Nonetheless, cloned meat represents an intriguing option for producing meat that minimizes environmental impacts and animal suffering.

See Also: Agribusiness; Agrofood System (Agrifood); Animal Welfare; Confined Animal Feeding Operation; Contract Farming; Organic Farming.

Further Readings

Atkins, Peter and Ian Bowler. *Food in Society: Economy, Culture, Geography.* London: Arnold, 2001.

Gregory, N. G. and T. Grandin. *Animal Welfare and Meat Production.* New York: CABI, 2007.

Hodges, John and In Han, eds. *Livestock, Ethics and Quality of Life.* New York: CABI, 2000.

Nierenberg, Danielle. *Happier Meals: Rethinking the Global Meat Industry.* Worldwatch Paper 171. Washington, D.C.: Worldwatch Institute, 2005.

Organisation for Economic Co-operation and Development (OECD). *Agriculture, Trade and the Environment: The Pig Sector.* Paris: OECD, 2003.

Simoons, Frederick J. *Eat Not This Flesh: Food Avoidances From the Prehistory to the Present.* Madison, WI: University of Wisconsin Press, 1994.

Steinfeld, Henning, et al. *Livestock's Long Shadow: Environmental Issues and Options.* Rome: Food and Agriculture Organization of the United Nations, 2007.

Harvey Neo
National University of Singapore

MECHANIZATION

The replacement of human work with work by machines is referred to as *mechanization.* Increasing mechanization in agriculture has allowed people to produce food in ever-greater quantities and with ever more consistent quality. Material elements of modern agricultural mechanization include rather low-tech plows and sophisticated and expensive combines, irrigation systems, and processing facilities and have led to the evolution of the safest and most plentiful food supply the world has ever seen. However, critics of highly mechanized agriculture point out that neither participation in this kind of agro-industrial production nor consumption of the food produced by this kind of agriculture is evenly distributed throughout the world. Instead, many people have too much to eat, and many more people still do not have enough. Though highly mechanized agriculture in general means that fewer people can produce enough food for many people to eat, some people have become concerned that this leads to a dangerous cultural and social disconnection from our food sources.

History of Agricultural Mechanization

In a basic sense, agricultural technology began with the birth of agriculture itself. Though we do not know exactly when and where people first began to plant seeds and cultivate their own crops, technological innovations have been ongoing in agriculture ever since that time. These innovations have included changes in systems of implementing and managing agricultural production, changes in the variety and type of crops planted, and innovations in the tools used to plant, nurture, harvest, and process agricultural products. All of these types of innovation in agriculture have occurred at different speeds and in different directions in different countries, and even in different regions of the same country.

Although all types of agriculture involve some level of technology, only certain types can truly be referred to as *mechanized*. Subsistence and shifting-cultivation systems of agriculture are some of the oldest agricultural systems and may include the use of technology in the form of tools such as hoes, rakes, spades, and machetes. However, it would be a stretch to call these kinds of agriculture mechanized, as the majority of the farm work in these systems relies heavily on human effort. The adoption of the use of the horse- or oxen-drawn plow represents a first step in agricultural mechanization.

Modern mechanized agriculture began as early as the 16th century in Great Britain. The widespread use of tools, as well as new practices such as crop rotation, improved crop yields. During this time, most farming still relied heavily on human and animal labor, including the widespread use of the plow. However, steady improvements in tools and farming methods were related to a subsequent population boom and freeing-up of much of the workforce, two factors frequently credited for the Industrial Revolution of the late 18th and early 19th centuries. With the Industrial Revolution, innovations in mechanization in European agricultural practices only intensified. During this time, major advances in technology included the tractor and the combine harvester, which greatly improved the speed at which farming tasks could be performed and led to an increase in the size of farms, as well as contributed to a less strenuous and safer work environment for laborers.

Mechanization and other agricultural technologies quickly spread throughout Europe and to the United States in the 19th and 20th centuries, especially after World War II. Advances in technology during World War II accompanied a shift in mainstream social conscience away from viewing the world as highly fragmented and toward understanding how interconnected the world is. With the new technology available, it became possible to produce enough food to feed more people than ever before, without the extensive human and animal inputs previously required. These shifts facilitated the move toward large-scale, highly mechanized monoculture production that became pervasive in Europe and the United States, as well as in other countries like Brazil, beginning in the mid-20th century.

The last half of the 20th century saw even greater improvements in agricultural mechanization, including improvements in technologies for irrigation, fertilization, and tillage. Farm machinery is sophisticated and expensive and relies on computers to control many aspects of agricultural production, such as the amount of seed or fertilizer to apply or when irrigation systems should be switched on. These advances have made large-scale agriculture exceedingly efficient, with much farming in the United States and other developed countries pushing the practical limits of crop yields. As a consequence of this efficiency, industrial and monoculture agriculture is the predominant method of food production in the developed world.

Benefits and Problems Associated With Mechanization

The benefits reaped from the mechanization of agriculture, particularly in terms of food safety and the sheer quantity and variety of food available globally, have been enormous. The speed with which modern farmers can plant, reap, and process crops for sale and consumption puts food on consumers' tables faster than ever before. As a result of improvements in processing, such as safe, large-scale canning facilities, and improvements in transportation, including refrigerated shipping via train, truck, airplane, and boat, consumers also have access to an unprecedented variety of food choices from around the world and throughout the year. Whereas previously, consumers could only eat what was fresh and local, today consumers can buy practically anything they want to eat at any given moment. Although there are isolated incidences of foodborne illness in developed countries, usually resulting from lax adherence to food-safety laws in processing facilities, the vast majority of processed and fresh food is clean and safe for consumption. People in both developed and less-developed nations go hungry, but it is not for lack of food available. Unlike in previous eras, most hunger in the world today can be accounted for by an inequitable distribution of resources and power between countries and within societies, instead of by the lack of global food security that existed before agricultural mechanization became widespread.

These benefits, however, have not come without costs. Increased mechanization has also meant an increased reliance on fossil fuels in farming. Farm machinery, including tractors, reapers, and combines, and processing and transportation equipment all rely on nonrenewable fuel sources. Moreover, machinery that applies fertilizers and pesticides to crops has made the application of these petroleum-based chemicals easier, and thus has encouraged increased use of them. In addition to the problems associated with reliance on fossil fuels, the intensive use of chemical inputs in agriculture has created a host of other problems. These chemicals have side effects dangerous to people, animals, insects, and other plants. Many of these side effects are unknown until the chemical has been widely used for many years.

Other disadvantages are related to the inequitable distribution of land, natural resources, and food products that has accompanied the mechanization of agriculture. Mechanization has allowed farmers to efficiently farm much larger tracts of land with less manpower than ever before, which is tied to land consolidation and changing socioeconomic landscapes in the countryside. Smaller, family-run farms have in many cases been bought by large, industrial-scale farms that have the resources to take advantage of economies of scale and wait out cycles of poor yields. This has had effects on the social structure of the countryside, as former farmers look for other sources of income, sometimes moving out of the countryside altogether. Industrial farms may also take more than their fair share of water for irrigation, leading to conflicts with local residents and with residents downstream of water sources. Industrial farms also are frequently criticized for polluting the air, water, and land without paying retribution for this pollution. Simply overfarming the land as a result of the intensive monocultures encouraged by mechanized agriculture may lead to an irreversible degradation of the soil in agricultural areas. Moreover, the products produced by this highly mechanized agricultural system are not evenly distributed. Even with the high capacity for food production created by mechanization, some citizens of both industrialized and unindustrialized countries are underfed.

The Future of Mechanized Agriculture

Even with the problems associated with mechanized agriculture, it is unlikely that this type of farming will be abandoned by industrialized nations. The efficiency and cost-effectiveness of mechanized agriculture are simply too great. In fact, many developing countries strive toward achieving the overall high level of food security found in the developed world by adopting such mechanized agricultural practices as they can. The 21st century can expect to witness continued advances in the mechanization of agriculture, including higher-tech farm machinery and even safer and more efficient processing facilities. Public attention to the problems tied to mechanization may encourage the adoption of more fuel-efficient and environmentally and socially sustainable agricultural practices. Tractors and combines that rely on renewable fuel sources, fertilizers, and pesticides made primarily of non-petroleum-based substances, as well as irrigation systems that dramatically reduce the amount of water required, are all real possibilities for the future of mechanized agriculture. As more countries move toward mechanized agriculture, the pressure for redistribution of resources to a more equitable state can also be expected to increase.

See Also: Agribusiness; Factory Farm; Fertilizer; Irrigation; Modernization.

Further Readings

Fitzgerald, Deborah. "Beyond Tractors: The History of Technology in American Agriculture." Review Essay. *Technology and Culture,* 32/1:114–26 (1991).

Knox, Paul L. and Sallie A. Marson. "Agriculture and Food Production." In *Human Geography: Places and Regions in Global Context,* 4th ed. Upper Saddle River, NJ: Pearson Prentice Hall, 2007.

Sassenrath, G. F., et al. "Technology, Complexity and Change in Agricultural Production Systems." *Renewable Agriculture and Food Systems,* 23/4:285–95 (2008).

Lisa Rausch
University of Kansas

Millennium Development Goals

The Millennium Development Goals (MDGs) are the expression of the strong commitment to universal development and poverty eradication made by the International Community in the United Nations (UN) Millennium Declaration in September 2000 and subsequently modified in 2005 at the UN World Summit. Presented within an operational framework, the goals offer a set of concrete targets that entail clear obligations for the developed countries to achieve generally by the year 2015.

Eight Goals

The eight major MDGs consist of the following: (1) eradicating extreme hunger and poverty through reducing the proportion of individuals earning incomes less than a U.S. dollar per day and increasing equal-opportunity and productive employment; (2) achieving universal primary education; (3) promoting gender equality and empowering women, particularly with reference to education; (4) significantly reducing child (< five years) mortality; (5) improving maternal health, including reducing the maternal mortality ratio; (6) reducing the spread of HIV/AIDS, malaria and other major diseases and increasing the accessibility of treatment options; (7) ensuring environmental sustainability including reducing the loss of biodiversity and environmental resources, halving the proportion of people without access to basic sanitation and clean drinking water, and integrating sustainable practices into governmental policies; and (8) developing a global partnership for development that includes measures for reducing tariff and international debt and establishing a nondiscriminatory trading and financial system.

Sustainable food/nutrition security, at the nexus of a number of issues from energy/water/health security to climate change, is emerging as one of the major challenges of the 21st century and occupies the central place in MDGs as far the developing countries are concerned. Hence, this article addresses the food security issues and their linkages to MDG 1, especially in developing countries in the context of burning debates on the sustainability of agricultural production.

MDGs and Food Security: The Policy Quagmire

In spite of their recurring political endorsement, the MDGs have been subject to a wide range of criticism, as, ironically, we have hunger and malnutrition even when we have adequate food stocks. Indeed, there has been much debate about the conceptual and methodological underpinnings of the MDGs and their predecessors, the International Development Targets, as targets that are easily set but seldom met. This is primarily a result of data gaps, lack of timely information, inconsistent indicators, frequent revisions, and negligence of indigenous traditional ecological knowledge. For instance, there are strong lobbies for agricultural biotechnology (BT) and genetic engineering (GE), supported by the state machinery and backed by the agricultural research institutes, as they are funded by international agencies and multinational companies, contrary to the accumulating realities on the ground.

The following are the inevitable consequences of the green revolution agricultural technologies (to some extent BT/GE too): pollution resulting from excessive use of fertilizers/pesticides, and the consequent degradation of soil/water/aquifers; loss of biodiversity; and heavy reliance on fossil fuels. In addition, farmers are forced to buy costlier hybrid seeds and other chemical inputs, whereas they were self-reliant earlier. As a result, food production has become unstable and/or unsustainable, and food and nutrient quality has become degraded, such that the chances of achieving the MDGs by 2015 are at risk. In addition, there are serious unanswered questions about the biosafety of agricultural BT. All these problems do not arise if we embrace agroecology.

MDG 1 (poverty reduction) is closely and inseparably linked to environmental/socioeconomic sustainability (MDG 7):

- Poor people's livelihoods and food security often depend on ecosystem goods and services that are affected by agricultural production methods.
- Poor people tend to have insecure rights to environmental resources and inadequate access to markets, decision making, and environmental information, limiting their capability to protect the environment and improve their livelihoods and well-being.
- Lack of access to energy services also limits productive opportunities, especially in rural areas. The lopsided subsidies by governments for nonrenewable energy sources and technologies send a negative signal to consumers.

The Road Ahead

With more than 800 million people in developing countries still suffering from chronic undernutrition, hunger, and food security (physical and economic access to sufficient safe and nutritious food to meet their dietary needs and food preferences), food sovereignty (the right of people to protect and regulate nutritious and culturally appropriate domestic agricultural production and trade to achieve sustainable development objectives) will have to remain the top priority for food policy for many years to come.

On the basis of this discussion, the following recommendations can be made for helping to achieve MDG 1:

- Democratic, de-institutional and participatory action research, involving innovative farmers, scientists, sociologists, and other civil society groups through networks
- Partnerships for forward (marketing) and backward (production) linkages for ensuring sustainable supply chains for producing and marketing certified/value-added organic farm products, simultaneously encouraging localization of distribution, trade, and marketing
- Participatory plant breeding and community seed banks for in situ conservation and regeneration and propagation of traditional seeds
- Land reforms, coupled with promoting cooperative farming for the landless in cultivable wastelands
- Horizontal transfer of socially equitable, culturally compatible, ecologically sustainable, self-reliant, and economically viable agro-eco technologies, based on an ideal blend of indigenous knowledge and modern scientific tools for enhancing total farm productivity, resilience, sustainability, equity, and stability by increasing crop and animal diversity (spatial/temporal) to minimize risks and increase synergism

See Also: Agrarianism; Agrarian Question; Agribusiness; Agricultural Extension; Agrodiversity; Agroecology; Agrofood System (Agrifood); Appropriationism.

Further Readings

International Assessment of Agricultural Knowledge, Science and Technology for Development. http://www.agassessment.org (Accessed January 2009).

United Nations Development Programme. "Millennium Development Goals." http://www.undp.org/mdg (Accessed October 2009).

Wagstraff, Adam and Mariam Claeson. *The Millennium Development Goals for Health: Rising to the Challenges.* Washington, D.C.: The World Bank, 2004

Gopalsamy Poyyamoli
Pondicherry University

MODERNIZATION

Modernization refers to a group of approaches popular in the social sciences during the 1950s and 1960s, particularly in the United States and the United Kingdom, which seek to explain the conditions necessary for successful economic, social, and political change in what were then referred to as "Third World" countries. Modernization was superseded in academic circles for a time by more radical theories of change such as dependency theory and World-Systems theory and by a range of postmodern approaches to development. However, it has reemerged recently in debates on, among other things, the relationship between environment and economic development.

Modernization thinking has its roots in 19th- and early-20th-century social theory, particularly in the work of Karl Marx, Max Weber, and Émile Durkheim, who sought to understand and explain the emergence of modern industrial society, considered as a new and unique societal form. For these theorists, although they did not use the term, modernization meant breaking with the nonindustrial past and adopting the values and institutions of the emerging industrial world based on technological innovation; the harnessing of inanimate or extra somatic sources of energy to serve human needs; scientific and secular ways of explaining society and nature; an increased social, political, and economic division of labor; and the belief in the application of the principle of instrumental rationality to the solving of human problems. Old ties of kinship, clan, ethnicity, and "tribe" were to be replaced by more impersonal, fragmented, and contractually based relations found in the new factories and workplaces of the growing cities of Europe and North America, by national territorial identities, and by new forms of political authority.

During the early Cold War period (1950–1970), Western academics and policymakers drew partly on these early theorists' work to develop modernization approaches. At this time, the United States and its allies competed with the Soviet Union for global political and economic dominance, and modernization approaches helped provide a scientific rationale and legitimacy for Western involvement and intervention into the economic and political affairs of Third World countries. The Soviet Union, its satellites, Communist China, and other aspirant communist societies also held their own notions of socialist modernization, which were considered by Western modernization theorists as aberrant or deviant models of development and change. However, both shared a belief in the importance of linking science and technology to the expansion of material production through rational control of society and nature, though they differed on the appropriate economic and political means to realize such goals. Attention here will be confined to Western approaches to modernization.

The Foundation of Modernization

Underlying all modernization approaches is the idea that the human society passes through particular stages of development, culminating in the industrialized societies of the West. These societies presented themselves as models to emulate through a combination of state regulation and free markets directed by modernizing political and scientific elites. Successful modernization required nonmodern societies to copy the examples of purportedly already modernized societies to raise their living standards and to move away from "tradition," defined residually as everything that is not modern. Early versions of modernization theory saw successful change as the total transformation of a traditional society's major institutions, which meant the dismantling of traditional institutions, considered internal obstacles to change in societies seeking to modernize. Later versions modified this view to allow for variations among societies in their capacity to modernize. However, the already modernized societies would show the poor, tradition-bound societies of the Third World the way to a stable, democratic, and prosperous market-based future.

Modernization theorists developed a mix of economic, political, and social aggregate indicators to assess the extent and type of modernization of the undeveloped (later called "developing") countries. These included gross national product; rate of economic growth; extent of private property ownership; degree of involvement in commercial farming; proportion of the population in "modern" employment; levels of urbanization, education, and literacy rates; level of malnutrition; and provision of modern healthcare. Traditional ways of living based on subsistence farming, craft production, village life, authoritarian and kin-based forms of rule, appeals to the past as the guide to proper conduct, communal forms of property ownership, and limited social mobility were to be replaced by commercialized and often highly concentrated farming, industrial production, an urbanized labor force supplied with mass-produced food and nonfood commodities, technological innovation, entrepreneurship, and an emphasis on individual and personal freedom to be realized through democratic political systems, private ownership, mass secular education, and a belief in constant and unending economic growth. Traditional societies were by definition poor or undeveloped societies, as they lacked the institutions of modern society, which had generated great wealth and rising standards of material life in the West. The developed societies could assist in modernizing the undeveloped by investment, aid, technology transfer, education, and nation-building. Modernization was a transitional phase in the development of societies from tradition to modernity. That is, there was an end point to societal evolution culminating in a movement of all societies toward so-called Western forms, captured in the notion popular in the 1960s that all industrial societies were becoming alike, and revived in the 1990s with Francis Fukuyama's end of history hypothesis.

Modernization theorists considered rural and agricultural life as one of low productivity, underemployment, and technological stagnation and as grounded in inadequate understanding of how to marshal resources to produce more food and nonfood cash crops for domestic and export markets. Agricultural modernization policies promoted various forms of production ranging from smallholder farming to plantation agriculture directed toward supplying export markets and feeding the growing landless rural and urban populations, themselves partly a product of the same modernization policies pursued by national governments and international development agencies. Traditional forms of management of natural resources based on common property regimes and usufruct rights were to be replaced by private and corporate ownership of such resources, regarded as a more efficient way of raising productivity than earlier ownership systems. More generally, traditional practices and institutions were to be swept away, as they were incompatible with modernity. The natural world was considered a realm of reality separate from humans and was a tabula rasa on which modernizing societies could write their own scripts.

How Modernization Grew

Modernization thinking flourished in a period of cheap fossil fuels, which underpinned technological improvements in agriculture and food production, and at a time when the environmental costs of modernization policies were considered of minor importance. A good example of this technocratic or productionist approach to agriculture was the "green revolution" of the 1950s and 1960s, when high-yielding varieties of barley, maize, rice, and wheat and chemical fertilizers and insecticides were introduced into many developing societies. The results of the green revolution were both positive and negative. On the positive side, crop yields increased, many farmers benefited from increased sales, and some countries were able to achieve self-sufficiency in foodstuffs. On the negative side, high demand for water and the use of chemical products led to water shortages and disputes over allocation, the use of chemical fertilizers increased pollution from irrigated agriculture, the increased costs of agriculture favored those with money, and the shift to monoculture systems of production reduced biodiversity. Modernization policies encouraged state-supported mega environmental and social engineering projects such as dam building, land clearance, draining of wetlands, and resettlement of populations. It was assumed that the nonhuman natural world should serve human interests and that the environment and the ecological services it provided were free goods.

Although productionist approaches to agriculture continue to dominate global food policies, they are being both modified and challenged by new models of agricultural change and food production. These include the growth of state and private regulation and the certification of the quality of food and food products; the promotion of more ecologically sensitive approaches to crop production, land and water use through crop diversification, reduced chemical inputs, and more organic farming methods; and the creation of more diversified income-generating opportunities for rural populations.

Early modernization approaches were criticized for being top-down, treating all nonindustrial societies as alike, failing to consider more than one pattern of development, and ignoring global structures of economic and political inequality, which were said to have created and ensured lack of development in the third world. Although no longer popular in development studies, there are some recent examples of revived modernization thinking in both academic and policy circles. For example, cultures of corruption are said to contribute to a lack of economic development in developing countries, global strategic insecurity is considered a result of civilizational differences and the lack of democratic political cultures in the developing world, and ecological modernization theory seeks to counter radical environmental critics of contemporary capitalism by arguing that continued economic growth and protection of the environment are compatible objectives, that these objectives can be achieved through better environmental planning, and that such planning avoids the need for major change in contemporary economic, political, and social institutions.

See Also: Cash Crop; Green Revolution; Productionism.

Further Readings

Dryzek, John S. *The Politics of the Earth: Environmental Discourses*, 2nd ed. Oxford: Oxford University Press, 2005.

Rostow, W. W. "The Stages of Economic Growth." *The Economic History Review*, 12/1 (1959).

Scott, James C. *Seeing Like a State: How Certain Schemes to Improve the Human Condition Have Failed.* New Haven, CT: Yale University Press, 1998.

Timmons Roberts, J. and Amy Hite, eds. *From Modernization to Globalization: Perspectives on Development and Social Change.* Malden, MA: Blackwell, 2005.

Bob Pokrant
Curtin University of Technology

MONSANTO

Headquartered in St. Louis, Missouri, Monsanto is a major multinational agricultural company producing agricultural inputs such as corn, canola, and soybeans. The company's controversial involvement in genetic engineering and its lobbying efforts to prevent product labeling of GM ingredients, with its lawsuits against seed-saving farmers and its pursuit of seed patents, have attracted much criticism and protest from around the world. Monsanto claims that its products support environmental sustainability in agriculture, but critics have raised questions around the environmental impacts of Monsanto's GM products.

History

Originally founded in 1901 by chemist John Francis Queeny, Monsanto began producing saccharin, an artificial sweetener, and later went on to manufacture sulfuric acid, plastics, chemicals, and other synthetic materials. In the 1940s, it produced the herbicide 2,4,5-T, which contained dioxin, a toxic chemical. By the 1960s, Monsanto produced polychlorinated biphenyl, a chemical that was later banned in the United States for its deleterious effect on human health and on the environment. During the Vietnam War, Monsanto also manufactured the herbicide Agent Orange, which was used by the American military to defoliate rainforests in Vietnam. Agent Orange contained high levels of dioxin, which led to health problems in Vietnam War veterans. These veterans later brought a lawsuit against Monsanto for damages. Throughout the 1980s, Monsanto focused on biotechnology research, leading to its first biotechnology product, Posilac, or recombinant bovine growth hormone (BGH). Also known as bovine somatotropin, BGH generated much controversy and protest.

BGH

Monsanto's BGH, a genetically engineered hormone injected into dairy cows to increase milk production, received approval from the U.S. Food and Drug Administration in 1993 and became commercially available in 1994. Critics state that Monsanto used its political leverage to gain regulatory approval of BGH. They further submit that BGH was not tested sufficiently before being approved. Additional concerns have been raised over the welfare of injected cows, as studies have shown an increased risk of udder inflammation and decreased fertility in addition to other health problems. Monsanto has claimed that these findings are false. Growing consumer concerns around BGH have led some retailers such as Whole Foods Market and Costco to carry BGH-free milk. These concerns also fueled efforts to label milk produced without the hormone. Claiming that BGH milk is no different from conventional milk, Monsanto attempted to ban or limit BGH-free labels and was successful in some states. Labels were now required to include a disclaimer stating the Food and Drug Administration's position that there is no difference between the two types of milk. Banned in Canada, Japan, Australia, New Zealand, and by the European Union, BGH was sold to Eli Lilly, a pharmaceutical company, in 2008. In its decision to sell BGH, Monsanto stated that it seeks to focus on its seed and pesticide products instead. Critics contend that the sale of the hormone to Eli Lilly was in part a result of the immense controversy surrounding the product.

GM Organisms

The controversy surrounding Monsanto's GM crops centers around two product lines. The first is Monsanto's Roundup Ready brand of seeds, which include soybean, canola, cotton, and corn. Roundup Ready seeds contain the soil bacterium gene Agrobacterium, a gene that confers resistance to Monsanto's Roundup Ready glyphosate herbicide. Glyphosate is an herbicide that destroys all green plants. Second, Monsanto produces a line of controversial crops that emit their own pesticide. These crops, which include cotton (Bollgard) and corn (YieldGard), contain *Bacillus thuringiensis* (Bt), a bacterium found in soil. The bacterium produces toxins that eliminate insects that threaten corn and cotton crops.

Monsanto asserts that its GM crops are beneficial for the environment and safe for human consumption. As part of its corporate responsibility statement, Monsanto outlines that sustainable agriculture is a key concern for its GM seed operations. Sustainable agriculture is characterized by Monsanto as an effort to increase crop yields to meet a growing world demand for food by using current levels of resources or less. Agricultural resources include land, water, and agricultural inputs such as pesticides and fertilizers. Because genetically engineered crops that contain Bt produce their own pesticides, Monsanto states that their crops require less pesticide use. Monsanto also asserts that its GM products are no different from their conventional counterparts.

Critics of Monsanto and GM foods claim otherwise. Roundup Ready plants have been criticized because they may transfer pesticide resistance to other plants such as weeds. Roundup Ready seeds have also been said to have spread to non-GM crops, as discussed further here with the case of farmer Percy Schmeiser. Critics also state that Bt crops may create new strains of disease that may be resistant to Bt. Although Monsanto states that its products increase crop yield, farmers and their advocates have cited crop losses and damage. Some have charged Monsanto with leveraging its political clout to influence international trade, for example, during the Uruguay Round of the General Agreement on Tariffs and Trade, a precursor to the World Trade Organization. The World Trade Organization has been said to have placed pressure on European countries to accept GM food. Opposition to Monsanto has emerged in various countries, with groups uprooting and burning Monsanto's GM test crops and fields. Monsanto's patenting of GM agricultural products has entangled the corporation in legal disputes sparking further controversy and protest.

The legal battle between Canadian canola farmer Percy Schmeiser and Monsanto brought attention to the stakes involved with patents and GM crops. Monsanto sued Schmeiser for growing Roundup Ready canola without Monsanto's consent. Monsanto claimed that Schmeiser infringed on Monsanto's patent rights. Schmeiser, in turn, sued Monsanto for contaminating his canola crop with GM seeds and for failing to prevent the spread of its seeds. According to Schmeiser, Monsanto's Roundup Ready canola seeds may have been blown onto his property from an adjacent farm or from trucks transporting the seed. The case was taken to the Supreme Court of Canada, which ruled in Monsanto's favor. Schmeiser was found to have violated Monsanto's patent by growing and saving patented seeds; however, he was not obligated to pay for damages to Monsanto because Schmeiser's profits did not differ from previous years. Monsanto has been involved with other legal disputes, but the Schmeiser case is notorious. It is also a significant event in Monsanto's complicated history and involvement in genetic engineering and chemical manufacturing.

See Also: Bt; Food Safety; Genetically Modified Organisms; Recombinant Bovine Growth Hormone; Roundup Ready Crops.

Further Readings

Brook, Mary M. "Monsanto." In *Encyclopedia of Environment and Society,* edited by Paul Robbins. Thousand Oaks, CA: Sage, 2007.

Monsanto Corporation. "Company History." http://www.monsanto.com/who_we_are/history .asp (Accessed December 2008).

Pringle, Peter. *Food, Inc. Mendel to Monsanto—The Promises and Perils of the Biotech Harvest.* New York: Simon & Schuster, 2003.

Shiva, Vandana. *Stolen Harvest.* Cambridge, MA: South End Press, 2000.

Tokar, Brian. "Monsanto: A Checkered History." *The Ecologist,* 28/5 (1998).

Karen Okamoto
John Jay College of Criminal Justice

Nanotechnology and Food

The use of nanotechnology in food is among the fastest growing in consumer markets. Nanotechnology includes a wide range of materials, systems, and processes that are used at the scale of 100 nanometers or less. Nanotechnology is increasingly being used in agricultural production, food manufacturing, food packaging, and foods themselves. Nanomaterials can help make agricultural production more efficient, enhance foods with nutrients and dietary additives, extend the longevity of packaged foods, and protect consumers from spoiled foods. Despite these benefits, the use of nanotechnology in food may have highly uncertain yet serious toxicological and ecological risks.

Industry is quickly moving to introduce nanomaterials in the food supply chain. Most large food companies, including Nestle, Altria, Kraft, Heinz, and Unilever, are researching nanotechnology. However, relatively little information is currently available regarding products on the market, and companies tend not to label or advertise products as containing nanomaterial. Analysts estimate that, for example, over 400 nanotechnology packaging applications may already be in use.

Specific applications include packaging that maintains the freshness and longevity of foods. Food packaging is the largest single application. "Smart packages" are being developed to extend shelf life by detecting spoilage and bacteria contamination, monitoring internal temperature and humidity, warning consumers through indicators, repairing package leaks, and releasing antimicrobial substances, colors, and flavors in response to triggers. Nanomaterials can also strengthen package barriers to gas, moisture, and ultraviolet light. DuPont is marketing a new nanoscale titanium dioxide light stabilizer. Clay nanoparticles also are used in plastic beer bottles to reduce breakage.

In turn, functional foods (or foods designed to provide health benefits to consumers) are becoming more important. Manufacturers are exploring the use of nanosized capsules to deliver active ingredients such as vitamins, omega-3 fish oil, minerals, and carotenoids. Such nanocapsules are said to increase bioavailability, solubility, and potency. The German company Aquanova has introduced a NoveSol capsule system. Other companies, such as Kraft, are interested in developing foods that can change their nutritional properties to reflect individual needs.

Industrial food production tends to process foods in ways that reduce their nutrition values. Nanotechnology may aid food processing by enhancing flavors, textures, and

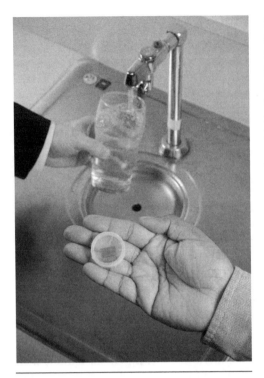

Although controversial, nanotechnology has the potential to improve detection of contaminated food and water. This chip uses nanotechnology-based biosensors developed by NASA to detect microorganisms that cause waterborne illness and infectious diseases.

Source: NASA Ames Research Center/Dominic Hart

nutrition. Unilever is developing a nano-emulsion ice cream with lower fat content that still tastes "fatty." Nanoparticles also may be used to selectively remove pathogens and chemicals from foods, and nanocapsules may be used to replace cholesterol in meat with plant steroids. Nanotechnology may also help reduce processing costs.

Nanotechnology is also being introduced to agricultural production. To date, most such applications center on the use of nanocapsules to deliver pesticides, herbicides, and plant growth regulators. Syngenta, for example, markets a PrimoMAXX plant growth regulator. Capsules can be engineered to control delivery rate and respond to triggers such as moisture in soil or contact with plant surfaces. Industry asserts that nanocapsules will strengthen active ingredients, reduce worker exposures, minimize harm to biodiversity, and lead to safer pesticide handling practices.

In time, nanotechnology may be used to automate farm management through remote sensing and integrated computer networks. Already, trials with wireless sensors in the form of "smart dust" are occurring on farms in the United States. As a result, inputs may be better matched to dynamically changing environmental conditions, reducing farming costs and increasing crop yields. Radio frequency identification chips are increasingly being used to track livestock.

However, government regulators and nongovernmental organizations are increasingly concerned about the use of nanomaterials in food packaging without adequate risk assessment. Two major reasons underlie this trend. First, there is emerging toxicological evidence that nanomaterials may pose health risks to humans and wildlife: Nanomaterials may be able to enter and circulate in human bodies more readily than larger-sized particles; they often have greater toxicity compared to their cousins, as seen in experiments with human tissues, cell cultures, and fishes; and the mechanisms of this toxicity remain poorly understood. For example, carbon nanotubes are being linked to inflammation and DNA damage in fishes and are being equated to asbestos fibers. However, research suggests that potential toxicity will depend greatly on the specific properties of a given nanomaterial, as well as its interactions with other materials. The charge, structure, reactivity, and agglomeration status may matter. Size alone is not sufficient.

Second, there is significant concern among many scientists, regulators, and nongovernmental organizations about the effects of chemicals used in food packages. Packaging material is in direct contact with food that humans ingest, thus implicating the most probable pathway of exposure. Some scientific evidence exists regarding the leaching out of packaging materials under conditions of everyday use and consumption. Indeed, regulators often assume that there will be some leaching out, and they set standards on the basis of whether the exposure exceeds a threshold of acceptable risk. The case of bisphenol-A (BPA) exemplifies the debate occurring around food packages. BPA is used widely in metal cans and plastic bottles and is strongly suspected to have endocrine-disrupting effects. As a result of consumer pressures and legal liability, Nalgene, a water bottle manufacturer in the United States, eliminated the use of BPA in its products in 2008. Similar issues may arise for nanomaterials in food.

Food manufacturers are beginning to commercialize food packages that contain nanomaterials without informing consumers or providing labeling. As a result, consumers may be exposed to uncertain health risks. Nanoscale silver, increasingly used as an antimicrobial substance in packages, is being associated with health risks. In turn, carbon nanotubes are being developed to pump out oxygen and carbon dioxide from packages, even though sizable evidence already points to their toxicity risks. In the agricultural arena, the use of nanocapsules may make pesticides far more inhalable by humans and may harm bees by substituting pesticide capsules for pollen. Little research has investigated whether pesticides in nanocapsules may have different toxicity compared with their larger-scale cousins. Nanotechnology may also have distributional justice effects: one possible use is to replace cotton with nanofabrics, which would displace many poorer farmers in developing countries.

Many nongovernmental organizations call for a moratorium on the use of nanotechnology in food until an adequate regulatory system is in place and nanomaterials have been assessed for their ecological and human health risks. Social scientists have identified significant consumer concern about the use of nanotechnology in food. For instance, a Hart Research Associates survey of 1,000 American adults in 2007 found that just 7 percent were prepared to buy foods produced with nanotechnology; 62 percent needed more data about the health risks and benefits.

Regulatory jurisdictions have varied in their responses thus far, following a similar divide in the genetically modified food case. In the European Union, the European Commission in mid-2008 requested that the European Food Safety Authority study the potential health effects of nanomaterials. In contrast, the Food and Drug Administration in the United States appears to have little interest in regulating nanomaterials in food, even if it had the regulatory authority and resources; the Food and Drug Administration has no capacity to evaluate industry claims about nanomaterial benefits. In conclusion, the use of nanotechnology in food may bring both promise and peril. Policy innovation and risk assessment are urgently needed, along with scientific innovation.

See Also: Agribusiness; Food and Drug Administration; Functional Foods; Pesticide.

Further Readings

Friends of the Earth. "Out of the Laboratory and Onto Our Plates: Nanotechnology in Food & Agriculture" (2008). http://www.foe.org/camps/comm/nanotech (Accessed February 2009).
International Risk Governance Council. *Risk Governance of Nanotechnology Applications in Food and Cosmetics.* Geneva: International Risk Governance Council, 2008.

Alastair Iles
University of California, Berkeley

NATIONAL ORGANIC PROGRAM

The National Organic Program (NOP) is the federal program that regulates certification of organic agricultural products in the United States. The NOP has regulated the use of the term *organic* on the labels of food and other products since October 2002. To label food or other agricultural products as organic, farms and processors must meet the organic certification standards created by the NOP and be inspected by a certifying agency accredited by the NOP. The NOP is administered by the Agricultural Marketing Service of the U.S. Department of Agriculture (USDA).

The NOP has three main functions: to regulate labeling of organic products, to set and update the standards for certification, and to oversee the certification of farms and processors by independent agencies. The regulations are published in the Code of Federal Regulations title 7 part 205 and known as the Organic Rule. The Organic Rule has been controversial since its inception, and understanding the history of its formation is key to understanding what it means when a product is labeled "organic."

The National Organic Standards Board

The USDA, headed by the secretary of agriculture, makes policy decisions about the Organic Rule. They do so with recommendations from a 15-member advisory council called the National Organic Standards Board. The board includes four farmers, two organic product processors, one retailer, one scientist, three consumer advocates, three environmentalists, and one certifying agent. It convenes several committees to make recommendations on specific regulations, including the Materials Working Group, the Policy Committee, the Crops Committee, the Livestock Committee, and the Certification Accreditation and Compliance Committee.

The NOSB plays an advisory role in most aspects of the NOP's rule making, meaning the USDA may decide whether or not to follow the NOSB's recommendations, with one exception. Decisions affecting the National List of Allowed and Prohibited Substances are made by the NOSB and are not considered advisory. Rather, in deciding which substances are allowed and disallowed, the NOSB's recommendations must become the rule.

Labeling

The word organic in the context of agricultural products is given a legal definition under the NOP. If a product is labeled organic or bears the USDA organic label, it must be produced in accordance with the Organic Rule and certified by an NOP-accredited certifying agency. A packaged product may be labeled with the words "made with organic ingredients" or claim that it is made with a specific ingredient, as in wine labeled "made with organic grapes," if at least 70 percent of the ingredients are certified organic. There is an exemption to the certification requirement for very small scale farmers or processors that have less than $5,000 in annual sales. They can claim that a product is organic without being certified by an accredited agency, but they must produce the product in accordance with the organic standards.

Standards

The substance of the Organic Rule is the organic standards—a long and detailed specification of the practices and materials that can legally be used in producing organic products.

Although originally conceived of as standards for producing organic food, the standards have been expanded to apply to cosmetics, clothing, and other agricultural products.

The rule defines standards not just for producers (farmers and ranchers) but also for *handlers* (a term that includes processors, shippers, and retailers) of organic products. At this time, all producers and only those handlers who transform a product, such as a food processor, must be certified. Certification is optional for shippers and retailers.

At the heart of the organic standards is the National List of Allowed and Prohibited Substances, popularly known as simply "the National List." In general, synthetic chemicals are not allowed in organic production, and naturally occurring substances are. The National List is a register of exceptions to that requirement. It lists natural materials that are prohibited because of their toxicity, such as arsenic and tobacco dust, and synthetic materials that are allowed because they are of low toxicity and considered essentially important; for example, synthetic insect pheromones used as a pesticide or vitamin C used as a preservative.

In addition to designating prohibited substances, the standards also regulate practices. For example, organic farmers must rotate crops, follow detailed regulations on making compost and applying manure, and prevent erosion, and organic ranchers are required to give livestock access to pasture and shade. Farmland must be managed organically for three years before it can be certified organic.

Planning and record keeping is another requirement of organic certification. A certifier verifies that an organic farmer has a detailed plan for maintaining soil fertility, rotating crops, preventing erosion, and monitoring pest populations. Producers and handlers are required have a long-term plan detailing their practices, the substances they use, and their monitoring systems to make sure their plan is being followed.

Certification

Certification is the mechanism for verifying that products marketed as organic were in fact grown and handled in accordance with the national organic standards. There are currently 95 certifying agencies accredited by the NOP. There is a great deal of diversity in the certifying institutions: 55 of them are domestic, and 40 of them certify organic products abroad for importation to the United States. They include nongovernmental organizations, private companies, and state agencies. Some have been certifying organic food since long before the federal standards were written, and others were created subsequent to the federal regulations. All are accredited and regularly audited by the NOP, which can order certifying agencies to change their policies and in some cases has revoked their accreditation.

The Organic Rule is relatively brief and is made up of deliberately general standards that can be adapted to the wide variety of business models, landscapes, and climates in which organic products are grown and processed. As a result, certifying agencies have a great deal of interpretive leeway when it comes to applying the rule, and not all agencies have the same interpretation. For example, the Organic Rule specifies that organic seeds should be used except "when an equivalent organically produced variety is not commercially available." Some certifiers interpret "not commercially available" to mean that farmers can use nonorganic seeds only when organic seeds are not available at all, whereas others interpret it to mean they can use nonorganic seeds if the organic ones are far more expensive. In some cases the varying interpretations come from differences in philosophy and tradition older than the NOP, because some of the certifiers have been certifying organic farms since long before a federal organic rule was proposed.

These differences in philosophy concerning certification have led some certifiers to create associations of certifiers that bring together several different certifying agencies for decision-making processes. The Accredited Certifiers Association, created in 2004, focuses on facilitating communication and standardizing practices between certifiers, whereas the Organic Materials Review Institute, created in 1997, researches agricultural input products as they come on to the market and decides whether they qualify as allowable organic inputs.

History

When the NOP wrote the legal definition of "organic" in the Organic Rule, it was superimposed on an existing organic agriculture movement with a long history and a diversity of ideas about what "organic" meant. This made for a controversial introduction of the national rule, and the controversy over what is and is not organic continues today.

The Organic Farming Movement and the Origins of Organic Certification

J. I. Rodale propelled the term *organic* into popular use in 1942, when he first published *Organic Farming and Gardening* magazine, which quickly gained a large readership. Rodale's definition included a sharp criticism of the industrialization of agriculture taking place at the time, especially the use of synthetic chemicals, and posited organic farming as an alternative. He drew ideas from English organic movement pioneers Sir Albert Howard and Lady Eve Balfour. Rodale, Albert, and Balfour all promoted nurturing soil with compost and manure and emphasized the connection between health of the agricultural ecosystem and health of humans.

Independent organizations began to offer certification of organic farms to differentiate their produce in the market in the early 1970s. One of the earliest and most influential organizations, California Certified Organic Farmers, was founded in 1973 and began certifying farms as organic. The state of California took notice and passed the California Organic Foods Act in 1979, giving the organic label a legal definition for the first time and allowing organizations to sue producers who falsely claimed that a product was organic. California Certified Organic Farmers brought the first major suit against Pacific Organics, a company that had been selling carrots grown with synthetic chemicals as organic. This and other misuses of the term *organic* encouraged many to call for a clearer law with a better mechanism for enforcement.

The Organic Food Production Act of 1990

Congress included the Organic Food Protection Act of 1990 as title XXI of the 1990 farm bill. The law gives three reasons for its existence: (1) to establish national standards governing the marketing of certain agricultural products as organically produced products, (2) to assure consumers that organically produced products meet a consistent standard, and (3) to facilitate interstate commerce in fresh and processed food that is organically produced.

The act mandated that the USDA give organic agriculture a legal definition, including a national list of allowed and disallowed substances, and implement a certification system to ensure consumers that products sold as organic were in fact produced according to federal standards. For the first time, the term *organic* was to have a single definition that carried the weight of federal law.

The Proposed Rule

In December 1997, the USDA staff published a draft rule for organic certification, and in accordance with standard procedure, it opened a 90-day public comment period. The public response was enormous and overwhelmingly negative, prompting the USDA to extend the public comment period and hold a series of hearings around the country. By April 1998, they had received nearly 300,000 comments, more than any other policy in history.

The huge volume of negative comments could be categorized into two kinds of objections. One had to do with the materials allowed, as the standards were seen as far too permissive. Particularly offensive to many commentators was the inclusion of provisions allowing genetically engineered plants, irradiation, and fertilizers derived from sewage sludge to be used in the production of organic products. The Organic Consumer Association and other groups representing consumers and organic farmers pointed out that those three technologies were highly controversial even in conventional agriculture and had never been a part of anyone's definition of organic agriculture before. These offensive technologies were dubbed "the Big Three" during the campaign for a more stringent rule.

The other kind of objection was more general and argued that the proposed rule was hostile to the principles and philosophy of the organic movement. Those objectors argued that the term *organic*, as defined by the movement of farmers and consumers that created it, was synonymous with small-scale family farms and a philosophical commitment to caring for the land. The proposed rule, they argued, would loosen the standards and allow big corporate growers to enter the certified organic market, confusing consumers and diluting the differentiated market that organic movement farmers had worked hard to create.

Why the USDA drafted such a contentious organic rule was the subject of intense debate. The proposed rule was in fact very different both from the independent organic standards existing at the time and from the USDA-appointed NOSB's recommendations. It is evident from the public comments and transcripts of the public hearings that many in the organic movement felt that it was an attempt by the USDA to co-opt the organic market. This view fit with the fact that the market for certified organic products was growing quickly, representing a lucrative opportunity for large-scale corporate food companies. Another theory is that the USDA, like any regulatory agency, was used to working with industries that want as little regulation as possible. In that view, it was a surprise for the USDA staff that organic farmers would demand more regulation.

The Final Rule

On March 7, 2000, the USDA published a revised rule, largely taking into account the specific concerns about substances and ignoring the more general concerns about scale and philosophy. The "Big Three" were disallowed, and several other regulations were tightened. The revised rule went into effect on October 21, 2002.

Present and Future Challenges

The rule continues to be updated as new questions arise and the scope of products that can be certified as organic expands. Recent contentious decisions include how to treat nonagricultural ingredients in certified organic processed products and exactly how much access to pasture livestock must have on certified organic farms.

The market for organic food in the United States, according to the most recent study by the Organic Trade Association, was $20 billion dollars in 2007 and was growing by an impressive 20.9 percent annually. It had grown from $1 billion dollars in 1990, the year Congress passed the National Organic Food Protection Act that created the NOP. By that measure, the NOP has been a resounding success.

The rapid growth of the organic sector has caught the attention of the largest agribusiness corporations, many of which now sell certified organic products. Many of these products are now grown on a large scale using intensive mechanization, subsidized irrigation water, and plenty of naturally derived, low-toxicity fertilizers and pesticides. Although these inputs are permissible under the Organic Rule, organic movement pioneers like Howard, Balfour, and Rodale would probably not recognize this as their vision of organic agriculture.

The most contested regulations in the NOP, including access to pasture, use of organic seeds, and regulation of nonagricultural substances, are controversial because they are difficult to implement on a large scale. They are part of a larger conflict between a growing industry and a movement for a less-toxic, smaller-scale food system. NOP decisions matter a great deal to the growing number of people involved in the organic market, including farmers, processors, retailers, and consumers. The NOP must balance the needs of a rapidly growing industry with the ideals and values that differentiate organic products from the conventional market.

See Also: Agribusiness; Beyond Organic; Certified Organic; Eco-Labeling; Farm Bill; Organic Farming.

Further Readings

Heckman, J. "A History of Organic Farming: Transitions From Sir Albert Howard's War in the Soil to USDA National Organic Program." *Renewable Agriculture and Food Systems,* 21/3 (2006).

Organic Food Protection Act of 1990, 7 USC. §6501-22 (2004). http://www.sarep.ucdavis.edu/Organic/complianceguide/national6.pdf (Accessed January 2009).

Shulman, S. W. "An Experiment in Digital Government at the United States National Organic Program." *Agriculture and Human Values,* 20/3 (2003).

U.S. Department of Agriculture. "National Organic Program." http://www.ams.usda.gov/NOP (Accessed January 2009).

U.S. Department of Agriculture, Agricultural Marketing Service. "National Organic Program" 7 C.F.R. part 205 (2008). http://www.access.gpo.gov/nara/cfr/waisidx_08/7cfr205_08.html (Accessed January 2009).

Vos, T. "Visions of the Middle Landscape: Organic Farming and the Politics of Nature." *Agriculture and Human Values,* 17/3 (2000).

Devon Sampson
Kathleen Elizabeth Hilimire
University of California, Santa Cruz

NITROGEN FIXATION

The process of nitrogen fixation involves taking nitrogen in its molecular form (N_2) and converting it into nitrogen compounds that help produce proteins. It was originally wholly a natural process, but chemicals are now used to engineer nitrogen fixation artificially. It is therefore the use of those chemicals on plants that differentiates between vegetables that are "green" and those that are not.

This process was discovered by the French agricultural chemist Jean Baptiste Joseph Dieudonné Boussingault (1802–87), who was born in Paris but spent the period until 1832 serving in the South American wars of independence, and then remained in the region, where he became increasingly interested in science and agriculture. As professor of chemistry at Lyons and then professor of agriculture at the Conservatoire des Arts et Métiers in Paris from 1839 until his death, Boussingault was able to show that the growing of legumes was made easier by fixing the level of atmospheric nitrogen. This was a major advance in microbiology that led to his being able to demonstrate that green plants absorb carbon dioxide. However, it was his experiment on nitrogen that encouraged the growth of leaves and stems, which is especially important in vegetables that are grown for their leaves, such as lettuce, cabbage, and Brussels sprouts. It also was shown by experimentation that a lack of nitrogen caused the plants to have stunted growth. Later, Martinus Beijerinck

At this farm in Hardin County, Iowa, in 1999, nitrogen was applied to cornfields in small doses during the season, rather than in one large dose, to avoid having excesses that might contaminate water supplies.

Source: U.S. Department of Agriculture, Natural Resources Conservation Service/Lynn Betts

(1851–1931), a Dutch microbiologist who established the Delft School of Microbiology, was able to add to Boussingault's discovery. The German chemist Justus von Liebig (1803–83) had already discovered that ammonium was beneficial to plants, and Beijerinck expanded this knowledge further by uncovering the fact that bacteria in the root nodules of certain leguminous plants perform nitrogen fixation naturally, which helps not only the growth of legumes but also the fertility of the soil over many years. His task, and that of later scientists, therefore, was the creation of artificial methods of increasing nitrogen fixation, which was expected to increase crop yields.

Obviously, as a natural process, nitrogen fixation in itself has been important in the growing of vegetables, but it was not long before chemical nitrogen fixation agents were used to help increase crop yields (and also occasionally in explosives through "fertilizer bombs"). This practice increased rapidly throughout the 20th century, and the largest source of fixed nitrogen in the Earth's ecosystem is now from the production of artificial fertilizers. This alone has major implications, because of the energy used to produce the chemicals, and for the entire ecosystems where it is used; this is why the debate dealing with "green food" is so important.

The making of artificial fertilizers to help with nitrogen fixation requires very high pressure, at about 200 atm, and also a temperature of at least 400 degrees C (752 degrees F) to establish what is known as the Haber Process (or the Haber-Bosch Process), which was invented by the German scientist and 1918 Nobel laureate Fritz Haber. The vast use of energy used in the manufacture of such fertilizers on such a wide scale has obvious implications for global warming, but the major concern is over the use of the fertilizer, especially its frequent overuse.

With nitrogen fixation agents, the problem is that they are intended to speed up the development of plants, and the use of synthetic nitrogen can cause problems. Indeed, the very large amounts of nitrogen fixation chemicals that are used have resulted in trace elements of some of the chemicals remaining in the crops, which are then ingested by humans, or in fodder fed to animals and then (through the animals) ingested by humans. Over time, these chemical traces can result in problems that had not been foreseen and that already are being held responsible for some allergies. Although organic fertilizer can be used to help grow crops, and indeed—as in the ash in swidden cultivation—has been used for several thousand years, chemicals have not been used for long, in evolutionary terms, and heavy use of chemicals has been practiced for an even shorter period of time. It is regularly mentioned that the pesticide agent DDT (dichlorodiphenyltrichloroethane) was used extensively and for long periods of time before Rachel Carson was able to demonstrate its toxicity in *Silent Spring* (1962), and even then she had detractors who criticized her work for years.

Another major concern about artificial nitrogen fixation involves the effect of the chemicals on the environment, not just on the crops in question. In 1965, following the discovery of the first dinitrogen complex, it was learned that there was a problem, in that the dinitrogen ligand could bond to a single metal or bridge two or more metals. This has the effect that concentrations of harmful chemicals can build up in the environment and leach through to the water table, which in turn can have major implications for flora, fauna, and humans in the region. This damage to the environment, especially when the synthetic nitrogen fixation agents have been used widely and over a long period of time, is certainly the reason why organic food growers have made a point of relying on organic fertilizers rather than chemical agents. They argue that the long-term potential damage to the environment is sufficiently important for a limit to be placed on synthetic nitrogen fixation agents.

See Also: Fertilizer; Legume Crops; Low-Input Agriculture.

Further Readings

Davidson, Eric A. and Sybil Seitzinger. "The Enigma of Progress in Denitrification Research." *Ecological Applications*, 16/6:2057–63 (December 2006).

Gaskell, George and Martin W. Bauer, eds. *Genomics and Society: Legal, Ethical and Social Dimensions*. Sterling, VA: Earthscan, 2006.

Giller, Ken E. and Kate J. Wilson. *Nitrogen Fixation in Tropical Cropping Systems*. Wallingford, UK: CABI, 1991.

Lerner, K. L. and B. M. Lerner. *Martinus Willem Beijerinck From World of Microbiology and Immunology*. Florence, KY: Thomas Gage, 2002.

Mosier, Arvin R., et al. "Policy Implications of Human-Accelerated Nitrogen Cycling,"
  Biogeochemistry, 57 (April 2002).
Postgate, John. *Nitrogen Fixation*. Cambridge: Cambridge University Press, 1998.
Unkovich, Murray, et al. *Measuring Plant-Associated Nitrogen Fixation in Agricultural
  Systems*. Canberra: ACIAR, 2008.
Werner, Dietrich and William E. Newton, eds. *Nitrogen Fixation in Agriculture, Forestry,
  Ecology and the Environment*. Dordrecht, Netherlands: Springer, 2005.

Justin Corfield
Geelong Grammar School, Australia

NORTH AMERICAN FREE TRADE AGREEMENT

NAFTA stands for the North American Free Trade Agreement—among Canada, the United States, and Mexico—that formed the largest regional free trade economy. The agreement represented the continuation of a trend established by the 1989 U.S.–Canada Free Trade Agreement. In proposing the expansion and its own inclusion in the regional economy, Mexico was able to rely on the precedent set by the U.S.–Canada Free Trade Agreement in designing and implementing extensive internal reforms. For the United States and Canada, the initial difference in NAFTA was the inclusion of Mexico.

The development of the NAFTA zone was in some respects a logical development of historical patterns of trade between the United States and its immediate neighbors to the north and the south. Canada and Mexico have traditionally been the first and second major suppliers of food to the United States, respectively. Although the trade between Canada and Mexico has been relatively smaller than each one's trade with the United States, the agreement advanced the harmonization of the markets of the three major trading partners. A further advance was in the formulation of the side agreements for labor and the environment. In NAFTA, food is considered in the context of trade and is then a commodity involving goods and services that are addressed through the regimes designed for foreign investment, labor, environment, and agricultural production systems.

The conclusion of the agreement and its implementation in 1993 led to broad consequences for the participating political systems and markets. Perhaps more so for Mexico than for the United States and Canada, governments at all levels in the region were required to reconfigure to equalize market access to eligible participants. In addition to reforming its investment laws, Mexico also changed its land ownership laws. The combination played a significant role in modifying the food industry in the region.

As sometimes occurs in planning for the future, some much-anticipated changes to the international flow of labor did not materialize in the manner it was expected. Although access to foreign direct investment and mobility rights were granted to professionals designated in NAFTA, the food industry was restructured more through investment than through changes in food sales. Sectors in which unrestricted access to local markets was established include food and beverages.

The influx of foreign investment in Mexico was facilitated by legal guarantees to the right to repatriate investments and profits, equal treatment to domestic and foreign investors, and indefinite continuity to these guarantees. A concurrent depreciation in Mexican currency and a strengthened dollar lowered costs in Mexican operations, amplified the value of the investments, and facilitated the transfer of food processing contracts to Mexico. The improved balance of trade in food sales was, however, insufficient to change Mexico's position as a net importer of foods because of other factors.

As a result of the foreign direct investment, changes to the trade in food were felt more extensively in Mexico, which experienced dramatic restructuring in its retail food distribution systems. The introduction of U.S.- and Canadian-style supermarkets and hypermarkets, along with lack of preparation by small local retail outlets in the larger urban areas, made it difficult for the smaller outlets to remain profitable.

Agriculture as part of the trade in food was also affected by the change in investment levels, along with reformed land ownership rules and the introduction of industrial style of farming. As is the practice in the United States and Canada, capital-intensive agricultural production includes equipment, management style, and technologically enhanced inputs such as fertilizers and genetically modified seeds. These inputs were used to replace traditional farming methods and seeds in larger fields. In Mexico, the aggravating factor for the displaced farmworker was the lack of some of the support systems available in the United States and Canada.

Food, of course, necessarily involves its consumption. The trade in food thus changed practices of food consumption, in that one part of the region generates demand, which directly affects the production practices and related working conditions, and hence food consumption, for the workers and their dependents at the site of food production. Those who participate in the food production, trade, and consumption system then are included in inherently homogenizing processes. For those who are not included in the system, the consequences are dictated by local support systems. By extension, the regime for labor and the trend toward capital-intensive flexible production combine to raise the significance of local support systems for those who are underemployed or unemployed. The overall effect is to increase the advantages of the lifestyle and incomes of those who invest capital over those of the producers of food and the losers in the emergent trading system.

The effects of NAFTA vary considerably by sector. For instance, it was determined that NAFTA had no apparent effect on the fishery sector in the short and medium term. In the longer term, the ability to adapt to changing environmental and consumer demands remains relevant to continuity. The dairy sector too was not susceptible, but for a different reason. As one of the most highly subsidized sectors, the dairy trade was addressed in a trilateral agreement that varied the tariffs by replacing them with transition provisions for trade-related quotas and duties for exceeding the quotas. The agreements also address access to supplies from non-NAFTA countries. Items subject to transition provisions include horticultural products (such as tomatoes), grains and oilseeds, livestock and meat, peanuts, sugar, cotton, and orange juice.

Over the long term, effects of NAFTA on U.S.–Mexico agricultural trade are more complex due to the multifaceted nature of the sector that includes the parties' relative underlying comparative advantage, economic growth, foreign investment, farm size and structure, labor markets and costs, and availability of new production inputs. The net effect is likely to homogenize the societies, with capital investment being the predominant influence in directing change.

In particular, NAFTA impacted the U.S.-Mexico agricultural trade in corn, fruits, and truck-crop vegetables. Following a 14-year transition period, U.S. corn exports to Mexico began unrestricted trading starting January 1, 2008. The yellow corn variety comprises the majority of U.S. exports, used primarily for animal feed to support the burgeoning Mexican poultry and hog industries. White corn, the variety used in making for tortillas and other foods for direct human consumption, also was increasingly exported into Mexico after installation of NAFTA but has seen a decline in recent years as new Mexican governmental policies favor domestically grown white corn. On the other hand, Mexican fruits and truck-crop vegetable (e.g., tomatoes) imports into the United States grew substantially. NAFTA encouraged increased agricultural trade by diminishing transaction costs and utilizing the comparative advantage of each country, thus increasing production efficiency and the cost-benefit ratio.

In the context of economic growth and a consumer economy, a redistribution of locations of productions and consumers may have been inevitable. With the region extending to the continent, the scale of redistribution to locations with lower costs of production, including employment costs and larger sales demand, became correspondingly greater.

The expectation that relocating production to Mexico would improve local circumstances has not been observed in reality. There has been little, if any, abatement in Mexicans wishing to relocate to places north of the Mexican border, and indeed the very persistence of borders would appear to some to be indicative of the reality.

Because air, water, and soil are all relevant to the production of food, changing environmental contexts have an immediate effect on the food industry. Food classified as "goods" and traded from Mexico to Canada has extended the role of the transportation sector and reconfigured the carbon footprint and cost structure of the industry. From a labor and environmental perspective, it is at least arguable that the trade in food under the NAFTA regime may not have been an unqualified success.

The linkage between trade and the environment in the NAFTA side agreement on the environment represents a historical development. Previously, the United States and Canada had declined to link international trade and environment in agreements, including the 1988 Canada–U.S. Free Trade Agreement. Events such as the 1991 U.S.–Mexico tuna fishing case at the General Agreement on Tariffs and Trade (GATT) panel illustrate the issue of not linking trade and environment. The case was brought by Mexico to the panel when the United States banned the import of tuna caught by Mexican fishing vessels using means prohibited to U.S. vessels. The regulations applicable to the U.S. vessels were established out of concern for the excessive dolphin deaths that were caused by the prohibited means of fishing. From an exclusively trade-based perspective, the GATT panel ruled that the U.S. prohibition was a trade barrier contrary to the GATT. The parties subsequently came to an understanding and resolved the matter.

In the case of food, trade liberalization reconfigured the industry for efficiencies and profitability. However, where the business model used in the United States and Canada replaced local traditional food trade systems, the higher concentrations and greater scale of production amplified negative labor and environmental impacts. In addition, shipping food from Mexico to Canada or vice versa raised the role of the transportation system and its carbon footprint. Finally, the uneven enforcement of regional rules further contributes to uneven environmental impacts. The continuing question is whether the benefits flowing from exposure to U.S. and Canadian agricultural and food trade investments have benefited all, or only a few.

See Also: Agribusiness; Land Reform.

Further Readings

Beltrán, L. F., et al. "Environmental, Economic and Social Effects Caused by NAFTA in the Fishery Food Companies of Baja California Sur, Mexico." *Sustainable Development*, 10/4 (2002).

Brandt, Deborah, ed. *Women Working the NAFTA Food Chain: Women, Food and Globalization*. Toronto: Second Story, 1999.

Chavez, M. "The Transformation of Mexican Retailing With NAFTA." *Development Policy Review*, 20/4 (2002).

"Commentary on Commodity Policy in an Era of Globalization: The Mexican Sugar Industry and Its Problems Under NAFTA." *Policy Studies Journal*, 31/1 (2003).

Goiling, Christine and Javier Calderon. "Foreign Direct Investment and Agricultural Trade." *Choices: The Magazine of Food, Farm & Resource Issues*, 14/3 (1999).

Scarpaci, Joseph L. and James J. Biles. "Globalization and Latin American Geography: Linking Scales of Analysis." *Journal of Latin American Geography*, 6/2 (2007).

Williams, Gary W. and C. Parr Rosson III. "Agriculture and the North American Free Trade Agreement." *Choices: The Magazine of Food, Farm & Resource Issues*, 7/4 (1992).

"World Watch Updates." *World Watch*, 18/1 (2005).

Zahniser, Steven and William Coyle. "U.S.-Mexico Corn Trade During the NAFTA Era: New Twists to and Old Story." U.S. Department of Agriculture, FDS-04D-01 May, 2004.

Lester de Souza
Independent Scholar

NORTHEAST ORGANIC FARMING ASSOCIATION

The Northeast Organic Farming Association (NOFA) is one of the oldest organic farming groups in the country. Started in 1972, its membership consists of farmers, gardeners and consumers concerned with food quality and sustainable agriculture, as well as health, energy use, and environmental stewardship. Its members tend to be small-scale, locally based, sustainable and certified organic farmers who market directly to consumers through farmers markets, community-supported agriculture (CSA) programs, and roadside stands, as well as small-scale wholesale markets, such as local restaurants and farm-to-school programs.

The NOFA Interstate Council is an affiliation of seven state chapters: Connecticut, Massachusetts, New Hampshire, New Jersey, New York, Rhode Island, and Vermont. The Council provides coordination between the state chapters, conducts an annual NOFA Summer Conference, and acts as an umbrella organization for collective projects. Each chapter offers educational conferences, fairs, workshops, apprenticeship programs, farm tours, loan programs, farm share programs (providing organic food to low income households), and publications to educate farmers, gardeners, consumers, and land care professionals within their state. Topics covered in their publications include farming practices, homesteading, livestock care, marketing, seed breeding, soil management, and beekeeping.

NOFA's Dairy and Livestock Technical Assistance Programs offer services and workshops designed to help organic dairy farmers improve their organic farm practices and increase their income. On-farm consultation services include business planning, energy auditing, and information about cost-cutting practices, including low-cost loan and cost-sharing programs. Advanced workshops are provided on quality, animal health and nutrition, grain and forage production, and farm budgeting, among other topics.

NOFA publishes a quarterly newsletter, *The Natural Farmer*, featuring articles on organic farming techniques, organic market conditions, certification issues, and environmental and safety developments that impact organic farmers, producers, and distributors. They offer a listing of CSAs that offer consumers prepaid subscriptions to a farm's produce for a season. Consumer "share holders" receive a weekly basket of items such as fruits, vegetables, herbs, flowers, cheese, and meat, depending on availability.

NOFA represents the organic small farm by visiting members of Congress in Washington, D.C., and lobbying for bills and resolutions that favor smaller organic farms. NOFA supports legislation that is scale-appropriate and risk-based—measures that mitigate product safety risk, or implement safety solutions that are based on actual risk assessments for the products and scale of farms, not on the industrial-scale food model.

Members of NOFA are typically organic farmers, so it is important to understand what constitutes an organic farm. In order to be certified organic, land must have no prohibited substances, such as synthetic fertilizers, pesticides, and genetically engineered organisms (GEOs), applied to it for a period of three years immediately preceding harvest of the organic crop. Operators must also continually prevent prohibited substances from having contact with production, handling operations, and products. The land must have distinct, defined boundaries and buffer zones to prevent unintended application of a prohibited substance to the crop, and it may not have contact with adjoining land that is not under organic management. Organic farmers may not feed mammalian or poultry slaughter by-products or manure to its mammals or poultry, including fats, animal proteins, or gelatin (made from hooves). Organic livestock slaughterhouses must also be certified. The regulations covering organic farming are far more stringent than those covering non-organic farming. Beyond the actual requirements, organic farm record keeping is extensive. Every certified organic operation must maintain records that verify its compliance with organic regulation and provide traceability. Organic farmers can turn to NOFA for technical assistance on how to comply with the additional requirements, and can find support from fellow NOFA members.

Good Agricultural Practices (GAPs) are a policy priority for NOFA. In the early 1990s, Cornell University's regional extension initiative GAPs project took a commonsense, nonregulatory, nonverification approach to farm food safety practices based on good sanitation practices and worker training. Its manual "Food Safety Begins on the Farm—a Grower's Guide" offers information on reducing risk in everything from worker hygiene and storage facility sanitation to manure management. In 1998, the USDA turned the Cornell recommendations into a regulatory checklist and certification program, officially titled "Food and Drug Administration's Guide to Minimize Microbial Food Safety Hazards for Fresh Fruits and Vegetables." GAPs are voluntary from a legal standpoint, but wholesalers and buyers are increasingly requiring GAP certification from their growers to protect themselves from legal liability, should contamination occur. Some of the GAP recommendations are impossible to comply with on small farms, such as keeping livestock two miles from the farmstead. NOFA and other agricultural groups have formed an east coast-based Leafy Greens Working Group. The group works in collaboration with food safety experts to develop food safety approaches utilizing GAP compliance, while incorporating provisions that are already part of the certified farmers' Organic Farm Plan.

See Also: California Certified Organic Farmers; Certified Organic; Community-Supported Agriculture; Factory Farm; Family Farm; Farmers Market; Food and Drug Administration; Genetically Modified Organisms; Integrated Pest Management; Organic Farming; Pesticide; Sustainable Agriculture.

Further Readings

Northeast Organic Farming Association. http://www.nofa.org (Accessed August 2009).

Rangarajan, A., et al. *Food Safety Begins on the Farm: A Grower's Guide.* Silver Spring, MD: U.S. Food and Drug Administration, 2000.

U.S. Food and Drug Administration. *FDA Issues Final Guidance for Safe Production of Fresh-Cut Fruits and Vegetables, 03.12.07.* http://www.fda.gov/bbs/topics/NEWS/new01584.htm. (Accessed August 2009).

U.S. Food and Drug Administration. *How the FDA Works to Keep Produce Safe.* http://www.fda.gov/ForConsumers/ConsumerUpdates/ucm094555.htm (Accessed August 2009).

Elizabeth L. Golden
University of Cincinnati

OGALLALA AQUIFER

The Ogallala aquifer is the largest underground source of water in North America. Also known as the High Plains aquifer, it underpins the economy of the Great Plains area and is believed to hold a "quadrillion gallons of water"—enough apparently to fill Lake Huron. The Great Plains occupy a sixth of the land mass of the United States and account for a fifth of its agricultural produce. The name Ogallala in the language of the Sioux means "spread throughout." The aquifer spreads through parts of eight states, stretching for more than 174,000 square miles, underlying South Dakota, Nebraska, Wyoming, Kansas, Colorado, Oklahoma, New Mexico, and Texas.

The Water

The aquifer, once believed to be an inexhaustible underground river, is in fact an ancient geological formation created around 6–10 million years ago as a result of erosion of the Rocky Mountains. Scientists have determined that most of the water currently in the aquifer is "paleo" or "fossil" water and was collected during the ice ages. Ogallala is classified as an "unconfined aquifer" and is made up of clay and gravel, which is thicker in the northern states of Nebraska and South Dakota, soaking up rainwater and melting snow very slowly through its surface. The water in the aquifer is replenished at less than one inch annually. Only 15 percent of the aquifer qualifies as being "saturated" and available for agriculture. At present, 16.6 million acre-feet of water are pumped or "mined" annually to satisfy the needs of farmers in the region.

The water level in the aquifer has dropped in some places by more than a hundred feet, and it is estimated that 9 percent of the water has been used in the last few decades. It is also believed that global warming will greatly reduce the amount of precipitation in the region and that water levels will fall faster in coming years. Another area of concern is pollution from pesticides and fertilizers that is slowly seeping into these once-pristine waters. A holistic approach to conserving our water resources is urgently needed. It is critical that the aquifer be managed equitably so as to keep the "bread basket" full for future generations.

The Food

It comes as a surprise to many when they learn that the "bread basket" or "green belt" in the Great Plains was once called the great American desert and deemed unsuitable for agriculture. Before the 1950s, it was very difficult to sustain agriculture here because of the long periods of drought the region endured. The plains lie in flat semiarid open country once dominated by tall prairie grasses teeming with buffalo, deer, and huge flocks of birds. The settlers in the 1860s followed a system of cultivation known as "sod busting," by which the drought-resistant native prairie grasses were decimated and replaced by more "useful" but water-hungry crops such as wheat and corn. This also led to the destruction of entire ecosystems on which Native Americans were dependent for their food supply.

When white settlers first cultivated the land, they were faced with long periods of drought, and agriculture was a very risky occupation.

In the late 19th century, wells were dug and water was pumped with the help of windmills and then later, in the 1900s, by steam-powered pumps, making agriculture somewhat more certain. Up until the 1930s, however, the region also known as the Dust Bowl was afflicted by the vagaries of the weather, most famously captured in John Steinbeck's novel *The Grapes of Wrath*. Thousands of people and their livestock starved when the earth dried up. It was only after World War II that farmers realized that they had a huge source of water right under their feet and that with the help of new machinery they could increase agricultural yields several hundred times. Farmers started mining for water and pumped as much as they wished, with almost no state or federal regulation. Central pivot sprinklers helped turn the great American desert into a polka-dotted landscape worth billions of dollars in farm revenues. To say that the entire nation now depends on the Ogallala aquifer for its food would not be an exaggeration. Not only do much of our corn, wheat, cotton, and soybeans come from the region, but the grain from the area also supports a huge meat and dairy industry. The federal government viewed the aquifer for many years as being central to the nation's food security.

The Future

The aquifer is spread over 184 counties in eight states and is therefore subject to a variety of federal, state, and county regulations. Seven states, however, contain only a third of the water among them, with the main reservoir sitting squarely beneath Nebraska. Although states like Nebraska and Kansas have been trying to conserve the aquifer by controlling the amounts of water pumped annually, Texas and New Mexico have until recently had almost no regulation. Texas even had a law that encouraged water capture to the extent that one could drain a neighbor's well without any legal repercussions. The commonly followed practice was "Use it or lose it." A Texas businessman, T. Boone Pickens, states on his company's website (http://www.mesawater.com) that the people living in the Texas Panhandle have received a God-given gift of water too large for them to ever be able to use and proposes the selling of this water to other parts of Texas. Farmers in the region are enraged by this scheme and believe that Pickens has no right to sell their water to the big cities. This, however, does raise the question of whether it is the farmers or businessmen such as Pickens who have a right to exploit a resource commonly held by millions of people for a short-term profit.

Environmentalists have pointed out that it is unnatural to have such a large amount of irrigated agriculture in such a dry region. The geography of the region has been ignored by farmers, and technology has made them care even less. Frank and Deborah Popper introduced the idea of a "Buffalo Commons" in the late 1980s. They believe that agriculture here is unsustainable and that human population will continue to fall in the region, and therefore it ought to be returned to its natural state, in which native grass prairies will once again allow buffalo and other native fauna and flora to grow freely. Opposition to the idea arose mainly from the fear of having to import food from other countries. This, however, has already happened, as a quick look at the produce in the grocery store will tell you. Instead of poisoning and depriving future generations of this large source of clean water, perhaps returning a large portion of the driest high plains to the buffalo is the only way to have green food in this artificial bread basket.

See Also: Agroecology; Corn; Farm Crisis; Irrigation; Rural Renaissance; Soil Erosion; Wheat.

Further Readings

Emel, J., and R. Roberts. "Institutional Form and Its Effects on Environmental Change: The Case of Groundwater in the Southern High Plains." *Annals of the Association of American Geographers*, 85/4:664-83 (1995).

Kromm, David E. "Ogallala Aquifer." *Water Resource Encyclopedia*. 2007. http://www .waterencyclopedia.com/Oc-Po/Ogallala-Aquifer.html (Accessed May 2009).

McGuire, V. L. "Changes in Water Levels and Storage in the High Plains Aquifer, Predevelopment to 2005." *Fact Sheet 2007-3029* United States Geological Survey. 2007. http://pubs.er.usgs.gov/usgspubs/fs/fs20073029 (Accessed May 2009).

Verchick, Robert R. M. "Dust Bowl Blues: Saving and Sharing the Ogallala Aquifer." *Journal of Environmental Law and Litigation,* 14/1 (1999).

Williams, Florence. "Plains Sense." *Mother Earth News* (January 15, 2001). http://www.hcn .org/issues/194/10194 (Accessed April 2009).

Ken Whalen
American University of Afghanistan

ORGANIC FARMING

Organic farming is a form of sustainable agriculture committed to growing food naturally with respect for ecological systems. Its methods promote diversity and emphasize closed nutrient cycles and healthy soil. Though organic farming is an agricultural science with specific international and domestic regulations, it also encompasses social values and economics. Organic farming is defined in contrast to conventional or high input agriculture, which depends on synthetic, often fossil-fuel-intensive, fertilizers, pesticides, fungicides, and other agrichemicals. Conventional agriculture is generally highly specialized, and monocultures, or farms only growing one crop, are common. Regulations surrounding organic agriculture vary from state to state and country to country, but organic farming generally does not allow genetically modified organisms, irradiation, or, in animal operations, the use of antibiotics or hormones, such as recombinant bovine growth hormone. Organic foods are defined by the U.S. Department of Agriculture (USDA) as food products

There are an estimated 30,000 organic farms in the United States. This organic farmer in San Juan Bautista, California, is consulting with a horticulturalist (right) about the condition of red chard planted among a variety of crops.

Source: U.S. Department of Agriculture Agricultural Research Service/Scott Bauer

produced without using most conventional pesticides, fertilizers made with synthetic ingredients or sewage sludge, bioengineering (genetic modification), or ionizing radiation.

History of Organic Farming

The terms *sustainable* and *organic* are relatively new, but the practices are not. In fact, organic farming in many respects represents a return to older, unconsciously agroecological land stewardship practices as a response to the introduction of agrochemicals in the early 20th century.

As plants began to be shipped around the world for cultivation and the global trade in food found its feet in the 19th century, diseases and foreign pests became serious problems for growers. Removed from their native habitats, natural checks and balances were dislodged, leaving plants more vulnerable to the ravages of nature's forces. Oranges, for example, are not native to California, and varietal importation has brought in many citrus diseases with no natural predators, such as the deadly cottony cushion scale. Chemical research from government agencies and university institutes offered a seemingly ideal solution to battling such pests: chemical warfare. With chemicals, diseases that lived in the soil could be eradicated, rather than forcing farmers to let the land lie fallow, or without crops, for years. At the same time, farmers, such as California's citrus growers and the cotton plantation owners of the south, began to focus on monoculture (single crop) production for the sake of efficiency. World competition was fierce, and a farmer could make more money by focusing on the tools, markets, and labor needed for only one crop. The result was more or less an ideal breeding ground for pests, with ample food sources and little, if any, predatory dangers. Farmers were more vulnerable than ever to diseases and pests and turned ever more readily to synthetic solutions. Today an estimated 5.6 billion pounds of pesticides are used every year worldwide. In turn, the large-scale use of pesticides, herbicides, and fungicides has created resistance to the chemicals, rendering them useless against certain genetic immunities. There has been a tenfold increase in pesticide use since 1945, and between 1946 and 1948, U.S. production of synthetic fertilizers grew from 800,000 tons to 17 million tons. However, during this same time, crop losses doubled. Internationally, over 250 varieties of plants were estimated to be herbicide-resistant in 2004, with the United States housing the majority of those plants.

Providing adequate nutrients for crops has been another challenge that conventional agriculture has met head-on with synthetic chemicals. With the advent of monoculture production, there were fewer nutrients cycling through the soil. Scientists such as German

Justus von Liebig argued that soil health was unimportant if artificial sources of nitrogen, potassium, and phosphorus—the trinity of macronutrients deemed most essential to plant health—were provided. However, it is now clear that sustained use of such inputs destroys the natural balance of the soil and results in a destructive cascade effect, negatively affecting the surrounding ecosystem health. For example, during the green revolution (which began in the 1940s but was first termed as such in the 1960s) agrochemicals and plants genetically designed to work with them were successfully promoted in combination, and the technological duo introduced to countries around the world with the hope of eradicating hunger once and for all. Although some crops, such as cereals, experienced increased production, it was realized that the input of energy into the development of the crops was greater than the energy outputs. In addition, the required use of chemicals negatively affected the soil and water quality, as well as worker health.

Organic farming represents an alternative to conventional agriculture for pest and disease control, fertilizer, and place-specific crops. The term *organic* has been used popularly since the early 1940s, a time period, as noted earlier, that was also witness to an exponential increase in the use of synthetic fertilizers. High fuel prices, a growing grassroots base of organic farming, and the publication of landmark books such as Rachel Carson's *Silent Spring*, which warned against the use of pesticides, eventually attracted attention to organic farming on a larger scale. The period from 1979 to 1990 may be described as the era of general recognition for organic farming at a national level in the United States, as well as around the world. In 1974, Oregon was the first state to define organic in a regulatory manner, but it was not until 1979 that the first state legislation was passed—the California Organic Food Act—after six years of activism by the young group California Certified Organic Farmers. Although it was a state-mandated program, there was no budgetary appropriation for enforcement. The Organic Foods Production Act of 1990 set guidelines for what would eventually become organic certification. The first publication of the USDA's Organic Rule was in 1998. In 2001, rules were written into law by the U.S. government for certified organic that regulate the food at all stages of production: farming, processing, packing, and retail. Certification also provides guidelines for transition times, offers a basic outline of allowable materials, and puts into place an enforcement and inspection protocol.

There have been many leaders of the organic movement. British agricultural scientist Sir Albert Howard is considered one of the fathers of the modern organic movement. His research in India as the director of the Institute of Plant Industry led him to stress the importance of compost to soil fertility as well as to regard farming as a systems science. Lady Eve Balfour, author of *The Living Soil* and the first president of the British Soil Association, an organization that continues to promote organic farming today, viewed soil as a living organism. This remains a fundamental platform for successful organic operations. In the United States, Jerome Rodale, originator of the still-published magazine *Organic Gardening* and author of *Pay Dirt*, which warned of the dangers of DDT (dichlorodiphenyltrichloroethane) before Carson's book, was an early organic success whose farm provided a model for others and whose publications served as the bible of organic farming. More recently, farmer and author Wendell Berry has written extensively about the cultural loss in the United States resulting from industrial farming in books such as *The Unsettling of America*.

Organic Farming Research

Although farmers have informally been practicing organic farming since the beginning of agriculture, official research into the economic, social, and environmental benefits of organic farming has been lacking. A University of Washington–sponsored report published in 1976 suggested that organic farming was economically feasible. A few years later, Dr. Youngberg published a report on organics titled "Time to Choose" under the auspices of the USDA. In 1980, after compiling 69 case studies, the USDA published a report titled "Report and Recommendations on Organic Farming" (available online through the National Agricultural Library), which recommended further government research on organic farming. For many years, there was no further action taken by the USDA. Almost 30 states had adopted definitions of organic nine years later when the National Research Council published a similar report titled "Alternative Agriculture." A follow-up by the National Research Council, "21st Century Systems Agriculture: An Update of the 1989 NRC Report 'Alternative Agriculture'" was released in 2009.

Though time and again research has suggested that there is much potential for highly productive and sustainable organic farming, there has been continued resistance to funding this research on any meaningful scale. Through the Sustainable Agriculture Research and Education program established in the 1980s, some funds have been made available for research, but the $40 million per year authorized for research has yet to be fully requested by the USDA. The Organic Food Production Act, in fact, intentionally avoided any mention of a research agenda for organics. Many people suggest that the reluctance to spend more resources on organic research is a result of the power of political lobbyists with economic interests in agrichemicals.

Principles of Organic Farming

Organic farms are extremely diverse, with different motivations. Farms include both small-scale operations marketing directly to consumers and large-scale operations targeting national and international markets. Although there exists a large degree of variation in the commitment to the principles of organic farming, some of the key principles include human health, environmental stewardship, animal welfare, consumer confidence, fairness, biodiversity, holism, sustainability, natural plant nutrition, natural pest management, and integrity. Generally speaking, organic farming takes as its tenets the on-site integration of farm inputs and the precautionary principle warning against new technologies unless there is assurance there will be no harmful consequences.

Together these principles attempt to achieve a farming system that emulates living ecological systems and cycles in which overall system health is emphasized, and the interaction of management practices is the primary concern. Maintaining biological and genetic diversity and replenishing soil fertility are two key goals. Organic farming attempts to produce high-quality, nutritious food that contributes to overall human health and well-being without the use of fertilizers, pesticides, animal drugs, and food additives that may have adverse health effects. Ideally, organic farming strives to adapt to local conditions, ecology, and culture with very few outside inputs. In addition, reuse, recycling, and efficient management of materials and energy contribute to the treatment of the farm as a living and dynamic system.

Other, more avant-garde principles that defined that early organic movement include equity, respect, justice, food sovereignty, and reduction of poverty. There are farmers, activists, and consumers who feel that not all of these principles are being met by today's large-scale organics movement.

Common Practices and Techniques

Soil diseases, weeds, and pests are the most notable challenges faced by farmers that, since the early to mid-20th century, have been tackled using synthetic chemicals. Organic farmers use other techniques to confront these problems, often mixing indigenous knowledge and the more modern science of agroecology. USDA regulations do allow certain botanical pesticides and fungicides to be used in organic farming, such as the microorganism *Bacillus thuringiensis* (Bt), copper, neem tree oil, pyrethrins, sulfur, and rotenone.

Continuously productive soil is a vital concern for organic farmers who do not use chemical fertilizers. Outside organic inputs, such as bloodmeal, bonemeal, guano, and kelp, are used as fertilizers to replace fossil fuel–based sources of nitrogen, potassium, and phosphorus. Compost, though, is a vital fertilizer that can be made on-site with no outside inputs and is thus essential to organic farming. Green manure or cover cropping also enhances soil fertility and is one of the most common organic practices. It involves planting leguminous plants, such as vetch and fava beans, to fix nitrogen from the air into the soil, where plants can use it. The crops are then turned under and the field planted with cash crops. Crop rotation is another technique considered a prerequisite for an efficient use of on-site resources. Crop rotation not only prevents quick soil depletion that can occur when growing the same crop year after year but also can aid in disease suppression and reduce the threat of pests as well.

Animal husbandry has historically been important to retaining on-site soil fertility. Livestock manure is a powerful fertilizer and speeds the composting process. If livestock are raised organically, they must reside in free-range, open-air systems and be provided with organic feed to meet regulations. Using rotational grazing and mixed-forage pastures for livestock operations and alternative healthcare for animal well-being are also common in organic farming.

Other holistic management practices such as planting beneficial habitat to attract "good" insects that then destroy the "bad" insects are key to successful organic farming. Such practices promote biological diversity on the farm and offer wildlife habitat. Choosing plant and animal species that are adapted to local conditions can reduce the threat of disease and insect invasion.

Society and Economy

In an attempt to lower input costs, decrease reliance on fossil fuels, and capture growing demand and premium market prices, more and more producers are turning to organic farming. In fact, it has become one of the fastest-growing segments of U.S. agriculture. Between 1997 and 2002, according to the USDA, the amount of U.S. certified organic cropland more than doubled. By 2005, all 50 U.S. states had some certified organic farmland. Today there are approximately 30,000 organic farms in the United States, which together represent about 2 percent of market share. U.S. organic sales have been growing at 20 percent or more per year for the past decade, currently exceeding $4 billion dollars a year, and the increases are expected to continue for at least another 10 years to meet consumer demand. In Europe, the annual growth in the market for organic products is 10–15 percent.

Organic produce is the "gateway" product and the number one organic seller. Other organic products, notably dairy and poultry, are rapidly expanding as well. The "Spring

2006 Organic Cotton Fiber Report" found a 55 percent annual growth rate in U.S. organic cotton sales from 2001 to 2005, with sales at $275 million.

Organic farming is part of a larger supply chain, which includes food processing, distribution and retail, and consumers. Restaurant Nora, the first organic restaurant in the United States, opened in 1979, and many other establishments, including Alice Waters' famous Chez Panisse in Berkeley, California, have followed suit with local, seasonal, and organic offerings.

Consumers choose to buy organic products for a variety of reasons, including health and nutrition, food safety and quality, and concern for the environment. Recent reports on climate change have accelerated interest in food that is not fossil fuel–intensive and have helped fuel the growing consumer demand for organic food. Though organic products generally cost more, many consumers are willing to pay the price. Of 3,500 people surveyed in Ohio, 39 percent said that they would pay 10 percent more for organically grown food.

Farmers markets have increased in popularity simultaneously with increased organic sales. A 2002 USDA survey of 210 farmers markets found that a third of the vendors were certified organic. In comparison, less than 1 percent of all farmers at the time in the United States were certified organic. A survey completed in 2002 by the Organic Farming Research Foundation discovered that 13 percent of organic produce is sold directly to the consumer (through community-supported agriculture or farmers markets). Despite the growing interest in direct sales, agriculture today is characterized by increasingly fewer but larger farms. Where 23 percent of the population farmed before World War II, only about 1 percent of the population does so today. The organics industry is not different, and organic production is increasingly consolidated.

Environmental Impacts

There is increasing evidence to suggest that conventional agricultural practices are environmentally unsustainable. By 1972, agricultural pollution, mainly from fertilizers, was found in 72 percent of waterways. Today, according to studies undertaken by the U.S. Environmental Protection Agency, agriculture is the largest nonpoint source of water pollution. The production of synthetic fertilizers for conventional agriculture uses egregious amounts of energy—it takes 2,200 pounds of coal to produce 5.5 pounds of usable nitrogen. Erosion has been a serious concern in the United States since the 1930s.

In response to these environmental concerns, the techniques adopted by organic farming mitigate many of the environmental disasters wrought by conventional agriculture. The elimination of toxic agrochemicals is, in and of itself, a huge environmental victory. The use of compost slows the intake of nitrogen and other nutrients, and therefore helps keep surface and groundwater clean, as well as maintains healthy microbial communities in soil. Other techniques practiced by organic farmers, such as lining waterways with grasses and contour farming, prevent unnecessary erosion. Finally, many organic techniques such as planting beneficial habitat and not using genetically modified seeds maintain biological and genetic diversity.

Health Impacts

The 1990 Organic Foods Production Act makes no health claims, in part to be politically palatable to representatives from heavily conventional agricultural states. Though there is little research into the actual health benefits of organic food, preliminary studies done in 1998 and 2003 show that children who eat organic food have lower to no pesticide metabolites (the bodily response to such chemical intake). Furthermore, links have been drawn between pesticide exposure and cancer.

Ultimately, it is difficult and expensive to test the benefits of organic food, as it requires a very controlled diet on the part of the subject and strictly followed growing techniques on the part of the farmers to scientifically link organic food to any health benefits that might be observed. Thus, it is much easier to argue that conventional farm practices are unhealthy than it is to argue that organic foods are healthier. For instance, methemoglobinemia, a disease caused by the ingestion of nitrogen compounds that reduces the ability of blood to hold oxygen, is common in the Midwest, where nitrogen compounds are heavily applied to agricultural land as fertilizer. In addition, an easily observed health benefit to organic farming is that farmworkers are not exposed to dangerous chemicals such as methyl bromide.

Controversies: Can Organic Farming Feed the World?

Although many people argue that conventional agriculture is unsustainable for the environment, human health, and rural economies, there are drawbacks to organic farming. Yields are consistently lower, and hand-weeding and other labor-intensive techniques are time consuming and expensive. Recent studies suggest that over the long term, organic techniques can result in higher overall yields, but there is not enough sustained research to conclusively argue this point. Earl Butz, former secretary of agriculture under Richard Nixon, famously said, "Before we go back to organic agriculture, somebody is going to have to decide what 50 million people we are going to let starve." Though not all research suggests this is true, as it stands, current conventional farming techniques produce more food with less labor.

In addition, there is wide debate over whether or not genetically modified foods harm the environment and human health or can provide a safe and effective way of combating disease, pests, and difficult climatic conditions such as drought. The debate is not simply two-sided. Some argue that genetic engineering is a safe and effective tool. Others adamantly disagree with introducing unnatural gene combinations into the environment with little research as to the possible effects on human health and the environment. Still others contend that genetic engineering can be a powerful aid in improving farming and feeding the world, but that current politics and the economics of the sector further the interests of only a select few. Critics of the genetic engineering market argue that current policies increase the economic dependency of farmers and perpetuate an agricultural system defined by monocultures and egregious use of synthetic fertilizers and pesticides.

See Also: Agroecology; Certified Organic; Low-Input Agriculture; National Organic Program; Sustainable Agriculture.

Further Readings

Conford, Philip. *The Origins of the Organic Movement*. Glasgow, Scotland: Floris Books, 2001.

Food and Agriculture Organization of the United Nations. "Organic Agriculture." http://www.fao.org/organicag/fram11-e.htm (Accessed January 2009).

Fromartz, Samuel. *Organics Inc., Natural Foods and How They Grew*. New York: Harcourt, 2006.

Gold, Mary V. "Organic Production/Organic Food: Information Access Tools." U.S. Department of Agriculture National Agricultural Library, June 2007. http://www.nal.usda .gov/afsic/pubs/ofp/ofp.shtml#orgs (Accessed January 2009).

Heckman, Joseph. "A History of Organic Farming—Transitions From Sir Albert Howard's War in the Soil to the USDA National Organic Program." *Renewable Agriculture and Food Systems*, 21 (2006).

National Sustainable Agriculture Information Service. http://attra.ncat.org (Accessed January 2009).

Organic Farming Research Foundation. http://ofrf.org/index.html (Accessed January 2009).

Stacy Vynne
Shannon Tyman
University of Oregon

ORGANOCHLORINES

Organochlorines are composed primarily of carbon, hydrogen, and chlorine. These chlorine-containing compounds are found in the environment as a result of human activities. Organochlorines such as chlorinated pesticides and polychlorinated biphenyls represent an important group of persistent organic pollutants that have caused worldwide concern as toxic environmental contaminants. The large-scale manufacture and distribution of organochlorines took place after the accidental discovery of dichlorodiphenyltrichloroethane (DDT) as an insecticide by Paul Hermann Müller in 1946. Uses of organochlorine pesticides (OCPs) take a wide range of forms, ranging from pellet application in field crops to sprays for seed coating and grain storage. Some organochlorines are applied to surfaces to kill insects.

The persistence of OCPs; their tendency to accumulate in soil, sediment, and biota; their harmful effects on wildlife; their widespread ecological damage; and global contamination and resistance by target insects resulted in their ban and restrictions in most countries, especially in the United States. Banned OCPs include DDT, aldrin, dieldrin, toxaphene, chlordane, and heptachlor. Despite the restrictions, these compounds are still detected in the environment and tissue samples. Biomonitoring studies continue to find them in food, blood, adipose tissue, and breast milk of humans. Body burdens have declined since these organochlorines were banned, yet virtually the entire population still carries detectable levels of the toxic chemicals. Chronic exposure to low levels of OCPs can cause a wide range of serious harmful effects in animals and humans. Those that are still being used include lindane, endosulfan, dicofol, methoxychlor, and pentachlorophenol. They are known animal carcinogens and potential human carcinogens. Examples of organochlorines are DDT, chlordane, lindane, aldrin, dieldrin, toxaphene, heptachlor, endosulfan, dicofol, methoxychlor, hexachloro benzene, mirex, pentachlorophenol, beta-hexachlorocyclo hexane, trans-nonachlor, heptachlor epoxide and pentachlorophenol, 2,3,5-trichloro phenol.

Organochlorines generally have low volatility, chemical stability, lipid solubility, and slow biotransformation and degradation. They are persistent and bio-concentrate and bio-magnify. They are hydrophobic compounds that tend to adsorb to suspended particulate matter and bottom sediments in aquatic ecosystems. These qualities are not environmentally desirable. The lipophilic nature, hydrophobicity, and low chemical and biological degradation rates of OCPs have led to their accumulation in biological tissues and the subsequent magnification of concentrations in organisms, progressing to the food chain. Polychlorinated biphenyls are very persistent in the environment and are among the industrial chemicals banned and included in the list of priority contaminants to be monitored regularly. They cause a variety of carcinogenic effects and neurological problems in organisms.

Soils can be contaminated in many ways by organochlorines. The most common ways include

- using more organochlorine pesticide than is recommended,
- spills during mixing and loading,
- tank overflows, and
- improper disposal of containers or surplus spray mixtures.

The capacity of the soil to filter, buffer, degrade, immobilize, and detoxify organochlorines is a function of the quality of the soil. Their presence and bio-availability in soil can adversely affect humans, animals, plants, and soil organisms. Spills on sand or sandy loam soils can lead to serious contamination of groundwater through leaching. Spills on clay soils remain on the soil surface longer and are more likely to spread to other areas because of surface runoff. Many pesticides degrade in the soil, but some persist for long periods of time. Organochlorines can enter surface and groundwater in several ways:

- natural processes including runoff, leaching, and erosion
- spray and vapor drift during application
- using more pesticide than is recommended
- container leaks and flooding while pesticides are in storage
- back siphoning of pesticides from the spray tank into wells during tank filling
- overflow of spray tanks during filling
- waste water from equipment cleanup
- improper disposal of excess spray mix, unwanted waste pesticides, and pesticide containers
- atmospheric fallout

Surface and groundwater can be contaminated if organochlorines are not handled and applied correctly. Once water is contaminated with pesticide, it may be toxic to fish and other aquatic life, wildlife, domestic animals, and humans. It may also cause damage to sensitive crops through irrigation or runoff. Many organochlorines that pollute the environment become incorporated into food webs and affect humans, who are final links in the food chain.

Organochlorine pesticides can pollute the air through spray and vapor drift during aerial spraying or when soil is eroded by the wind. Pesticides in the air may return to Earth via rain or snow, far away from where the pesticide was used. This is the reason why pesticide residues have been found in the snow and in the tissues of animals in the arctic and Antarctic.

The toxicity level of an organochlorine pesticide depends on the deadliness of the chemical, the dose, the length of exposure, the health status of recipient, and the route of entry or absorption by the body. People can be exposed to OCPs via the following routes:

- eating fatty foods such as milk, dairy products, or fish that are contaminated with them;
- eating foods imported from countries that still allow the use of persistent pesticides;

- passing pesticides through the placenta to the unborn child or by breast feeding; and
- inhalation and dermal contact.

OCPs contribute to many acute and chronic health effects including cancer, neurological damage, birth defects, tremors, headache, dermal irritation, respiratory problems, and dizziness. They are also suspected endocrine disrupters and are highly toxic to the nervous system, particularly during the early stages of development. Prenatal exposure to organochlorines has been associated with neurological effects such as learning deficits and behavioral changes in infants. They also have been linked with many forms of cancer. Animal studies have shown the potential for reproductive and developmental effects and disruption of normal hormone function. Children may be especially vulnerable to OCP exposure because they consume larger amounts of food and water relative to their body weight than adults. Children's developing organ systems are more sensitive, and their bodies have thus limited ability to detoxify OCPs. There are many different pesticides in use with very different modes of action and levels of toxicity.

The World Health Organization estimated that there are 3 million cases of acute, severe poisoning annually, with 220,000 deaths. OCPs should be chosen such that they give the needed pest control with the least risk of harm to nontarget species and the environment. The adverse environmental effects of pesticides used in public health can often be mitigated through proper selection and application procedures. As a general rule, the minimum of pesticide should be applied by the most efficient method at the most suitable time to achieve the required goal.

See Also: DDT; Pesticide.

Further Readings

Ize-Iyamu, O.K., et al. "Concentrations of Residues From Organochlorine Pesticide in Water and Fish From Some Rivers in Edo State Nigeria." *International Journal of Physical Sciences*, 2/9:237–41 (September 2007).

Kasozi, G. N., et al. "Organochlorine Residues in Fish and Water Samples From Lake Victoria, Uganda." *Journal of Environmental Quality*, 35:584–89 (2006).

Osibanjo, O., et al. "Regionally Based Assessment of Persistent Toxic Substances, Sub-Saharan Africa Regional Report." UNEP/GEF, 2002.

UNEP/FAO/WHO. "Assessment of Chemical Contaminants in Food. Report on the Results of the UNEP/FAO/WHO Programme of Health-Related Environmental Monitoring." Nairobi, UNEP/FAO/WHO, 1988.

Akan Bassey Williams
Covenant University

ORIGIN LABELING

Country-of-origin labeling is a practice increasingly demanded by consumers in an age when food crises seem more and more prevalent and more and more widespread as tainted produce, once discovered, is found to have been distributed nationwide or farther. The history of food legislation in the United States shows an increasing amount of required information in food labeling to inform and in some cases warn consumers—from ingredient labeling (and regulations as to what ingredients may be called) to nutritional information to allergy warnings. The 2002 Farm Bill required retailers to label fresh (but not processed) beef, pork, and lamb with country-of-origin information; peanuts, poultry, and produce were later added to this list. As a result of delays and continued debate, so-called COOL (country-of-origin labeling) was not implemented until September 30, 2008.

Though the inclusion of COOL in the Farm Bill was not as contentious as nearly every other item—most congressional debate centered, as usual, around subsidies, the necessity or lack thereof, and the specifics thereof—various contradictory legislation was introduced from 2001 to 2008, including measures that would make COOL labels voluntary for meat, and others that would make them mandatory for all uncooked/unprocessed food products and would implement the labels effective almost immediately. The reasoning for supporting or opposing COOL varied. Some proponents said that COOL would give domestic products a deserved competitive advantage, since domestic food would be presumed to be fresher (even apart from any desire to Buy American on the part of the consumer). Further, the average American can be assumed to have some basic understanding of the food safety requirements of his country and to have some faith in them; in contrast, although there are requirements placed on imported food, the same direct oversight is not in place, and when a food safety crisis centers around an imported product, consumers become more motivated to buy domestic products instead. For this reason, many supermarkets were already implementing limited COOL labeling—if nothing else, drawing attention to domestic products except in such unusual cases where foreign origin was a selling point (New Zealand lamb, Kobe beef from Japan, volcano oranges from Italy)—in response to the same consumer demand that elsewhere manifested itself as pressure from congressional constituencies.

Bovine spongiform encephalopathy—mad cow disease—is the best known recent example of an international food safety crisis that affected food trade because of the severity of the disease (and the variant Creutzfeldt-Jakob disease, which can be contracted by humans who consume brain or spinal cord matter from an infected cow). Though U.S. regulations meant to prevent mad cow infections have been shown to be exceptionally poor and ill enforced, fears have centered predominantly around beef from Britain (where more than 150 people have died as a result of mad cow) and Canada (historically a major importer of beef to the United States and where mad cow was found beginning in 2003). Though little British beef is imported to the United States, the crisis made consumers want to be sure of exactly where their beef came from—and after television talk show host Oprah Winfrey's overcautious and later recanted declaration that she would avoid all beef just to be safe, it was in the U.S. beef industry's best interests to differentiate between domestic and foreign products.

Opponents argue that such labeling caters to xenophobic fears; that instead of distributing false pamphlets about the despicable practices of immigrants or hanging "No Irish" signs outside their businesses, modern xenophobes characterize foreign products as dirty/tainted/unclean/toxic, playing up crises started by foreign products and downplaying those of domestic origin. These opponents criticize U.S. Department of Agriculture food safety regulations as weak, easily subverted, and poorly enforced, and claim that the existence of COOL labels impugns foreign products and bolsters the reputation of domestic ones, acting as an unfair trade barrier precisely at a time when World Trade Organization negotiations are trying to get countries to reduce and eliminate trade barriers. Implementation costs have been estimated at about $500 million per year, with a $4 billion initial cost—though there is much argument over the accuracy of those figures, with proponents of COOL calling them too high, and some opponents declaring them incomplete.

See Also: Department of Agriculture, U.S.; Doha Round, World Trade Organization; Farm Bill; Food Safety; Mad Cow Disease.

Further Readings

Becker, Geoffrey S., et al. *Mad Cow Disease Bovine Spongiform Encephalopathy.* Hauppauge, NY: Nova Science Publishers, 2008.
Gale Reference Team. "Consumers Say They Want Country-of-Origin Labeling." *Food and Drink Weekly*, 13/26 (July 2007).
Pollan, Michael. *The Omnivore's Dilemma: A Natural History of Four Meals.* New York: Penguin, 2007.

Bill Kte'pi
Independent Scholar

PEASANT

> The peasant—remote, conservative, somewhat archaic in his ways of dressing and speaking, fond of expressing himself in traditional modes and formulas—has always had a certain fascination for the urban man. In every country he represents the most ancient and secret element of society. For everyone but himself he embodies the occult, the hidden, that which surrenders itself only with great difficulty: a buried treasure, a seed that sprouts in the bowels of the earth, an ancient wisdom hiding among the folds of the land. (Paz, 1985)

In 2007, the absolute number of people living in urban centers had overtaken for the first time in history the number of people living in the countryside. The estimates for 2010 are that there will be around 3.3 billion people in rural areas and 3.5 billion in urban communities. These dramatic demographic changes regarding the rural–urban population distribution are quite recent. In 1970, the total world population was 3.7 billion, with 2.4 billion rural and 1.3 billion urban. During the same period, the change in the agricultural/nonagricultural population was even more dramatic. In 1970, the agricultural population stood at 2 billion people and the nonagricultural population at 1.7 billion. According to Saturnino Borras, by 2010 this will be radically reversed, with a 2.6 billion agricultural population versus 4.2 billion nonagricultural.

The World Bank states that, even as the number of urban dwellers overtakes the number of rural population, the percentage of poor people in rural areas continues to be higher than that in urban areas: Three-fourths of the world's poor today live and work in the countryside. In 2008, world poverty remained a largely rural phenomenon.

Although decreasing in relative terms, the absolute number of rural dwellers remains very significant. Many of these rural dwellers might be peasants, others not. The current usage of the term proposed by Ron Johnston et al.—"peasants work and live on family farms which function as relatively corporate units of production, consumption and reproduction"—sounds comprehensive to most of us. But additional elements will be needed for a more comprehensive understanding of the meaning of "peasant."

Peasants have existed, and have been defined, under a variety of economic, political, and cultural circumstances, like feudalism, capitalism, and communism, spanning vast periods of history. This article outlines only a few approaches by a few authors in the vast field of peasant studies.

The term *peasant* is derived from 15th-century French *païsant*, meaning one from the *pays*, or countryside, but it ultimately originates from the Latin *pagus*, or outlying administrative district (when the Roman Empire became Christian, these outlying districts were "pagan"; that is, not Christian). The term *peasant* was in common use in 15th-century England, referring to individuals working on the land and living in the countryside, without any apparent negative connotation. It was in the 19th century that it became a term of abuse, and nowadays it is still sometimes used in a pejorative sense for impoverished farmers. One of the strongest examples of a negative connotation is Karl Marx's concept of "the idiocy of rural life." Marx considered the peasantry to be disorganized, dispersed, and incapable of carrying out change. He expected that this class would tend to disappear, being displaced from the land and joining the proletariat. Mao Zedong, in contrast, gave the peasantry a heroic and revolutionary connotation.

Peasant studies date back to the first half of the 20th century, and many authors since then have employed different frameworks to define and to understand peasants. Some authors described peasants by their relationship to the outside world, the best known maybe being Alfred Kroeber, who wrote in 1948 that "peasants are definitely rural—yet live in relations to market towns; they form a class segment of a larger population which usually contains also urban centers. They constitute part-societies with part-cultures." These "part-societies" are more differentiated than, for example, tribal societies, both socially and culturally. Lloyd Fallers in the same vein states that peasant society is not completely isolable, completely capable of self-sufficiency, but neither is it so completely knit into a larger fabric by crisscrossing occupational structures as is the modern society.

Economic behavior is another important analytical approach to analyzing peasant societies. Frank Cancian in 1989 outlined three schools of peasant economic behavior: homogeneity theorists, heterogeneity theorists, and differentiation theorists. Homogeneity theorists claim that peasants have a special sociocultural system, which is different from other people and those peasants are resistant to no-peasant people and economic change. Socioeconomic differences between households are small, and people are uniformly poor; the rich people are nonpeasants. There also are local customs that generate and maintain a homogeneous population.

Eric Wolf, one of the homogeneity theorists, coined the term *closed corporate community*. He refers to a peasant as an agricultural producer in effective control of land who carries on agriculture as a means of livelihood, not as a business for profit. Examples of these closed corporate communities can be found in Mesoamerica and central Java. These communities maintain a body of rights to possessions, such as land, but also put pressures on members to redistribute surpluses at their command, preferably in the operation of a religious system, and induce them to content themselves with the rewards of "shared poverty." These societies strive to prevent outsiders from becoming members of the community and place limits on the ability of members to communicate with the larger society. This means that they are both corporate organizations, maintaining perpetuity of rights and membership, and closed corporations, because they limit these privileges to insiders and discourage close participation of members in the social relations of the larger society.

Wolf's statement emphasizes the isolated, local aspects of many peasant communities—the inward-looking part of their nature. Often the closed corporate community developed in reaction to events in the larger society, as during conquest and colonization. In other places, the closed corporate community is a result of internal colonization. Polarization is crucial to the appearance of these societies. They are closed toward the larger society.

George Foster uses "cognitive orientation" as the explanation for much peasant, but finally human, behavior. The model of cognitive orientation he uses to account for peasant behavior is the "image of limited good." By "image of limited good," Foster refers to the idea that much of peasant behavior is patterned in such fashions as to suggest that peasants view their social, economic, and natural universes—their total environment—as one in which all the desired things in life such as land, wealth, and health, among others, exist in finite quantity and are always in short supply, at least for peasants. Not only do these and all other "good things" exist in finite and limited quantities, but in addition, there is no way directly within peasant power to increase the available quantities. The image of limited good explains much peasant behavior that helps to keep the community homogeneous. The goods to be distributed are very few, so peasants fear that the balance can be attacked. They hesitate to take leadership, they gossip, and they do not sanction fellow villagers changing their relative economic standing in the community. They also maintain local institutions that restore the balance when economic differentiation does occur. According to Foster, peasants live tense and atomistic lives; they are distrustful, suspicious, and hostile, and they trust nobody except close family members. Individualism and suspicion mean that peasants will not cooperate with each other for their mutual benefit; the image of the limited good means they do not think it is worth trying to improve their economic lot, and the envy and hostility mean that those who try and succeed will soon suffer negative sanctions from their neighbors.

Alexander Chayanov considers the peasant economy as a special type. He sees peasants as independent from the larger society. The peasant family is a producing unit with no connection to the market economy, which makes their economic behavior distinctive.

Peasants, according to Chayanov, are primarily concerned with providing for their own families. Their strategy is as follows: when the family has many small children, the peasant parents work very hard to support all of them. They expand the size of their fields to meet the needs of the family. Later, when the children grow up and can contribute to their own support by work in the fields, the parents work less hard. Eventually, the children leave and the size of the family farm is again reduced. In this model, the consumer/worker ratio defines how hard the workers have to work to support the family, and it describes the variation in their level of effort that Chayanov saw as a distinctive feature of peasant economy. The Chayanov peasants have no infinite needs, which keep work effort constant, and they are not connected to socially defined infinite needs, which makes its members work hard all their lives. The structure of a peasant society is basically egalitarian. The family cycle is the explanation of having more or less land, being more or less rich. Wealth differences among peasant families are products of their temporary positions in the family cycle. Because all families are going through the same cycle, thus they are essentially homogenous in socioeconomic terms.

All three authors see peasant social and economic organization as different from modern, industrial free market (capitalistic) organization, which encourages the individual accumulation of wealth and leads to socioeconomic differentiation and heterogeneity.

Heterogeneity theorists, in contrast, define peasants as similar to the rest of the people; like the rest of society, peasants adapt their economic behavior when opportunities are good. The emphasis of this school is on the differences between people within peasant communities, which offer a key to understanding the internal dynamics and the responses to economic change programs. These open peasant communities, in contrast to closed corporate communities, depend on cash crops and have a high degree of market integration. According to Wolf, the open community comprises a number of subcultures, as opposed to the closed peasant corporate community, of which the peasantry is only one, although very important, subculture. The open community emphasizes continuous interaction with the outside world and ties its fortunes to outside demands. The open-ended community permits and expects individual accumulation and display of wealth during periods of rising outside demand, and it allows this new wealth much influence in the periodic reshaping of social ties.

Differentiation theorists, often using a political economy approach, also see peasants in relation to the larger economic society and analyze the history of transformation from independent producers to rural residents who must sell their production and labor on the market. In the last few decades, because of a closer integration into the larger economic system, peasants have become less self-sufficient, less distinctive from their environment, and more directly dependent on the larger system.

Although agriculture remains quite important to the livelihoods of almost half of the planet's population, rural households have increasingly diversified their ways of earning a living. Labor has become more mobile, and labor migration has taken multiple directions and characters: rural–urban, rural–rural, urban–rural, in-country and international, permanent and cyclical. Similar to the migration to which it is related, the growing "pluriactivity" of rural households and the increasing connections between the city and countryside complicate the question of defining the modern peasant. Moreover, there are alarming environmental and climate-related problems facing the rural world today. If temperatures rise by more than 3 degrees Fahrenheit, yields of major crops like maize may fall by 20–40 percent in parts of Africa, Asia, and Latin America.

See Also: Family Farm; Farm Crisis; Rural Renaissance.

Further Readings

Borras, Saturnino M. "Agrarian Change and Peasant Studies: Changes, Continuities and Challenges—An Introduction." *Journal of Peasant Studies,* 36/1 (2009).

Cancian, Frank. "Economic Behavior in Peasant Communities." In *Economic Anthropology,* edited by S. Plattner. Palo Alto, CA: Stanford University Press, 1989.

Chayanov, Alexander. V. *The Theory of Peasant Economy.* Manchester, UK: University of Wisconsin Press, 1986.

Fallers, Lloyd A. "Are African Cultivators to Be Called 'Peasants'?" *Current Anthropology,* 2/2, 1961.

Foster, George M. "University of California, Berkeley." *American Anthropologist,* 67/2 (1965).

Johnston, Ron J., et al. *The Dictionary of Human Geography.* Oxford: Blackwell, 2000.

Kroeber, Alfred L. *Anthropology.* New York: Harcourt, Brace, 1948.

Paz, Octavio. "The Sons of Malinche." In *The Labyrinth of Solitude and the Other Mexico,* translated by Lysander Kemp et al. New York: Grove, 1985.

Wolf, Eric. 1957. "Closed Corporate Communities in Mesoamerica and Java." *Southwestern Journal of Anthropology,* 13/1 (1957).

World Bank. *World Development Report 2008: Agriculture for Development.* New York: Oxford University Press for the World Bank, 2007.

Birgit Schmook
El Colegio de la Frontera Sur (ECOSUR)

PERMACULTURE

Permaculture is variously practiced and defined around the world. In a sense, permaculture is being practiced in any relatively self-reliant community or home. However, it was in the mid-1970s that the term was explicitly coined and explained by Bill Mollison and David Holmgren. In their first book on the subject, Mollison and Holmgren outlined a system designed to create sustainable human settlements inspired by observations of the patterns within natural systems. The term *permaculture* was understood as a combination of the words permanent and agriculture and was described further as a means of evolving a system of plants and animals to meet human needs. Since its initial definition, the term has been elaborated and expanded (e.g., by including cultures as well as agricultures, finances as well as foods). At its most basic, though, the purpose of permaculture design remains the same. Permaculture design is a way of thinking about, creating, and supporting interrelationships of various elements (house, garden, chicken, etc.) to form sustainable human settlements.

Permaculture, though expressly about design and engaging in landscapes, has underlying ethical principles that cannot be ignored in its application. The ethical guidelines of permaculture can be divided into four main principles. The first is referred to as care of the Earth. This means that within permaculture systems, living and nonliving elements and their needs must be not only considered but also given support and opportunities to thrive within the system. The second ethical principle is care of people. This means that the system designed should provide for the basic needs of the people within the system, expanding from the individual to wider communities, including considerations such as food and shelter, but also social contact, satisfying work, and so on. This principle suggests that although permaculture systems may be human-centered (i.e., they serve human needs), people's use should be careful and frugal (and in accordance with the first ethical principle). The third and fourth principles are the limitation of consumption and population and the redistribution of resource surpluses (time, energy, materials, etc.). These two principles are really elaborations of how to achieve the first two ideals, but they are important nonetheless. In applying these broad principles, one might (for example) limit the use of fossil fuels, designing a low-energy system using renewable energy sources (solar, wind, water, bio-energy, etc.) or waste whenever possible. By doing this, consumption would be limited (low-energy, no fossil fuels), and care of the Earth would be furthered (less overall negative environmental impact and interference), while still meeting the needs of the people within the system for energy. Sharing the stored energy surplus, your knowledge about the systems, or any extra materials from the project with your community would reflect the fourth principle. Although the ethical underpinnings of permaculture, and its relations with environmental ethics more broadly, remains a limited area of exploration, these four ethical guidelines do provide a starting point for the main concentration of permaculture writings: design.

Permaculture design focuses on the connections between various elements within a system. So, in designing a system, one might ask, "How does this bee interact with the buckwheat or pears?" or "How does that black walnut tree relate to the vegetable garden?" or "How do the algae and water lilies in the pond relate, and what does this have to do with my water use?" Permaculture design can be broken down into principles and techniques. Although design principles are held to be universally applicable, techniques vary depending on the particulars of a system. In other words, the design principles guide thinking and action, but how those principles are applied and achieved is site specific. Similar to the definition of permaculture, the principles of design vary somewhat in the literature, both in specificity and in content, but these differences do not necessarily result in the contradiction of various approaches. Bill Mollison and Reny Mia Slay outline several design principles that relate to elements and energy within a system under consideration. In general, they state that each element should be placed in the best possible relation with other elements and should perform many functions within the system; each system function should be supported by many elements; species diversity should be built into the system; cooperation with natural patterns (such as succession) will enhance the system; and energy should be efficiently used, recycled where possible, and emphasize biological resources (vs. fossil fuels). More recently, David Holmgren has suggested a set of 12 design principles:

1. Observe and interact

2. Catch and store energy

3. Obtain a yield

4. Apply self-regulation and accept feedback

5. Use and value renewable resources and services

6. Produce no waste

7. Design from patterns to details

8. Integrate rather than segregate

9. Use small and slow solutions

10. Use and value diversity

11. Use edges and value the marginal

12. Creatively use and respond to change

There are similarities between these two sets of design principles; for example, in their treatment of energy and of diversity, and the Holmgren model makes explicit some of the assumptions of the first principles such as "observe and interact" and "use small and slow solutions." Though not necessarily contradictory, these principles of design do differ and suggest that permaculture is not only still an evolving field but also one in which there is variance in approach and understanding even between student and teacher (as in the case of Holmgren and Mollison).

One of the most important design principles of permaculture relates to each element's relative place within the system. In applying the principle of maximizing positive relations between elements, one must first examine each element in relation to its characteristics and behaviors, needs, and products (including waste). Then one can consider how to best integrate the element into the system as a whole. If one were considering integration of bees for honey, then reflection is necessary on (among other things) the particular breed's aggression, nectar preferences, climate tolerance, average flying distance, disease and pest vulnerabilities, likely disturbance by people or animals, and diversity of nectar available at production times. It is not only a consideration of how to get honey for consumption but also of how to get the best honey, provide the best habitat, and support the best living for the bees as possible in relation to the rest of the system. Given the complexity of these considerations and the evolving nature of the system (and the permaculturalist's knowledge), much permaculture design occurs through zone planning so that those elements we use most frequently and intensively are closest to the dwelling and are designed first, whereas those used least intensively and designed last are farthest away. For example, in zone one around a rural house, one might have herb and vegetable gardens, a greenhouse, workshops, and so on, whereas zone five would be primarily a wilderness corridor; the zones in between might include animal forage and an orchard of decreasing intensive production. In this example, one can see how conscious system design is more necessary in zones of intensive use, though this does not suggest that design can be neglected between zones or in zones of less use. Of course, how the particular system evolves and what it looks like depends on the context of the site, the possible elements, and the permaculturalist's observations and knowledge.

From its origins in Australia, permaculture has spread around the globe. Though still relatively limited in scope, permaculture projects, both individual and collective, can be found in global South and global North countries. On the initiative of dedicated individuals, and some nongovernmental organizations, permaculture is spreading. Although some permaculturalists are self-taught, many have taken a permaculture design course offered by an experienced (often certified) permaculture teacher. Although efforts have been made to create a standard curriculum that could be adapted to local circumstances, most of the time teachers create their own course, though they often use the same permaculture books as their guides. The potential of permaculture design lies in its offer of a way of thinking systemically, but in a way that is adaptable to local situations and always evolving. It is not the only model for designing more sustainable human settlements, but it is one that synthesizes solutions to various concerns and one that is well worth considering as we face many ecological and social challenges to human and nonhuman survival and thriving.

See Also: Agrodiversity; Agroecology; Biodynamic Agriculture; Holistic Management; Intercropping; Low-Input Agriculture; Sustainable Agriculture.

Further Readings

Holmgren, David. *Permaculture: Principles and Pathways Beyond Sustainability*. Hepburn, Australia: Holmgren Design Services, 2002.

Mollison, Bill. *Permaculture: A Designer's Manual*. Tyalgum, Australia: Tagari, 1988.

Mollison, Bill and David Holmgren. *Permaculture One: A Perennial Agriculture for Human Settlements*. Melbourne, Australia: Transworld, 1978.

Mollison, Bill and Reny Mia Slay. *Introduction to Permaculture*. Tyalgum, Australia: Tagari, 1991.

Whitefield, Patrick. *The Earth Care Manual: A Permaculture Handbook for Britain and other Temperate Climates*. East Meon, UK: Permanent, 2004.

Catherine Phillips
Bishop's University

PESTICIDE

A pesticide is a material used for the mitigation, control, or elimination of plants or animals detrimental to human health or economy. Specifically, a pesticide is a chemical—liquid, granule, or gas—used to kill or control pests such as insects, weeds, bacteria, fungi, rodents, and worms. Pesticides play significant roles in increasing food production and

This agronomist is collecting broccoli samples from a test plot in Salinas, California, to test them for pesticide residue.

Source: U.S. Department of Agriculture Agricultural Research Service/Scott Bauer

eliminating diseases. However, only a few environmental issues have aroused the concern of the public as much as pesticides because exposure to them can be harmful to humans. The ill effects experienced may result from short- or long-term exposure or low- or high-level exposure through dermal absorption, inhalation, and oral ingestion. Frequent application of pesticides has led to the development of resistance to them. Scientists thus face the challenge of developing new environmentally friendly pesticides that overcome the rapid development of resistance by pests. Pesticides can be classified according to chemical class (e.g., organochlorine, carbamate, organophosphorus, and chlorophenoxy compounds) or according to their intended use (e.g., insecticide, herbicide, rodenticide, fungicide, fumigant, and acaricide). Insecticides act by poisoning the nervous system of target harmful insects. They include organochlorines, organophosphates, carbamate esters, pyrethroids, and botanical insecticides. Herbicides target weeds and include chlorophenoxy compounds and bipyridyl derivatives. Rodenticides target rodents and include zinc phosphide, fluoroacetic acid and derivatives, Naphthyl Thiourea, and anticoagulants. Fungicides target fungi, and they include hexachlorobenzene, organomercurials, pentachlorophenol, phthalimides, and dithiocarbamates. Examples of fumigants are phosphine, ethylene dibromide, and dibromochloropropane. Acaricides target acarose (a mite).

When a pesticide is introduced into the environment by application, disposal, or spill, it is influenced by many processes. These processes determine a pesticide's persistence, its movement, and its ultimate fate. The fate processes can have both positive and negative influences on a pesticide's effectiveness or its effect on the environment. The fate processes can be beneficial. They can move a pesticide to the target area or destroy its potentially harmful residues. Sometimes they can be detrimental, leading to reduced control of a target pest, injury of nontarget plants and animals, and environmental damage. Different soil and climatic factors and different handling practices can promote or prevent each process. An understanding of the fate processes ensures that applications are not only effective but also environmentally safe. Fate processes could be grouped into four major types: degradation, transfer, adsorption, and bio-accumulation/bio-magnification. These physical and chemical properties of pesticides influence their environmental risk.

Pesticide Degradation

Pesticide degradation refers to the breakdown of pesticides in the environment. The rate at which the degradation occurs is measured by the pesticide's half-life. A pesticide with a long half-life is described as persistent. Persistence is good for long-term pest control, but the pesticide can cause environmental damage over a long period of time. Pesticide degradation is usually beneficial, as pesticide-destroying reactions change most pesticide residues in the environment to nontoxic or harmless compounds. However, degradation is detrimental when a pesticide is destroyed before the target pest has been controlled. The rate of pesticide degradation is affected by many environmental factors including temperature, moisture, and pH. The three types of pesticide degradation are microbial, chemical, and photodegradation.

Microbial degradation is the breakdown of pesticides by fungi, bacteria, and other microorganisms that use pesticides as a food source. Pesticides are broken down into basic compounds such as water and carbon (IV) oxide. Most microbial degradation of pesticides occurs in the soil. Soil conditions such as moisture, temperature, aeration, pH, and the amount of organic matter affect the rate of microbial degradation because of their direct influence on microbial growth and activity. The frequency of pesticide application is also a factor that can influence microbial degradation. Rapid microbial degradation is more likely when the same pesticide is used repeatedly in a field. Repeated applications can actually stimulate the buildup of organisms that are effective in degrading the chemical. As the population of these organisms increases, degradation accelerates and the amount of pesticide available to control the pest is reduced. The possibility of very rapid pesticide breakdown is reduced by using pesticides only when necessary and by avoiding repeated applications of the same chemical.

Chemical degradation is the breakdown of pesticides by processes that do not involve living organisms. It is the chemical reaction that occurs between the pesticide and other chemicals in the environment, resulting in the splitting of pesticides into less hazardous compounds. The adsorptive capacity, pH, temperature, moisture, mineralogy of the soil, and physical and chemical properties of the pesticide determine the rate and type of chemical reactions that occur. One of the most common pesticide degradation reactions is hydrolysis, a breakdown process in which the pesticide reacts with water.

Photodegradation is the breakdown of pesticides by ultraviolet or visible light, particularly sunlight. Photodegradation can destroy pesticides on foliage, on the surface of the soil, and in the air. Pesticides vary considerably in their stability under natural light. Factors that influence this kind of degradation include intensity of the sunlight, characteristics of the application site such as soil type and vegetation, application method, and physical and chemical properties of the formulated pesticide. Pesticide losses from photodegradation can be reduced by incorporating the pesticide into the soil during or immediately after application and by using adjuvants in the pesticide formulation to protect the active ingredient from photodegradation.

Pesticide Circulation

Pesticide transfer is sometimes essential for pest control, as some pesticides need to circulate for effective utilization. Too much movement, however, can move a pesticide away from the target pest, leading to reduced pest control, contamination of surface water and groundwater, and injury of nontarget species, including humans. Pesticides can be transferred through natural processes such as volatilization, runoff, leaching, absorption, and crop removal.

Volatilization is the conversion of a solid or liquid into a vapor. Once volatilized, a pesticide can move in air currents away from the treated surface. Vapor pressure is an important factor in determining pesticide volatility. The higher the vapor pressure, the more volatile the pesticide. Environmental factors tend to increase volatilization. These include high temperature, low relative humidity, and air movement. A pesticide tightly adsorbed to soil particles is less likely to volatilize. Soil conditions such as texture, organic matter content, and moisture can thus influence pesticide volatilization. Volatilization can result in reduced control of the target pest because less pesticide remains at the target site.

Runoff refers to the movement of water over the land surface or a sloping surface. It occurs when water accumulates on the land surface faster than it can infiltrate the soil. Pesticides can be carried in the water itself or bound to eroding soil particles. The amount of pesticides in runoff water is a function of site-related factors such as the slope of the land and moisture content of the soil. Climatic factors such as temperature and the amount and timing of rainfall relative to the pesticide application are also of influence. Climatic factors affect the persistence of pesticides in the environment, thus influencing the availability of pesticides for transport by runoff. Other factors of note are the pesticide-water-soil interactions, such as the solubility and adsorptivity of the pesticide and the erodibility and texture of the soil.

Leaching is the movement of pesticides through the soil rather than over the surface. Pesticides can leach downward, upward, or side to side. Leaching depends on the pesticide's chemical and physical properties such as adsorption, solubility, and persistence. A pesticide held strongly to soil particles by adsorption is less likely to leach. A pesticide that is rapidly broken down by a degradation process is less likely to leach because it may remain in the soil for a short time. Soil factors that influence leaching include permeability, texture, and organic matter. The more permeable a soil is, the greater the potential for pesticide leaching. A sandy soil, for example, is much more permeable than a clay soil. Monitoring weather conditions and the amount and timing of irrigation can help minimize pesticide leaching.

Absorption is the movement of pesticides into plants and animals or structures such as soil and wood. Absorption of pesticides by target and nontarget organisms is influenced by environmental conditions and physical and chemical properties of the pesticide and the soil. Once absorbed by plants, pesticides may be broken down, or they may remain in the plant until tissue decay or harvest. Similarly, desorption is the release of pesticides from soil, wood, or other substances.

Pesticide Transfer

Crop removal transfers pesticides and their breakdown products from the treatment site. Most harvested food commodities are subjected to washing and processing procedures that remove or degrade much of the remaining pesticide residue. Although we typically associate harvesting with food and feed products, it is easy to forget that pesticides potentially can be transferred during such operations as tree and shrub pruning and turf grass mowing.

Pesticide adsorption is the binding of pesticides to soil particles and organic matter. Positively charged pesticide molecules, for example, are attracted and bound to negatively charged clay particles. The amount of adsorption in the soil depends on the type of soil and the soil conditions such as temperature, pH, moisture content, and the characteristics of the pesticides. Soils high in organic matter or clay are more adsorptive than coarse, sandy soils, in part because a clay or organic soil has more particle surface area onto which pesticides can bind. Moisture affects adsorption. Wet soils tend to adsorb less pesticide than dry soils because water molecules compete with the pesticide for the binding sites. Pesticides vary in their adsorption to soil particles. Some pesticides bind very tightly, whereas others bind only weakly and are readily desorbed into the soil solution. Pesticide adsorption leads to reduced pest control; hence, higher application rates are recommended when the chemical is applied to adsorptive soils.

Bio-accumulation is the ability of some pesticides to build up in the body tissue of animals. Accumulation occurs if the pesticide is metabolized or excreted from the body slowly. As the amount of pesticide increases, it becomes more harmful to the person or animal. Pesticide buildup can cause death or long-term damage. It can also build up in the food chain—a process called *bio-magnification*. Bio-accumulation and bio-magnification also occur in aquatic systems. Fishes, for example, are affected when their water habitats or food sources are contaminated. The extent of damage to the fish depends not only on the properties of the pesticide but also on the species of fish, its age, size, and its position in the food chain.

The use of pesticides introduces some risk to the environment. The degree of risk depends on four factors; namely, persistence, mobility, nontarget toxicity, and volume of use. Thus, environmental risk is minimized when any of the risk factors is close to zero. A pesticide that remains active in the environment for a long period of time is described as *persistent*. The more persistent a pesticide is, the higher the risk it poses to the environment. *Mobility* refers to the ability of the active ingredient of the pesticide to move away from the site of application through the soil, water, or air. The more easily the pesticide is able to move away from the site of application, the higher the risk to the environment. *Nontarget toxicity* refers to the unintended harmful effect of the pesticide on any organism other than the pest. The risk to the environment increases if the nontarget toxicity of the product is high. *Volume of use* describes the total amount of pesticides used in the environment. The larger the amount of product used in the environment, the higher the potential will be for environmental damage.

Pesticide degradation generally results in a reduction in toxicity. Pesticides are inherently toxic. However, some pesticides have breakdown products (metabolites) that are more toxic than the parent compounds. Pesticides could be classified according to their potential toxicity to humans and animals. Chemical structures differ within categories as well as between categories. Thus, toxicity to humans can vary widely within each group.

See Also: DDT; Organochlorines.

Further Readings

Ademoroti, C. M. A. *Environmental Chemistry and Toxicology*. Idaban, Nigeria: Foludex, 1996.

Kielly, Gerard. *Environmental Engineering*. New York: McGraw-Hill International Edition, 1998.

Akan Bassey Williams
Covenant University

PLANTATION

Plantations are large farms growing a single crop that is sold for cash. They are established in areas that can produce a marketable cash crop. They often use large numbers of laborers, which enables the crop grown to achieve benefits from economies of scale.

Plantations are classified according to either the type of crop that is grown or the way in which labor is used. Large farms that raise cattle or grow some type of cereal such as wheat are usually excluded from the term *plantation*. Orchards are sometimes identified as plantations, especially if they grow nuts.

Large plantation-style farms existed in the Roman Empire, where lands owned by a wealthy businessman were worked by slaves. Large quantities of labor were needed because caring for vineyards and olive trees to make wine and olive oil requires a lot of hand labor. Mediterranean *latifundia* were a type of plantation introduced into Latin American countries.

In Anglo-American history, the term *plantation* was first used in Northern Ireland (Ulster). The Irish had been defeated by the English and Scots with great loss of life. Colonists (Ulstermen) were planted in the northern counties. From there the agricultural system and many descendants moved to America, where tobacco, indigo, rice, and ultimately cotton were the crops of the plantation days of the antebellum South. The Southern plantations were worked by slaves from Africa as well as by European indentured servants.

With the end of slavery at the end of the Civil War in 1865, all labor became free labor, but the slave system persisted in the form of sharecropping, used to raise cotton in fields that previously had been worked by slaves. Using "free" labor, plantations persisted into the 1900s because people on these plantations were trapped by the sharecropping system. The plantation owner would provide farmland, house, tools, seeds, perhaps draft animals, and other supplies in exchange for labor. Both the plantation owner and the sharecropper would get a portion of the crop and use it to pay debts at harvest time. If there was a good harvest, the system was beneficial; however, if the harvest was poor, sharecroppers would be in debt after the harvest. It was not until about 1900, when the boll weevil forced diversification of crops and mechanization from the 1920s and afterward, that the system almost ended.

Slave plantations were used extensively in the days of European colonialism in the Americas. Sugarcane was grown on plantations in Brazil (Portuguese *engenho* or "engine"), in Louisiana, and on many Caribbean islands such as Cuba, Haiti, and Jamaica. Since large gangs of labor are often needed for agricultural crops, the end of slavery did not end plantation labor. Hawaiian pineapple, Central American bananas, and chicle are only three of the crops grown in the 1800s and 1900s that were grown on plantations.

Many plantations have been opened in tropical or semitropical countries. Plantation crops grown in Egypt and India include cotton. Tea, coffee, tobacco, cacao, sisal, hemp for rope, rubber, chicle (gum), indigo, and other crops have commonly been grown on plantations in semitropical or tropical areas.

Modern plantations grow a great many crops, but each is usually a monoculture. The single crop is sold for cash and is then distributed to distant markets. The laborers used on modern plantations are increasingly well educated and employ machinery in the production of their respective crops.

Industrial plantations are devoted to a single crop, usually trees. Timber plantations are common in temperate zones today. Trees grown may be Douglas fir trees in Washington State or pine trees in Georgia. The goal of timber companies in this type of industrial timbering is to raise a volume of wood as quickly as possible. This means that each tree is looked on as a stack of dollar bills (or other currency in other countries), rather than as a part of nature.

In many of the northern states and Canada, Christmas trees are planted on plantations. What is central to tree plantations is that the natural forest is replaced by a tree monoculture. Tree plantations are common in Europe as well as North America. In southern Asia (e.g., Malaysia), rubber, oil palm, coconuts, bananas, and teak trees are all grown on plantations.

When all of the trees on a tree plantation are planted in a large area at the same time, they can be managed in case insects or diseases threaten. The whole planting can be harvested economically at the same time. Because these are commercial trees, they are often seedlings from a special variety of a rapidly growing and high-producing parent tree.

Tree plantations such as pine trees are planted in specific varieties for specific geographic locations. They are also selected for their favorability for quick growth on a particular soil. The land may have a "site index" that has been determined by its capacity to grow a pine tree or some other tree. For example a site index of 90 for loblolly pine trees means that that piece of land will grow a 90-foot-tall pine tree in 50 years.

In the past the problems posed by plantations arose from slavery or from using indentured laborers. Today modern plantations are more likely to pose environmental problems. They have been criticized because their monoculture often exhausts the soil and because a monoculture is likely to attract a concentration of pest or to be more vulnerable to disease. Additional concerns include the use of hybrids and genetically modified trees.

Many new plantations have been developed where natural forests or jungles grew before. In places such as Indonesia or Brazil, the diversity of tropical forests is being destroyed for the sake of a monoculture cash crop. Also, in Indonesia, rubber plantations have had a negative effect on the natural environment by reducing biodiversity and destroying the natural habitat. For example, pine plantations may produce a cash crop, but deer do not eat pine needles. The clear-cutting of sections of a tree plantation is a threat to specialist species that need stands of mature trees for nesting or for food.

Environmental plantations are devoted to preserving a watershed area. They may be put into areas that need erosion control or windbreaks and are used to recover native species of fauna and flora. Among the proposals in the Kyoto Protocol were several promoting tree plantations as a way to lock large quantities of carbon from carbon dioxide into growing trees.

See Also: Agribusiness; Agrodiversity; Cash Crop; Labor; Sustainable Agriculture.

Further Readings

Carrere, Ricardo and Larry Lohmann. *Pulping the South: Industrial Tree Plantations and the World Paper Economy*. London: Zed, 1996.

Evans, Julian. *Plantation Forestry in the Tropics: Tree Planting for Industrial, Social, Environmental, and Agroforestry Purposes*. New York: Oxford University Press, 2006.

Konings, Piet. *Labour Resistance in Cameroon: Managerial Strategies and Labour Resistance in the Agro-Industrial Plantations of the Cameroon Development Corporation*. Portsmouth, NH: Heinemann, 1993.

Pons, Frank Moya. *History of the Caribbean: Plantations, Trade and War in the Atlantic World*. Princeton, NJ: Markus Wiener, 2007.

Royce, Edward Cary. *Origins of Southern Sharecropping*. Philadelphia: Temple University Press, 1994.

Savill, Peter, et al. *Plantation Silviculture in Europe*. New York: Oxford University Press, 1998.

Andrew Jackson Waskey
Dalton State College

PRODUCTIONISM

Within agriculture, productionism is the ideology and suite of associated practices that privileges agricultural productivity over other values or goals that might be associated with agricultural landscapes or communities. Although productionism remains the dominant perspective among most agricultural professionals—from farmers to agricultural scientists—there is a growing recognition that other values may also be important to consider in today's society. The ideology of productionism is prevalent, though by no means universal, among agricultural professionals around the world. However, most scholars trace the ideology back to European roots—to the Enlightenment, the Protestant Reformation, and the Industrial Revolution.

Productionism is the idea that agricultural yield, especially of staple food and feed crops, is the most important goal of farming and the measure by which producers are judged. Other social goals, long-term conservation or environmental goals, and even economic goals are secondary concerns. Productionism can be understood as both a hegemonic ideology as well as a suite of technical and socioeconomic practices supporting the ideology. Although productionism is not limited to agriculture, it is arguable that it is within agriculture that productionism is embodied in its most encompassing form. That is to say that although yield or production of goods remains one of the primary goals of most material-based industries, there are usually other goals, especially those of profit and efficiency, that are also central.

The roots of productionist ideology and practice can be traced to the development of modern Western thought regarding science, technology, and morality. Early productionism emerged in the West out of an interaction between agricultural communities and landscapes, with the changing view of the world provided by the Protestant Reformation, the Enlightenment, and the Industrial Revolution.

The philosophical and spiritual debates over the nature of salvation and morality that led to the Protestant Reformation also fed productionism by linking productivity with morality. Max Weber's early sociological work examining the effects of the Protestant work ethic and ideas about predestination, published in the early 20th century, though not specific to agriculture, can help us understand the very deep, and in many ways spiritual and moral, roots of productionist values. Weber traced the development of capitalism to the Protestant work ethic, and in his day would likely have exempted farming from the suite of capitalist enterprises he was interested in. However, his analysis of how the work ethic leads to productivity and accumulation is central to understanding agricultural productionism.

For productionism to have taken hold, farmers must have a reason for producing more than they are already producing. A peasant farm family's current level of production was sufficient for their subsistence needs; for example, they may have been uninterested in expending the extra effort to produce excess. However, during the 15th–17th centuries, Protestant religions took hold in most of northern Europe. Protestant reformers were especially concerned about the perceived hedonism and corruption of the Church, and many protestant groups adopted an ascetic Puritan ethic and endorsed the virtue of hard work, arguing that God was best served if one's talents and labor were spent on being productive. Hard work in itself became a virtue, and sloth or leisure a vice. One of the best ways to demonstrate that one was working hard, and was therefore a virtuous person, was to produce a great deal of whatever it was one was engaged in producing. This production led to the accumulation necessary for capitalist investment.

Many Protestant groups also believed in predestination—that an individual's status as saved or damned was determined by God before that individual was born, and therefore nothing he or she did while alive could change salvation's outcome. On the surface we might expect a belief in predestination to mitigate against the Protestant work ethic; however, in practice the two beliefs served to reinforce the value of productionism. Although predestination indicated that one could not influence one's salvation for good or ill, it was still believed that the morally superior individuals would be saved and the degenerates would be damned. Therefore, embracing morally upstanding practices marked one as saved rather than damned, and individuals strove to demonstrate their virtue through productivity.

Two fundamental tenets of the European scientific Enlightenment provide the basis for the development of productionist hegemony within agriculture. First, the purpose of scientific research was to enhance society's understanding of the natural world. The natural world, including crops, livestock, pests, and water systems, was conceptualized as a machine, the understanding of which required reductionist scientific inquiry and could result in positivist conclusions about how the natural world functions. The natural world was knowable by breaking it down into its constituent parts and was separate from the human and social world. Second, the Enlightenment embraced the normative goal of science that remains dominant, though often unspoken, within most of the scientific establishment today—science's ultimate purpose is to improve the human condition. The Enlightenment's framing of science as both reductionistic and humanistic adds another layer of complexity to productionist agriculture.

The Industrial Revolution can be thought of in many ways as the marriage of capitalism (rooted in the Protestant work ethic) and the Enlightenment. Science and technology were applied to the problem of production, with production already being acknowledged as a measure of both human well-being and virtue. Efficient use of labor and capital toward the end of production was the goal of a great deal of very successful scientific effort, and in agriculture this meant increased yields and increased overall production.

It is worth noting here that although agriculture certainly adopted technological advances, it did not centralize and resort to hired labor in the same way that other industries did, and in many ways and in many places, agriculture remains recalcitrant to these capitalistic forces. As a relatively decentralized industry, and one in which its workers are related to the work through family ties and so are often not paid a wage, efficient use of labor has been less a concern within agriculture than in other industries. This has reinforced the emphasis on productionism over efficiency or profit, as labor as a cost of production remains uncounted and farmers remain willing to exploit themselves and their families toward the end of higher production.

The post–World War II green revolution introduced a suite of agricultural technologies that very dramatically increased yields in many parts of the world, combining hybrid seeds, irrigation, and petroleum-based fertilizers. At the time of the green revolution, and continuing to today, the productionism ideology's explicit motivations have been less about generalized virtue and have been much more specifically focused on human population growth and the belief that population growth was and would continue to be so significant that all possible resources must be allocated to practices and technologies that would result in at least equivalent increases in production yields of staple food crops. Indeed, world population increased from just under 2.5 billion people in 1950 to over 6 billion people in 2000. World food production more than kept pace, in large part as a result of the green revolution technologies, as well as conversion of wildlands.

International trade has also played a role in reinforcing productionist tendencies in modern agriculture. In the 1970s, U.S. secretary of agriculture Earl Butz famously advised American farmers to "get big or get out" and to "plant fencerow to fencerow." At the time, American grain was selling at high prices on the international market, and similar to many other countries with a large agricultural sector, U.S. policy took advantage of productionist ideologies to increase exports and corner emerging markets.

Productionist ideology and practice have been criticized for a variety of reasons. By privileging yield over other goals, critics contend that productionist agriculture has resulted in a variety of negative unintended consequences for the environment, agricultural communities, consumers, and farmers themselves. Environmentally, focus on sustaining high yields, often of one crop, over many years often results in depletion of soil fertility and organic matter, and eventually soil erosion. Pesticides and herbicides used to protect yields are also linked to negative environmental effects, especially on wildlife, as well as human health impacts. Wildland conversion results in reduced natural biodiversity, as well as decreased carbon storage. Industrial agriculture is also the second-largest user of fossil fuels and thereby contributes to climate change.

In an industry focused on production of large quantities rather than on particular qualities of a product, producers who have the capital to farm larger areas will be more successful than smaller-scale farmers, and indeed we are seeing a loss of medium-scale producers. Agricultural communities lose population and suffer the social and economic consequences. Though consumers are the imagined beneficiaries of productionism in that staple foods are cheap and plentiful, the abundance of inexpensive raw products fuels much of the processed food industry, which has been linked to rising levels of overweight and obesity around the world.

For farmers, an ideology of productionism can undermine their long-term viability. Economically, the emphasis on yield and quantity of production often comes at the cost of inefficient use of resources such as labor and inputs. For example, rather than measuring yield per unit of labor, many producers focus only on yield as their measure of success. Pursuing ever-higher yields can also result in decreased ability of the land to support crop growth over time. Productionism remains a prevalent ideology within agriculture; however, there is some evidence that values around farming might be integrating additional environmental and social values beyond yield. For example, the focus on multifunctional agriculture—agriculture that meets a variety of needs, including environmental and cultural needs, not only food production—broadens the acceptable goals of farming. The increasing popularity of sustainably produced products, local agriculture, and agritourism also speak to both consumers' and producers' changing ideas about the purpose of agriculture.

See Also: Agribusiness; Green Revolution; Rural Renaissance.

Further Readings

Jenkins, Willis J. *Ecologies of Grace: Environmental Ethics and Christian Theology.* New York: Oxford University Press, USA, 2008.

Lappe, Frances Moore and Anna Lappe. *Hope's Edge: The Next Diet for a Small Planet.* New York: Tarcher, 2003.

Thompson, Paul. *Spirit of the Soil: Agriculture and Environmental Ethics.* New York: Routledge, 2005.

Ann Finan
Australian National University

PROLETARIANIZATION

Proletarianization, a term coined by Karl Marx, is the process through which producers are separated from the means of production, such as land. This can occur forcibly through "extra-economic" means, such as coercion or other exercises of political power, or through market mechanisms as independent producers are outcompeted by others. Through these processes farmers lose their land and are forced to become workers, or proletarians, who sell their labor as a necessary survival strategy. Debates over proletarianization in agriculture, and the forces that delay it, have occupied scholars' investigations of the transition from a countryside dominated by peasant farm families to one dominated by capitalist units of production. Considerable debate remains on whether the agrarian transition that occurred in Europe can be generalized to areas where the transition is ongoing, and whether the final outcome of the transition—complete proletarianization—has actually occurred in agriculture in the core capitalist economies.

Primitive Accumulation as Proletarianization

According to Marx, the capitalist mode of production expands through two main processes, primitive accumulation and capital accumulation. Primitive accumulation, a form of privatization, is the process of separating producers from inputs—land, tools, machines, and other resources—that are necessary for subsistence production. Primitive accumulation is a precondition for capital accumulation. Primitive accumulation in Europe and its colonies frequently meant taking land away from peasants by force. Expropriation of traditional lands created commodities of labor and land, simultaneously making a labor and land market.

Enclosure, starting in England in the 16th century, was the first political project that turned peasants into proletarians. English peasants, as in many areas of the world, existed as subsistence producers with traditional usufruct rights to land for farming and shared rights to common areas for wood collection, hunting, and other subsistence activities. Enclosure transformed the English countryside from a society based on the two classes of the feudal mode of production, peasants and lords, to one with three classes, known as the triad of capitalist agriculture: workers (proletarians), tenant farmers, and landowners. In this triad, farmers rent land from landowners, employ workers, and use family members' labor. The most efficient farmers were better able to maintain access to rental lands and outcompeted other units of production.

Proletarianization through primitive accumulation has progressed unevenly around the world with the spread of capitalism. It has been incomplete in part because of resistance from those whose resources and attendant ways of life are lost by elite expropriation. Colonialism involved attempts by colonial powers to transform independent producers from a subsistence orientation to production for market. This involved expropriation and taxation, which required producers to generate income by producing cash crops. Marxist scholars have shown that proletarianization and the more general expansion of capitalist relations of production, in combination with extreme weather events, caused massive famines in the colonized world.

Proletarianization and the Agrarian Question

Proletarianization as a concept is inextricably tied to the "agrarian question" of 19th- and early-20th-century Marxism. The debate centers on the question of the peasantry's place within the capitalist mode of production and the "delayed" transition from peasants to proletarians. Marx and Friedrich Engels thought agrarian petty commodity producers were doomed. In their view, the peasantry would be squeezed so hard by capital and onerous forms of indebtedness and be outcompeted by mechanized capitalist agriculture that it would disappear and join the urban proletariat. Following Marx, Lenin proposed that concentration and class differentiation would occur in agriculture as in other economic sectors, so that the less-efficient producers would be outcompeted, leaving farming to a small number of large-scale, productive farmers and freeing up a vast proletarian labor force for industry.

Karl Kautsky opposed Vladimir Lenin's differentiation thesis. Kautsky, on seeing an increase in the number of peasant holdings in Europe even as capitalism expanded, argued that the laws of capitalism play out differently in agriculture. Kautsky's *The Agrarian Question* explained the persistence of petty commodity producers in agriculture to refute those who saw their continued existence as reason for abandoning Marx's explanation of capitalist development. For Kautsky, the particularities of agriculture that delayed the agrarian transition included (1) land, as the major factor of production, is not reproducible like capital, so capital could penetrate agriculture while maintaining traditional forms of property, and (2) economies of scale did not necessarily favor large-scale farm units, in part because intensive peasant production often produced the same goods for less (because of the ability of the family to engage in underconsumption during hard times) and because peasant farm units provided labor to nearby large farms. Kautsky maintained that these and other reasons explained the delay in the ultimate proletarianization of the peasantry.

Along similar lines, agricultural economist Alexander Chayanov pointed to the resiliency of the peasantry by focusing on the farm unit. Chayanov's concept of "self-exploitation" and his focus on demographic factors supported his argument that the peasantry occupied a more secure place in agrarian societies than Marxist scholars believed. Susan Mann and James Dickinson expand on these Marxian and Chayanovian arguments to explain other obstacles to a complete capitalist transformation of agriculture. Agrifood scholars such as Harriet Friedmann have drawn on these explanations to examine cases of agribusiness removing itself from direct production in agriculture to focus on credit, inputs, contracts, processing, and sales. Similarly, a host of scholars, largely in the developing country contexts, have drawn on these debates to argue that one cannot take Marx's theory of capitalist development, which was based on the empirical situation of Britain and Europe, as a philosophical historical theory. They instead emphasize that local circumstances play a role in influencing how proletarianization proceeds as peripheral agrarian economies are integrated into capitalism.

The Class Position of Family Farmers in Capitalism

Given the lack of a completed proletarianization even in advanced industrial economies, the class position of the farmer remains an area of debate among scholars. Richard Lewontin and Michael Watts argue that through contracts with agribusiness such as food processing corporations, contract farmers essentially become "propertied proletarians," as the factors of production they own can be used for a single purpose and they are strongly controlled by the contract. This suggests that capital does not need to fully modify the farm ownership structure along capitalist lines for capital accumulation to proceed. Similar arguments rely on the concept of appropriationism, which is the process through which petty commodity producers increasingly purchase off-farm inputs to substitute for inputs that they once produced on-farm to remain competitive. Appropriationism allows for capital to flow to capitalists even through noncapitalist ownership relations and the non-capitalist social relations of production by appropriating surplus value. Contract farming can involve an extreme form of appropriation in which all of the inputs for agricultural production are supplied by the contracting firm.

The Other Side of the Coin: Consumer Proletarianization

Josiah Heyman, drawing on the feminist revision of Marxist thought from Rayna Rapp, has proposed the concept of "consumer proletarianization." This idea refers to localities and householders "that lose the traditional devices, raw materials, skills, and social relations needed to produce their daily existence: to heat their houses, cook their food, cover their roofs, and so on. . . . [H]aving lost the main means of self-provisioning, consumers must purchase commodity inputs from the capitalist economy—appliances, construction materials, grocery store food, manufactured clothing, and so on." This concept has yet to be applied in agrifood studies but holds promise in understanding households' positions vis-à-vis the capitalist agrifood system.

See Also: Agrarian Question; Appropriationism; Commons; Concentration; Contract Farming; Labor; Peasant.

Further Readings

Friedmann, H. "World Market, State, and Family Farm: Social Bases of Household Production in the Era of Wage Labor." *Comparative Studies in Society and History,* 20/4:545–86 (1978).

Goodman, D. and M. Redclift. *From Peasant to Proletarian: Capitalist Development and Agrarian Transitions.* Oxford: Basil Blackwell, 1981.

Heyman, J. M. "The Organizational Logic of Capitalist Consumption on the Mexico-United States Border." *Research in Economic Anthropology,* 15:175–238 (1994).

Lewontin, R. C. "The Maturing of Capitalist Agriculture: Farmer as Proletarian." In *Hungry for Profit: The Agribusiness Threat to Farmers, Food, and the Environment*, edited by F. Magdoff, et al. New York: Monthly Review, 2000.

Magdoff, Fred, et al., eds. *Hungry for Profit: The Agribusiness Threat to Farmers, Food, and the Environment.* New York: Monthly Review, 2000.

Watts, M. "Living Under Contract: Work, Production, Politics, and the Manufacture of Discontent in a Peasant Society." In *Reworking Modernity: Capitalisms and Symbolic Discontent*, edited by A. Pred and M. Watts. Brunswick, NJ: Rutgers University Press, 1992.

Wood, E. M. "The Agrarian Origins of Capitalism." In *Hungry for Profit: The Agribusiness Threat to Farmers, Food, and the Environment*, edited by F. Magdoff, et al. New York: Monthly Review. 2000.

Ryan E. Galt
University of California, Davis

Public Law 480, Food Aid

Public Law 480, or the Agricultural Trade Development and Assistance Act of 1954, was enacted by the Eisenhower administration as a means for relieving American farmers of surplus production. The act allocated the majority of surplus commodities to "friendly nations," and the rest for humanitarian relief or famine purposes. By 1961 the Act became commonly known as the Food for Peace Program, and by 1966 the emphasis had changed from selling commodities to foreign food aid for poor nations.

Although touted as a successful foreign aid program by the U.S. government, Public Law (P.L.) 480 has not been without controversy. It has been a target of farmers in developing countries, who accuse the United States of "dumping" farm commodities in foreign countries and undermining any opportunities for international agricultural competition.

History of P.L. 480

P.L. 480 was signed into law on July 10, 1954, by President Dwight D. Eisenhower. In signing the legislation, Eisenhower stated that the legislation's purpose was to "lay the basis for a permanent expansion of our exports of agricultural products with lasting benefits to ourselves and peoples of other lands." In 1961, President John F. Kennedy renamed it "Food for Peace," stating, "Food is strength, and food is peace, and food is freedom, and food is a helping to people around the world whose good will and friendship we want."

P.L. 480 was modified by the Federal Agriculture Improvement and Reform Act of 1996, known informally as the Freedom to Farm Act, or the 1996 U.S. Farm Bill. The changes allow for agreements with private entities in addition to foreign governments. In addition, the 1996 bill broadened the range of commodities available for use under P.L. 480.

The act comprises four parts, titles 1, 2, 3, and 5, with titles 1 and 2 being the largest. Title 1 is managed by the U.S. Department of Agriculture and involves the sale of farm commodities overseas on the basis of long-term, low-interest loans and in local currency. Sales are then reinvested in development projects for low-income countries to improve their agricultural production.

Title 2 is the largest and, many would argue, the most important part of the bill. It is managed by the U.S. Agency for International Development (USAID). Surplus foods are donated for humanitarian needs and distributed by other voluntary U.S. organizations such as the Catholic Relief Fund. Food is most often channeled into food for work programs or child nutrition programs. Title 3 is dedicated to food for development, and Title 5 to Farmer to Farmer programs. The latter two titles are also managed by USAID.

Food aid, however, remains tied to the mandate it is given in Congress. P.L. 480 has three principal political and economic objectives: (1) to promote economic development, particularly in agricultural production; (2) to develop and expand markets for U.S. agricultural commodities; and (3) to promote the foreign policy of the United States.

P.L. 480, therefore, is not purely altruistic in nature, but is instead a powerful economic and political tool. The implications are that food aid may not be distributed in a manner that is directly proportional to the need of the recipient country.

Support for this theory can be found in P.L. 480's early history. The American agro-industrial complex grew out of the restructuring of the economy after World War II. In the

1950s, greater mechanization, widespread chemical use, and the consolidation of farmland created large surpluses of food commodities. At the same time, there was growing fear of a pan-Asian communist revolution that later spread to Latin America. P.L. 480 could help address both problems: Farmers would be paid for and relieved of their surplus commodities, and recipient countries could cheaply pay for agricultural food aid in their own currency and develop stronger relations with the United States.

Today, about $2 billion in taxpayer money is used every year to purchase food commodities from U.S. producers and to ship them overseas (using U.S. carriers). The process can take up to six months, and with rising fuel costs, the quantity of food that tax dollars can buy has diminished. To save costs, in 2008, President George W. Bush petitioned Congress to modify the farm bill so that 25 percent of the food aid budget would be spent on food purchased in other countries. Tax dollars would have greater purchasing power overseas, and shipping costs could be reduced. But Congress turned him down.

Other countries such as Canada have modified their food aid programs to give cash instead of farm commodities. The United Nations' World Food Program, which provides emergency food aid and is responsible for distributing close to half of the food from USAID, has also decided to buy most of its stocks from developing countries. By doing so, World Food Program has gained logistical efficiency and has saved temporal and shipping costs in its operations. Farmers in developing countries also benefit from these direct market opportunities.

Criticisms

Critics of P.L. 480 point out the United States' program for food aid does more harm than good. By dumping farm surpluses in poor countries, the United States is creating a cycle of dependency, where recipient countries have little economic incentive to develop their own agricultural sectors. In addition, by disproportionally supplying wheat products, the United States has systematically modified local food tastes. In places where cassava, rice, or yam represented a traditional food staple, people have substituted and demanded wheat-based products instead, further undermining local agriculture.

P.L. 480, others argue, distorts global agricultural markets. Although USAID earmarks millions of dollars yearly to develop agricultural projects in developing countries, food aid can make it even more difficult for small farmers to compete on the international (or even local) market.

In spite of the economic obstacles that food aid can create in developing countries, political issues arise as well. Critics of the current food aid paradigm claim that in poor nations, food can be a powerful tool for coercing governments. Or, conversely, recipient governments can use food aid to coerce their own people. Either way, food aid is not distributed according to actual need. It is turned into a political instrument. In reproach, a United Nations initiative called the Right to Food has earned support from numerous countries. The Right to Food positions food as a right for all people and not as a means to political ends.

Taxpayers also complain that the P.L. 480 program is inefficient and that by allowing for overseas purchases of food, more farm commodities can be bought and shipping costs can be cut.

See Also: Department of Agriculture, U.S.; Farm Bill; Food Security.

Further Readings

Newman, Brian. "A Bitter Harvest: Farmer Suicide and the Unforeseen Social, Environmental and Economic Impacts of the Green Revolution in Punjab, India." *Food First Development Report*, 15 (January 2007).

Salzman, Avi. "U.S. Food Aid: We Pay for Shipping," *BusinessWeek* (July 9, 2008).

U.S. Department of Agriculture. "The 1996 Farm Bill." *Agricultural Outlook* (1996).

Vengroff, Richard and Yung Mei Tsai. "Food, Hunger, and Dependency: P.L. 480 Aid to the Third World." *Journal of Asian and African Studies*, 17:3–4 (1982).

Anna Carla Lopez
San Diego State University

RECOMBINANT BOVINE GROWTH HORMONE

Recombinant bovine growth hormone (rBGH) is a genetically modified hormone that increases milk production in dairy cows. Also called recombinant bovine somatotropin, or rBST, the hormone rBGH was the first genetically engineered commercial product marketed in the United States by Monsanto. Use of the hormone has sparked international controversy and public debate over its effects on human health, livestock health, and the environment. These controversies occur within the increasingly complex context of industrialized U.S. farming practices, agribusiness, government regulatory structures, and consumer interests. Perhaps in part as a result of the controversies surrounding it, rBGH use in the United States is decreasing; in 1998, about 30 percent of cows in the United States were being injected with rBGH, but by 2007, only 17 percent of U.S. cows were injected with the hormone.

rBGH augments natural processes in dairy cows. A cow's pituitary gland naturally produces somatotropin, a hormone that triggers lactation. To produce rBGH, the synthetic hormone, scientists remove a portion of the cow's DNA that includes the somatotropin code and then implant it in the DNA of the bacteria *Escherichia coli*. After some bacterial growth occurs, the rBGH can be extracted from the bacteria. Although rBGH mimics hormones produced naturally by cows, the synthetic version contains a slightly different set of amino acids. To use rBGH, dairy farmers inject the hormone into cows every two weeks. The rBGH triggers the release of another hormone, insulin-like growth factor 1 (IGF-1), which spurs additional milk production. Although IGF-1 also naturally occurs in cows, rBGH encourages higher levels of IGF-1 production to increase milk production anywhere from 5 to 20 percent.

Early trials of rBGH focused on testing the efficacy and economics, rather than the safety, of the synthetic hormone. Cornell University researcher Dale Bauman conducted the first scientific trials of rBGH on 30 cows from 1982 to 1985 with the support of Monsanto Company, the sole U.S. retailer of rBGH. The study found that it indeed increased milk production. At the same time, Robert Kalter conducted an economic analysis of the hormone. The Kalter study predicted that introducing rBGH into the market would force 30 percent of dairy farmers out of business within five years. These reports, which focused solely on the effectiveness and economics of rBGH, formed the basis for many future decisions about the substance. Before federal approval, Monsanto conducted a safety study in which rBGH was tested on 30 rats for 90 days. Using that study as evidence, the U.S. Food and Drug Administration (FDA) approved rBGH in 1994, declaring its use in dairy production to be safe for humans. In 1994, Monsanto began selling rBGH under the trade name Posilac to dairy farmers. Early opponents were not concerned with the effectiveness of the product; they were concerned about how introduction of rBGH into the market would affect family farms, livestock, human health, and the environment.

Human Health, Livestock, and Environmental Impacts

The approval and use of rBGH has been riddled with controversy, both for its potential human health impacts and for its effects on cow health and on the environment. The science around these issues has been strongly contested, creating uncertainty about the effects of rBGH. For example, some studies have asserted that rBGH and IGF-1 appear in milk. The effect of increased concentrations of IGF-1 in humans is largely unknown, but it has been linked to increased prevalence of several types of cancer including prostate and breast cancer, as well as to an increased prevalence of multiple births. rBGH use can also increase the fat and decrease the protein content in milk. Consumers are also concerned about the tendency of rBGH use to increase somatic cell counts, or pus from infected udders, in dairy products. Cows injected with rBGH are more prone to infections that require antibiotics. These antibiotics may find their way into milk consumed by humans, contributing to broader concerns of increasing bacterial resistance. Some consumers also worry that rBGH in milk may stimulate an immune response or allergic reaction in some people. These concerns were not considered in the FDA approval process.

The use of rBGH also presents potential adverse health effects on dairy cows. Because rBGH prolongs lactation in cows, it also prolongs the period during which cows are weakened and susceptible to infection. The risk for clinical and subclinical mastitis, or inflammation of the udders, is particularly heightened. These conditions often require antibiotics, including penicillin, amoxicillin, and erythromycin, which may leave residue in the cow's milk. Injecting rBGH into cows can also increase the incidence of cystic ovaries and digestive disorders in cows. These health effects may be quite significant: A 1998 report by Health Canada attributed the use of rBGH to a 25 percent increased risk of mastitis and a 50 percent increased risk of critical lameness in cattle.

The use of rBGH also raises several environmental issues including the proliferation of antibiotics, the overall impact of industrialized dairy farming, and the effect on climate change. As described earlier, the use of rBGH increases the need for antibiotic applications in the dairy industry, thereby contributing to broader problems of bacterial resistance in the environment. Some have also suggested that rBGH contributes to a shift from small family farms to industrialized dairy farming, with the latter activity contributing to a myriad of environmental problems including soil erosion, water pollution, and land degradation. In addition, industrialized dairy farming may contribute to climate change as a result of increased methane emission resulting from corn-fed cows as compared with grass-fed cows. Recently, however, rBGH proponents have been touting the hormone's use as a technique to reduce greenhouse gas emissions by suggesting that the resulting increased milk efficiency will decrease the number of methane-producing cattle. Dairy cows do produce about a fifth of U.S. methane emissions, but critics have questioned the claims that rBGH reduces these emissions. Some critics suggest that cows injected with rBGH need additional feed inputs to meet the higher milking demands, thereby negating any potential emission reductions. Other studies have shown that rBGH use either does not affect green-house gas emissions or may slightly increase them.

Labeling and Corporate Politics

Until recently, the Monsanto Company has been the only U.S. retailer of rBGH. In some ways, Monsanto needed to overcome a large obstacle in marketing rBGH: Since 1950, U.S. dairy farmers have produced more milk products than Americans can consume, so the introduction of rBGH feeds an already-flooded market. To create a need in the market, and to overcome consumer concerns, Monsanto engaged in controversial corporate politicking with the U.S. regulatory agencies, the media, and any groups attempting to label consumer products as rBGH-free. For example, the connections between FDA officials and Monsanto reveal the corporate politics: the FDA official involved in prohibiting rBGH labeling was Michael R. Taylor, a former partner of King and Spaulding, a firm that brought forth lawsuits on Monsanto's behalf. This "revolving door" created some consumer skepticism about the safety of rBGH. Some also suggest that Monsanto played a role in the firing of Dr. Richard Burroughs, a senior FDA scientist who claims he was fired when he raised concerns about the safety of rBGH.

Further controversy has arisen over the extensive public relations war Monsanto has waged against rBGH naysayers. For example, in 2001 two Fox News investigative journalists in Tampa, Florida, were fired after they reported on consumer concerns about rBGH. The station initially delayed airing the story because Monsanto threatened to sue the station. Concerns over the human health, livestock, and environmental impacts of rBGH have sparked additional controversy over consumer labeling of rBGH and non-rBGH milk products. Public surveys have shown that most Americans would prefer to have milk produced using rBGH to be labeled as such. Initially, however, the FDA did not allow retailers to label milk as being rBGH-free for two reasons: the FDA does not regulate rBGH and cannot keep track of the accuracy of labeling claims, and according to the FDA, there is "virtually" no difference between the rBGH-free milk and milk with rBGH. The agency also argues that rBGH is not an additive and therefore does not require labeling.

Monsanto has vehemently opposed state and corporate efforts to label or ban the use of rBGH in consumer products. For example, the state of Maine had been applying the "Quality Seal" to dairies that did not use rBGH. In 2003, Monsanto attempted to stop this practice by suing Oakhurst Dairy, Maine's largest dairy operation, for implying that there is a difference between rBGH milk and milk produced without the artificial hormone. A settlement allowed Oakhurst to label if they also included a statement that the FDA has found no significant difference between milk produced with or without rBGH. It is worth noting that the FDA has a limited mandate and only can consider a fraction of the many reasons why consumers might oppose rBGH. In 1994, shortly after the introduction of rBGH, the state of Vermont attempted to require labeling of dairy produced using rBGH, which initiated a protracted battle with Monsanto and agribusiness. Six trade groups, including the International Dairy Foods Association, brought a lawsuit against Vermont claiming that the labeling law challenged First Amendment rights. This resulted in a scaling back of Vermont's rBGH labeling program from mandatory to voluntary. In 2008, a group called the American Farmers for the Advancement and Conservation of Technology formed to try to stop these rBGH labeling efforts. The group has been largely funded by Monsanto and an agriculture public relations firm.

Another big player in the rBGH labeling controversy is the Oregon-based Tillamook County Creamery Association, which banned the use of rBGH by its 147 dairy farmer members in 2004. This ban followed the creamery consumers' concerns about the effects of rBGH, although it caused controversy both within the cooperative and with Monsanto. Before the vote to ban rBGH use in the cooperative, Monsanto sent letters to dairy farmer members asserting their support of the farmer's freedom to choose the way they produce milk. After the initial vote, Monsanto urged a revote, which upheld the previous decision. Although Tillamook does not allow rBGH use by its members, the creamery does not label its consumer products as rBGH-free.

Facing increased consumer concern over rBGH, some retailers have banned the sale of dairy products produced with the synthetic hormone. In 2007, grocery chains Kroger and Safeway banned the sale of dairy produced with rBGH. In 2008, Starbucks and Wal-Mart (for store-brand dairy only) followed suit. Many countries have also banned the use of rBGH, including Japan, Australia, Canada, and the countries of the European Union. Health groups including the American Nurses Association have expressed concern about the human health effects and have supported labeling efforts. Future regulation of rBGH in the United States is uncertain, but Monsanto will not play a role in that future; in August 2008, Monsanto sold its Posilac division to Eli Lilly & Co for $300 million.

See Also: Agribusiness; Dairy; Genetically Modified Organisms; Monsanto.

Further Readings

Ehrenfeld, D. "The Cow Tipping Point." *Harper's Magazine* (October 2002).
Mills, L. N. *Science and Social Context: The Regulation of Recombinant Bovine Growth Hormone in North America*. Montreal, Canada: McGill-Queen's University Press, 2002.

Kate Darby
Arizona State University

RETAIL SECTOR

The food system consists of broad sectors that include production, processing and manufacturing, distribution, and retail. The retail sector is the part of this system that is closest to the consumer. Although processors/manufacturers were once quite powerful, exerting tremendous influence over retailers and consumers, this dynamic is shifting toward greater retailer power. The leading retailers make up a large and increasing percentage of sales for the largest food processors. Wal-Mart, for example, sells more than 10 percent of the products made by Dean Foods, General Mills, Kellogg, Kraft, Campbell's, Tyson, and Pepsi. This gives Wal-Mart the leverage to demand low wholesale prices, as well as exacting packaging and delivery requirements, from its suppliers. Leading retailers are also more effectively encouraging consumers to behave in ways that enhance their profits, using means such as store design and shelf placement. Their increasingly sophisticated marketing strategies also include discounts for customers who use loyalty or rewards cards—these enable the collection of massive amounts of data and highly targeted advertising efforts.

A shopper in Maputo, Mozambique, selecting packaged meat at a supermarket butcher counter in 2002. The supermarket format has spread to much of the world, and Western companies are looking to developing countries for new audiences.

Source: World Bank

Food retailing was dominated by supermarkets from the 1950s to the 1990s. The supermarket format originated in the United States and is characterized by: (1) a reliance on customers to self-service and line up to check out; (2) a larger store size and higher number of food items (currently 15,000+) than most other formats; (3) low markups from wholesale prices; and (4) high sales volumes (currently $2 million or more annually). Many of these innovations were facilitated by the development of a national highway system and refrigerated trucking, which lessened dependence on railroads and encouraged more self-distribution by retailers. By shipping directly from manufacturers, supermarkets were able to negotiate price discounts based on volume.

Until recently, more than 75 percent of food sales in the United States were made through supermarkets and grocery stores, and the average person was estimated to spend 2 percent of their life in these retail outlets.

Supermarkets and grocery stores are facing increasing competition from other retail formats including hypermarkets/supercenters (a combination supermarket and department store, pioneered by the French firm Carrefour in the 1960s), wholesale buying clubs, convenience stores, inexpensive department stores (e.g., dollar stores), and specialty retailers (e.g., natural/organic markets, ethnic markets). They are also affected by a trend of increasing expenditures for food consumed outside the home, particularly at fast food restaurants. The differentiation unfolding in the retail sector reflects a shift from mass production and consumption to a model more highly targeted to particular groups of consumers (based on age, socioeconomic status, ethnicity, or other differences), or even to individuals.

The entrance of Wal-Mart into food retailing has had a dramatic effect on the retail sector. Wal-Mart began selling groceries in a supercenter format in 1988, and in little more than a decade it became the leading food retailer in the United States. The corporation also operates a chain of wholesale buying clubs (Sam's Club) and introduced a chain of smaller-format grocery stores (Wal-Mart Neighborhood Market) in 1998. Wal-Mart's involvement in food retailing helped to trigger a wave of consolidation in the 1990s, particularly in North America and Europe, as other retailers merged or made acquisitions to remain competitive. It is estimated that the top 3 firms control 80 percent of grocery sales in Australia, the top 4 firms control 75 percent of sales in the United Kingdom, the top 5 firms control 50 percent of sales in the United States, and the top 30 firms control 33 percent of sales globally.

Food retailing is an increasingly international business, with the leading firms in North America and Europe rapidly dominating markets in Latin America, Africa, and Asia. This interest in expansion is a result of slowing sales in highly industrialized nations. Much of their growth in foreign countries has occurred through acquisitions of existing supermarket chains but has also included opening new stores. Even firms that do not sell globally do source food from all over the world to ensure an adequate supply of food (such as out-of-season fresh produce) and, increasingly, to secure the lowest possible prices for a given commodity.

By establishing their own distribution networks and supply chains, retailers exert a high level of control over the food production process. These companies develop their own standards for growing, harvesting, and processing fresh produce, often requiring more stringent safety practices than government regulations. Control over processed food production is increasing as well, through the growth of private label or store brand products. Retailers enter into contracts with manufacturers to package products with the store brand that are typically sold at lower prices than comparable name-brand products. The quality of these private label foods is much improved compared with early generic products, and sales are growing much faster in this segment than name-brand items.

These trends have resulted in relatively low prices for consumers, as retailers have engaged in cost-cutting strategies and increased the efficiencies of their operations. Another benefit is the increasing diversity of food that is available year-round. In addition, the convenience of ready-to-eat meals and markets open 24 hours a day has enabled consumers to reduce the amount of time spent on planning and preparing meals.

The retail sector has also faced criticism, however. The low prices for consumers, for example, are partially at the expense of producers, who lack sufficient bargaining power and may sell at below the cost of production. Access to low-priced, nutritious food is not equitably distributed either. A number of retail food chains in the United States have shut down inner-city stores and opened new locations in growing suburbs. Some urban areas may have abundant convenience stores and fast food restaurants but few options for fresh produce. Retailers have also been accused of encouraging sprawl, excessive energy consumption in the production and transport of food, mistreatment of animals, antiunion activities, and in some instances, price fixing. Recent concern with rising obesity rates has led to claims that food retailers encourage the consumption of unhealthy but profitable foods.

Retailers have implemented a variety of efforts to attempt to address these criticisms. U.S. companies including Hannaford and SuperValu have introduced nutrition labeling programs on their store shelves to help customers identify foods that are more or less healthy than others. In the United Kingdom, Tesco is labeling some of its private label products with their carbon footprint—the amount of carbon dioxide produced in their production. Wal-Mart has touted its commitment to reducing energy consumption and reducing waste. Ninety percent of the leading supermarket chains in Europe have removed genetically engineered ingredients from their stores, and many of the largest retailers in the United States have banned milk produced with a genetically engineered growth hormone for their store-brand dairy products.

See Also: Concentration; Fast Food; Supermarket Chains; Wal-Mart.

Further Readings

Martinez, Steve W. "The U.S. Food Marketing System: Recent Developments, 1997–2006." U.S. Department of Agriculture, Economic Research Service (May, 2007). http://www.ers .usda.gov/Publications/ERR42 (Accessed January 2009).
Progressive Grocer. http://www.progressivegrocer.com (Accessed January 2009).

Seth, Andrew and Geoffrey Randall. *Supermarket Wars: Global Strategies for Food Retailers.* New York: Palgrave Macmillan, 2005.

Philip H. Howard
Michigan State University

RICE

This Japanese farmer is transplanting young rice seedlings into a rice paddy. Rice has been cultivated in Japan since at least 1000 B.C.E.

Source: iStockphoto.com

The term *rice* refers to the grain-producing species of grass with narrow leaves that grows about 60 to 180 centimeters tall. There are 20 wild species and two cultivated species of rice. Rice is native to Africa, India, and China. The word *rice* has Indo-Iranian origins. Rice is a major staple food, and rice production is a major source of income and employment for millions of people in 144 countries worldwide. Rice is a superior source of energy among the cereals. Polished rice, however, has shown vitamin A and mineral deficiencies, causing weak immune systems in children in countries where rice is a major staple food. Parboiled rice, in contrast, is nutritionally superior to polished rice. In south and southeast Asia, rice is cultivated under submerged conditions in paddies. The practice of intensive rice monoculture in Asia has caused problems of waterlogging, salinity buildup, and nutrient depletion. Rice cultivation is one of the most important sources of atmospheric methane. Rice is used in making noodles, infant foods, snack foods, breakfast cereals, and fermented products. Rice straw is an important cattle feed throughout Asia.

Origin

Rice belongs to the Poaceae or Gramineae family and genus *Oryza*. The genus *Oryza* includes 20 wild species and two cultivated species. The wild species are generally found in humid tropics and subtropics. Of the two cultivated species, African rice (*Oryza glaberrima*) is native and confined to west Africa, whereas Asian rice (*Oryza sativa*) is commercially grown in all continents. The reason for the parallel evolution of the above two cultigens in Africa and Asia is the origin of genus *Oryza* in Gondwanaland in the supercontinent Pangaea before it fractured and drifted apart. *Oryza*'s origin in Gondwanaland is also confirmed by the presence of closely related wild species of rice in Australia, Central America, and South America. The wild rice species of North America is *Zizania palustris*. Traditionally, this species was self-propagated and harvested only by

Native Americans in the Great Lakes area. Today, it is grown in Minnesota and northern California.

There are two major subspecies in *Oryza sativa*: the sticky, short-grained *japonica* (or sinica) variety, and the nonsticky, long-grained *indica* variety. *Japonica* is generally cultivated in dry fields in east Asia, whereas *indica* is mostly cultivated submerged under water in tropical Asia. Although the origin of domestication of Asian rice is still controversial, the scientific consensus concludes that *Oryza sativa indica* was domesticated in eastern India and *japonica* in southern China. A recent archaeological study suggests a single domestication of rice in the lowlands of China. The Yangtze was probably the site of the earliest rice cultivation.

China introduced *japonica* rice to Korea before 1030 B.C.E. Rice cultivation in Japan began in the late Jomon period in about 1000 B.C.E. From the Indian subcontinent and mainland southeast Asia, the *indica* rice spread to Sri Lanka before 543 B.C.E., to the Malay Archipelago, to the Indonesian islands between 2000 and 1400 B.C.E., and to central and coastal China.

The Middle East acquired rice from south Asia in 1000 B.C.E. The expedition of Alexander to India (c. 327 B.C.E.) introduced rice to the Romans; Persia introduced rice to Europe through Egypt in the 4th and the 1st centuries B.C.E. The Turks brought rice from southwest Asia into the Balkans. Moreover, in Africa, Indonesian settlers introduced Asian rice to Madagascar. Countries in west Africa obtained Asian rice through European colonizers. In the Western Hemisphere, the Caribbean islands also received rice from their European colonizers in the 15th and 16th centuries, whereas Central and South America received rice seeds from Spain between the 16th and 18th centuries. Rice cultivation in the United States began around 1609 in Virginia. California began growing the *sinica* variety of rice in 1909. Rice was introduced into Hawaii by Chinese immigrants. Cultivation of African rice (*O. glaberrima*) remains confined to west Africa, but African rice has been rapidly displaced by Asian rice.

Production

In Asia, rice is often grown submerged in paddies. Edible fish and frogs are sometimes grown in the same paddies. Rice plants generally grow up to 1.8 meters tall. Farmers in Asia initially raise rice plants in nursery beds, then transplant young rice seedlings into level and puddle fields. Transplanting practices and the use of water buffalo and spike-tooth harrow began during the Han period (23–270 C.E.) in China and subsequently diffused to Southeast Asia. Although transplanting is a labor-intensive operation, it provides efficient water utilization, better weed control, uniform maturation of the plants, and higher grain yield. In Asia, women usually do activities related to planting, weeding, harvesting, processing, management, and preservation of rice seeds.

In the southern United States, rice is not transplanted, but seeds are dribbled into dry soil and briefly irrigated and drained. The seeds are allowed to germinate, and water is reintroduced when the seedlings are established. In the southern United States, slave laborers from Senegambia and Sierra Leone brought the knowledge of rice cultivation. In northern California, pregerminated seeds are dropped from airplanes into cool water several inches deep. Rice in California was initially cultivated in small amounts by Chinese laborers during the California Gold Rush.

During the first half of the 20th century, production of rice increased greatly as a result of the increase in the area of rice production and the use of nitrogen fertilizers. In the late 1960s, the green revolution increased global rice production. World rice production in

2007 was approximately 645 million tons. Nearly 114 countries grow rice across an area from 53 degrees latitude north to 35 degrees south. China and India grow more than half the world's rice production. Asian farmers produce about 90 percent of the total global rice production. The top five rice-producing countries include China, India, Indonesia, Bangladesh, and Vietnam. The scientists at the International Rice Research Institute in the Philippines and in other institutions are conducting research to create new, higher-yielding rice varieties to increase rice production.

Environment

Rice cultivation is one of the major sources of atmospheric methane. The contribution of rice paddies to global methane is estimated at 100 plus or minus 50 Tg per year. Cultivating rice under submerged conditions enables organic matter to accumulate in soils, which contributes to carbon sequestration. The continuous flooding of rice fields decreases the rate of soil nitrogen mineralization and increases salinity buildup.

Consumption

Rice is a staple food for nearly one-half of the world's population. China and India are the largest consumers of rice in the world. Rice provides 60 percent of the food intake in southeast Asia and about 35 percent in east Asia and south Asia. Myanmar (Burma) has the highest per capita rice consumption (217 kilograms (kg) per person per year), followed by Vietnam (211 kg/person/year) and Bangladesh (182 kg/person/year). In Bangladesh, Cambodia, Indonesia, Laos, Myanmar, Thailand, and Vietnam, rice provides 50–80 percent of the total calories consumed. Only about 5 percent of the world's rice enters the international market, as most of the rice is consumed in the producing countries.

Nutritional Value

Rice is a superior source of energy among the cereals. Only oat (68 percent) ranks above rice (65 percent) in protein quality among cereals. However, milling of brown rice into white rice results in a nearly 50 percent loss of the vitamin B complex and iron, and washing milled rice reduces the water-soluble vitamin content. Rice is low in sodium and fat, and rice is cholesterol and allergen free.

In countries where rice is a major source of calories, about 125 million children are suffering from vitamin A deficiency, which might lead to blindness and death. The rice consumed is generally polished rice that contains no beta-carotene or other forms of provitamin A. Use of polished rice has shown to have increased beriberi because of the lack of thiamine (Vitamin B_1). An effort was made in the 1950s to enrich polished rice with free vitamin derivatives. After the 1950s, nutritional intakes of the masses in Asia generally improved, and with dietary diversification, beriberi receded as a serious threat.

Parboiling rough rice is another way to keep beriberi at bay. Parboiling permits a smaller loss of vitamins, minerals, and amino acids in rice. Parboiled rice is popular among the low-income people in south Asia and parts of west Africa. In 1999, scientists at the Institute of Plant Sciences at the Swiss Federal Institute of Technology developed genetically engineered rice known as Golden Rice. In Golden Rice, two genes were inserted into the rice genome by genetic engineering. This intervention leads to the production and accumulation of β-carotene in the grains. Golden Rice has a Humanitarian Use License.

As long as a farmer does not make more than $10,000 per year by selling Golden Rice, no royalties need be paid to Syngenta, the company that sponsored the research.

Culture and Rice

Rice plays important roles in Asian cultures. In the Shinto religion, the emperor of Japan is the embodiment of the ripened rice plant. There are festivals dedicated to rice and rice cultivation in China. The Japanese refer to rice as "mother." Villagers in Senegal honor their special guests with a rice meal. In India, uncooked rice mixed with turmeric or saffron is very auspicious and is used in religious rituals.

Depending on the region, rice is eaten with fish or meat or with legumes such as beans, lentils, and chickpeas. The combination of rice and fish in the diet is common in Asian societies. The combination of rice and legumes is characteristic of cuisines from Cajun to Mexican to the Middle East to southern Europe. In Colombia, "rice and beans" is a national food. White boiled rice is the primary way in which rice is eaten in Asia. Rice is also used in making a variety of dishes. Dosa (rice pancakes) and idli (made from black lentils and rice) are commonly eaten in southern India. Other rice dishes in India include bisi belebath, biryani, curd rice, puffed rice, khichidi, and so on. In the Chinese culture, the common rice dishes are fried rice, steam rice soup, zongzi, rice noodles, and so forth. Japanese sushi is available throughout the world; damburi, onigiri, chazuke, and kayu are some of the Japanese rice dishes. Bibimbap, bokkeumbap, and boribap are rice dishes from Korea. Other rice dishes include goto from the Philippines; jambalaya, a Creole dish; and kateh, a Persian rice dish. Ketupat and lumpia are Malaysian rice dishes, lemang is an Indonesian/Malaysian dish, and bhuna kichuri is common in Bangladesh. Amazake, horchata, rice milk, and sake are some of the rice beverages found in Asia.

Economy

Much of the rice is consumed where it is produced. *Indica* rice accounts for 75–80 percent of the global rice trade, followed by *japonica* (10–12 percent) and aromatic rice such as basmati and jasmine (10 percent). For many small farmers in Asia and Africa, the plunge in rice prices is a major cause of poverty and hardship. Although the global rice market has liberalized, most rice-producing nations heavily support rice producers through market price interventions. The European Union, United States, Japan, and Korea rely heavily on direct payment or market price support. The United States and the European Union procure and transfer large quantities of rice through food aid on concessional terms. Developing countries of south and southeast Asia have some form of price support system such as public procurement for public distribution of food grains (India) or for market interventions (Vietnam, Thailand, China, and Indonesia).

See Also: Agrofood System (Agrifood); Famine; Green Revolution.

Further Readings

Carney, J. *Black Rice: The African Origins of Rice Cultivation in the Americas*. Cambridge, MA: Harvard University Press, 2001.

Chang, Te-Tzu." Rice." *The Cambridge World History of Food*, 1st ed., edited by K. K. Kiple and K. Ornelas. Cambridge, UK: Cambridge University Press, 2000.

Grew, Raymond. *Food in Global History*. Boulder, CO: Westview, 2000.

Jiang, Leping and Li Liu. "New Evidence for the Origins of Sedentism and Rice Domestication in the Lower Yangzi River, China." *Antiquity*, 80 (2006).

Mew, T. W., et al. "Rice Science: Innovations and Impact for Livelihood." Manila, Philippines: International Rice Research Institute (IRRI), 2003.

Wassmann, R., et al. "Methane Emission From Rice Paddies and Possible Mitigation Strategies." *Chemosphere*, 26/1–4 (1993).

Shrinidhi Ambinakudige
Mississippi State University

ROUNDUP READY CROPS

With increasing pressure for higher crop yields, there has been extensive use of the broad-spectrum herbicide Roundup, which is manufactured in the United States by Monsanto and makes use of the active ingredient glyphosate. Since 1980, Roundup has been the best-selling herbicide around the world. To use the herbicide, Monsanto has been involved in the production of seeds that have been genetically engineered so that they are tolerant to glyphosate; these are known as Roundup Ready crops. This allows farmers to plant their seeds—purchased from Monsanto—and then use Roundup to prevent the emergence of weeds. With Roundup heavily used by private individuals in garden maintenance and also by local governments in removing weeds from roadsides, glyphosate has been found to be effective in removing grasses and also woody and broadleaf plants.

Essentially, much of the debate about green food centers on the nature of genetically modified (GM) crops and whether these are safe for human consumption over long periods of time. The debate over GM crops hinges not only on whether they are safe for consumption but also on whether they should be labeled as "GM crops" when sold in stores; the other factor is that most of the seeds sold by Monsanto are capable of crops for only one season. This means that each year the farmers have to buy seeds from Monsanto, making them dependent on the company, which can, after it has achieved a monopoly position, increase the price at which it sells the seeds.

At this time, the available Roundup Ready crops include soybeans, maize/corn, canola, sugar beet, and cotton—the latter obviously not for consumption and therefore having far fewer problems with environmental activists. Roundup Ready wheat and alfalfa are currently in development. The arguments for Monsanto's promoting Roundup Ready crops essentially rest on the fact that crop yields are higher and that the crops themselves are safe. They also argue that there is no major effect on the environment.

All these debates are also central to the Roundup Ready crops, and all are keenly contested. According to Monsanto, the crop yields produced are between 7 and 11 percent higher than would otherwise be the case. In contrast, some environmentalists have claimed that there is a 6.7 percent lower yield. However, many would argue that whichever is the case, the yield is not important because the crops themselves are contaminated by the use of Roundup.

Clearly, as an herbicide Roundup is toxic and not meant for human consumption, and hence its use on food crops has brought the issue of its use in conjunction with food crops into focus. Monsanto initially countered by arguing that Roundup is "practically non-toxic" to mammals, birds, and fish. There are claims that Roundup is far safer for use with crops than many other herbicides, and the research has tended to bear this out. That some types of clover, often found in gardens, are not affected much by Roundup shows that some plants have, or have developed, a resistance to the herbicide. But Monsanto has gone further and has argued that Roundup does not pollute the environment, and at one stage the company even went so far as to state that it was safer than table salt—the latter, obviously, in large quantities is detrimental to the growing of many crops, as in the apocryphal story of the Romans sowing the ruins of Carthage with salt to prevent the locals from growing any crops there afterward—a story believed now to be untrue.

Environmentalists and consumer rights advocates have countered by claiming that Monsanto was inaccurate when it claimed that Roundup was not dangerous, and that it was classed in the European Union as being "dangerous for the environment" and also "toxic for aquatic organisms." It certainly seems poisonous to amphibians, especially to many frog species. This led to a court case that saw Monsanto, in 2007, fined $19,000 in France for false advertising after the chairman (by the time of the court case, former chairman) of Monsanto Agriculture France had claimed that Roundup was biodegradable. Many also point to the fact that glyphosate has been used as an herbicide by the U.S. government to destroy coca crops in Colombia as evidence of the harmful nature of Roundup.

Furthermore, the case of Rachel Carson's campaign against DDT (dichlorodiphenyltrichloroethane) is often highlighted, whereby DDT was found to have a dramatic effect on wildlife and polluted the ecosystem in a far wider manner than had been thought to be the case before Carson's *Silent Spring* (1962) was published. Her facts were challenged for years but have now been found to have been correct; DDT has since been banned. This has been why the effect of Roundup on overall ecosystems is regarded by many environmentalists as being as important as the crops being produced, especially in relation to the water table and the possibility of polluting drinking water.

Overall, given the nature of the debate over GM foods, many of the consumer-led arguments have been that the food products resulting from GM crops should be labeled as such in supermarkets and stores to allow purchasers to choose whether or not they want to buy (and hence consume) GM products. Monsanto and companies connected with GM products have fought this, and this is also the case with the sale of Roundup Ready crops: Arguments have been made against any of the crops grown being labeled as GM products, with arguments made that the use of many synthetic fertilizers and the like for the production of crops are also not stated at the point of sale.

See Also: Corn; Fertilizer; Genetically Modified Organisms; Monsanto; Soybeans; Weed Management.

Further Readings

Benbrook, Charles. "Evidence of the Magnitude and Consequences of the Roundup Ready Soybean Yield Drag From University-Based Varietal Trials in 1998." Ag BioTech InfoNet Technical Paper Number 1. http://www.biotech.info.net (Accessed July 2009).

de la Perriere, Brac, et al. *Brave New Seeds: The Threat of GM Crops to Farmers.* New York: Zed, 2000.

Fukuda-Parr, Sakiko. *The Gene Revolution: GM Crops and Unequal Development.* London: Earthscan, 2007.

Shipitalo, M. J., et al. "Impact of Glyphosate-Tolerant Soybean and Glufosinate-Tolerant Corn Production on Herbicide Losses in Surface Runoff." *Journal of Environmental Quality*, 37/2:401–08 (2008).

Justin Corfield
Geelong Grammar School, Australia

RURAL RENAISSANCE

The Earth reached a noteworthy milestone in 2007. For the first time in history, more of its human population lived in urban areas than in rural areas. In the United States that milestone was passed almost a century earlier, in the 1920s. The United Nations projects that by 2050, approximately two-thirds of the world's population will reside in urban areas. Urban immigration is often accompanied by smaller, counter emigrations to rural settings. The term *Rural Renaissance* broadly describes those ex-urban movements.

In the strictest sense, Rural Renaissance describes a period in the United States from the late 1960s to the mid-1970s. During this period, for the first time since before the Great Depression, practically all indicators of population and economic growth pointed to rural areas instead of urban centers. However, the term *Rural Renaissance* more broadly has described a variety of counter-urban movements throughout the 20th century in the United States, as well as in Australia, Japan, and a variety of European nations. In the 21st century, the term *Rural Renaissance* has experienced renewed use to describe revitalized hopes and opportunities for life in rural places.

Although the primary direction of migration in the United States during and following the Great Depression was toward the nation's urban centers, at that time there were also counter movements promoted by men such as Ralph Borsodi that encouraged empathizers to return to the land to live simply and self-sufficiently. Although the majority of the people moving were immigrating to urban areas to find employment, a minority was optimistically moving to rural areas in an attempt to sustain themselves off the land. That movement continued through and following World War II.

A significant moment in the United States' Rural Renaissance occurred in 1954 with the publication of Scott and Helen Nearing's influential *Living the Good Life*, a book that significantly contributed to the hope for a Rural Renaissance that peaked at its demographically identifiable height in the 1960s and 1970s. Together with that book, periodicals like *Whole Earth Catalog* and *The Mother Earth News* provided the principles and practicalities for the back-to-the-land movement.

Following the migration of people and their accompanying wealth to the perceived higher quality of life in rural areas during the late 1960s and 1970s, rural employment growth expanded by almost twice the urban growth rate. Although the numbers of people employed by farming and agriculture continued a two-decade decline, the hiring by manufacturing and service industries relocating to rural areas more than offset the losses. In addition, rural retirement communities multiplied as women and men of means left cities and sought a higher quality of life and a lower cost of living in more scenic rural atmospheres.

From its earliest time, the Rural Renaissance movement has been spurred by the need many people felt to recover a better, simpler life, free from the problems of urban living.

Fueling the height of the rural population shift of the early 1970s were sentiments described by some as feeling out of touch with nature, disgust with rampant consumerism, frustration with perceived shortcomings of government, and urban deterioration. In addition to principle and sentiment, the population shift of the early 1970s was fueled by the practical reaction against the era's widespread pollution and energy shortages. By the latter 1970s and early 1980s, the movement had additional proactive motivations such as the increased health value of noncommercially produced foods.

Although most vivid during the period of the late 1960s and 1970s, the Rural Renaissance is more nearly an ongoing phenomenon than an event unique to a particular time or location. The scale of migration to rural areas has waxed and waned in a variety of regions and eras. However, the latent interest by some urban dwellers to relocate to or to invest in rural lands, endeavors, and lifestyles has endured.

By the 1980s, the scales of population and economic growth had tipped back toward urbanization, making the early 1970s an anomaly in the trends of population movement in the United States. Many who embraced the back-to-the-land movement returned to urban and suburban life because of the stressful economic realities of rural homesteading. Widespread talk of a Rural Renaissance in the United States faded.

In spite of the numerous liabilities and obstacles of rural life, for the final decades of the 20th century, the ever-present latent sentiment toward a simpler, more connected, more satisfying lifestyle associated with rural places continued to take root in the hearts and dreams of a small but steady group of Americans who swam against the flow toward increased urbanization. In that sense, aspirations for a Rural Renaissance have lived on. Even for urbanites with no aspirations of relocating, the later 20th century saw many rural property values soar as investors purchased rural real estate, especially farms and forest land.

In the 21st century, a number of conditions make rural immigration more economically viable and more desirable than ever. Expanding broadband Internet access increasingly enables global connectivity and enhances rural life. The growing awareness of the health benefits of food produced outside the present industrial agriculture system fuels a growing market for organically grown and locally produced foods that originate in rural areas. The current emphasis on both production and consumption of renewable energy supplies grown in rural areas has renewed the motivation in some to embrace the possibility of a Rural Renaissance. Finally, some expectation for a Rural Renaissance is kindled by the hope that urbanites who experience sentiments toward a renewed connection to nature and life's basics, yet who are not inclined to relocate from urban and suburban settings, will sustain the emerging incidence of agricultural tourism. This would further open the possibility of additional revenue for those who make their homes rurally while providing an opportunity for urban residents to experience a taste of rural life.

An additional new factor affecting a future American Rural Renaissance, the scale of which can only be projected by early trends, is the migration of retirees from the "baby boom" generation. Following the Rural Renaissance of the 1970s, the rural communities that experienced the most consistent, predictable growth were communities attractive to retirees. At this time, as this unusually large demographic segment of the United States reaches retirement age, it is yet to be seen whether they will follow trends of previous generations and relocate from more urban to more rural settings.

A renewed emphasis on rural development at the level of the federal government has additionally spurred hopes and possibilities for a Rural Renaissance. The Rural Agenda for the Barack Obama presidency calls for increased economic opportunities for family farmers; increased capital available to rural entrepreneurialism and small business development; and increased funding for law enforcement, education, health care, and infrastructure, each of which strengthens the economic viability and quality of life in rural areas. In 2008, the U.S. Department of Agriculture Rural Development adopted the phrase *Rural Renaissance* to describe the current reality of rural opportunity and the U.S. Department of Agriculture's goal of empowering individuals, businesses, and communities to seize emerging rural opportunities.

Whether motivated by aspirations for increased quality of life, lower living costs, cleaner air, and availability of food, or for the fleeting availability of jobs in the 1970s, the latent dream of "a home in the country" goes on in the United States. Whether the dream is ever to be realized in its envisioned fullness is yet to be discovered. However, in the contemporary, as in the past, hope for an elusive Rural Renaissance endures.

See Also: Department of Agriculture, U.S.; Family Farm; Sociology, New Rural.

Further Readings

Dorr, Thomas. "Promoting a Rural Renaissance." U.S. Department of Agriculture. http://www.usda.gov/oce/forum/2008_Speeches/PDFSpeeches/Dorr.pdf (Accessed January 2009).

Galston, William and Karen Baehler. *Rural Development in the United States: Connecting Theory, Practice, and Policies.* Washington, D.C.: Island, 1995.

UN-HABITAT. "State of the World's Cities 2008/2009: Harmonious Cities." http://www.unhabitat.org/pmss (Accessed January 2009).

Richard B. Gifford
University of Arkansas Community College at Morrilton

S

Salmon

Salmon are anadromous fish, meaning that they migrate from oceans to freshwater streams to reproduce. There are six distinct species of Pacific salmon and one species of Atlantic salmon. The six Pacific species are: (1) chinook, also known as king, Tyee, and spring; (2) chum, also known as dog and keta; (3) coho, also known as silver; (4) pink, also known as humpie; (5) sockeye, also known as red; and (6) steelhead.

These large, red-fleshed fish are an increasingly important foodstuff for the residents of the United States and the European Union. Over half of this demand is met through the production of a million metric tons of farmed Atlantic salmon annu-

About 1 million metric tons of Atlantic salmon like this are produced every year through aquaculture.

Source: U.S. Department of Agriculture, Agricultural Research Service/Troutlodge, Inc.

ally. Wild salmon are culturally and economically important for Native Americans in the Pacific Northwest, as well as for many relative newcomers to the region. Atlantic salmon were once similarly important to the Northeast, but habitat destruction long ago reduced their prevalence. Salmon can migrate vast distances through the ocean and then return to the location of their birth, often far inland, to spawn. Small numbers of Atlantic salmon still return to the rivers of northeastern United States and Canada, though they are best known to modern food shoppers as farmed salmon, raised in cold oceanic waters around the world. Wild salmon, once prolific along both coasts, have suffered a long and slow demise throughout most of their range.

Salmon farming, an example of aquaculture, typically takes the form of raising hatchery-born Atlantic salmon in floating nets in oceans around the world. Significant producers include northern Europe, Chile, Canada, and the United States. Although salmon aquaculture helps meet the increasing demand for salmon, it introduces a number of environmental problems. For instance, the densely populated nets become breeding grounds for diseases affecting wild salmon, antibiotics used in aquaculture affect other organisms, and vast quantities of fish must be caught to feed to the carnivorous farmed salmon. Also worrisome, some research indicates that farmed salmon contain significant amounts of pollutants and toxins, and therefore may constitute health risks to consumers.

Wild salmon once migrated up the rivers of the Pacific and Atlantic coasts to spawn in great numbers. Historically, salmon "ran," meaning that they migrated through or reproduced in all accessible streams and rivers from Southern California to northern Alaska. Early European settlers to the Pacific Northwest reported massive runs of salmon. Some chinooks in the Columbia River, known as "June Hogs," were as large as 70 pounds. Some steelhead and coho migrated inland as far as Nevada. The range of all species has been severely reduced because of a number of factors, and many of the remaining runs are precarious at best. Because salmon are genetically unique in each stream, they are regulated by the Endangered Species Act at the level of "Evolutionarily Significant Units" rather than species or population.

Their complex lifestyle makes salmon vulnerable to diverse threats in multiple political jurisdictions, complicating efforts to protect them. Adults spawn in freshwater, the eggs hatch there, and the young (smolt) stay there for a month to several years, depending on the type of salmon. When ready, the young fish make their way out to sea. Most salmon spend the majority of their lives at sea, where they take advantage of high levels of biological productivity. When they reach sexual maturity, salmon return to the same stream where they were born to reproduce. The salmon returning to a certain stretch of stream only spawn with others that return to that same stream at the same time, leading to the development of genetically distinct stocks adapted to the conditions of their birthing streams.

On returning to their birth stream, female salmon thrash back and forth, using their bodies to create depressions in the sediment, known as redds, where they lay their eggs. Males swim alongside and fertilize the eggs. Females then cover the eggs with a layer of gravel. After spawning, adult Pacific salmon die; Atlantic salmon can occasionally return to sea and repeat the cycle. In healthy runs, the bodies of thousands of migrating salmon fertilize inland ecosystems throughout their distribution. The large fish bring nutrients from the ocean, providing food to human societies alongside wild scavengers and predators. The nutrients from decaying bodies fuel the juvenile salmon as they prepare to make their own migrations out to sea. The health of salmon populations in the rivers of the lower 48 United States and parts of Canada has been severely reduced, changing the ecosystems and the lifestyles of the human inhabitants of the region. Many Alaskan salmon runs remain healthy.

There is a strong cultural affinity for salmon in the Pacific Northwest, a relationship that has gone back thousands of years for the Native peoples in the region. Salmon are still the center of modern Native American culture for the tribes that depend on them, and many perform ceremonies to honor the return of the first salmon each year. Urban dwellers in Seattle, Portland, and other western cities take pride in the salmon as a symbol of their regions, and wild salmon fisheries provide both cultural meaning and significant income to communities from Oregon to Alaska.

There are multiple causes of the decline of wild salmon populations, which makes it easy to point fingers but hard to find real solutions. This is especially true because many of the causes are directly tied to the lifestyles and economic well-being of stakeholders in the issue:

- Overfishing is part of the problem, and regulations intended to protect remaining runs may inadvertently reduce genetic diversity by encouraging anglers to keep only the largest fish.
- Logging can destroy riparian habitat and cause sediment to flow into streams, which smothers developing eggs.
- Development and urbanization lead to runoff of harmful chemicals and destroy the streams on which salmon depend.
- Ranching and farming can have similar effects, especially when cattle are allowed to graze in riparian areas. Farmers also occasionally use water to the extent that salmon do not have enough to migrate.
- Dams—for irrigation, river navigation, and electricity generation—prevent upstream migration of salmon returning to spawn. For young salmon attempting to migrate out to sea, the slackwater behind dams makes downstream migration difficult. Fish ladders have been installed at many dams to allow salmon passage, with some success. In addition, in many locations, juvenile salmon are trucked around dams to improve downstream migration. The countless dams without these measures prevent salmon access to vast areas, and thereby have made many runs go extinct. Of all of the causes of salmon decline, dams have been among the most contentious because they represent significant barriers to recovery, yet many people derive their livelihoods from them and consider them to be important symbols of progress.
- Salmon hatcheries are commonly used to boost populations, but among other problems, they have adverse effects on the genetic diversity.
- Salmon farming, as previously mentioned, spreads disease and can have adverse effects on genetic diversity.
- Predators, such as killer whales and sea lions, eat large quantities of salmon but are protected under the Marine Mammal Protection Act.

There is significant passion in disputes over salmon. In 1974, a federal judge decided that Native Americans had a right to half of the catch because of their historical reliance on salmon. Commercial fishermen, who harvested the vast majority of salmon but who blamed Native Americans for the demise of the fishery, resorted to violence. According to historian Blaine Hardin, the judge received death threats, and Native American anglers were shot. State and local police also got into the fray, using tear gas in an illegal raid of a Native American fishing camp.

See Also: Aquaculture; Fisheries; Sustainable Fisheries Act.

Further Readings

Harden, B. *A River Lost: The Life and Death of the Columbia*. New York: W. W. Norton, 1998.

Hites, R. A., et al. "Global Assessment of Organic Contaminants in Farmed Salmon." *Science*, 303/5655:226–29 (2004).

Lichatowich, Jim. *Salmon Without Rivers: A History of the Pacific Salmon Crisis*. Washington, D.C.: Island, 1999.

Wolf, E. C. and S. Zuckerman. *Salmon Nation: People and Fish at the Edge*. Portland, OR: Ecotrust, 1999.

Mark Neff
Arizona State University

SALMONELLA

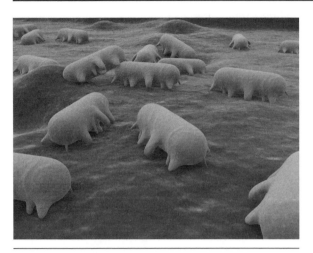

A highly magnified image of *Salmonella* bacteria reveals their rod-shaped form. Since the early 1990s, *Salmonella* has been the most common cause of reported food-borne illness, with the most common sources being milk and eggs.

Source: iStockphoto.com

Salmonella is a one-celled, gram-negative, rod-shaped, microscopic bacteria family containing over 2,300 different strains (also called serotypes). U.S. scientist Daniel E. Salmon's laboratory discovered the first strain, *Salmonella cholerae-suis*, in 1885. *Salmonella* can live in the intestinal tracts of both humans and animals and can be transmitted through contact with feces and contracted through the consumption of tainted foods. According to the U.S. Centers for Disease Control and Prevention, *Salmonella* has become the most frequently reported cause of food-borne illness since the early 1990s, with *Salmonella enterica* serotype *typhimurium* and *Salmonella enterica* serotype *enteritidis* the most frequently reported U.S. strains. Persons infected with *Salmonella* may develop moderate to severe cases of salmonellosis, with symptoms lasting from four to seven days on average.

Common symptoms of salmonellosis include diarrhea and abdominal cramps that may be accompanied by nausea, vomiting, fever, chills, or headaches. Dehydration is also a common threat and cause of hospitalization. Most people recover with or without the use of antibiotics, but the disease can be life threatening in certain vulnerable populations, such as infants, children, the elderly, pregnant women, and those whose immune symptoms are already compromised by other illnesses or chronic conditions. A minority of those patients who develop salmonellosis later develop Reiter syndrome—a condition that can cause joint pain resulting in arthritis, eye irritation, and painful urination. Food-borne illnesses such as salmonellosis cost the United States billions of dollars annually for medical expenses related to treatment.

Prevention of *Salmonella* contamination can occur at all points along the production, distribution, preparation, and consumption process. Prevention is critical because food contaminated with *Salmonella* cannot be identified by taste, smell, or appearance. The U.S. Department of Agriculture's Food Safety and Inspection Service (FSIS) sets the standards for food processing companies to help ensure the safety of commercial meat, poultry, and egg products. They established industry guidelines in the 1996 Pathogen Reduction/ Hazard Analysis and Critical Control Point System. FSIS randomly samples consumer products, testing them in FSIS laboratories for the presence of *Salmonella*. The U.S. Department of Agriculture has also partnered with the U.S. Food and Drug Administration, the Centers for Disease Control and Prevention, and others to form FoodNet to study the effectiveness of current preventative measures in food production and consumption.

Food safety guidelines for commercial and domestic food preparation are readily available. Food preparers should cook foods to safe internal temperatures, as measured with a food thermometer, and thoroughly wash raw fruits and vegetables. Leftovers should be frozen or refrigerated promptly at proper temperatures and reheated thoroughly; reheated liquids should come to a boil. Frozen food should not be thawed at room temperature. Preparers should also wash their hands and preparation surfaces thoroughly after contact with raw meat or poultry, as remaining juices can transfer present *Salmonella* to other surfaces or foods in a process known as cross-contamination. *Salmonella* that may be present can survive the food preparation and cooking process if such safety guidelines are not followed.

Milk and eggs are the most common sources of *Salmonella*, but it has also been found in meat, poultry, dairy products, seafood, fruits, vegetables, and water. *Salmonella* outbreaks have affected a wide variety of consumer products, including frozen pot pies, cantaloupe, peppers, frozen chicken dinners, peanut butter and peanut products, and pistachios. Petting zoos have also been linked to *Salmonella* outbreaks, as children may come in contact with contaminated animal feces and place their fingers in their eyes or mouths. Outbreaks generate widespread publicity in the news media and frequently result in recalls and public avoidance of suspected products.

Large- and small-scale outbreaks of *Salmonella* can also occur as a result of acts of bioterrorism, or the use of pathogenic microorganisms as biological weapons. The first known U.S. case of *Salmonella* bioterrorism involved the Rajneeshee religious cult in Oregon. The cult deliberately caused an outbreak of salmonellosis just before county Election Day so that cult members would be among the only voters well enough to go to the polls, thus allowing them to influence the election's outcome as they desired. The group's actions came to light after the Federal Bureau of Investigation searched their compound as part of an unrelated criminal investigation. Agents found a vial of *Salmonella enterica* serotype *typhimurium* identical to the strain linked to the salmonellosis outbreak. Cult members admitted that they had deliberately tainted 10 local restaurant salad bars and one city water tank. Publicity surrounding the outbreak revealed the need for adequate preventative measures and plans to respond to such attacks in the future. *Salmonella* has also played a role in efforts to combat another emerging bioterrorist threat, *Yersinia pestis*, the causative agent for the plague. Live *Salmonella*–based oral vaccines have shown the ability to protect against the pneumonic form of plague, meaning they could be useful for large-scale public immunizations if such a weapon were used.

The U.S. Department of Agriculture, FSIS, and Centers for Disease Control and Prevention work with local public health officials to investigate *Salmonella* outbreaks to determine the affected food products as well as the source of the contamination. Determining the products and source is a difficult process, and investigations can take months or more to complete. It is also very difficult to determine whether an outbreak was accidental or a deliberate act of bioterrorism. Thousands of people can be contaminated before the investigation results in a public warning and recall of affected products. Delays also mean that products pulled from shelves may be safe, because the contaminated products have made their way through the food supply chain. Recalls result in economic losses and disruptions in the food supply chain, as production facilities shut down for cleansing and reinspection and businesses and grocery stores recall items for safety even if they have not been directly linked to an outbreak.

See Also: Department of Agriculture, U.S.; Food and Drug Administration; Food Processing Industry; Food Safety; Supply Chain.

Further Readings

Bell, C. and Alec Kyriakides. Salmonella: *A Practical Approach to the Organism and Its Control in Foods*. Malden, MA: Blackwell Science, 2002.

Brands, Danielle A. and I. Edward Alcamo. Salmonella *(Deadly Diseases and Epidemics)*. Philadelphia: Chelsea House, 2006.

Guthrie, Rufus K. *Salmonella*. Boca Raton, FL: CRC Press, 1992.

Nestle, Marion. *Safe Food: Bacteria, Biotechnology, and Bioterrorism*. Berkeley: University of California Press, 2003.

Marcella Bush Trevino
Barry University

SEED INDUSTRY

The seed industry is dominated by a monopoly of multinational corporations that control the majority production and distribution of seed crops. The top 10 seed companies in the world in 2006 accounted for more than half of the commercial seed market worldwide. Many of these corporations also dominate the combined pesticide, food, pharmaceutical, and veterinary product industries, satisfying every aspect of the commodified agricultural industry. Such aggregate companies include the world's top three: Monsanto, Dupont, and Syngenta. Monsanto is the largest and most powerful of these corporations, proclaimed in 2007 to be the world's number one ranking agrochemical firm and the world's largest seed company, alone accounting for 20 percent of the world's commercial seed market. Vandana Shiva, physicist, ecologist, and activist, points out that this global corporate domination of the seed industry favors the rich while endangering the livelihoods of the poor. This monopoly also diminishes global food biodiversity and security—in 2002, 90 percent of the world's food supply was estimated to come from only 15 species of crop plants and 8 species of livestock among the estimated 10 million species of plants and animals in the world.

The history of the seed industry builds from seed existing in the public domain in the 19th century to the Green Revolution of the 1960s to increased commoditization and the implementation of global trade policies. In the 1800s, seed in the United States was sourced from a few seed catalogues or peddlers, through farmer seed saving and exchange, or through support of the government, which established research stations to develop and distribute quality seed to farmers. Influenced by the rediscovery of Mendel's work in 1900, farmers learnt to read "good" seeds by considering both their visible qualities and invisible hereditary composition. However, some groups believed government research was hindering the seed industry's advancement and lobbied for change. The development of hybrid seed in corn in the 1910s through the 1930s spawned the nascent seed industry. Since hybrid seed does not breed true, farmers had to purchase new seed every year. Future Vice President Henry Wallace founded Pioneer Hi-Bred in the 1920s and today it is the second largest hybrid seed producer, owned by Dupont. Hybrid seeds were quickly accepted —90 percent of corn planted in the United States by 1945 was hybrid. The green revolution emerged in the latter half on the 20th century based on the assumptions that technology was a superior substitute for nature and that by using a combination of high-yielding variety crops with high commodity potential inputs, such as artificial fertilizers and pesticides, soil could be reconditioned to grow the same crop in large quantities in different locations. This process favored high commodity inputs over low commodity potential, such as traditional agricultural practices that require increased chemical inputs, technologies, and financial investment. The transition from heirloom to hybrid seeds further commoditized agriculture, producing goods suitable for sale on the world market. Thus, the green revolution replaced local biodiversity with the large-scale production of monocrops.

Global trade policies, such as the World Trade Organization's Trade-Related Intellectual Property Rights Agreement, were introduced as a result of the argument that increased globalization of the economy without sufficient patent systems was acting as a barrier to trade. The concept of patents, a form of intellectual property, was originally introduced to encourage people to invest time, money, and energy in the innovation process while protecting them from theft. In exchange for this protection, creative work would be made public to benefit society. However, corporations concerned with securing patents to protect economic profits could be threatened by local companies producing cheap imitations. These patents thus restricted local innovation of plant breeding, and hence severely limited seed biodiversity. This change in patent protection was predicated on previous changes led by the United Nations Convention on Biological Diversity and Plant Breeders Rights, which were originally introduced to protect seed biodiversity. However, the Union for the Protection of Plant Varieties, which previously prohibited the production, reproduction, or offering of material from seed without authorization from the plant variety owner, was revised in 1991, changing the concept of "novelty" from physical novelty to commercial novelty. This new definition meant that a plant is novel only if it has not been previously marketed, making Plant Breeders Rights ineffectual in defending farmers' rights to save the seed of their crops. The emergent result extended the corporations' ability to lock countries into bilateral and pluri-lateral treaties requiring them either to patent plants and animals, to join Union for the Protection of Plant Varieties, or to provide legal protection for biotechnological inventions. These treaties favor corporations, who own the technologies to patent.

The consequences from the emergent seed industry have affected both the developed and developing worlds. In developed countries, the main consequence of the global seed industry is the consolidation of small, independent family-owned farms into large farming corporations as a result of increased agricultural costs, greater corporate control by agrofirms, and the threat of contamination of crops by genetically modified breeds, such as illustrated by the Percy Schmeiser case. Schmeiser battled Monsanto in court, claiming his canola fields were contaminated by Monsanto's crops, whereas his opponents stated Schmeiser had stolen the seed. Schmeiser lost. Even more daunting was the threat of Monsanto's Terminator and Traitor biotechnologies, as Terminator prevented seed from reproducing, and Traitor required a chemical input to activate certain traits. As a result of international outrage, Monsanto announced in 1999 that it would abandon its plans to commercialize both these technologies.

However, it is in developing countries where the main consequences of the green revolution, global trade policies, and increased corporatization are felt. These trade policies have negative economic consequences as they draw developing countries' farmers deeper into the neoliberal system, espousing the virtues of expanded export markets, lower trade barriers, and improved productivity through higher technology, favoring large-scale monoproduction over the small and diverse farming systems of the North and the South, enforced by powerful World Trade Organization trade sanctions. These trade policies in effect strip independence and power from the farmers of the developing world, championing trade liberalization by removing power from local constituencies to centralized corporate control. This imposed neoliberalism adversely affects seed biodiversity because with more farmers in the Third World growing more green revolution and biotechnology seeds, there is less attention devoted to promoting traditional crops, which consequently become endangered.

The social consequences of the imposed commoditization of agriculture include a change in the structure of social and political relationships of Third World societies, with relationships formerly based on mutual village obligations replaced with relationships between individuals, banks, and agribusiness. Farmers in developing countries are also threatened by biopiracy, the theft of biological resources and knowledge without recognition. As patents necessitate economic compensation for use, indigenous and peasant farmers must pay to access their original resources every year. Hence, patenting by corporations reduces the very food security that biotechnology companies claim can feed the world. Furthermore, the potential threat to Third World food security is emphasized by the fact that Monsanto dominates the major global food industries of corn, cotton, and wheat industries—the very species that replaced the diverse traditional crops grown by the Third World before the onset of the green revolution.

It is also very evident that there are increasing environmental costs from this industry, which fosters short-term economic profit and monocrops while neglecting environmental externalities. These costs include the hazards of unstable food systems, increased air pollution and greenhouse gas emissions from industrialization and transportation, and greater amounts of pesticides and herbicides polluting streams, soils, and oceans. In addition, the potential dangers of genetically modified crops are not yet fully known, but a wide range of activists, scientists, and the general public fear that genetically modified foods could contaminate, and hence dominate, the majority of crops while possibly causing health implications.

However, many grassroots groups are emerging to challenge current policies and reclaim power over seed resources. There are also many groups that continue to practice alternative food cropping methods that promote seed biodiversity. Together, these groups represent both the developing and developed worlds, encompassing a network of actors fighting to reclaim farmers' rights to their livelihoods and support a more resilient and sustainable environment through increased biodiversity. Groups advocating change include Navdanya, the International Forum on Globalization, the Third World Network, the International Federation of Organic Agriculture Movements, and the Action Group on Erosion, Technology and Concentration. Practices such as seed saving, community-supported agriculture, and the purposeful cultivation of heirloom seed varieties provide a distributed and decentralized network of millions of actors producing food on a small, local scale. These methods promote dynamic seed biodiversity, growing seeds in response to cultural preferences and climate change, in contrast to the seed conservation method of establishing large-scale frozen seed vaults. These seed biodiversity groups aim to raise awareness of these issues while practicing alternatives to seed industry corporatization to achieve greater food sovereignty.

See Also: Agrofood System (Agrifood); Crop Genetic Diversity; Food Sovereignty; Green Revolution; Monsanto.

Further Readings

Action Group on Erosion, Technology and Concentration (The ETC Group). "Who Owns Nature? Corporate Power and the Final Frontier in the Commodification of Life" (November 2008), http://www.etcgroup.org/en/materials/publications.html?pub_id=707 (Accessed January 2009).

A SEED Europe: Action for Solidarity, Environment, Equality and Diversity. Website. http://
www.aseed.net (Accessed January 2009).

Busch, L., et al. *Plants, Power and Profit: Social, Economic and Ethical Consequences of the NewBiotechnologies.* Cambridge, UK: Blackwell, 1992.

Fowler, C. and P. Mooney. "Shattering: Food, Politics and the Loss of Genetic Diversity." Tucson: University of Arizona Press, 1990.

Gorelick, S. "The Farm Crisis: How We Are Killing the Small Farmer." *The Ecologist*, June 2000.

GRAIN. "TRIPS-plus Rights." http://www.grain.org/rights/tripsplus.cfm (Accessed January 2009).

Kloppenburg, J. *First the Seed: The Political Economy of Plant Biotechnology.* Cambridge, UK: Cambridge University Press, 1988.

Manno, J. "Commoditization: Consumption Efficiency and an Economy of Care and Connection." In *Confronting Consumption*, edited by T. Princen, et al. Cambridge, MA: MIT Press, 2002.

Rural Advancement Foundation International. "Seed Industry Consolidation: Who Owns Whom? RAFI's Seed Industry Consolidation Chart." RAFI Communique (July/August 1998). http://www.etcgroup.org/en/materials/publications.html?pub_id=404 (Accessed January 2009).

Shiva, V. *Stolen Harvest: The Hijacking of the Global Food Supply.* Boston: South End, 1999.

Ferne Edwards
Australian National University

SEWAGE SLUDGE

Sewage sludge, also called *sewage biosolids,* is the end product of sewage treatment. Sewer systems using water to carry away wastes have been used in a variety of civilizations since ancient times. The sewage water carries wastes either in solution or in suspension to a place for depositing. It oddly enough is mostly pure water, but the wastes it contains are such that it contains significant quantities of fecal or other biological materials in sufficient quantities to sicken and kill humans, and often animals as well.

Since the Industrial Revolution and the growth of cities, of industries, and of chemically based agricultural production, it has become extremely important that water used in one location not be dumped untreated into a river or stream so that the sewage will not pollute the water for the next downstream user.

Storm water runoff can be classified as surface, commercial, industrial, and agricultural runoff. Surface runoff comes from streets and roads, where it has been polluted with all manner of waste including animal wastes. Commercial runoff usually comes from commercial parcels, where storm water cleans the surfaces of buildings, parking lots, and other facilities. Industrial runoff can include chemicals, pathogens, heavy metals, and other pollutants that can end up on farmland or in watersheds, where the food chain is affected.

In addition to the surface runoff from the several sources identified above, some of which may enter a sewer system to end in a sewage treatment plant, there is the normal mass of human and animal wastes that end up at a sewage treatment plant. In the water that reaches a sewage treatment plant are human body wastes, wash water from laundries (home and commercial), food wastes, and other waste products that may be flushed down a drain or toilet. Among the items flushed are aged medicines and wastes from hospitals for humans and animals. These often include waste that has pathogens and semiused antibiotics or other medicines.

Sewage treatment plants use a variety of methods for treating sewage. Septic tanks are usually in-ground systems that use anomic bacteria to reduce home sewage to sludge. Biofilters can be added to these to make the system cleaner; however, many have to be pumped occasionally. Others use aeration, which can mean that local ponds can be used to treat the sewage in a natural way, or aeration can be used in a system that then gathers the sludge and transfers it to a new location to make room for the plant to continue operating. Other systems include sequential batch reactors and Diamond Aeration systems.

As sewage is treated, the solids settle out in several stages of the treatment process. They are treated by anaerobic digesters to destroy all pathogens before the sludge can be dumped into landfills. This stage produces significant quantities of methane gas, which can be processed for use in a power system. After sterilization, water is removed from the sludge. Removing the water reduces the cost of landfill dumping and is often accomplished with large centrifuges and the use of polymers.

Usually, sewage water is a complex chemical mixture, whether in liquid or solid form. There are usually chemicals from a number of household products, including ammonia, nitrates, phosphates, acids, and a variety of carbon compounds including oils, and other chemicals such as trihalomethanes that were used in past disinfections. To treat sewage water, a variety of systems are used.

Some systems are simple and local such as an aeration pound for a small community. Others are giants such as the Deer Island Sewage Treatment Plant, which cost nearly $4 billion. It serves Boston and protects Boston Harbor from pollution from Boston's city sewers and from storm water runoff. It operates in compliance with regulations established by state and federal officials including the U.S. Environmental Protection Agency.

The newer systems or those being planned for the near future are being built with new and emerging technologies that make them more effective and more efficient. They use methods that give a higher rate of coagulation, that clarify bacteria and other media, and that rotate much more effectively while using significantly reduced amounts of energy. The energy savings and the efficiency of sewage treatment are accompanied by a much cleaner sludge. The sludge is then turned into compost for use as organic fertilizer.

Historically, human waste has been collected from outhouses or other sources. It is taken to gardens and fields, where it is used as manure. This use of raw sewage is risky because it may contribute to the spread of diseases. Animal waste from horses, sheep, cows, and chickens has also been long used as organic fertilizer. However, the amount of composting is important because if it is used when fresh, it exposes areas to odors and to pathogens. If composted too long, however, some of its potency as a fertilizer may be lost.

With the development of huge agribusiness for slaughter of animals and poultry, the amount of waste available for disposal has created challenges for animal sewage sludge treatment. For example, there are around a million horses in south and central Florida. Disposing of the resulting horse manure is becoming a Herculean task that has prompted the building of a horse sewage treatment plant in central Florida. Other areas are also developing animal sewage treatment plants. The resulting sludge can be sold as organic compost; however, there are limits to the amount of sludge that can be applied to farmland, regardless of whether it is human or animal. If an excess is applied, it runs off in heavy rains into streams to pollute the watershed.

Many farms have on-site aerobic composting units. These units for horse farms, poultry operations, dairy farms, schools, or packing houses can produce usable sludge. However, spreading sludge over forests, fields, and gardens is being challenged by individuals and groups that see it as too much of a good thing, even if the sludge is a clean organic fertilizer. In response, systems are in place, with more still in development, that can burn equine waste in a raw state as a fuel source for electrical power generation.

Using waste as a biochar in a waste-to-energy system is one method of handling sludge. The ash in biochar systems contains heavy metals that can be removed or used as an aggregate in concrete. Others are experimenting with ways to use sludge as a biomass for producing biogas through anaerobic digestion. Others are working with pyrolysis, which uses chemical decomposition to produce syngas.

Synergies from biochar, biogas, and pyrolysis systems reduce the methane released into the atmosphere, where it becomes a greenhouse gas. Instead, the sludge becomes a fuel in a closed carbon cycle.

See Also: Composting; Confined Animal Feeding Operation; Fertilizer.

Further Readings

Birkett, Jason and John N. Lester, eds. *Endocrine Disrupters in Wastewater and Sludge Treatment Processes.* Oxford: Taylor & Francis, 2002.

Bruce, A. M., et al., eds. *Disinfection of Sewage Sludge: Technical, Economic and Microbiological Aspects.* New York: Springer, 1982.

Casey, T. J., et al., eds. *Methods of Characterization of Sewage Sludge.* New York: Springer, 1984.

Davis, R. D. and P. L'Hermite, eds. *Utilisation of Sewage Sludge on Land: Rates of Application and Long-Term Effects of Metals.* New York: Springer, 1983.

Lue-Hing, Cecil, et al. *Municipal Sewage Sludge Management: A Reference Text on Processing, Utilization and Disposal.* London: Taylor & Francis, 1998.

Pawlowska, Lucjan, ed. *Management of Pollutant Emission From Landfills and Sludge.* Boca Raton, FL: CRC Press, 2007.

Andrew Jackson Waskey
Dalton State College

SLOW FOOD MOVEMENT

Slow Food is a nonprofit, member-supported organization whose aim is to raise awareness of the social, environmental, and economic impact of food choices. Founded by grassroots efforts in 1989 in Italy by Carlo Petrini, Slow Food includes over 100,000 members from 132 countries. The movement is based on the principle of ecogastronomy—the awareness of the link between the environment and the production, preparation, and consumption of food. Embedded in Slow Food's philosophy is the belief that individuals have a fundamental right to pleasure as well as a responsibility in consumption. Slow Food's headquarters are in Bra, Italy, and its symbol is the snail, because it moves and eats slowly.

The slow food movement began to organize in 1986 as a result of a protest against the building of a McDonald's in the Piazza di Spagna just below the Spanish Steps in Rome. The term *slow food* was used as a natural counter to *fast food,* as well as to what organizers felt was an increasingly fast life. Poet Folco Portiari was charged with writing the Slow Food Manifesto, first published in Italy in 1987. Slow food became an international movement in 1989, when the manifesto was presented and signed at the International Slow Food Congress in Paris.

Central to Slow Food's philosophy is the support of good, clean, and fair food. That is, food should taste good and be prepared in a way that is healthy to individuals and to the environment, and food producers should receive fair compensation for their work. Slow Food believes consumption cannot be divorced from production and distribution and celebrates food traditions and diversity. Committed to sustainable agriculture, Slow Food opposes the standardization and industrialization of food. The organization seeks to highlight the consequences of the modern diet including such issues as obesity, hunger, waste, and the effect on rural economies.

The mission of Slow Food is to

- defend biodiversity by preserving traditional and local produce, grains, and animal breeds that are disappearing;
- provide education on a variety of topics from farm to table, such as the taste of food, food production, and preparation; and
- connect producers and consumers or "coproducers"—Slow Food considers consumers to be part of the production process, and therefore refers to them as *coproducers.*

Slow Food attempts to fulfill this mission through numerous activities and publications that have emerged over the past 20 years.

Slow Food Editore (publishing) was launched in 1990, publishing a variety of books and periodicals to support the mission of the organization. Examples of such publications include food and restaurant guides, studies and surveys, educational manuals, and magazines (e.g., *Snail* in the United States).

Since 1996, the Salone del Gusto (Halls of Taste) have been held every two years in Turin. A large food and wine fair, the Salone allows the public to sample a wide variety of products from around the world while simultaneously promoting good, clean, and fair food.

The International Ark of Taste, a catalog of foods in danger of extinction, began in 1996. To date, over 800 products have been catalogued worldwide. The U.S. Ark of Taste includes over 200 foods. To be included in the Ark, a product must be considered of outstanding quality (in terms of taste), linked historically to a region or culture, at risk of extinction, and produced in limited quantities.

Slow Food Presidia—or local projects—began in 1999 as a means to improve and to protect artisan food production. The Presidia's goal is to assist producers of a product within a particular region in developing the infrastructure necessary to compete in the global food industry.

The Foundation of Biodiversity was established in 2003 and supports projects that seek to defend biodiversity, promote sustainable agriculture and food traditions, and protect small producers. Although projects are funded around the world, priority is given to those in developing countries. The foundation is funded by Slow Food, for-profit companies, and donations.

Also in 2003, Slow Food established the University of Gastronomic Sciences in Piedmont and Emilia-Romagna, Italy. The academic institution is a private university, the objective of which is to create an international research and training center, and which offers undergraduate, graduate, and master's degrees in such areas as food science, food communications, food culture, and Italian tourism.

Some believe that 2004—when the first Terra Madre (Mother Earth) meeting was held in Turin, Italy—marked the turning point in enhancing Slow Food's profile. Terra Madre is a meeting of food communities, bringing together thousands of producers, consumers, activists, and academics from around the world to discuss critical issues facing the world's food supply and production and to develop strategies for sustainable agriculture. Held every two years in Turin, many feel Terra Madre was the catapult that moved Slow Food from what many considered a gourmet association to a significant political movement. It was at the first Terra Madre that Slow Food's principle of good, clean, and fair food emerged. Since its inception, Terra Madre has established smaller events in countries around the world, including Terra Madre in the United Kingdom, Brazil, Austria, Kenya, Sweden, France, Tanzania, and Cuba.

Slow Food also sponsors a variety of food and beverage festivals around the word, including Cheese, a biennial cheese fair in Bra; Slow Fish, a fish festival in Genoa; and country-specific events such as Slowbier, a beer fair in the Oberfranken region of Germany.

Slow Food USA began in 2000 with its office in New York. It is now the second-largest Slow Food association in the world, with over 17,000 members and 200 local chapters. Alice Waters, notable owner of Chez Panisse restaurant in Berkeley, California, and strong advocate of sustainable agriculture, is an International Governor of Slow Food and visionary of Slow Food USA. She organized Slow Food Nation, a subsidiary nonprofit organization of Slow Food USA. In September 2008, the Slow Food Nation four-day festival brought over 85,000 people from 44 states and 10 countries together in San Francisco for tastings, workshops, lectures, and food markets. Slow Food USA also supports Slow Food in Schools, teaching youth about foods, biodiversity, the environment, and ecofriendly food production and consumption.

Critics of Slow Food claim it to be an elite organization—one of wealthy gourmands without actual interest in solving the world's food problems. Despite the attempts to negate this criticism, this has been particularly problematic for the organization, as many local and organic foods are financially out of reach for many people. Others argue that Slow Food's antiglobalization and antiagribusiness ideals may help enhance the taste of food, but they will not help to feed the world's poor. Furthermore, others argue that the organization has not established a thorough evaluation system to assess the effect of the movement and its activities.

Organizationally, Slow Food is led by the International Executive Committee, which is elected every four years. The International Executive Committee consists of the President's Committee and the International Council, which includes representatives from countries with at least 500 members. At this time, Slow Food International has over 100,000 members across 132 countries and more than 100 Slow Food chapters around the world.

See Also: Agrodiversity; Fast Food; Food Security; Sustainable Agriculture.

Further Readings

Andrews, Geoff. *The Slow Food Story: Politics and Pleasure*. London: Pluto, 2008.
Petrini, Carlo. *Slow Food: The Case for Taste*. New York: Columbia University Press, 2003.

Petrini, Carlo and Gigi Padovani. *Slow Food Revolution: A New Culture for Eating and Living*. New York: Rizzoli, 2006.

Slow Food. http://www.slowfood.com (Accessed May 2009).

Slow Food USA. http://www.slowfoodusa.com (Accessed June 2009).

Marcia Thomas
New York University

SOCIOLOGY, NEW RURAL

The term *New Rural Sociology* was first used by Howard Newby in 1983. He referred to "Sociology of Agriculture" as the New Rural Sociology. During the time, there were concerns within the rural sociology community about making the subject attractive and influential to wider researchers. Although Rural Sociology was originally associated with agriculture during the 19th century and early 20th century, the association disintegrated by the 1950s. Rural sociologists were focused on the rural–urban continuum during that period. However, the farm crisis of the 1970s, suburbanization, and the decreasing number of small farms developed interest among rural sociologists in reconnecting the subject with agriculture. It was in this environment that the 50th anniversary meeting of the Rural Sociological Society in 1978 formed a task group to look at the possibility of a new field—the sociology of agriculture. During the 1980s, in an attempt to transform the field, many universities renamed, merged, and even dropped the rural sociology programs. It was against these events that Newby suggested some changes to rural sociology by including the study of agriculture in rural communities.

What is Rural Sociology?

Rural sociology is the study of geographical areas with low-density population. In the beginning, rural sociology decided to be associated with rural society rather than with agriculture. In the earlier stages, rural sociology did not include other sectors such as development, farmers, agriculture, and so on. Traditionally, rural sociologists were concerned with rural sociability, kinship, customs, traditions, demography, education and occupations of rural youths, and community change. They viewed agriculture as an obstacle affecting the above variables. Rural sociologists never considered rural agriculture as their core subject, except for some few studies on diffusion-adoption of technological innovations after World War II. The majority of the research after World War I focused on the rural-urban continuum developed by Pitirim Sorokin and Carle C. Zimmerman during the 1930s. This new research area was very empirical and involved rigorous quantitative methods.

In the past, rural sociologists were secluded even in the land-grant institutions, which sustained the field. These institutions did not support research on agriculture by rural sociologists. Rural sociologists were expected to be applied and to influence policymakers on rural issues. However, even the administrators in land-grant universities hired experts from other areas to study agricultural issues. For example, they hired applied scientists for agriculture-related research and economists to study issues related to agricultural economy. Some of the setbacks of rural sociology during the period were its inability to produce knowledge that is of use to sociology, to other social sciences, or to the land-grant institutions, and that it was not symbolically recognized by general sociology. This created a kind of crisis within the rural sociological society.

At the same time, traditional rural sociology was challenged by the reverse migration of urban people into rural areas and the inability of traditional occupations like agriculture, forestry, and extractive industries to support the rural population. The introduction of the tractor transformed the rural basis of American society; with the migration of urban people into rural areas, a new field of "suburban studies" emerged, mostly of interest to nonrural sociologists. The farm crisis of the 1970s, the decreasing number of rural farms, and the integration of rural agriculture with the greater network of agribusiness had an effect on policies, which tried to reestablish the relationship between agricultural change and rural social change in the United States. In an attempt to avoid the crisis within rural sociology, some rural sociologists suggested renaming the subject. One of the strongest proposals was to rename it "Sociology of Agriculture" or "New Rural Sociology."

The revitalization of the connection between rural sociology and agriculture at the 1978 Rural Sociological Society meeting instigated new interests among the rural sociologists. New Rural Sociology attracted mostly young faculty and graduate students. The development was like going back to the 1930s, when agriculture was the major area within rural sociology. The attraction to the New Rural Sociology increased with the globalization of agribusiness and the connection of developing nations with the global market system. Thus, the New Rural Sociology was created under the theme "Sociology of Agriculture."

History of the Sociology of Agriculture and Rural Sociology

Even though sociology of agriculture is rural sociology's oldest specialty area, it was never used before the mid-1970s. The history of sociology of agriculture can be divided into three eras. The first era consisted of research on agriculture by rural sociologists before the 1930s. Some of the earliest work on agriculture in rural areas was conducted during the late 19th century. The research focused on different agricultural practices, the production system, and the behavior of communities. The earlier work also looked at the relationship between agriculture and community development and prosperity in rural America. In the early 20th century and after World War I, extensive rural social surveys across the nation were undertaken by rural sociologists. The unit of the survey was communities with a population between 250 and 2,500. The survey looked at farm value, farm per household, and the per capita contribution to village churches. Similar surveys were carried out across the nation until the 1950s. The census of 1930 also recorded data on crop types, farm size, and economic characteristics of farmers. In addition to survey research, specialized studies about tenure, part-time farming, low-income farmers, effect of federal policies, cooperatives and marketing systems, black laborers, and other structural components of agriculture were carried out. By the late 1940s and early 1950s, the study of agriculture and rural communities began to disintegrate. There was an increasing trend of studying agriculture separately and not relating to rural communities. This breakdown marked the beginning of the second era of the sociology of agriculture.

The second era of the sociology of agriculture is the period between the 1950s and the 1970s. During this period, a new cohort of rural sociologists emerged who focused on the social-psychological component of farming communities. These researchers viewed farmers as actors responding to stimuli from new agricultural technologies, the mass media, and educational and occupational opportunities. As a result, this era was characterized by studies on diffusion and adoption of agricultural innovations and their effect on peasant communities. To summarize, there were seven categories of research on the diffusion-adoption theme as part of the research about agriculture:

1. Characteristics of farm practice adopters

2. The social psychology and role of aspirations and motivations in farm practice adoption

3. Group influences on adoption

4. Role of communication and media in adoption

5. Conceptual clarifications of the diffusion-adoption studies

6. Methodology of diffusion-adoption studies

7. Criticisms of diffusion-adoption research

This period was also characterized by the dominance of the quantitative method in studying rural communities, indicating a strong influence from sociology. Later in the early 1970s, rural sociologists renewed the study of farmers in rural areas, which was called the "New Sociology of Agriculture" or the "New Rural Sociology."

The new sociology of agriculture was diverse theoretically, as it attempted to combine broad, macrosocial theories with falsifiable theoretical formulations and testable hypothesis together with its greater representation of historical and qualitative research methods rather than the dominant quantitative methods in the 1960s. Sociology of agriculture tried to fill the vacuum created with the demise of the rural–urban continuum within rural sociology. It tried to build a theory of society by connecting social with spatial. This new field was more theoretically informed, holistic, critical, and radical than traditional rural sociology. The new rural sociology focused on two aspects of agriculture: the political economy of agriculture, and the environmental aspect of agriculture. Thus, the sociology of agriculture reunited with rural sociology after decades of separation.

The Status of Rural Sociology

The new rural sociology or sociology of agriculture is primarily about the economic relationship between small farmers and the wider economic system. The earlier work of new rural sociologists was influenced by the works of Karl Marx and Max Weber on political economy. Research focused on the labor–capital relationship within the global capitalist system, the structure of agriculture, state agricultural policy, agricultural labor, regional inequality, and agricultural ecology. The other areas of studies under the new rural sociology were as follows:

- *Industrial agriculture:* study of production, processing, and distribution of agro-products and labor structure within these industries
- *Wage labor in agriculture:* migrant laborers in agriculture and the effect of mechanization on labor market
- *Small farms and part-time farming:* although the nature, social aspect, and socioeconomics of small farms and part-time farming are basic topics within rural sociology, three things in the late 1970s and early 1980s increased attention on small farms and part-time farming: concern for the status and the prospects of small farms, the increasing number and stability of small farms, and the diversification of economic sources through off-farm jobs by small farms

- *Gender and agriculture:* research on gender and agriculture falls into four major categories: descriptive studies on women's involvement in on-farm, household, and off-farm tasks; assessment of changes in women's involvement in on-farm, household, and off-farm tasks through longitudinal studies; exploration of the interrelationship between on-farm, household, and off-farm roles of men and women; and the effect of institutions of capitalism and patriarchy on the involvement of men and women in the workforce.

There were mixed reactions to this proposed (trans)formation of new themes within rural sociology. Some of the proponents suggested that rural sociology only be associated with agriculture and for all other fields not related to agriculture to be separated from rural sociology. There were also suggestions to reduce rural sociology to "Sociology of Agriculture." People who supported the conversion of rural sociology to sociology of agriculture insisted rural sociology be replaced by analyses of labor processes and social relations of production across agricultural commodities. In contrast, there was strong opposition to this transformation because this economic reductionism threatened rural sociology and suggested a broader definition of rural sociology.

The Future of Rural Sociology

Right now, sociology of agriculture is one of the subfields within rural sociology, just like demography, natural resources, community development, and international development. To become a prominent field in the future, the new rural sociology must adapt to the changing pattern in the agricultural system nationally and globally. To attract more people, it must add more areas into its basket, such as the following:

- Research on the changing structure of farmland ownership
- Research on the structure and effect of agribusiness
- Relationship between agriculture and rural well-being
- Organic farming and its effects on agriculture
- Recruitment and management of wage labor
- Social movements and social control
- Impact assessment of various agricultural policies

In the last three decades since its creation, the new rural sociology has not been able to establish itself as an alternative to rural sociology. Although agriculture is always attached to rural life, it was abandoned in the process but now is reunited with the field, having the strongest relationship. Even though the new rural sociology might be in its dormant stage, the recent food crisis and changes in the global agricultural system from climate change promise the need for the field. In its latest edition, sociology of agriculture is widely published as sociology of agriculture and food. There is even an international journal by the same title. Some departments offer courses on sociology of agriculture to capture increasing interest among graduate students. We can only hope that sociology of agriculture remains and expands within rural sociology and that the mistakes of the past are not repeated. Rural sociologists have a challenging task to embrace agriculture as the core subject of research within their interests.

See Also: Agrarianism; Agribusiness; Rural Renaissance.

Further Readings

Buttel, Fredrick H., et al. *The Sociology of Agriculture*. Westport, CT: Greenwood, 1990.

Friedland, William H. "The End of Rural Society and the Future of Rural Sociology." *Rural Sociology*, 17/4 (1982).

Newby, Howard. "The Sociology of Agriculture: Towards a New Rural Sociology." *Annual Review Sociology*, 9 (1983).

Krishna Roka
Penn State University

SOIL EROSION

This scientist with the U.S. Department of Agriculture Agricultural Research Service (standing) consults with a farmer in the Lower Rio Grande Valley who demonstrates the effects of severe wind erosion on his land.

Source: U.S. Department of Agriculture Agricultural Research Service/Jack Dykinga

Soil erosion caused by wind and water is the most destructive phenomenon worldwide because it is like a slow death. Civilizations have been lost as a result of the scourge of soil erosion. Barren hills in central India, parts of Greece, Lebanon, and Syria are some of the civilizations lost because of washed-away deep and productive topsoil, a result of ignorance or mismanagement of soil. Soil erosion is a major environmental concern in industrialized nations, as well as in developing countries. A good-quality (fertile) soil is essential to meet the objectives of attaining self-sufficiency in food and keeping the food price low. Excessive soil erosion is responsible for topsoil loss and subsequent decrease in agricultural production. According to a study conducted by R. Morgan (1986), the rate of soil erosion in developing and developed countries is at an alarming level—more than the allowable limit of around 5 tons/hectare/year. Countries like the United Kingdom, the United States, and Belgium have succeeded in arresting the soil loss close to the allowable limit, but soil loss from cultivated and bare land is still very high in all the countries of the world. The soil erosion rate is higher in the developing countries, where the farmers are less engaged in soil conservation programs and practices.

Soil erosion is a natural phenomenon of detachment and transportation of soil particles by natural agents like water, snow, or wind. Not only does it reduce productivity, but it also has much more destructive effects such as siltation in reservoirs; deposition of infertile materials on good land; harmful effects on water supply, fishing, and power generation; and above all, increased flood potential of rivers. However, soil erosion has become more ominous since 1960. The burgeoning world population is responsible for this because farmers of the world are now desperate to grow double or triple the amount of food they once grew, using intensified cultivation. They achieve this by expanding the land for cultivation by clearing forests and other bushy areas, plowing grasslands, and performing slash-and-burn culture on mountain or hill slopes. Incessant population growth is paving the way for overgrazing of rangelands and overexploiting timber resources. Thus, more and more land is becoming devoid of natural vegetation, making it an excellent medium for soil erosion reduction. This is a downward cycle of deterioration. The soil erosion process accelerates as a result of such land degradation, and food production diminishes; as a consequence, people become desperate to grow more food, clear more forested land, and cultivate grasslands, indirectly encouraging overgrazing by cattle.

There are two types of soil erosion—*geological erosion* and *accelerated* or *man-made erosion*. Canyons, buttes, rounded hills, river valleys, deltas, plains, and pediments are examples of geologic erosion. Accelerated erosion is the consequence of livestock overgrazing, forest clearing for agriculture, hillside plowing, and creating impervious surfaces like roads and buildings by tearing up pervious land cover. Accelerated erosion is 10 to 1,000 times as destructive as geologic erosion.

Four types of soil erosion, such as sheet, rill, gully, and stream bank erosion, occur by water. The major factors that contribute to soil erosion were initially defined by W. H. Wischmeir et al. (1958) through their Universal Soil Loss Equation; that is, *A (Soil loss per unit area) = $R*K*L*S*C*P$*. These are rainfall erosivity (*R*), soil erodibility (*K*), slope length (*L*), slope gradient or steepness (*S*), cropping management (*C*), and conservation practice (*P*) factors. *R*-factor represents the driving force for sheet and rill erosion, and it varies spatially over the Earth as a result of rainfall intensity. *K*-factor represents the inherent susceptibility of soil to erosion. Infiltration capacity and structural stability are the two most significant soil characteristics that determine *K*-factor. Topographic factors *L* and *S* contribute less toward erosion if slope gradients are low and the slope length is long. Rill and sheet (interrill) erosions are consequences of high steepness and short slope length. *C*-factor is a major contributor toward soil loss. Pervious land cover like forests, grass, and dense crops (legumes) reduce soil loss, whereas bare land and impervious land cover like urban and semiurban development and row crops (corn, wheat, sugarcane, etc.) increase soil erosion amount. Strip cropping, contour bonding and planting, terracing, straight-row farming with up and down slopes, slope tilling (ridge tillage), stubble mulching, and chisel plowing are some of the positive conservation practices (*P*) factor to reduce soil erosion.

More than half the erosion in the United States and other countries comes from agricultural land. Therefore, it is essential to follow better cropping management and conservation practice methods. The other half of erosion is caused by deforestation, urban growth, shifting cultivation, and so on. It should be the practice of governments and other organizations to encourage large-scale afforestation to counter these causes. Some other soil erosion control measures are as follows. For construction activities, schedule excavation activities in low rain periods, clear as small an area of vegetation as possible during construction, cover disturbed soil always using vegetation and other materials, control flow of runoff, and trap runoff before releasing water off-site.

Soil erosion by wind is a menace in many countries. The factors affecting wind erosion are soil moisture, wind velocity and turbulence, soil properties, and vegetation. Wind erosion is predicted through the equation *E = f(ICKLV)*, where *I* = soil erodibility,

C = climate, K = soil, soil-ridge-roughness, L = width of field, and V = vegetative cover factors. Among these factors, C and V are very important. Wind erosion can be reduced with the provision of windbreaks or shelterbelts that reduce wind velocity. If the field is not kept fallow in dry seasons, wind erosion can be reduced by a great deal. Perpendicular to wind direction, cropping is a deterrent to wind erosion. Tillage conducted at the time of enough soil moisture conditions could reduce wind erosion by a great deal.

Soil conservation would not only reduce soil erosion but will enhance food production in the world. Thus, the world will be a better place to live, as people would have more peace of mind.

See Also: Agricultural Extension; Cover Cropping; Grazing.

Further Readings

Blaikie, Piers. *The Political Economy of Soil Erosion in Developing Countries.* New York: Longman Scientific and Technical, 1985.

Brady, N. C. and R. R. Weil. *Elements of Nature and Properties of Soils,* 2nd ed. New York: Prentice Hall, 2004.

Morgan, R. *Soil Erosion and Conservation.* Hong Kong: Longman Scientific and Technical, 1986.

Panda, S. S. "Planning Soil Conservation Measures on a Watershed Basis Using Remote Sensing and Geographic Information System: A Case Study." Master's Thesis, Asian Institute of Technology, Bangkok, 1996.

Wischmeir W. H., et al. "Evaluation of Factors in Soil Loss Equation." *Agricultural Engineering,* 39:458– 62, 474 (1958).

Sudhanshu Sekhar Panda
Gainesville State College

SOIL NUTRIENT CYCLING

Soil nutrients are essential for plant growth and subsequent food production. These nutrients are of macro- and micronutrient categories based on their required amount for plants. Primary macronutrients required in large amount are nitrogen (N), phosphorous (P), and potassium (K). Secondary macronutrients required in moderate amounts are calcium (Ca), magnesium (Mg), and sulfur (S). The nutrients like iron (Fe), manganese (Mn), zinc (Zn), copper (Cu), boron (B), molybdenum (Mo), and chlorine (Cl) are required in very small amounts for plants and are considered micronutrients. Carbon (C), hydrogen (H), and oxygen (O) are not soil nutrients but are required for plant growth, and they are available through air, soil, and water. Other nutrients in soil that are not directly essential for all plants but are required for other purposes, such as sodium (Na), silicon (Si), and nickel (Ni), are vital for some plants and may stimulate growth in other plants; cobalt (Co) is a vital nutrient necessary for N fixation in soil by legume plants to support other crop growth; and selenium (Se), arsenic (As), and iodine (I) are not essential for plants as such but are required for humans and other herbivores who eat plants.

These soil nutrients are available in the soil as a result of the weathering of soil minerals; decomposition or breakdown of dead plants, animals, and microbes in the soil; application of fertilizer, manure, compost, lime, and other biosolids; fixation of N in soil by legume plants; and movement of soil from other locations by wind and water erosion, and so on. Decomposers present in soil, such as bacteria, fungi, and termites, process and break down decaying matter to form soil nutrients. The decomposition process varies from tropical climate to temperate climate as the amount of decomposer varies with soil and atmospheric temperature. Therefore, soils in the Amazon forest and other tropical areas are rich in soil nutrients. These nutrients are also lost in soil as a result of soil erosion, leaching or percolation, volatilization, and crop uptake. In slope land, soil nutrients can be lost with water runoff.

It is essential for a soil to have high nutrient concentrations at the time of high crop growth rate, but the soil should have a high storage capacity to retain nutrients in it at low crop need. Cation exchange capacity of soil is the property that helps in attracting and holding positively charged ions from leaching. A high cation exchange capacity helps in storing the nutrients (K^+, Ca^{2+}, Mg^{2+}, NH_4^+), and other micronutrient trace metals are held together by soil clay and negatively charged organic matters, known as a reservoir of plant nutrients.

Biological cycle and chemical transformations are two basic plant nutrient cycling processes. The soil food web and the soil–plant–animal system are part of the biological cycle that makes the soil fertile. Soil organic matters store plant-available nutrients. Plants use those nutrients to grow food. Humans and animals consume these foods, and their excreta, including plant by-products, are used to prepare compost and manure. The manure and compost is used in soil to increase the soil nutrients amount. This cycle is known as the soil nutrient cycle.

In chemical transformation, soil nutrients are absorbed through dissolution and precipitation, surface adsorption, and exchange, for example. N is the most important nutrient required for plant growth. N exists in both chemical and physical form. The transformations between these forms to make N available for plant growth is conducted as a cycle, known as the N-cycle. The N-cycle is part of the soil nutrient cycling. It is like a maze, rather than a simple, circular cycle. Nitrification, denitrification, mineralization, and N-fixation are the chemical transformations performed in soil by a diversity of soil-inhabiting organisms. Through physical transformations, N moves freely between soil and the atmosphere, thus being available in soil to plants. N, like many other soil nutrients, is also added to soil with chemical fertilizers. These fertilizers get broken down to individual nutrients for plant uptake. N is easily lost in soil, as it is very soluble and one of the most mobile plant nutrients in the soil. It gets lost through leaching, denitrification, and runoff.

Soil nutrient cycling is not 100 percent efficient because it involves some losses or leaks from the cycle. In the case of a natural ecosystem, there are possibilities of imbalanced nutrient cycling because of an uneven amount of runoff, pest attacks on plants, erratic solar radiation, and so on. In farm conditions, farmers sometimes mismanage the field to produce a tremendous imbalance in the soil nutrient cycle. If too much of a crop is grown in the field, the soil nutrient loss becomes vigorous because the soil cannot compensate for the amount of naturally producing nutrient uptakes by crops. Cash crops mostly decrease the amount of soil nutrients availability because the harvests mostly get exported out of the farm. Therefore, the soil nutrient cycling cannot take place. The negative balance is compensated for with the application of fertilizers. There are several farming practices followed worldwide to take care of the imbalance of the soil nutrient cycling.

Crop rotation is used in most parts of the world. Legumes, such as soybeans, green/black and other grams, and other lentils are grown in fields after wheat, paddy, or corn is cultivated to replenish the depleted N. Some of the other legumes are alfalfa, clover, peas, lupins, mesquite, carob, and peanuts, for example. The legume plants help in fixing atmospheric N at their roots as the result of the presence of a certain bacteria, rhizobia. In many cases, green crops like fava beans, mustard, clover, fenugreek, lupin, sun hemp, alfalfa, tyfon, buckwheat, velvet beans, and so on are grown between two cash crops. These green crops are typically plowed into the soil as green manure. They increase the soil nutrients amount, provide protection to the soil, and even bring nutrients from subsoil that is generally unavailable to shallower root crops. Mixed crop and livestock management in a farm setting is another means to maintain balance in the soil nutrient cycle, as the soil–plant–animal system is the best way for cycle preservation. Concentrated livestock management is generally considered to be a net nutrient importer. Manures and composts from high-density livestock farms are imported to put in cash-crop production farms to preserve the balance of the nutrient cycle. This can be termed manure management.

Soil and water conservation practices reduce soil erosion and thus reduce the loss of soil nutrients. Proper tillage of the crop field breaks down soil aggregates and thus increases soil aeration, which helps in organic matter decomposition. Improving drainage in the field reduces soil erosion and decreases N losses from denitrification, which is substantial on waterlogged soils. Proper tilling also decreases the leaching ability in soil so that the soil nutrients do not leach into the subsoil and beyond. Root growth management in plants increases the nutrient supply to plants as more roots interact with large volumes of soil. Soil acidity management through liming is another option for preserving the soil nutrient balance. The ideal soil pH for most crops is 6.3–6.8. In this range, soil nutrients availability is well balanced, and it is optimum for an active and diverse microbial population in soil. Liming or application of limestone is the most common method to increase the pH level in soil. As discussed earlier, fertilizer application based on soil tests can recompense soil nutrient loss and preserve the soil nutrient balance.

See Also: Cash Crop; Crop Rotation: Fertilizer.

Further Readings

Ashman, M. R. and G. Puri. *Essential Soil Science: A Clear Concise Introduction to Soil Science*. New York: Wiley Blackwell, 2002.

Brady, N. C. and R. R. Weil. *Elements of Nature and Properties of Soils*, 2nd ed. Upper Saddle River, NJ: Prentice Hall, 2004.

Sudhanshu Sekhar Panda
Gainesville State College

SOYBEANS

Soybeans (cultivated variety, *Glycine max*; wild variety, *Glycine soja*), are native to Asia and were originally cultivated in north China. Soybeans are one of the world's most important crops because of their high protein content and their high production of edible oil. Along with rice, wheat, barley, and millet, soybeans have been an important part of the Chinese diet for many centuries. They did not arrive in Europe until the 18th century, and the first known report of their cultivation in the Americas was not until the early 19th century. Today, most of the world's soybeans are produced in the Americas, with the United States and Brazil being the leading producers. Soybeans are used to produce a variety of foods for human consumption, but their main use today is the manufacture of feed for livestock and soybean oil. Soybean cultivation in the Brazilian Amazon basin and the manufacture of soy-based biodiesel have been at the center of important debates about conservation, agriculture, and renewable energy since the end of the 20th century.

The soybean was domesticated by the Chinese by the beginning of the 11th century B.C.E., during the Zhou Dynasty (1100 B.C.E.–256 C.E.), although the domestication process likely began several hundred years earlier, during the Chang Dynasty (ca. 1700–1100 B.C.E.). The earliest depictions of the archaic Chinese character for soybean (*shu*) can be found throughout the *Book of Odes*, a collection of Chinese poems written between the 11th century and the 7th century B.C.E. Though there is no explicit mention of

Mature soybeans in the pod. Soybeans are used in a wide variety of foods, but it is their use in livestock feed and for soybean oil that makes them one of the world's most important crops.

Source: U.S. Department of Agriculture Agricultural Research Service/Scott Bauer

the cultivation of soybeans in this book, scholars have interpreted these first depictions as evidence of domestication, because of the somewhat intimate knowledge of the soybean that the authors of the book must have had to create the pictograph for the soybean. The first specific reference to cultivated soybeans was not until 644 B.C.E., when it was recorded that the Shan-Jung tribe paid a tribute to the House of Zhou with soybeans. By the 1st century C.E., the soybean had spread to other parts of China as well as to the Korean Peninsula, closely mirroring the spread, rise, and fall of ancient Chinese dynasties. Soybean cultivation began to take off in Japan and southeast and central Asia starting in the 1st century C.E. The spread of soybeans throughout Asia was closely related to the establishment of land and sea trade routes, including the Silk Road.

Soybeans first arrived in Europe by way of the Netherlands and France in the middle of the 18th century. They were introduced in the United States by the turn of the 19th century. The first seeds were likely sent from France to the Philadelphia Society for Promoting Agriculture by the U.S. ambassador to France, Benjamin Franklin, who established a seed exchange program between the two countries during his ambassadorship in the late

18th century. The first specific mention of soybeans in U.S. literature was an 1804 recommendation for their cultivation in Pennsylvania from the president of the Philadelphia Society for Promoting Agriculture. The soybean is known to have arrived in South America by 1882.

Before World War II, soybean crops in the United States were primarily used as forage crops for livestock or cut for hay. The disruptions to trade routes as a result of the war, however, caused the United States to look for alternatives to importing a large percentage of its edible fats and oils. Soybean plantings expanded drastically as the United States strove to meet its food and food oil needs during the war. In addition to providing ample amounts of oil and meal, the soybean was particularly suited to large-scale production in the United States because it grew easily in conditions similar to corn, which was already being planted on a large scale. It also worked well as a rotation crop with corn and with rice. After World War II, soybean production in the United States continued to rise dramatically, with the geographic extent of the soybean's cultivation expanding in the south and the Corn Belt. The United States overtook China as the world's leading soybean producer in 1955 and produced more than 75 percent of the world's soybeans in the 1950s, 1960s, and 1970s.

Brazil began producing soybeans in earnest at the end of the 1960s, and the period of rapid expansion of this crop between the 1960s and the end of the 1970s is commonly referred to as the Brazilian "Soybean Boom." Though still second to the United States in soybean production, Brazil has been the world's leading exporter of soybeans since 2004. Chinese production of soybeans remained somewhat steady between 1945 and 2004, rising slightly since then, but not nearly enough to compensate for its rising demand for soy products. Argentina and India round out the world's top five soybean-producing countries.

Soybeans were originally used in Asia for their seeds, which can be eaten a number of different ways: fresh, fermented, or dried. Whole, fresh soybeans, simmered until they are soft and eaten out of the shell, are known as edamame in Japan. Fermented soybeans also can be used to make miso, which is a bullion-like product that can be used in soup and sauce preparation, as well as soy sauce, a common condiment for many kinds of Asian foods, even in the West, and tempeh, a food original to Indonesia that is a popular main dish and meat substitute throughout the world.

Popular nonfermented soy foods include tofu, soy milk, yuba, and sprouts. Tofu was first eaten in large quantities by the Chinese during the Sung Dynasty (960–1127 C.E.) It is now an important part of the Japanese and Korean diets as well, gaining in popularity throughout the world as a meat and dairy substitute food, partially because of its highly digestible nature. Soy milk originated before the Sung Dynasty and is a popular substitute for cow's milk in Asia and in other parts of the world. Yuba, a protein-rich food that is less well known outside Asia than many other soy foods, is made by boiling soy milk. Yuba can be served dried, fresh, or somewhere in between. Soybean sprouts are simply germinated soybeans, which are high in vitamin C and amino acids and are often eaten fresh; for example, as toppings on salads. Soy foods are rich in nutrients and protein and are inexpensive to produce, but they are also a common source of food allergies. As more processed products, including food ingredients, began to be produced from soy in the last half of the 20th century, soy allergies began to receive more attention from healthcare professionals and concerned citizens.

Soybeans are made primarily into two products in the West: crude oil and defatted soybean meal. Crude soybean oil is made by crushing the seeds to extract the oil; the remaining matter is the defatted meal. Both of these products are then further processed into other final products. The first step in processing soybean oil is called degumming. The gum is then processed into a product known as soy lecithin, a commonly used emulsifying and stabilizing agent found in many processed foods, including beverage mixes, margarine, shortening, candies, and baking products. It is also found in nonfood products including paints, insecticides, and cosmetics. The degummed oil can undergo further processing and be used as an edible oil in cooking oils, salad dressings, and other products. Biodiesel can also be produced from this oil—a relatively new use for the oil that may increase in popularity as more alternatives to fossil fuels are sought. Almost all of the defatted meal left over from extracting soybean oil is used in livestock and poultry feed and pet food, though soy flour for human consumption is also made from a small percentage of this meal.

Although the United States still produces the most soybeans overall, no country in the world has more potential land for soybean cultivation than Brazil. By some estimates, Brazil has an additional 100 million acres of land suitable for soybean production that are not yet being used. Much of this land is located in the Amazon rainforest and *cerrado* (tropical savannah) biomes of Brazil—land that requires considerable amounts of fertilizer to maintain production. In particular, deforestation of the Amazon rainforest related to soybean production has received considerable attention as a result of the efforts of international environmental groups, including Greenpeace. In 2006, the largest soy traders in Brazil signed an agreement to certify that no soy grown in Brazil would come from newly deforested areas in the Amazon biome. Though it has not yet been thoroughly tested by time, this agreement may be a notable milestone in effective consumer activism because it was the result of large-scale soybean buyers reacting to demands from consumers and environmental groups, and not of requirements set by the Brazilian government.

See Also: Cash Crop; Grain-Fed Beef; Legume Crops; Vegetarian.

Further Readings

Banck, Geert A. and Kees de Boer. *Sowing the Whirlwind: Soya Expansion and Social Change in Southern Brazil*. Amsterdam: CEDLA, 1991.

Boerma, H. R. and J. E. Specht, eds. *Soybeans: Improvement, Production, and Uses*, 3rd ed. Madison, WI: American Society of Agronomy, 2004.

Gibson, Lance and Garren Benson. "Origin, History, and Uses of Soybean (Glycine Max)." http://www.agron.iastate.edu/Agron212/Readings/Soy_history.htm (Accessed January 2009).

Hymowetz, T. and C. A. Newell. "Taxonomy of the Genus Glycine, Domestication and Uses of Soybeans." *Economic Botany*, 35/3 (1981).

Raffensperger, Lisa. "Soy Moratorium Working to Protect Amazon." http://earthtrends.wri .org/updates/node/277 (Accessed January 2009).

Lisa Rausch
University of Kansas

Substitutionism

Substitutionism refers to industrial sector efforts to reduce agricultural products to industrial inputs for food manufacture or to entirely replace farm-based products with industrially produced substitutes. David Goodman, Bernardo Sorj, and John Wilkinson developed the parallel concepts of appropriationism and substitutionism in their theory of agro-industrial development described in *From Farming to Biotechnology*. Examples of substitutionism include the industrial addition of synthetic additives to foodstuffs during processing to increase product shelf life or to reach other taste or nutritional criteria not possible through farm-based production. Examples of agricultural product replacement include the production of latex instead of rubber, or margarine instead of butter, which can be made with the cheaper, industrially interchangeable inputs of animal fats or vegetable oils. Together, the concepts of appropriationism and substitutionism explain the industrialization of agriculture through the application of science and capital investments to discrete rural labor and biophysical processes in agricultural production.

Although substitutionism generally occurs in the postharvest, downstream agricultural product processing or involves the replacement of farm products altogether, appropriationism describes the processes whereby industrial products stand in for on-farm rural and biological production processes. Examples of appropriationism include the use of industrially produced nitrogen fertilizer as a purchased input in place of farm-based crops or animal rotation or the use of animal manure for crop nutrient supply.

The first and most essential component of substitutionism, established between 1870 and 1914, was the emergence of industrial intermediaries between the site of agricultural production and food consumption. Initially, industry acted between agricultural production and consumption to scale up food processing activities historically performed by the farm or consuming household.

The technologies applied to food processing in this stage of substitutionism were not new, but labor moved to the factory, and the scale of both food-processing operations and product standardization increased. For example, industrial flour milling outcompeted more localized, decentralized milling as a result of its economies of scale and its ability to supply large quantities of high-quality, standardized flour that industrializing bakeries demanded.

With industrial presence established between farm production and final consumption, industrial investments and role in food production processes expanded by transforming the composition, form, or packaging of foodstuffs. Food processors drew on the large volumes of standardized food products becoming available, like flour, and applied new processing technologies to "add value" by recomposing foodstuffs into more complex products. Food manufacturers then competed by marketing products as having differentiated qualities imparted through a company's food-processing techniques. As food products became increasingly associated with a food processor's technical approach and brand name, rather than with a particular place or farm, the association between farm labor processes and food products began to break down, and larger portions of the profits derived from food sales accrued to industry.

Other iterations of substitutionism created entirely new industrial products that performed the same function as rural products but took on a different form, unlike industrial flour milling, which maintained the same basic milling methods and end product. Milk, for example, was historically preserved through rural cheese and butter production; the industrial sector developed new, less perishable products to preserve milk: dehydrated milk powder or condensed milk, preserved with sugar.

The advent of refrigeration and rapid food-freezing technologies in the 1930s marked another similar version of substitutionism. In this case, new food industry technologies were applied to outdo their rural product counterparts. For example, quickly freezing vegetables for transport and later consumption maintained product flavor and quality better than traditional, rural methods of food preservation like pickling or dehydrating.

Substitutionism also refers to industry efforts to reduce agricultural output into simple, interchangeable industrial inputs. Margarine production, for example, entailed a more radical qualitative shift from the rural product butter, but unlike powdered milk, it was not produced as an industrial method of preserving a less stable rural product. Instead, margarine manufacture replaced butter with a product that reduced agriculture's role to the supply of cheap, raw materials (animal fats and then vegetable oils) that could be used interchangeably as market and technical concerns dictated.

This process of substitutionism also applies to more recent practices of dividing a simple agricultural product, like corn or soybeans, into their protein, starch, and other components. The fractionation of natural products into simple component parts allows industry to variably recompose food products to fit specific flavor, texture, color, nutritional, or other criteria. Processing also makes single agricultural products more versatile to industry. Corn, for example, can be de- and recomposed into animal feed, seed, human food, or ethanol or used in other industrial applications like the manufacture of bioplastics. Taken a step further, substitutionism means entirely replacing agricultural products with nonagricultural industrially produced versions that satisfy the same function. Whereas margarine still relies on agricultural products, albeit in simplified forms, some products' manufacturers have succeeded in rendering rural production processes and their products entirely unnecessary. The advantage for the industrial sector here is that it can avoid reliance on the more time-consuming biological production processes inherent in agriculture. For example, innovations in organic chemistry led to the production of latex in place of rubber, nylon and other synthetic fabrics in place of natural ones, and the use of mineral rather than plant sources for dyes. In another example, a mixture of lime and chlorine replaces sunlight as a means for bleaching cotton.

More recently, substitutionism has entailed the creation of "designer" or "health foods." Such foods might include potato chips manufactured with olestra, which reduces the consuming body's ability to absorb fat contained in the chips. Another example might be "functional foods" or "nutraceuticals" that contain industrially added nutrients to make the food more nutritious or otherwise desirable than the original agricultural product. The practice of adding inulin to yogurt to increase shelf life and the consumer's absorption of the yogurt's calcium serves as an example here. Nondairy coffee creamers and the proliferation of soy-based, industrially produced meat alternatives ranging from nuggets of textured protein to strips of fake bacon are other more recent examples of substitutionism. These foods are completely industrial in the sense that the typical farm or household is nowhere near capable of producing them, and the majority of the profits involved in their manufacture accrue to industries seeking to fulfill consumers' nutritional or other market criteria not achievable with agricultural products.

Consumers, however, are increasingly suspicious of more advanced forms of substitutionism that rely on chemical additives to ensure food longevity or durability. For example, recent controversy over the use of synthetic materials in products that are certified organic demonstrates consumer concerns about the role of industrial processing in food provisioning. For another example, bagged salad mixes have become one of the fastest-growing produce categories, but recent *Escherichia coli* outbreaks have raised consumer doubts about the ability of large-scale industrial processors to safely deliver even minimally processed agricultural products on a mass scale.

See Also: Agribusiness; Appropriationism; Corn.

Further Readings

Goodman, David, et al. *From Farming to Biotechnology: A Theory of Agro-Industrial Development*. Oxford: Basil Blackwell, 1987.

Goodman, David and Michael Redclift. *Refashioning Nature: Food Ecology and Culture*. London: Routledge, 1991.

McMichael, Philip. *Food and Agrarian Orders in the World-Economy (Contributions in Economics and Economic History)*. Santa Barbara, CA: Praeger, 1995.

Sean Gillon
University of California, Santa Cruz

SUGARCANE

These washed stalks of sugarcane are moving to a mill for processing. Over 1 billion tons of sugarcane are produced every year in Asia and the Americas.

Source: U.S. Department of Agriculture Agricultural Research Service/David Nance

Sugarcane has always been sweet, but its cultivation, usage, and meaning have evolved and expanded over time and space over eight millennia. Sugarcane is a tall, bamboo-like plant that falls under the genus *Saccharum*, but it is difficult to classify into specific species, as species within the genus easily crossbreed among each other, and species themselves also cross easily with other species in similar genera. Hence there are many species and varieties of sugarcane, although sugar production is generally limited to four specific species or their hybrids. Early hybridization between domesticated and wild sugarcane varieties resulted in two main categories of sugarcane: the "thin" varieties, which favor cooler tropical or subtropical regions, and the "thick" or "noble" varieties, which favor warmer tropical zones. The sugarcane plant itself is made up of a shallow root system, a jointed stalk, and leaves that grow out of the stalk. It is the sucrose stored in the stalk that is harvested for sugar production.

Sugarcane cultivation originated in what is now New Guinea around 8,000 years ago and began to spread into southeast Asia and India about 1000 B.C.E. The Arab conquests of North Africa and Europe were responsible for the spread of the plant cultivar into these regions, first to

Egypt and then to the Middle East in the 7th century C.E., finally entering Spain at the beginning of the 8th century. Portuguese colonizers further expanded the reach of sugarcane into the Canary Islands, the Azores, and then into West Africa in the 15th century. From there the plant made its first foray into the Americas, entering what is now the Dominican Republic in 1493. It spread into Central and South America in the mid-16th century and later into the West Indies. The development of the West Indies was centered on sugarcane production until the independence movements in the early 19th century, and production depended on imported African slave labor. As a consequence to the growth of sugar plantations in the new world, sugar developed into a major commodity during the industrial revolution and initiated changes in European and North American diets. Today over one billion tons of sugarcane are produced annually in North and South America, East Asia, and the Indian subcontinent.

Sugarcane's early use in southeast Asia was restricted to chewing, and it was cultivated strictly for this purpose. It is believed that the technique of boiling the cane juices to make solid sugar was not discovered until the 1st century B.C.E. in the Indian subcontinent. Sugar production and its usage as a sweetener were unknown outside south Asia in ancient times, and travelers to the region from Europe, where honey was the only sweetener, would marvel at the vast fields under cultivation. Sugar itself is produced by crushing and extracting the juice from the cane, and then heating and evaporating the clear juice to produce sugar crystals. This raw sugar is then further refined to produce white and other kinds of sugar. Sugarcane juice is consumed as a beverage by itself, or fermented and distilled into rum, Brazilian *cachaça*, or other cane liquors. Bagasse, the fibrous waste material left over after crushing the cane, is also used as a fuel source in sugar production as well as a source of paper pulp. Sugarcane is moreover used as a source of ethanol—an alcohol fuel made by distilling the cane juices. Ethanol is available as a by-product of hard sugar production, but it is increasingly being produced in dedicated processing plants for fuel production, especially in Brazil.

Sugarcane sits squarely in the middle of current debates around food versus fuel production. Proponents of expanded sugarcane-based ethanol production cite its properties as a renewable energy source and its competitive cost in comparison with gasoline. They look to Brazil as the leading example, which began mandating ethanol mixed into gasoline (or flex fuel) for the transport sector in the 1970s. Brazil greatly increased its sugarcane and ethanol production after September 11, 2001, amid rising world oil prices; the result has been that the majority of Brazilian cars are currently flex fuel vehicles.

In contrast, there are various critiques of expanding sugarcane production for ethanol. First is the argument that the process of transforming cane into ethanol requires fossil fuels and also releases large amounts of carbon emissions, as does the burning of the ethanol itself for fuel in vehicles. Environmentalists also point out the increasing demand for freshwater associated with expanded processing facilities, as well as the wastewater generated by such facilities. Critics point out the dangers of changes in land use associated with increased demand for sugarcane, including intensification of existing sugarcane cultivation that involves higher rates of agrochemical usage, often citing the wholesale takeover of the Florida Everglades by commercial sugar production. Encroaching sugarcane monocultures into former areas of food production can also potentially create problems with food security. Mechanized harvesting is increasingly in use, which eliminates the need to burn the cane for harvesting, greatly improving air quality in areas of production. Debates have recently arisen on the environmental and economic sustainability of other countries duplicating Brazil's experience with sugarcane ethanol. The jury is still out, however, on whether sugarcane for ethanol production is the lesser of two evils or a boon to sustainable development.

See Also: Agribusiness; Cash Crop; Corn.

Further Readings

Goldenberg, José. "Ethanol for a Sustainable Energy Future." *Science*, 315 (2007).
Hollander, Gail M. *Raising Cane in the 'Glades: The Global Sugar Trade and the Transformation of Florida*. Chicago: University of Chicago Press, 2008.
Kriz, Kay Dian. *Slavery, Sugar, and the Culture of Refinement: Picturing the British West Indies, 1700–1840*. New Haven, CT: Yale University Press, 2008.

Heather R. Putnam
University of Kansas

Supermarket Chains

The supermarket occupies a niche similar to the general store in earlier decades: It is the store where you buy the things you need to buy most frequently, and most of the nonfood items will either answer that description or will represent high-margin impulse purchase items (such as DVDs and toys) that offset the discounted loss leaders. Supermarket chains are a major part of the food industry, particularly when considered in tandem with the distribution centers serving them, which are usually owned by a common parent company. Although locally owned and "mom and pop" grocery stores have not been displaced, in the United States and other developed countries where supermarkets have become the norm, haggling and other forms of price negotiation have died out to such a great degree that Americans abroad are often uncomfortable attempting to haggle even when it is expected of them.

The transnational supermarket chain enjoys such economies of scale that it can provide extraordinary discounts on food—which, in time, are no longer seen as discounts but are, rather, received by the consumer as the appropriate price of the good, even if it is priced as a loss leader (and thus below its true value), which can make it difficult for nonchain stores to compete. One result of this, an example of the phenomenon of the disappearing middle, has been that more and more nonchain grocery stores are boutique stores that do not compete directly with the chain supermarkets because they sell a distinctly different product: meat from a full-service butcher, regionally made cheeses and milk from local cows, local farm produce, or food products with a particular value-added tag like "organic," "vegetarian," "gourmet," or "imported." Similarly, kosher and halal butchers are still able to thrive, as are "ethnic" grocery stores that specialize in a particular country's cuisine and comforts, or a particular clientele. Even these stores' customers likely still shop at the chain supermarket for 50-cent bananas, six-packs of soft drinks, and toilet paper.

Though the line between a supermarket and a nongrocery store has blurred in recent years as Wal-Mart and Target have added supermarkets to many of their stores, supermarkets typically sell fresh food like meat, produce, dairy, and baked goods; canned, frozen, and other packaged foods; and basic nonfood staple goods such as pet supplies, cleaning and basic automotive supplies, greeting cards and magazines, batteries and lightbulbs, and vitamins and over-the-counter health remedies. Depending on local laws and the chain, they may also carry alcohol (sometimes only beer and wine, sometimes liquor as well) and may have an in-store pharmacy or bank. The parking lot may include a gas station. Increasingly common are in-store eating areas—prepared foods like salads and fried chicken are taxed differently—and self-service soda fountains or cashiered in-store coffeehouses (much of Starbucks's expansion before the recent contraction was in such locations). Unlike other stores of similar size, a supermarket is nearly always on a single floor. It is usually located near a residential area or in the part of town commuters pass through on their way home from work, which is not only convenient but encourages impulsive shopping trips, particularly in the time after rush hour and before dinner, when shoppers are likely to be hungry and prone to more extravagant purchasing.

Because they offer so many goods, supermarket chains are masters of using products as loss leaders. A loss leader is sold at a price that is so low, no profit is made, which works to the store's benefit only if the shopper purchases other things at the same time. Making staple products like bread and bananas loss leaders is a good way to get shoppers to come back every week, and supermarkets can depend on making their profits on the other purchases that shoppers will make. Supermarkets do not need to increase the prices of other goods to make loss leaders practical; they can charge the same price for greeting cards as the drugstore down the street, for instance, so long as that price is a profitable one. Using loss leaders to attract business is sufficient. Another inventory characteristic of supermarkets is stock rotation, which constantly moves the products with the earliest expiration dates to the front of the display area to increase the odds that they will sell before that date arrives.

Supermarkets and the Food Industry

Supermarkets as they exist today—having taken the place of those general stores—are a result of changes in the food industry. In decades past, most food did not arrive packaged for the consumer—except for canned goods, once they were introduced—and thus had to be repackaged in the store, either in advance of purchase or to order (with dry goods sold in bulk, as some limited number of goods are still sold today). Butchers cut much of their meat to order and ground their ground beef in the mornings. Everything had to be laboriously packaged and weighed. Furthermore, customers used to have to be waited on; the idea, common today in stores of most sorts, of customers wandering the aisles and filling their hands or a cart with products is an early-20th-century development. Today, these processes are streamlined to such a degree that more and more supermarkets have self-checkout lanes requiring no cashier—the customer can even weigh goods sold by the pound himself, using the register's built-in electronic scale. More and more food is packaged before it arrives in the store—not just the flour prepackaged in paper containers instead of arriving in barrels or burlap sacks, but the canned goods, the meat already ground and steaks already cut (many supermarkets' butchers should more properly be called meat salesmen), often even corn on the cob already shucked and shrink-wrapped on Styrofoam trays.

The result is a significant savings in labor costs. Shelves can be stocked during working hours or, more conveniently for customers (and allowing for narrower aisles), at night while the store is closed. A small staff of cashiers and stock boys is sufficient for huge daily sales volumes. The supermarket and the supermarket chain developed side by side; the first "self-service" grocery store (with customers helping themselves rather than being waited on) was the Piggly Wiggly chain in the 1920s, and when the Depression struck, the additional innovation of buying in bulk and using loss leaders to attract business was adopted by other grocery store chains like Kroger and Safeway. All three chains are still around today and are among the largest in the country. Kroger was the first chain to construct a store as an "island" surrounded on all sides by a parking lot, to accommodate the significantly greater number of customers that a supermarket could serve simultaneously, relative to the old-style slower stores. Of course, all of this developed at the same time that the automobile was becoming more common, and supermarket chains spread alongside the suburban sprawl, introducing Americans to national product brands like the sugared cereals that sponsored children's radio dramas and the frozen ready-to-bake meals that quickly became known as "TV dinners," perfect for heating up and watching around the newest appliance.

Winning Customer Loyalty

Because chains compete principally with each other, they use a variety of loyalty programs to entice customers to continue shopping with them. The oldest of these is the trading stamp program, extremely popular throughout the 1950s and 1960s, when every purchase of a certain amount of money rewarded the shopper with a certain proportional number of trading stamps, which could be saved up and redeemed for "prizes" at a later date—like portable radios, roller skates, and so on. The prizes, worth a small fraction of the total amount of money that had to be spent to earn the stamps, were not meant to entice shoppers to spend more—only to spend it "here." More recent loyalty programs include seasonal incentives—many stores award a shopper who exceeds a particular amount of spending within a particular month with a seasonally appropriate prize, like a free turkey at Thanksgiving, a free ham at Christmas, or an assortment of grilling items in July—and loyalty cards, which make the shopper eligible for specific discounts on selected items (which change periodically, usually every week).

Because supermarket chains represent, in a sense, extraordinarily large buyers, they are accused of monopsonistic practices. A monopsony is the converse of a monopoly, a situation in which there is one big buyer and many potential sellers, giving the buyer power to cause an anticompetitive environment. For instance, if a major supermarket chain were to elect to stop carrying a particular brand of cereal, that cereal would face significant profit losses; the threat of this gives the chain (and the distributor owned by its parent company) bargaining power in dealing with the owner of that brand. The practice of slotting fees— fees charged by supermarket chains for placing a product in a highly visible spot, such as at the end of an aisle, or at eye level—is often criticized as unethical and anticompetitive, and the general public is not widely aware of it.

See Also: Agribusiness; Disappearing Middle; Supply Chain; Wal-Mart.

Further Readings

Allen, Gary J., et al., eds. *The Business of Food: Encyclopedia of the Food and Drink Industries.* New York: Greenwood, 2007.

Burch, David and Geoffrey Lawrence. *Supermarkets and Agri-Food Supply Chains: Transformations in the Production and Consumption.* Northampton, MA: Edward Elgar Publishing, 2007.

Werbach, Adam. *Strategy for Sustainability: A Business Manifesto.* Cambridge, MA: Harvard Business Press, 2009.

Bill Kte'pi
Independent Scholar

SUPPLY CHAIN

A supply chain is an organization of a network that connects producers, marketers, distributors, and consumers in a complex manner to move a commodity from producers to consumers. More precisely, a supply chain is a system of organization in which goods move from producers to consumers through the exchange of payment, credit, and capital among actors; price signals, pricing behavior, and value added; the dissemination of technology; and the flow of information across the chain. A supply chain contains various actors, from producers to consumers, technology, regulations, activities, resources, information, and forms of governance. The supply chain—also known as the global commodity chain—is currently one of the most widely used approaches to studying interfirm relations in the global agrofood system. Within this framework, however, there is a proliferation of overlapping names and concepts, such as global value chains, value system, production networks, value networks, commodity system analysis, and so forth, as various authors have developed the idea in different ways.

The supply chain recognizes that different configurations of actors may influence capabilities and relative bargaining power and subsequently may affect outcomes along the chain. In sum, the term *supply chain* defines product chains in terms of actors and their objectives, the structure of its markets, the strategies or instruments that actors use to influence this structure, and finally, by the product itself, as different product chains may display different characteristics. Supply chain analysis, therefore, has become a useful analytical tool in identifying the central role that global buyers such as Wal-Mart, Nike, Gap, Darden, and Lyons play in organizing activities within commodity chains.

Governance in the Supply Chain

Supply chain analysis facilitates the mapping of the world division of labor. It was elaborated and popularized by, among others, Gary Gereffi and his colleagues in the sixteenth Political Economy of the World System conference. According to them, the chain consists of sets of interorganizational networks clustered around one commodity or product, linking households, enterprises, and states to one another within the world economy. Their concept contains four important elements: First, they emphasized the fact that chains frequently involve cross-border coordination of the activities of independent firms. Second, they emphasized the issue of governance, drawing attention to the fact that large retail and brand-name companies create interfirm networks characterized by a high degree of coordination. Third, they highlighted the increasing role played by international buyers, retailers, and brand-name companies in the trade of labor-intensive manufactured products such as garments. Finally, in recognition of these elements, they initially postulated two types of commodity chain: buyer driven and producer driven.

In a buyer-driven commodity or supply chain, large retailers, marketers, or brand-name companies play a pivotal role in establishing and driving geographically dispersed production and distribution systems without necessarily owning any themselves. The United Kingdom–Africa horticulture value chain, for example, exhibits several characteristics of a buyer-driven supply chain, in which powerful lead farms (supermarkets) govern supply networks that cross several African countries, defining not only what is to be produced but also how and under what conditions it is to be produced. These supermarkets increasingly determine the production imperatives of horticultural farms upstream in the chain and, indirectly, the employment strategies they adopt. Producer-driven chains are typical of capital- and technology-intensive industries in which transnational manufacturing firms drive the chain, controlling the core technologies and the production facilities, often through vertical integration.

The supply chain analysis has become an important framework for analyzing economic development, as well as the evolution and complexity of export-oriented industries in the context of globalization, proving fruitful for delineating the power relations or governance within chains and their prospects for broad-based growth. The point of departure for supply chain analysis is the fact that some firms directly or indirectly influence the organization of global production, logistics, and marketing systems. Through the governance structures that such firms create, the decisions that they make have important consequences for the access of developing-country farms to international markets and their range of activities. The governance structures also generate changes in local landscapes.

Gereffi and his colleagues used their model mainly in the export-oriented manufacturing industries. The basic supply chain frame has been used and elaborated, with substantial variance, by many agrofood scholars suggesting that powerful buyers increasingly govern and drive production definition/specification, production processes, and enterprise participation in the international supply chain. Some of them, however, linked their discussion of supply chain to complementary use of earlier traditions of producer-consumer networks, such as commodity systems (sometimes termed *production chains*, *food chains*, or *filieres*).

Although agrofood scholars' extension of Gereffi's producer-/buyer-driven analogy to the agrofood sector seems quite insightful, we can discern some critical scrutiny of and challenges to a simple dichotomous characterization of producer-driven versus buyer-driven chains. First, the nature of globalization's effect on the agrofood system is different from that of other export-oriented manufacturing industries. As a result of some inherent characteristics and dynamics, most agrofood industries are not characterized by the vertically integrated transnational production that Gary Gereffi and his colleagues found in the apparel industry. In the agrofood commodity chain, the nature of lead firms varies significantly. Among buyer-driven agrofood commodity chains, some are driven by large supermarket retailers, but others are dominated by processors, international traders, or global branders, and some are also partly influenced by environmental and labor rights movements.

Second, in the traditional supply chain model, economic actors are given primacy, and political conditions are treated as contextual. However, in agrofood supply chains, economic, social, and political forces are central for chain governance. Although economic actors are important, agrofood chains sometimes contain nonhuman actants such as viruses and bacteria and other elements of nature. Even apparel, cars, and so on may include nonhuman actants, but they are more likely to be actively considered in agrofoods because of the ties to metabolic process (agri, bodies, etc.) and the influence of science studies. Therefore, the agrofood supply chains should include nature to permit a more complex understanding of the ways in which biology and the organic temper the political economy of food system. In this sense, we suggest that the question of risk, perishability, seasonability, sustainability, and nonidentity of production and labor time are necessary to grasp the commodity-specific dynamics of production systems but should be linked with food consumption as integral elements of the global food supply chain.

The supply chain approach has been widely used by agrofood scholars, albeit differently, mainly to explore production-consumption networks that traditionally have tended to privilege analyses of buyers and, to some extent, suppliers. Recent studies, however, extended this analysis to explore local labor relations as well as gender issues.

See Also: Agrofood System (Agrifood); Commodity Chain; Food Safety; Labor; Supermarket Chains.

Further Readings

Dolan, C. and J. Humphrey. "Governance and Trade in Fresh Vegetables: The Impact of UK Supermarkets on the African Horticulture Industry." *Journal of Development Studies*, 37/2 (2000).

Gereffi, G. and M. Korzeniewicz, eds. *Commodity Chains and Global Capitalism*. Westport, CT: Praeger, 1999.

Humphrey, John and Hubert Schmitz. "Governance in Global Value Chain." *IDS Bulletin*, 32/3 (2001).

Islam, M. S. "From Pond to Plate: Towards a Twin-Driven Commodity Chain in Bangladesh Shrimp Aquaculture." *Food Policy*, 33/3 (2008).

Ponte, S. "The 'Latte Revolution'? Regulation, Markets and Consumption in the Global Coffee Chain." *World Development*, 30/7 (2002).

Raynolds, L. "The Globalization of Organic Agro-Food Networks." *World Development*, 32/5 (2004).

Vandergeest, P. "Certification and Communities: Alternatives for Regulating the Environmental and Social Impacts of Shrimp Farming." *World Development*, 35/7 (2007).

Md. Saidul Islam
Nanyang Technological University

Sustainable Agriculture

The concept of sustainable agriculture—a food and fiber production system that does not destroy the ecological, human, and animal foundations on which it is built—is not a new idea. Well before the rise of the industrial agricultural system that developed in the wake of World War II and the green revolution of the 1960s that brought mechanized, high input agriculture from the United States to countries around the world, people advocated for a food production system that fostered health in the soil, in animals, and in people. Over the last 20 years, sustainable agriculture has gained a significant foothold around the world in terms of consumer interest, public policy, and number of farm acres. Many challenges face the continued spread of sustainable agricultural practices, however, and its future may lie in synergistic associations with other movements for environmental sustainability, adaptive management, land tenure, and social justice.

As is often the case with ideas that have long histories, there are diverse definitions of sustainable agriculture; however, most approaches share several key concepts. The care of the soil is certainly a foundation stone, and historians as well as farmers have argued that the fall of many a civilization was caused by agricultural abuse of the soil. Arguments have been made by proponents in industrialized countries since the 1920s that the use of industrially produced agricultural chemicals such as fertilizers and pesticides destroys soil life and structure—the very basis on which sound agriculture depends. Basic tenets of sustainable care of soil include inputs of organic composts, rotational planting and cover cropping to build soil, and minimizing tillage and other soil disturbance.

Proponents of sustainable agriculture also advocate for farming practices that take advantage of natural cycles, rhythms, plant characteristics, and animal behaviors. Thus, a sustainable farmer knows a great deal about the ecology and biology of soil, plants, and animals and will use that knowledge to achieve production goals. Certain crops are grown sequentially (rotational planting) or next to each other (companion planting) to take advantage of natural pest repellents in a certain plant or soil-building abilities in another. Sustainable farmers might take advantage of the natural proclivity of pigs to root to turn a compost pile, for example, or will rotationally pasture chickens and beef cattle together because the chickens clean up insects found in and under cow patties. A sustainable cotton farmer will rely on a seasonal frost to defoliate plants and expose cotton bolls, rather than a synthetic chemical.

Another key concept in sustainable agriculture is the idea of a closed system, or at least a production system that emphasizes the use of locally available resources and minimizes both waste and inputs from off-farm. The manure from cattle and other livestock on the farm might be composted along with plant materials to produce fertility for the soil and to improve soil tilth. This leads to another idea common in sustainable agriculture—that a sustainable farm will harbor a diversity of organisms and often produce multiple agricultural products.

Examples from traditional indigenous agricultural systems from around the world have often been upheld as models of agricultural sustainability. These are long-lived systems that developed independent of imported seeds, fertilizers, and chemicals and that are informed by deep, local knowledge about the environment on and around a farm. Traditional agroforestry, for example, involves the deliberate maintenance of trees and other woody perennials in fields and pastures, which helps secure the future of the surrounding forest. Some indigenous farmers encourage wild species of crop plants in their fields to bolster the long-term genetic vigor of the crop plant. Indigenous farming practices are often interwoven into indigenous cultural institutions and are central to sustaining a culture. Homegardens in countries around the world have been found to harbor increased biodiversity, to be naturally resistant to pests and disease, and also to play important roles in cultural sustainability and food sovereignty. As the result of changes in land tenure and shifts in economies to a focus on cash crops, however, many traditional ways of farming have been lost or have become economically unsustainable.

All Sustainable Agriculture Is Not the Same

The definition and deployment of concepts in sustainable agriculture can vary widely. Rudolf Steiner, who lectured on biodynamic agriculture in the 1920s to a group of German farmers, for example, argued strongly for the model of a self-contained farm system. Other authors, such as the British botanist Sir Albert Howard, who brought ideas about composting from India to Britain in the 1930s and 1940s, described a broader agricultural system that was linked to the expanding urban presence and would help recycle waste from cities. J. I. Rodale, who published Howard's writings in the United States, placed a strong emphasis on the connection between healthy soil and personal health.

In Japan in the 1930s, Masanobu Fukuoka developed a noninvasive farm system eschewing tillage, weeding, and external inputs including even animal manure, arguing that all needs could be met within a balanced microsystem. Fukuoka argued further, as did Rudolf Steiner, that the ultimate goal of farming is not crop production but the cultivation of human beings as an integral part of the spiritual whole of nature. These core concepts of an integrated farm system, diversity, and the spiritual development of the farmer, lend themselves to the idea of a sustainable farm as necessarily smaller scale. A farm must not be so large that a farmer cannot, as author Wendell Berry has described it, carefully consider all associated, interacting parts.

Despite this plethora of ideas many decades in development, sustainable agriculture has had to wait a long time to experience wider acceptance, and this acceptance has been very partial. Organic agriculture, initially a marginalized and derided agricultural sustainability movement, was pulled into the mainstream during the 1980s by consumer fears of pesticide contamination of food and genetically modified organisms in the food system. Opportunities for more sustainable agriculture also arose as a result of the U.S. farm crisis in the 1980s, which threw many conventional farmers into bankruptcy.

The spectacular development of the organic food industry since 1985 indicates the potential as well as the pitfalls of growth in markets for products of sustainable agriculture. Markets for organic products expanded, especially in the United States, Britain, and European countries, and the number of organic acres farmed began to grow rapidly in the mid-1980s at rates that would be sustained for over 15 years. The U.S. 1990 Organic Food Production Act authorized federal standards for organic production, and a contentious decade later, the country passed its first set of federal standards regulating the organic production process. Despite strong gains, however, only about 5 percent of U.S. vegetable acreage, 2.5 percent of fruit and nut acreage, 1 percent of dairy cows, and 0.7 percent of layer hens were under organic management in 2005.

The global market for organic products reached a value of over US$46 billion in 2007. In 2008, more than 31 million hectares of agricultural land were being managed organically in 119 countries. At this time, the world's consumption of organic food and beverages is concentrated in Europe and North America, but the production of certified organic products is scattered worldwide. International trade in organic products—the Japanese market for organic soybeans, for example—was important for U.S. farmers as early as the 1980s, but as markets have continued to expand in the United States and Europe, more developing countries are producing for organic markets. The United States ranked fourth behind Australia, China, and Argentina in certified organic land in 2006. In Australia and

Argentina, pasture and rangeland are the dominant use of the organically managed land. Although the domestic market for organic products is expanding modestly in China, most of that country's organic production—as well as organic production in countries such as Bolivia, Chile, Uruguay, and Ukraine—is slated for export. According to the International Federation of Organic Agriculture Movements, Latin America and Africa are seeing strong growth in organic acreage, but in terms of certified organic farmland as a proportion of national agricultural area, Austria (13.4 percent) and Switzerland (11 percent) top the list.

New Opportunities

The market-driven success of organic production has also created new opportunities for sustainable agriculture in other institutions. Numerous North American colleges and universities offer educational and degree programs in organic and other sustainable agricultural technologies. In addition, many university extension programs now support organic farming. In 1988, the U.S. federal government created the Sustainable Agriculture Research and Education Program, a nationwide research and education grants program. In addition, many farmers' organizations, originally formed to certify organic farms, have now dedicated themselves to sustainable agriculture education and outreach.

The success of the organic sector has not come without costs, and critics have raised many issues about organics that reflect generally held concerns for the sustainable agriculture movement. People have pointed out that some organic growers, for example, are only pursuing price premiums rather than embracing the core intent of sustainable production. Such farmers, critics complain, pursue a kind of "input substitution," in which allowable substances are used in the place of prohibited ones, but no attention is paid to the farm system as a whole or to diminishing off-farm inputs and waste. In addition, the deployment of the organic model across a range of settings has revealed critical oversights in the original concept, particularly the issue of farm labor. Concerned mostly with small-scale farms, and with very little to say about self-exploitation of farm families, the sustainable agricultural literature has been poorly prepared to offer guidance on the issue of sustainable agricultural labor, particularly at a large scale. This has been an issue in terms both of domestic farmworkers' rights and of fair trade concerns regarding working relations with farmers in other countries.

Other issues of agricultural sustainability that have emerged in the wake of the rise of organic agriculture include humane treatment of farm animals, overall biodiversity (including wild animals) of a farm, and relations with neighbors and surrounding communities. In response, some people have worked to augment existing certification programs such as the U.S. National Organic Program and Demeter, the certification body for biodynamic farming. In addition, a number of other market-based certification programs have emerged that have been specifically designed to cover shortcomings in organic certification. Food Alliance, for example, includes requirements for safe and fair working conditions and the health and humane treatment of animals, the Rainforest Alliance provides for wildlife protection and improved farm labor conditions, and numerous free trade certifications advocate the payment of a fair price as well as social and environmental standards in the production of a wide variety of goods.

Thus, the consumer-driven interest in products of sustainable agriculture has created enormous opportunities for more sustainable ways of producing fiber and food. Markets are not always safe terrain for holistic ideas, however, and supporters of organic farming and other alternative agricultural approaches have been frustrated to see commitments to core concepts, such as the balance of a farm and humane treatment of animals, dissolve in the context of a narrow consumer focus on personal health, intense competition for profits, and the larger vertical integration of agriculture.

In addition, the concept of sustainable agriculture continues to be challenged as new interests pull and tug at the definition. Supporters of genetically modified organisms, for example, have argued that this technology is critical for food system sustainability because genetically modified organisms require the use of fewer chemicals and can be modified to grow in new environments with less water.

Despite these challenges, the thinking behind sustainable agriculture has no doubt improved as a result of being ground-tested in so many different situations. Moreover, on-the-ground activities are supported by a broad range of resourceful and committed people. Rather than the development of an ideal model, the future of sustainable agriculture instead resides with the ability of supporters to develop synergies with others working on issues such as social justice, fair trade, environmental health, and animal rights, and to change policy and market contexts to support their common interests.

See Also: Biodynamic Agriculture; Fair Trade; Green Revolution; Organic Farming.

Further Readings

Altieri, Miguel A. *Agroecology: The Science of Sustainable Agriculture.* Boulder, CO: Westview Press, 1995.

Feenstra, Gail, Chuck Ingels, and David Campbell. "What Is Sustainable Agriculture?" http://www.sarep.ucdavis.edu/Concept.htm (Accessed July 2009).

Ikerd, John E. *Crisis and Opportunity: Sustainability in American Agriculture.* Omaha, NE: Bison Books, 2008.

National Sustainable Agriculture Coalition. http://sustainableagriculture.net (Accessed July 2009).

Mrill Ingram
University of Wisconsin, Madison

Sustainable Fisheries Act

The Sustainable Fisheries Act (SFA), enacted in 1996, reauthorized and significantly modified the Magnuson Fisheries Conservation and Management Act of 1976, the federal fisheries management statute. The SFA renamed that statute the Magnuson-Stevens Fishery Conservation and Management Act ("the Magnuson-Stevens Act") in honor of Senator Ted Stevens (R-AK), a key advocate for improved fisheries management. The Magnuson-Stevens Act, as amended by the SFA, was modified further in 2007 when the act was reauthorized. The SFA modifications to the Magnuson-Stevens Act were based on a new congressional finding that "[c]ertain stocks of fish have declined to the point where their survival is threatened, and other stocks of fish have been so substantially reduced in number that they could become similarly threatened as a consequence of (A) increased fishing pressure, (B) the inadequacy of fishery resource conservation and

Sea turtles like this one, which died after becoming entangled in a fishing net, are included in the Sustainable Fisheries Act's definition of bycatch.

Source: National Oceanic and Atmospheric Administration

management practices and controls, or (C) direct and indirect habitat losses which have resulted in a diminished capacity to support existing fishing levels (16 U.S.C. 1801)." The SFA addressed three major problems: overfishing, bycatch, and protection of essential fish habitat.

Overfishing

The most important problem addressed by the SFA is overfishing, which has devastated fish in U.S. waters. A stock is overfished when it has fallen below the biological threshold for sustaining that fish stock. Overfishing occurs when the fish stock mortality rate is high enough that, in the long term, the fish stock will be overfished. As of the end of 2008, of 173 fish stocks, 124 stocks are not overfished, 4 are approaching an overfished condition, and 45 stocks are overfished. There is insufficient data and analysis to establish the overfished status of 57 stocks. Of 188 fish stocks with adequate data, 148 stocks are not being overfished, but 40 are being overfished. An additional 42 fish stocks have insufficient data and analysis to determine whether they are being overfished.

The SFA rewrote National Standard 1 of the Magnuson-Stevens Act to require that fishery management plans (FMPs) prevent overfishing while achieving an optimal yield from each fishery on a continuing basis. The SFA requires the plans, which are prepared by regional fishery management councils and approved by the secretary of commerce, to establish specific scientific criteria for determining whether a stock is overfished or whether the rate of harvest and other causes of mortality are causing overfishing to occur. Once these criteria are identified, the National Oceanic and Atmospheric Administration Fisheries Service (NOAA Fisheries, also known as the National Marine Fisheries Service) monitors the condition of fish stocks. If a fishery is overfished or approaching an overfished condition, NOAA Fisheries notifies the regional council. The regional council must revise its FMP within a year with the goal of rebuilding overfished stocks as soon as possible and no longer than 10 years, absent special circumstances.

Bycatch

The SFA addressed a second significant problem: bycatch. The term *bycatch* as defined by the SFA refers to fish killed by fishing efforts that are not sold or kept for personal use, including fish that are discarded as too small for economic or regulatory reasons. Although sea turtles and kingfish are included within the SFA definition of bycatch, marine mammals and sea birds are not. Any bycatch minimization efforts for those species that regional councils include in FMPs are based on authority granted under the Endangered Species Act, the Marine Mammal Protection Act, the Shark Fishing Prohibition Act, and state laws. Under National Standard 9, FMPs must minimize bycatch to the extent practicable. Bycatch minimization measures include at-sea observers, bycatch caps, gear restrictions, and closure.

Essential Fish Habitat

Protection of essential fish habitat (EFH) is another important issue addressed by the SFA. Congress defined EFH as "those waters and substrate necessary to fish for spawning, breeding, feeding or growth to maturity." Pursuant to the SFA, regional councils must identify EFH in accordance with NOAA EFH regulations and adopt measures to minimize fishing impacts on EFH "to the extent practicable." The SFA also requires all federal agencies to consult with NOAA Fisheries on actions that may adversely affect EFH to determine measures that can be taken to conserve EFH.

In NOAA's view, habitat is the area used by fish throughout their life cycle, including habitat for spawning, feeding, nursery, migration, and shelter. NOAA also recognized that fish may change habitats with changes in life history stage, seasonal and geographic distributions, abundance, and interactions with other species. The EFH regulations provide standards for identifying EFH and guidance encouraging councils to distinguish EFH from all potential habitats. The EFH regulations also provide federal agencies with guidance on evaluating fishing activities that may adversely affect EFH, including what information councils should provide in the evaluation and standards for deciding when councils must act to minimize adverse effects. Finally, the EFH regulations set forth the EFH consultation procedures. The EFH consultation procedures reiterate NOAA Fisheries' preference for combining EFH consultations with other environmental reviews (e.g., NEPA, Clean Water Act) to promote efficiency, to use streamlined procedures for developing General Concurrences (to eliminate the need for individual consultations on actions with minimal impacts to EFH), and to clarify that very brief EFH assessments are appropriate for relatively simple actions.

See Also: Fisheries; Salmon.

Further Readings

Magnuson-Stevens Act. http://www.nmfs.noaa.gov/habitat/habitatprotection/efh/stat_reg_a .htm (Accessed February 2009).
NOAA Fisheries. Essential Fish Habitat Regulations, 50 CFR Part 600. http://www.nmfs .noaa.gov/habitat/habitatprotection/pdf/efh/archives/efhfinalrule.pdf (Accessed February 2009).
NOAA Fisheries. Fish Stock Sustainability Index Reports. http://www.nmfs.noaa.gov/sfa/ domes_fish/StatusoFisheries/2008/4thQuarter/Q4_2008_FSSI_Summary_Changes.pdf (Accessed February 2009).

NOAA Fisheries Service. http://www.nmfs.noaa.gov (Accessed February 2009).

NOAA Fishery Management Councils. http://www.nmfs.noaa.gov/councils.htm (Accessed February 2009).

Sustainable Fisheries Act. http://www.nmfs.noaa.gov/sfa/sustainable_fishereries_act.pdf (Accessed February 2009).

Susan L. Smith
Willamette University College of Law

SWIDDEN AGRICULTURE

Swidden agriculture—a more expansive form of slash-and-burn agriculture—is a type of shifting cultivation that entails the cutting and drying of all vegetation from a forest patch until fire can properly take hold. Following burning, the resultant ash provides the swidden farmer with a sufficient nutrient base from which to farm for up to four years. The burning process also removes potentially damaging pests from the plot. After the nutrients have been exhausted from the swidden plot, the farmer moves to an adjacent plot of forest land and repeats the entire process of cutting, drying, burning, planting, harvesting, and abandoning again.

Swidden agriculture has been practiced on most forest types throughout the globe but currently is almost exclusively used in tropical forests. Tropical forests are not suitable for the type of conventional agriculture commonly practiced in temperate climates because of their soils. Tropical forest oxisol and ultisol soils are very old and largely devoid of essential nutrients required for farming crops. Nutrient cycling in tropical biomes is vastly different than other forest types. In the tropics, the nutrient base is bound up almost exclusively in living trees and plants and in the decaying wood and vegetation that litter the forest floor—which can exceed a meter in depth. The swidden farmer cuts and burns a forested plot to capture a portion of the nutrient base bound up in the vegetative material to nourish future crops.

In tropical forest systems, with few cultivators compared with vast expanses of forest land, and with small farmed plots, slash-and-burn agriculture is relatively benign, as the surrounding forests quickly encroach on the farmed plot following abandonment. Indigenous Amazonian populations have practiced this type of agriculture for centuries with minimal deleterious effects. This method of farming, when practiced with properly spaced small plots, can even increase biodiversity, as it mimics natural forest openings created when large overstory trees die and fall, taking with them large sections of understory vegetation. Small openings within the forest interior allow early successional species that cannot tolerate the high-shade conditions of old growth tropical forests to reestablish themselves. Furthermore, forest openings provide edge effects that increase the variety of plant and animal food resources for some tropical forest herbivores and carnivores.

However, when human populations expand in tropical environments and the swidden plots are large, the continued practice of this form of agriculture can be devastating to the biological integrity of tropical forests. When colonialists practice swidden agriculture in its more expansive form, it takes long periods of time for these lands to recover for a variety of reasons. Because tropical forest soils are nutrient poor and have been further impoverished following successive rotations of conventional cropping, they cannot easily support tropical revegetation. Furthermore, the loss of a protective cover of forest detritus, root systems, and vegetation subjects tropical soils to bleaching by the strong tropical sun, which can destroy essential rhizomal systems. Frequent tropical rains can also remove remaining topsoil from the abandoned patch. When a combination of insults is foisted on tropical forests, it might take from many decades to more than a century before a swidden-farmed plot will return to its preharvest state.

A major vector for facilitating the expansion of swidden agriculture is the construction of roads to reach mineral supplies—often petroleum—in the forest interior. When viewing aerial photographs of tropical forests with road crossings, the observer will notice a series of herringbone patterns lying adjacent to tropical forest road systems. Such patterns represent the main and arterial arms of forest plots converted to swidden agriculture. From these roads, over time, agriculturalists will expand their land clearing away from the road, progressively clearing up more pristine tropical forest land.

Lamentably, many swidden-farmed and abandoned plots will not, in the near future, reach a preharvest state as a result of cattle ranching. At this time, cattle ranching is a common end point to the slash-and-burn, harvest-and-abandonment agricultural cycle. Some swidden farmers transition to cattle ranching, but more commonly, more affluent cattle ranchers take possession of abandoned farmland and convert it to cattle pasture. The economics of raising cattle over continue swidden agriculture makes economic sense for several reasons. Cattle ranching is not nearly as labor intensive as the backbreaking work that goes into subsistence agriculture—the animals keep tropical forest from returning, so only a few hands are need to monitor a large number of cattle. Second, unlike agriculture, which is at the mercy of Mother Nature's fury, a cattle rancher holds a physical asset that can be immediately sold to recoup an investment. Moreover, during hard times, a cattle rancher can always slaughter a steer for food. Finally, cattle are a relatively market-safe commodity that holds its value despite fluctuations in the local economy.

The fact that swidden agriculture is practiced almost exclusively in tropical biomes makes it as a major threat to the planet's biodiversity resources. Tropical forest ecosystems are the most biodiverse of any ecotone on the planet. Although tropical forests represent less than 2 percent of the planet's surface, they support greater than 50 percent of the world's species diversity. They also store large amounts of carbon, second only to the vast expanse of the world's oceans. Alterations in carbon levels and the long-term removal of vegetation with the capacity to remove more carbon from the atmosphere through photosynthesis can have deleterious environmental effects, contributing to global warming. Unfortunately, tropical forests are also one of the most fragile ecosystems, as they do not quickly recover from swidden agriculture.

To many, the plight of the swidden agriculturalist boils down to a battle between the basic human right for self-sufficiency versus the protection of the planet's most valuable natural resources. There have been long-standing arguments in scientific and policy arenas about whether swidden agriculturalists are being unfairly blamed for biodiversity destruction. Most swidden agriculturalists are poor subsistence farmers who do not destroy tropical forests to enrich themselves. They are merely surviving, one farm plot after another, with an unintended aftereffect of habitat destruction. But the forests are coming down, and species are in peril.

The political economics that surround swidden agriculturalists are complex. These farmers find themselves in their current states as a result of a combination of poverty, food insecurity, poor education, the exhaustion of mineral resources, political dynamics, and high population densities in other areas. By and large, swidden farmers live in poverty, practicing subsistence farming—they do not sell their crops and have few material posses-sions. Given the remoteness of their lives, access to education is not an option. Thus, their knowledge of farming is passed down from generation to generation, without input from modern agricultural scientists. They are also constantly on the move, as minerals are depleted from agricultural soils following successive crop rotations.

There are instances in which government programs incentivize the large-scale move-ment of inner-city poor to the forest frontier as a method of ameliorating the threat of political unrest. In the case of Brazil, the cities of Rio de Janeiro and São Paulo witnessed a large movement of their peasant population to the Amazon when the Brazilian govern-ment offered free tracts of land to willing migrants who maintained occupancy for five years. Initially, swidden agriculture was practiced by many of these new farmers, but cattle ranching quickly ruled the day on most of these gifted plots of land.

In the case of rural-to-frontier migration, as seen in northern Guatemala, subsistence agriculturalists are pushed out of their communities, as high local fertility rates have led to land scarcity and insufficient resources to support a family. Because these farmers do not possess sufficient education or employable skills useful in an urban setting, they choose to avoid the city and migrate to the forest frontier. On arriving in the forest frontier, they promptly locate and illegally squat on a piece of suitable forest land, often in protected national parks or reserves, to undertake swidden agriculture.

In summary, swidden agriculture is a method of shift cultivation that has been embroiled in a larger fight for poverty rights and environmental protection. Answers to current problems posed by this form of agriculture cannot be found with changes in agri-cultural technology. An effort must be made to provide swidden agriculturalists with alternative employment opportunities—short of cattle ranching—that will entice these families from the forest interior to places where access to education, health care, and nutri-tional resources are provided.

See Also: Agribusiness; Cash Crop; Grazing.

Further Readings

Carr, David. "A Tale of Two Roads: Population, Poverty, and Politics on the Guatemalan Frontier." *Geoforum*, 37/1 (2006).
Kellman, Martin and Rosanne Tackaberry. *Tropical Environments: The Functioning and Management of Tropical Ecosystems.* New York: Taylor and Francis, 1997.
Loker, William. "The Human Ecology of Cattle Raising in the Peruvian Amazon: The View From the Farm." *Human Organization,* 52/1 (1993).
Myers, Norman. *Conversion of Tropical Moist Forests.* Washington, D.C.: National Academy of Sciences, 1980.

Jason Davis
University of California, Santa Barbara

TRADE LIBERALIZATION

Trade liberalization is the process of reducing or completely eliminating any government policies and practices that interfere with the free flow of goods and services among nations. Such policies and practices include not only tariffs (and subsidies that create the same artificial price disparity between imports and domestic goods) but also nontariff barriers, such as quotas or certain trade regulations. Although essentially an extension of capitalism—extending the idea of the domestic free market to an international market of free trade—trade liberalization nevertheless often finds opposition among some of those who are otherwise free market policy proponents, because of the long association of such policies with protectionism. Trade liberalization is the essential function of the World Trade Organization (WTO), and the Doha developmental round thereof has repeatedly stalled since 2001 because of disagreement over the liberalization of international agricultural trade.

Tariffs, subsidies, price supports, and other government-created and government-imposed devices that either increase the price of imports or decrease the price of domestic goods or exports all interfere with the natural prices that would result from unfettered supply and demand and generally increase the cost to the consumer. Over the course of the General Agreement on Tariffs and Trade, which preceded the WTO in the post–World War II years, most of the developed world has abandoned or significantly reduced its protectionist policies. The principal exception is agricultural trade; it is still common in much of the world to artificially lower the prices of domestic agricultural goods as protection against foreign competition.

Free trade not only allows for fair competition and fair prices but also enables specialization among countries. Different countries have different resources and can produce goods at different prices; if all nations can trade among each other without price interference, and if the costs incurred by trade (such as transportation) are sufficiently low or sufficiently offset by economies of scale, then in an ideal situation, everyone can enjoy the highest-quality goods at significantly lower prices. This and other general laissez-faire arguments were the initial arguments in favor of free trade in the 18th and 19th centuries, and they were largely persuasive. Since then—perhaps in particular since the development of modern economics and the history of American politics, which has taken capitalism for granted—moral arguments in favor of free trade have become common. Though the argument that government restrictions on trade abridge or even abrogate the sovereign rights of the consumer dates to the middle of the 19th century, it has been especially common in the United States with the rise of libertarianism (which is promoted by, but not synonymous with, the Libertarian Party). The right of a worker to exchange the fruits of his labor for the fruits of another's labor, with no one but that other worker having the right to interfere, is considered inalienable by this school. This extends naturally not only to the liberalization of trade and an unfettered domestic economy but also to minimal or nonexistent prohibition of particular goods and services, up to a point (it is more reasonable to construct a libertarian argument in favor of legalized alcohol or legalized prostitution than it is one in favor of legalized murder for hire).

Classical liberalism—features of which are shared by both liberals and conservatives today—argues that commerce between nations decreases the chance of war between those nations. War is principally motivated by economics, historically, even when other motives are cited to rally the troops. Economic interdependence between nations counteracts the economic gains of winning a war between them. This is a difficult theory to prove, even if it seems intuitively true: We see so much evidence of war between nations with economic relations, including not only both world wars but the two wars between the United States and Iraq (and the war between Iraq and Iran in the 1980s). The argument that commerce discourages war would have to suppose that these wars represent a small fraction of the wars we would see if countries were less interconnected by trade. It seems difficult to imagine the world wars being a small fraction of anything.

There is, though, the compelling and intriguing McDonald's example. The golden arches of McDonald's have long been the icon of globalization, just as McDonald's was instrumental in the creation of national restaurant chains in the United States. McDonald's has locations from Argentina to Zimbabwe, in countries as disparate as Aruba and Iceland, Botswana and India. In Western Europe, only Vatican City lacks a McDonald's. *The Economist* has annually published the "Big Mac Index," which compares the cost of a Big Mac in local currency around the world to assess the purchasing power of those currencies, since 1986. The company would seem to benefit from trade liberalization in order to operate globally with minimal complication. But even apart from this, there is what has semi-jokingly been called "The Golden Arches Theory of Conflict Prevention." American journalist Thomas Friedman was the first to point out that "no two countries that both had McDonald's had fought a war against each other since each got its McDonald's," and although it was somewhat tongue in cheek and a mild exaggeration, the exaggeration truly is mild: the exceptions are the U.S. invasion of Panama in 1989, the bombing of Serbia, the conflicts between India and Pakistan, and the war between Russia and Georgia; in addition, the introduction of McDonald's has no effect on existing conflicts like those between Israel and Lebanon. Compared with the great number of wars and the great number of possible combinations of multinational conflicts among McDonald's nations, that list seems quite short.

In light of this theory, consider the fanfare with which the first McDonald's in Russia was greeted. It opened in Moscow in 1990, almost immediately after the cessation of the Cold War, and in a very real and genuine way was as symbolic a gesture of that cessation as had been the dismantling of the Berlin wall. Some notable names on the list of the countries that lack McDonald's include Iraq, Iran, Libya, Syria, and North Korea. Cuba is an intriguingly ambiguous case: the one McDonald's in Cuba is located on Guantanamo Bay Naval Base, on land owned by Cuba but operated by the U.S. Navy, and Cuban citizens are forbidden to go there. Iraq is home to MaDonal, a hamburger stand using a nearly identical logo and menu to McDonald's, which was founded by Suleiman Qassab, a Kurdish Iraqi who worked at a McDonald's in Vienna before returning home to open MaDonal after the McDonald's Corporation refused him permission to open an official McDonald's location, as they have refused many others in Iraq.

Friedman himself later upgraded the theory as the "Dell Theory of Conflict Prevention," in reference to the computer chain. The Dell theory states: "No two countries that are both part of a major global supply chain will ever fight a war against each other as long as they are both part of the same global supply chain." Opponents of globalism often complain that too much power is placed in the hands of corporations rather than of states (and their electorates). Friedman's Dell Theory turns that complaint on its head, crediting corporations and the connections they create with the prevention of war.

"Free Trade" Agreements

Many so-called free trade agreements, such as the North American Free Trade Agreement, do not in fact create a situation of free trade. In fact, they often create exactly the situation that trade liberalizers seek to prevent. The North American Free Trade Agreement, for instance, opens many trade doors between the United States, Canada, and Mexico, but because of the agricultural subsidies in the United States, those open doors allow American farmers to sell corn in Mexico at less than the cost of producing it. Corn is a staple crop of Mexico, and the effect here is far different from the specialized production of countries that free trade enthusiasts envision: The U.S. corn is not necessarily better than Mexican corn, but it is much cheaper, simply because of additional money paid to the farmers by the government (and the American taxpayer) to keep them in the corn business. Because no such arrangement exists for the Mexican corn farmers, they cannot possibly compete. Even the economies of scale that make foreign strawberries cheaper than local strawberries are not at work here, only the effect of agricultural subsidies, which is precisely why the WTO opposes them. At the same time, it is easy to see how U.S. agriculture has become dependent on subsidies, and that fewer farms could afford to operate without them—and to a greater extent, perhaps, than some countries, the United States considers itself a farming nation. It is no coincidence that the nation's farmland is concentrated in "the heartland," and surviving damage to the heart does not create a willingness to endure that damage.

The WTO officially condemns dumping—selling goods at below fair value in foreign markets when a higher value is charged in the domestic market—but does not ban it, and even the condemnation addresses only dumping activities that specifically injure the relevant industry in the foreign country. This is, it must be admitted, a fairly weak position. It is difficult, though, to reconcile antidumping policies—which could easily fall under protectionism—with free trade advocacy. It is specifically when dumping is made possible by government intervention, such as agricultural subsidies, that the free trader can safely oppose it as an anticompetitive practice. The United States is not the only country with such subsidies; most developed nations have them to some degree, and the European Union's Common Agricultural Policy sets such subsidies as well. The reason the European Union and the United States are so often criticized for their agricultural subsidies is because they represent such huge markets that theirs are the subsidies with the most effect on the world agricultural markets.

Other free trade agreements, which have varying degrees of actual free trade, include the Association of Southeast Asian Nations Free Trade Area; the African Free Trade Zone, which as of 2008 unified the East African Community, the Common Market for Eastern and Southern Africa, and the Southern African Development Community; the Caribbean Community; the Central European Free Trade Agreement; the Dominican Republic–Central America Free Trade Agreement; the Economic Community of Central African States; the Economic Community of West African States; the Greater Arab Free Trade Area; Mercosur; and the South Asia Free Trade Agreement. Even without including the WTO, a significant part of the world is either part of a free trade agreement or is in the process of adopting one.

Opposition to Trade Liberalization

The pragmatic, economic argument for free trade is that it leads to a situation in which goods are produced for the best price, to everyone's benefit. Socialists would argue that another way to describe that situation is one in which goods are produced by the lowest-paid workers to the benefit of capitalists (owners).

Others oppose trade liberalization because of the effect on domestic industries that would be threatened by their inability to compete with cheap imported goods. The Mexican corn farmers mentioned above are in that position. Even without agricultural subsidies, though, many countries would find themselves in that position when the costs of labor are significantly lower overseas, or when production costs are affected by, for instance, stricter safety regulations in one country than in another. Asian manufactured goods have long been cheaper than their American analogues, for instance, principally because of an extraordinary disparity in the cost of labor. Even aside from wage differences caused by varying costs of living, not all countries have strong labor unions requiring expensive benefits or reasonable production standards.

See Also: Disappearing Middle; Doha Round, World Trade Organization; Farm Bill; Fast Food; North American Free Trade Agreement.

Further Readings

Akwuole, Kenneth. *Trade Liberalization as Instrument for Development: The United States Trade Policy Towards Sub-Sahara Africa.* Saarbrücken, Germany: VDM Verlag, 2009.

Djurfeldt, Goran, et al. *The African Food Crisis: Lessons From the Asian Green Revolution.* New York: CABI, 2005.

McCalla, Alex F. and John Nash. *Reforming Agricultural Trade for Developing Countries: Key Issues for a Pro-Development Outcome of the Doha Round Negotiations (World Bank Trade and Development Series).* Washington, D.C.: World Bank Publications, 2006.

Bill Kte'pi
Independent Scholar

United Farm Workers

Established in 1962 by César Chávez, Dolores Huerta, Larry Itliong, and others, the United Farm Workers (UFW) is a labor union representing and advocating for farmworkers' rights and environmental and social justice concerns. It is presently organizing in 10 states across the United States. Through farmworker strikes, social movement organizing tactics, marches, and other nonviolent actions, the UFW has raised awareness about and helped to improve the working conditions of impoverished, mainly Latino and Latina farmworkers. Their organizing efforts, both past and present, also connect the struggles of disenfranchised farmworkers with the environmental and health impacts of pesticides, making the UFW one of the earlier groups to advocate for environmental justice. Some also credit the UFW for taking early action against harmful pesticides when government regulations were inadequate, and before the emergence of the environmental movement.

In 1962, César Chávez resigned from his position as the executive director for the Community Services Organization in California, a group associated with Saul Alinsky's Industrial Areas Foundation, which worked with Mexican Americans. Chávez left the Community Services Organization to pursue his dream of organizing farmworkers. Excluded from mainstream labor organizations and unprotected under the National Labor Relations Act, farmworkers endured low wages, poor working conditions, and inadequate access to drinking water and sanitary restrooms. Chávez began speaking to farmworkers, encouraging them to join his newly formed group, the National Farm Workers Association.

In 1965, the Agricultural Workers Organizing Committee, an affiliate of the American Federation of Labor and Congress of Industrial Organizations, which represented predominantly Filipino-American farmworkers, initiated a major and now well-known strike against grape growers in Delano, California. The Agricultural Workers Organizing Committee invited Chávez's mainly Latino National Farm Workers Association to also participate in the strike. A year later, the Agricultural Workers Organizing Committee and the National Farm Workers Association joined forces to form the UFW. Chávez became the leader of this new organization. UFW garnered support from labor, church groups, and student and civil rights activists. Senator Robert F. Kennedy also came to support the strike. Committed to nonviolent forms of protest by way of Mahatma Gandhi and Martin Luther King, Jr., Chávez organized a significant march from Delano to the state capitol in

Sacramento. He also conducted the first of many fasts in 1968 to bring attention to the farmworkers' cause.

By 1968, the UFW began to take action against pesticides in response to two major cases of farmworker poisoning by an increasingly popular type of pesticide: organophosphates. Organophosphates came to replace pesticides such as DDT (dichlorodiphenyltrichloroethane) in California's fields by the mid-1960s. This new type of pesticide, though less persistent in the environment, was highly toxic and dangerous to farmworkers' health. Although it was considered a fairly better alternative to DDT because it did not bioaccumulate in the environment, some farmworkers became violently ill after exposure to the pesticide, and some were hospitalized. There was little concern during this time for farmworkers' health problems resulting from pesticide exposure. In fact, government regulation of pesticides was said to be inadequate and favored growers' interests. The U.S. Department of Agriculture was responsible for overseeing pesticides before 1970, but the department, some charge, was merely a lobbying instrument for agribusiness. As result, the UFW decided it needed to campaign around pesticide use and exposure.

From 1969 to 1970, the UFW launched an impressive and effective pesticide campaign that linked pesticide regulation to their unionizing efforts. The Delano table grape boycott and strike had focused to this point on the plight of farmworkers. Now, the UFW framed the boycott in terms of a larger health and environmental concern for consumers and environmentalists, in addition to that of farmworkers. This expanded support for the boycott on a national level as popular awareness around pesticides, food safety, and environmental issues began to emerge. Soon, concerns surrounding pesticide use prompted the major grocery chain Safeway to conduct independent laboratory tests of their grapes. The tests revealed that residues of a highly toxic pesticide were present at levels higher than fit for human consumption. Safeway soon canceled its contract with their grape suppliers. This public relations disaster for the table grape industry, along with the broad support for farmworkers from consumers, labor, religious leaders, Democrats, environmentalists, and student, Latino, and civil rights activists led table grape growers to sign contracts with the UFW in 1970.

This major victory for the UFW translated into 150 contracts being signed with table grape producers, covering over 20,000 workers. The first of its kind, the UFW table grape contracts required growers to prohibit, limit, or regulate the use of certain pesticides. Organophosphate pesticides were still permitted, but growers were now responsible for periodically examining workers' health and for maintaining records of pesticide application. They were also required to designate sprayed crops as off-limits for a specified period of time. By including these regulations in the union contract, the UFW was able to supplement weak and inadequate government regulation and oversight of pesticides. The contracts also increased workers' wages and bonuses. Growers furthermore agreed not to use DDT and other toxic pesticides.

The struggle against pesticides, however, continued. Confident after their 1970 victory, the UFW then sought to pursue contracts with California lettuce growers. However the growers, fearful of the militant UFW, signed contracts with the Teamsters Union instead. The Teamsters' contracts did not require growers to monitor and regulate pesticide use, making these contracts more attractive to them. The Teamsters also took over the hard-won table grape contracts the UFW signed in 1970. This led to a seven-year-long battle between the two unions that ended with the Teamsters giving way to the UFW. During this time, however, the UFW forged important links with environmental organizations such as Friends of the Earth. Together they pressured the U.S. Environmental Protection Agency to limit the use of toxic pesticides.

The UFW continued its organizing efforts against harmful pesticides into the present. In 1986, Chávez launched the "Wrath of Grapes" campaign countering pesticide poisoning of farmworkers and their children. Continuing his tradition of nonviolent organizing, 61-year-old Chávez also fasted in 1988 to bring attention to the detrimental effect of pesticides. The UFW continues to sign new contracts and organize around pesticide issues. Its struggles, originating in the 1960s and articulated by one of its enduring slogans, "Si, Se Puede!" ("Yes, We Can!"), a slogan that was used during President Barack Obama's campaign, resonates powerfully into the present.

See Also: Agribusiness; Bracero Program; DDT; Labor; Pesticide.

Further Readings

Gordon, Robert. "Poisons in the Fields: The United Farm Workers, Pesticides, and Environmental Politics." *The Pacific Historical Review,* 68/1 (1999).
Levy, Jacques E. *César Chávez: Autobiography of La Causa.* Minneapolis: University of Minnesota Press, 2007.
Shaw, Randy. *Beyond the Fields: César Chávez, the UFW, and the Struggle for Justice in the 21st Century.* Berkeley: University of California Press, 2008.
United Farm Workers of America. "History." http://www.ufw.org/_page.php?menu= research&inc=research_history.html (Accessed January 2009).

Karen Okamoto
John Jay College of Criminal Justice

Urban Agriculture

Urban agriculture is any agricultural venture that produces a diversity of food, fuel, and/ or livestock in response to the daily demands of consumers within a town, city, or metropolis, primarily using local natural resources and recycling urban wastes. Urban agriculture can occur on either an individual basis (e.g., in a private garden in one's own yard or on a privately owned farm) or a collective basis (e.g., in a community garden or other collective agriculture venture). According to the United Nations Development Programme, urban agriculture contributes to about one-third of agricultural production of vegetables, eggs, meat, and fish worldwide. Similarly, 30 percent of U.S. agricultural production occurs within metropolitan areas.

In 2007, the world's urban population exceeded the number of people living in rural areas. According to the United Nations Food and Agriculture Organization, by 2030, two-thirds of the world's population will be living in cities. At this time, in the United States, almost 80 percent of the population lives in urban centers. Many low-income urban dwellers live in neighborhoods that are underserved by supermarkets or other venues that sell food, making access to fresh foods difficult to impossible. Despite technological advances in industrial agriculture, hunger and food insecurity in both rural and urban areas persist. With predicted surges in urban populations, food security challenges will only increase, especially for the very poor, those with chronic illnesses, the elderly, the homeless, single mothers, and those with limited transportation and mobility. Urban agriculture, which

These students in a Philadelphia, Pennsylvania, elementary school learned about urban agriculture by creating their own garden in 1997.

Source: U.S. Department of Agriculture Natural Resources Conservation Service/Bob Nichols

may at first seem like a contradiction in terms, offers a strategy for self-provision of fresh produce.

Throughout history, food production in urban areas has increased as a response to times of social and economic crisis. From the early 1890s through the present, vacant lot gardens in the United States have arisen out of food insecurity and unemployment brought about by economic depression, war, and social unrest created by population, industrial, or institutional changes. During World War II, urban agriculture in the United States (i.e., victory gardens) produced over 40 percent of fresh produce that was consumed—almost all available city land was cultivated, including rights of way, city parks, schoolyards, vacant lots, and people's yards. In the 1980s, urban agricultural production accelerated dramatically worldwide. For example, surveys in Moscow in 1970 and 1991 show a shift from 20 percent to 65 percent of families engaged in urban agriculture.

Urban agriculture is often tied to policy decisions to build sustainable cities because of its numerous environmental, economic, and social benefits. Urban agriculture closes urban nutrient loops by using local natural resources and recycling urban wastes via composting, gray water systems, and vermiculture. In addition, urban agriculture has the potential to turn vacant lots into productive, green, diverse ecosystems. It therefore improves the quality and health of the urban environment through greening and pollution reduction. Urban agriculture also conserves resources and reduces the cost of food by reducing the need for transportation of produce. Urban agriculture is used as a strategy for food security and poverty reduction throughout the world. Through urban agriculture programs, people are empowered to grow their own food, either in their own yard or as part of a community garden. Thus, people are no longer reliant on the current market structure for access to fresh produce.

In impoverished countries throughout Africa, Asia, and Latin America, urban agriculture is an especially important strategy for improving food security. Poor families are able to sell produce and livestock raised near their homes—thus decreasing poverty and raising money to purchase other staples—and at the same time grow food to feed their own families. Some international urban garden programs improve food security for those most in need. For example, Gardens for Health, a community garden organization in Rwanda, develops easily replicable gardens throughout the country to target malnutrition and HIV/AIDS simultaneously. Cuba is an example of a country with a highly successful urban agriculture program. As a direct result of the U.S.-imposed embargo, which caused fuel shortages and therefore severe deficiencies in the transportation sector, a growing percentage of Cuba's agricultural production occurs in urban areas. In Havana, 90 percent of the city's fresh produce comes from local urban farms and gardens.

In the United States, urban agriculture takes many forms. Community gardens are perhaps the most familiar type of urban garden and consist of lots divided into smaller plots tended by individuals or households. However, one can also find school gardens, entrepreneurial gardens, and healing and therapy gardens located in every region throughout the United States. For the most part, urban agriculture programs in the United States have the primary goal of communal food production. However, they are also a way to mitigate poverty, improve neighborhood aesthetics, provide space for the study of nature and ecology, and develop a sense of pride and public citizenship.

Despite its contributions to the urban environment and food security, there exist numerous barriers to urban agriculture. For example, the use of untreated wastewater can result in the spread of diseases. Similarly, cultivation on contaminated land or along roadsides can expose food and consumers to toxic substances. As well, agriculture and urban development are considered by some to be incompatible, as urban land availability is limited. However, urban agriculture tends to use land that has been vacant or unused because it is otherwise unattractive for urban development. It uses available land in people's backyards, on right-of-ways, and in vacant lots. The use of marginalized land and vacant lots for urban agriculture is often limited by the lack of access to resources (such as water) and lack of security of tenure. Several urban agriculture programs worldwide have managed to overcome these obstacles, especially those with support and funding from local, state, and national governments.

See Also: Community Gardens; Family Farm; Food Security.

Further Readings

Koc, Mustafa, et al. *For Hunger Proof Cities: Sustainable Urban Food Systems.* Ottawa, Canada: International Development Research Center, 1999.

Smit, Jac, et al. *Urban Agriculture: Food, Jobs, and Sustainable Cities.* New York: United Nations Development Programme, 1996.

Hilary Melcarek
University of California, Santa Cruz

V

Vegan

A vegan is a pure vegetarian who chooses to eat a totally plant-based diet, avoiding any food taken from an animal, such as flesh, eggs, or dairy. Veganism is typically motivated by ethical concerns over animal rights or welfare but can also be motivated by health and environmentalism. Ethically motivated vegans often embrace it as a lifestyle rather than just a diet, boycotting all animal products, including clothing and products tested on nonhuman animals. Donald Watson, founder of England's Vegan Society, coined the term *vegan* in 1944 to differentiate vegans from "ovo-lacto" vegetarians, who will eat nonflesh animal products such as eggs and

This vegan pie was made with peas, mushrooms, vegan gravy, and mock meat produced from soy protein. Vegans can get most necessary nutrients from a plant-based diet, with the exception of vitamin B-12.

Source: Wikipedia/Peter Halasz

dairy. Because veganism is stricter, there are fewer vegans than vegetarians. For example, vegans make up no more than 1 percent of North Americans, whereas 2–3 percent of the population is likely to be vegetarian. National percentages for Europeans are similar, but the United Kingdom is an exception, with vegetarianism being twice as popular there. Worldwide statistics on veganism do not exist, but a portion of Buddhists, Hindus, Taoists, and Krishnas worldwide practice some version of vegetarianism, with India possessing the largest concentration of vegetarians—several hundred million—based on class and religious beliefs.

When it comes to ethical and spiritual dilemmas related to animal farming, killing has been the primary contention, which is why so many spiritual principles exist to regulate how and if animals can be slaughtered morally. Before the advent of industrialized animal farming, or "factory farming," nonfatal animal products such as eggs and dairy were less of an ethical issue. However, modern egg and dairy industries kill the hens and cows who work for them once individual production dies down to less-profitable levels. This results in increased suffering for these animals, more so than for animals raised primarily for meat, as egg-laying hens and dairy cows endure a year or more of intense production in confinement before being killed for low-grade meat.

Hens are typically raised in dark warehouses, where they are confined in cramped, stacked cages. The tips of their beaks are seared off at birth to prevent them from pecking their cage-mates in a futile attempt to establish a pecking order in these crowded conditions. Calcium depletion causes bone fractures, especially during the rough trip from crate to slaughter. Cows used for dairy are increasingly raised on dry lots instead of pastureland. Constantly impregnated to keep their unnaturally engorged udders lactating, they are attached to milking machines several times daily for the majority of their pregnancies. When a mother delivers, she is typically only allowed to nurse her calf for a few days before the farmer separates them, despite the mournful bellowing of both. Male calves are either killed immediately or crated for veal, and female calves become the next generation of dairy producers.

A minor proportion of eggs and dairy come from small, free-range farms that may be more sustainable and humane than factory farms. Welfare proponents may find these products ethically acceptable because they reduce suffering. However, vegans tend to adhere to an animal rights philosophy that goes beyond a concern for suffering toward a concern for the rights of the individual animals to live freely and control their own bodies, eggs, milk, and offspring. This philosophy eschews domestication and breeding animals in captivity for human use and boycotts all animal agriculture products, including clothing such as leather, fur, wool, and silk and by-products such as gelatin, casein, and whey. Depending on strictness, some vegans avoid honey and refined sugars (which use animal bones in processing). Ethically motivated vegans also tend to boycott any products tested on nonhuman animals, primarily cosmetic and cleaning products.

Vegans can enjoy good health on a plant-based diet, which is the principal motivation of some, but they should supplement vitamin B-12, as that is primarily obtained from animal products. In addition to produce, vegans consume plant-based proteins such as whole grains, nuts, and legumes. Vegans who grew up eating animal products often choose to eat a variety of plant-based substitutes, such as milks, yogurt, and frozen desserts made from soy, rice, or nuts, and "mock-meats" made of seitan (wheat gluten), tofu (soy), tempeh (fermented soy), or textured vegetable protein.

Environmentalism may be a primary or secondary motivation of vegans, as eating lower on the food chain is more ecological in many cases, depending on location and what local foods can be grown sustainably. Rather than using land to grow plants for nonhuman animal feed, vegans eat plants directly, therefore using food calories more efficiently, using fewer resources (such as water and energy), and causing less pollution (greenhouse gases, excrement, chemicals, etc). Although some scientists propose that a diet that includes grass-fed meats is more sustainable and actually causes fewer animals to die (as crop farming kills small field animals), other scientists claim that since a vegan diet more efficiently uses land to feed more people, it causes one-fifth as many animal deaths. Aspiring vegans can obtain some information from environmental organizations but will still get greater support through animal rights and vegetarian organizations.

See Also: Animal Welfare; Factory Farm; Vegetarian.

Further Readings

American Dietetic Association. "Vegetarian Diets." http://www.eatright.org/cps/rde/xchg/ada/hs.xsl/advocacy_933_ENU_HTML.htm (Accessed June 2003).

Davis, Steven. "The Least Harm Principle May Require That Humans Consume a Diet Containing Large Herbivores, Not a Vegan Diet." *Journal of Agricultural and Environmental Ethics,* 16 (2003).

Food and Agriculture Organization of the United Nations. "Livestock a Major Threat to Environment." http://www.fao.org/newsroom/en/news/2006/1000448/index.html (Accessed November 2006).

Francione, Gary. *Rain Without Thunder: The Ideology of the Animal Rights Movement.* Philadelphia: Temple University Press, 1996.

Matheny, Gavin. "Least Harm: A Defense of Vegetarianism From Steven Davis's Omnivorous Proposal." *Journal of Agricultural and Environmental Ethics*, 16 (2003).

Maurer, Donna. *Vegetarianism: Movement or Moment?* Philadelphia: Temple University Press, 2002.

Singer, Peter and James Mason. *The Ethics of What We Eat: Why Our Food Choices Matter.* Emmaus, PA: Rodale, 2006.

Vegan Outreach. http://www.veganoutreach.org (Accessed February 2009).

Vegan Society. http://www.vegansociety.com (Accessed February 2009).

VegNews. http://www.vegnews.com (Accessed February 2009).

Carrie Packwood Freeman
Georgia State University

VEGETABLES

Following industrialization and improvements in agricultural science, the production of vegetables has become an industry of intensified agriculture. With greater health awareness and demand for vitamin and minerals from natural sources, there is an increase in demand for fresh vegetables in people's diets. From subsistence or small-scale traditional vegetable gardens, there has been a shift to commercial farms situated away from urban centers, with the produce later transported to consumers. "Conventional farming practices" in the vegetable industry usually refers to intensive cultivation aided by the use of synthetic fertilizers, agrochemicals (pesticides, herbicides, fungicides, etc.) and high energy consumption to maximize production and profit. However, increasingly, these practices pose concerns of food safety as well as environmental degradation. Thus, there is now a growing movement worldwide calling for sustainable agricultural practices as an alternative to these conventional farming practices.

Some of the key environmental concerns in vegetable production include soil erosion and soil quality, fertilizer and pesticide contamination, water availability, and food safety and quality.

Commercial vegetable production is a labor-intensive crop system, but at the same time, it is also highly lucrative. Thus, many farmers choose to heavily intensify their vegetable production to maximize crop yield per area of arable land. This intensification also means repeated tillage and excessive use of fertilizers, pesticides, and irrigation water. These are the practices that are frequently followed in conventional vegetable farming, which result in considerable environmental impact both on the farm and off the farm.

The term *tillage* generally refers to the activities involved in the preparation of land for cultivation. However, most of the tillage activities involve turning the soil through plowing for preparing seedbeds and weed management. This can damage the soil structure and also result in soil loss as the loosened soil gets carried away by water or wind erosion. This reduces soil fertility and soil quality and, in turn, lowers the productivity of the land. Often this results in farmers using more chemical fertilizers to compensate for nutrient loss in the soil. The soil that gets carried away by water ends up as sediment in reservoirs, reducing the capacity of hydroelectric dams, clogging waterways, and causing additional and costly problems.

It needs to be emphasized, however, that tillage is not necessarily bad, and it has often been highlighted that the negative effects of tillage are usually the result of malpractice. There are, of course, no-tillage or reduced-tillage systems being advocated as alternatives to conventional tillage practices. However, the effects of tillage differ according to climate; in different regions of the world, tillage may cause considerably less impact on the environment. In fact, in certain regions, tillage may be required to encourage crop growth. Nonetheless, in the humid tropical regions where much of the world's vegetables are grown, tillage is noted as one of the severest causes of soil erosion as well as of promoting pests and weeds, which results in excessive use of chemicals.

The Concern About Chemicals

This excessive use of chemical pesticides poses a large concern for both human and environmental health. Chemical use in pest control covers a broad range from general insecticides, herbicides, and fungicides to specifics that target specific pests such as nematodes that feed on plant roots. However, pesticides also affect wildlife and humans at the same time. Pesticides are a health hazard to the farmworkers who apply the chemicals on the crops. At the same time, pesticide residuals on vegetables is one of the key concerns in consumer food safety standards. These chemicals also pose a concern to biodiversity, killing the wildlife along with the agricultural pests.

The heavy reliance on chemical fertilizers and pesticides in conventional vegetable production also results in irrigation runoffs carrying these chemicals into the waterways and contaminating water supplies. Chemical runoffs can lead to algal blooms in the receiving waterways, resulting in eutrophication, a process in which the increase in algal growth results in oxygen depletion, degraded water quality, and the death of fish and animal populations. The inefficient and excessive use of inorganic nitrogen fertilizers in conventional vegetable production also leads to acidification of soils. In many intensive vegetable systems that aim at maximizing yield over a small acreage of land, high amounts of nitrogen fertilizers are applied. In cases of excessive application and poor management, soils in such vegetable fields become saline. Some farmers practice the flooding of vegetable plots to wash away the salts accumulated, and this waste fertilizer pollutes the groundwater. In terms of human health impacts, there is still much controversy surrounding the effect of consuming high dietary nitrate in vegetables. However, the environmental impact of excessive nitrogen in soil, water, and air is clear, especially as vegetables are also known to heavily accumulate nitrates and leave residuals both in the leaf blades for consumption and in the vegetable fields. This makes the fields more vulnerable to potential water contamination and soil acidification. Furthermore, the production of these inorganic fertilizers is often tied to the heavy use of fossil fuel for heat production in fertilizer production. Thus, many environmentalists have spoken up against these inorganic fertilizers.

Growing Vegetables Without Soil

Vegetables can also be grown in a soilless medium via hydroponics, a nutrient film technique system. In these systems, plant roots are partially submerged in a thin "film" of circulated liquid nutrient solution, and the vegetables are grown above the nutrient reservoir tank. These soilless systems rely purely on combinations of inorganic fertilizers to recreate the nutrients required by the crops.

Other than growing in water, soilless mediums can also be any inert material, either organic or artificial. Common mediums include rockwool, rice hulls, peat moss, perlite, or aggregate combinations. These growing mediums are used in "bag culture," in which vegetables are grown in bags filled with these inert mediums. Fertility is controlled via commercially available soluble fertilizers fed to the bags via a drip or injector system. This popular method, known as fertigation, uses irrigation lines to apply this liquid fertilizer to the bags of crops. Crops such as tomatoes, cucumbers, and even lettuce can be grown via these soilless mediums. One of the chief advantages of such soilless cultures is the elimination of soil-borne diseases, and thus a reduced need for pesticides and fungicides.

Most soilless farms, as well as some soil-based vegetable farms, are cultivated in a greenhouse setting. The term *greenhouse* can refer to any structure that provides climate control for cultivation. This prevents the vegetable crops from being exposed to variable weather conditions, allows heating or cooling of the growing environment, and also reduces pest numbers, thus reducing the need for pesticides. Greenhouse production is increasingly popular, as it is seen as a key contributing factor to increased yield in crops like tomatoes and cucumbers in North America. However, ill-managed and excessive use of soluble nutrients can likewise contribute to water contamination by inorganic fertilizers through overflows, discharges, and runoffs.

In the 1960s and 1970s, as a response to the increasing soil erosion, pesticide use, and groundwater contamination in conventional vegetable production, a growing movement for sustainable agriculture evolved. This umbrella concept hosts a variety of alternative farming methods that aim at more ecofriendly practices.

One of the more moderate alternatives is low-input agriculture, in which synthetic fertilizers or pesticides are still used, but in minimal quantity. In general, any system that reduces purchased chemical inputs can be called low-input farming. For example, no-till vegetable production systems reduce the need for herbicides to kill cover crops. This often uses integrated pest management to achieve this goal. Integrated pest management is an environmentally sensitive method that involves understanding the life cycle of pests and their interaction with the environment combined with existing pest control methods to manage pests with the least possible harm to people and the environment. This includes using biological pesticides or natural predators to manage pests.

The Organic Way

Similar to low-input agriculture, organic vegetable production uses integrated pest management, as well as relying strongly on mixed cropping, crop rotations, and organic manures to fertilize crops and optimize the nutrient cycle. Organic production can also occur in soil cultures in greenhouses, which reduces pest levels. Most important, organic farming's emphasis is on managing soil fertility using various methods without using inorganic fertilizers. Organic manure improves the fertility and water retention capacity of the soil. Much emphasis is also given to ensuring that nutrient intake by the plants is optimized so as to prevent excessive nutrients input and loss. Nutrients are recycled and used various times in different forms. One way of doing this is to practice crop rotation, in which a variety of different crops are grown in the same area. Different crops require different combinations of nutrients, and by strategically planning the sequence of crops grown per harvest, this can ensure that soil nutrient intake is balanced and optimized. Other ways of optimizing interaction between crop and soil nutrients include intercropping (different crops in alternate rows, contour strips, or no particular arrangement on the same land), mixed cropping (one main crop and a few other subsidiary crops on the same land), or undersowing, which aims to cover the ground beneath the crop with a layer of vegetation to prevent weed growth.

These crop diversification methods reduce pest incidence, as different vegetable crops usually attract different pests, and a high diversity of crop varieties confuses the pests and limits the spread of epidemics. Plant diseases and their severity are also reduced, as it is harder for the diseases to spread with crops of other family groups planted in between their target crops. Certain intercropping design uses flowering subsidiary crops to attract beneficial insects that are natural predators to police the pest situation. Unlike monocultures, a diverse vegetable field also encourages agro-biodiversity, creating habitats for beneficial insects and soil organisms. Strip intercropping, especially on hillsides, help to prevent soil erosion. Undersowing also reduces the amount of herbicides and manpower needed to remove weeds by leaving less space for weeds to grow. At the same time, relay growing of a second, faster-growing crop beneath a slow-growing crop, sometimes as it nears harvest, allows for multiple harvests while managing fertility, pests, and weeds. Some also grow legumes as intercrops, as these plants can fix nitrogen from the air and convert them into nitrates and other nitrogen compounds through bacteria in their roots. When the plant dies, these nitrogen nutrients are released and transferred to other crops grown, through fertilizing the soil.

In addition to using crop rotation and diversification to manage soil fertility, much effort is also invested in creating topsoil and organic fertilizer. Manure and composting are the common methods of fertility maintenance. Manure includes animal manure by animals reared on farms. There is also chicken or buffalo dung that is collected and sold commercially to vegetable farmers as fertilizers. Green manure refers to the adding of plant material into the soil as fertilizers: Undecomposed green plant tissues are plowed or turned into the soil to provide organic material and to increase topsoil (humus) and additional nitrogen. Plants for green manure can also be planted as an intercrop or as undersowing but are usually buried as manure before they produce any crop. They are usually fast-growing plants grown for their green leafy material, which is high in nutrients and protects the soil from erosion. Green manure can also be used as a mulch to prevent weed growth and to keep the soil moist to reduce irrigation.

Composting is also a key method in fertility management. Instead of undecomposed plant or animal waste, composting uses decomposition by microorganisms to convert this organic matter into a humus-like product. When composting, the odor and potential health problems posed by using manure will be eliminated as the manure is decomposed and broken down. The final humus product is not only good for fertilization but also as a soil-conditioning agent, which eliminates the need for inorganic fertilizers. The use of fresh animal manure for fertilizer often poses a threat of bacterial contamination and human health risk. By fermentation or the aerobic digestion of this animal manure in a compost system using enzymes and other microbes, the risk is reduced, as the final compost product is a stable and safe fertilizer.

Vegetables and Food Safety

Vegetables are quite often eaten fresh or even raw and unprocessed, unlike grain crops and other staples. The food safety aspect of vegetables often comes under extra scrutiny to prevent human casualties. A good reason for this was evident in the North American *Escherichia coli* outbreak in raw spinach and iceberg lettuce in 2006, which resulted in several deaths. Irrigation water was contaminated with animal wastes, which resulted in bacterial contamination of the vegetable produce. To ensure the cleanliness and food safety of vegetables, the California Leafy Greens Handler Marketing Agreement was established in 2007, which imposes a mandatory food safety audit by the government on leafy green producers who voluntarily sign on to the agreement.

In most temperate climates, production of vegetable crops is limited by season. As a response, growers have turned to climate control through greenhouse cultivation. However, many retailers have turned to imported out-of-season vegetables from tropical countries or to vegetables produced in cooler highlands in tropical countries that can produce year-long supplies of temperate zone vegetables. However, this highland cultivation is on marginal lands that are more susceptible to soil erosion and contribute to environmental degradation. These imported vegetables also travel long distances, often by air, and thus expend large amounts of fossil fuel in transport. Thus, there is a strong movement in organic agriculture to eat local and seasonal vegetables. In the United Kingdom, there has even been talk of banning imported vegetables for environmental reasons. However, this would also mean the loss of livelihood for large populations in developing countries engaged in export-oriented agriculture. Some have also argued that it is more efficient and environmentally sound for some countries to import organic vegetables than to produce their own locally.

In affluent countries, the demand for "quality" vegetables is often limited to either "clean" vegetables in terms of food safety standards or the aesthetic quality of the vegetables. However, the demand for aesthetic quality in organic vegetables, which is often synonymous with safe, pesticide-free products, is a contradiction for many growers. Conventional vegetables are often aesthetically perfect, as a result of the use of chemical pesticides. Organic vegetables, in contrast, often have blemishes on the leaves or even earthworms in the harvested crop. Consumer education is often needed to change perceptions of the "quality" of vegetable appearances.

Some perceive that organic vegetables are just conventional vegetables grown without pesticides or chemical fertilizers. However, there is a strong element of community, diversity, decentralization, and farmer livelihood in the original organic movement. With the mainstreaming of organic vegetables, there is increased large-scale, monoculture, industrialized production of "organic" vegetables. This has sparked debates about how that element contradicts the principles of organic agriculture, which began as a response to conventional industrial farming methods. As a response, there is now consideration of "Beyond Organic" to address the ruling in 2005 by the U.S. Food and Drug Administration that allows certain synthetics to be allowed in food labeled "organic."

See Also: Beyond Organic; Fertilizer; Food Safety; Integrated Pest Management; Low-Input Agriculture; Organic Farming; Pesticide; Soil Erosion.

Further Readings

EIRI Consultants and Engineers and Engineers India Research Institute. *Handbook of Organic Farming & Organic Foods With Vermi-Composting and Neem Products*. Delhi, India: Engineers India Research Institute, 2006.

Kristiansen, Paul, et al. *Organic Agriculture: A Global Perspective*. Collingwood, Australia: CSIRO, 2006.

Nonnecke, Ib Libner. *Vegetable Production*. New York: Van Nostrand Reinhold, 1989.

Wells, A. T., et al. "Comparison of Conventional and Alternative Vegetable Farming Systems on the Properties of a Yellow Earth in New South Wales." *Agriculture, Ecosystems and Environment*, 80:47–60 (2000).

November Peng Ting Tan
National University of Singapore

VEGETARIAN

A vegetarian is a person who chooses to eat lower on the food chain, avoiding any animal flesh, such as red meat, poultry, or fish. This choice may be spawned by concerns over the environment, animal rights/welfare, personal health, world hunger, and/or spirituality.

Different levels of vegetarianism exist, including veganism on the stricter end of the dietary spectrum, and flexitarianism or pescetarianism on the looser end. Flexitarianism, or semivegetarian, is the term for someone who avoids meats as much as possible but may include them occasionally for convenience or necessity. Pescetarians also have some dietary flexibility, as they allow themselves to consume fish and other animals from the sea but avoid land animals. Some vegetarians dislike inclusion of any semivegetarianism in the category of vegetarian because vegetarianism, by definition, means the abstinence of any animal flesh. Because vegetarians do eat some animal products that do not require killing, such as eggs and dairy, they are sometimes referred to more specifically as "ovo-lacto" vegetarians. Those who eat a strictly plant-based diet and avoid all animal products, including eggs and dairy, are called vegans.

Because of varying definitions of vegetarianism, exact numbers are difficult to obtain. Although approximately 7 percent of the North American population self-identifies as vegetarian, millions of these people eat some meat, as flexitarianism is gaining popularity. So the percentage of North Americans who actually abstain from animal flesh is likely closer to 2–3 percent, or approximately seven million people, with approximately a million of those people being vegan. The typical Westerner most attracted to vegetarianism is a young, white, middle-class, atheist female. Statistics for Europeans show a similar low national percentage of vegetarians; the United Kingdom has the highest percentage, at 6 percent. Although worldwide statistics do not exist, several hundred million Indians are vegetarian for class and religious reasons, and a portion of Buddhists, Hindus, Taoists, and Krishnas worldwide practice vegetarianism.

One reason that meat is popular is the common modern assumption that humans have always hunted other animals and are biologically predisposed to a meat-based diet; however, some anthropologists theorize that early *Homo sapiens* were primarily herbivorous, based on humans descending from great apes, who mainly eat plants. Many Westerners today mistakenly assume that vegetarianism is a fad that originated during the hippy movement of the 1960s, but the practice of consciously avoiding animal foods has been around since ancient times and is an ethical cornerstone of several Eastern religions. Writings from the last several millennia indicate that ethical and spiritual rationales have predominated as a motivation for vegetarianism, although its healthfulness is also touted.

Ancient vegetarian proponents include Buddha, Mahavira, Lao Tzu, Plato, Plutarch, and Porphyry. In fact, through the 18th century, Western vegetarians were often called Pythagoreans, after the most well-known vegetarian, Pythagoras. For centuries after, the hegemony of a humanist Christianity kept vegetarianism from growing. However, vegetarian writings surfaced again in the 18th century, largely in resistance to scientist Rene Descartes's popular instrumental theory suggesting that nonhuman animals were soulless automata with little consciousness.

Nineteenth-century authors, such as Dr. Anna Kingsford, Henry Salt, and Leo Tolstoy, often based vegetarianism on a concern for human peace, but they showed a concern for animal welfare by including vivid descriptions of slaughterhouse violence. In addition to promoting an ethical justification, 19th-century physician William Alcott was one of the only early writers to propose a sustainability argument regarding the inefficiency of using land to grow animals and not plants. However, his argument was motivated by humanitarianism more than environmentalism, as it claimed vegetarianism would enable more humans to exist.

Twentieth-century provegetarian writings expanded on all previous notions of kinship, sympathy, and a virtuous character to include animal rights and environmentalism. Prominent writers included Mahatma Gandhi, Tom Regan, Peter Singer, Carol Adams, Frances Moore Lappé, and John Robbins. Modern ethical rationales for vegetarianism relate either to animal welfare concerns over the intensive confinement and suffering of farmed animals in agribusiness or to animal rights concerns over the ethicality of enslaving and killing other animals. Regan and Singer are perhaps the most noteworthy animal ethicists who helped usher in the animal rights movement of the late 20th century.

Regan argued that no morally relevant aspect of humankind exists that separates them from other beings with a conscious interest in living. Under this view, it does not matter if humans kill a nonhuman in an idyllic farm, a factory farm, or the woods—it is considered morally inconsistent to take away the life of a nonhuman animal when one would not have taken the life of a human. Similarly, Singer argued that animal agribusiness, whether free-range or intensive, is a speciesist practice, meaning it discriminates based on species. Agriculture controls and sacrifices the lives (major interests) of farmed animals to satisfy humans' taste (minor interests) for flesh, milk, and eggs; therefore, according to Singer, to recognize the interests of other sentient beings, humans should make it a general principle to stop killing animals for food except when necessary for survival.

Indeed, humans can survive on a vegetarian diet and meet their nutritional needs if given access to a variety of plant-based proteins (legumes and nuts), grains, and vegetables. In addition, because plant foods contain fiber but do not contain any cholesterol, nor as much saturated fat as most animal foods, the American Dietetic Association recognized the role of a plant-based diet in preventing illnesses common to Americans, such as obesity, heart disease, hypertension, diabetes, and certain cancers. Animal-based foods are related to issues of both nutritional excess (diseases of industrialized nations) and deficiency (hunger in some less industrialized countries). A critical health crisis exists worldwide, as millions of people die annually of hunger-related causes in part as a result of inequitable food distribution. The United States alone is said to produce enough plants to feed the hungry worldwide, but it inefficiently uses most of its crops, particularly grain and soy, to fatten farmed animals.

Environmentalists have noted that confined animal feeding operations and all the plant crops required to feed tens of billions of farmed animals worldwide (fish included) cause pollution and use significantly higher amounts of resources such as soil, water, land, and energy than do the components of a plant-based diet. Studies from organizations such as the Worldwatch Institute and the United Nations have linked animal agribusiness and meat consumption to the most devastating modern environmental issues, such as deforestation, erosion, water scarcity and pollution, air pollution, extinction, disease epidemics, and global warming. The United Nations concluded that a meat-based diet is a major contributor to global warming because raising livestock generates close to 20 percent of the world's greenhouse gas emissions, proving even more damaging than the transportation sector.

Ethical rationales for vegetarianism, such as concerns for human and nonhuman animal welfare and the environment, are believed to create more long-term commitment. Although many people are motivated to go vegetarian for health reasons, they may be more likely to succumb to convenience or dietary fads. For broad appeal, organizations that promote vegetarianism often tout a variety of rationales based on notions of both personal and social responsibility.

Major nonreligious provegetarian organizations include the International Vegetarian Union, North American Vegetarian Society, Vegetarian Resource Group, Physicians Committee for Responsible Medicine, People for the Ethical Treatment of Animals, Farm Sanctuary, and the Farm Animal Rights Movement. Although major environmental organizations are generally supportive of a plant-based diet, they tend not to make vegetarianism a campaign priority; when food is highlighted, they often emphasize the solutions of eating local and organic foods, avoiding factory farm products, and reducing one's consumption of certain animal species, especially grain-fed cattle and threatened fish populations. Overall, the attempt of provegetarian organizations to make production and consumption of animal products an ethical issue, or at least a problem, represents an important challenge not only to mainstream food industry discourse but also to sociohistorical norms and basic ideals about who it is morally acceptable to eat and what costs are involved in choosing a meat-based diet.

See Also: Animal Welfare; Confined Animal Feeding Operation; Factory Farm; Vegan.

Further Readings

American Dietetic Association. "Vegetarian Diets." http://www.eatright.org/cps/rde/xchg/ada/hs.xsl/advocacy_933_ENU_HTML.htm (Accessed June 2003).

Food and Agriculture Organization of the United Nations. "Livestock a Major Threat to Environment." http://www.fao.org/newsroom/en/news/2006/1000448/index.html (Accessed November 2006).

Lappé, Frances Moore. *Diet for a Small Planet.* New York: Random House, 1991.

Mason, James. *An Unnatural Order: Why We Are Destroying the Animals and Each Other.* New York: Continuum, 1997.

Maurer, Donna. *Vegetarianism: Movement or Moment?* Philadelphia: Temple University Press, 2002.

Regan, Tom. *The Case for Animal Rights.* Berkeley: University of California Press, 1983.

Robbins, John. *Diet for a New America.* Novato, CA: New World Library, 1998.

Singer, Peter. *Animal Liberation.* London: Random House, 1990.

Singer, Peter and James Mason. *The Ethics of What We Eat: Why Our Food Choices Matter.* Emmaus, PA: Rodale, 2006.

Walters, Kerry S. and Lisa Portmess. *Ethical Vegetarianism: From Pythagoras to Peter Singer.* New York: SUNY Press, 1999.

Well-Fed World. http://www.wellfedworld.org/resources.htm (Accessed February 2009).

Worldwatch Institute. "Meat: Now, It's Not Personal!" *World Watch Magazine* (July/August 2004).

Carrie Packwood Freeman
Georgia State University

VERTICAL INTEGRATION

Vertical integration is an industrial business tactic that has increasingly been applied to the agricultural industry. It is also known as vertical combination, vertical expansion, or vertical acquisition. Vertical integration occurs when a single corporation gains control over all aspects of a product, from the supply of raw materials and other inputs needed for production to the transportation, distribution, and sale of the finished product. Horizontal integration is a related tactic in which a corporation dominates a particular sector or aspect of production.

Vertical integration exists to varying degrees within individual corporations. It can include control of input production, such as farm machinery, seed, feed, fertilizer, pesticides, grain silos and elevators, poultry hatcheries, or livestock operations, control of transportation, distribution centers and retail outlets, or both. Vertical integration can be achieved through a variety of methods, including mergers and acquisitions, the use of legally owned subsidiaries, control of contracted farmers or growers, and the use of tactics such as overproduction, rice reductions, and informal agreements to reduce the number of competing firms.

Historically, U.S. agriculture was a subsistence-based enterprise consisting mainly of small, self-sufficient family farms. These farms traditionally grew a diversified array of crops, raised animals, and provided most of their own inputs, such as seeds, tools, and labor. Family farms produced mainly for their own consumption, with any small surpluses sold or traded within the local community. The development of the Industrial Revolution in England and the United States in the early 19th century changed the practice of agriculture as well as of business. Most factories were located in or near cities and towns, giving rise to ever-larger urban populations that no longer produced their own food, placing more commercial demand on farmers.

As farming thus became increasingly commercialized, many industrial business practices such as vertical integration were adopted by agribusiness. The earlier pattern of subsistence agriculture was slowly overtaken by commercial agriculture, or agribusiness. Farms grew in size, leading to the clearance of large amounts of land for planting or grazing. Initially, many smaller local companies were competitive in a marketplace, but gradually fewer larger firms came to dominate. This system arrived in agriculture later than other industries, but had largely taken hold by the mid-20th century. Examples of large agribusiness corporations include ConAgra, Tyson Foods, Phillip Morris, Altria, Monsanto, Cargill, Maple Leaf Foods, Smithfield, Philip Morris, British Nutrition, Chiquita, and Imperial Foods.

Vertical integration within agribusiness gave rise to a new business arrangement between large corporations and farmers and growers known as contract production. Under this system, the corporation pays the farmer or grower on a piece rate basis and controls all major aspects of the production process. Contract production is most common in the poultry sector, where the corporation provides a grower with chickens, feed, and other necessary inputs, as well as specifies building designs and equipment. Farmers and growers function as hired labor reliant on large corporations rather than as independent businessmen.

Many farmers and growers remain trapped in a cycle of debt because they borrow capital to raise buildings and must periodically update buildings and equipment. They also have little to no choice of contracting corporations as fewer and fewer corporations operate within a single geographic area and contracting corporations enter into informal agreements not to use growers contracted to competing firms. Once independent family farmers are now commercial farmers, dependent on middlemen to provide inputs and transport, distribute, and market their goods. Due to vertical integration, these middlemen often consist of subsidiaries of the same large corporation.

Agricultural corporations use vertical integration as one strategy in their effort to improve their economic power and market shares. Large, vertically integrated corporations can lower their financial risks through contract production and other forms of outsourcing and can withstand financial losses in a particular sector or country for prolonged periods of time because these losses will often be offset by gains in other sectors or countries under their control. Vertical integration has allowed for the production of the agricultural surpluses needed to feed the populations of largely urbanized and industrialized countries. Vertical integration also helps agricultural corporations improve the quality of their products through standardization of inputs.

Vertical Integration Challenges

Vertical integration also has negative connotations, as it can aid in the creation of monopolistic corporations that use both legitimate and illegal tactics to acquire competing firms and eliminate competition. Larger corporations can afford the expense of the additional inputs required in commercial agriculture. Larger corporations can also afford to increase production and lower prices over a sustained period of time in order to force smaller companies who cannot afford such losses out of the market. Smaller companies, which lack the capital and resources to remain competitive, either go out of business or are acquired by larger corporations, further decreasing competition.

Vertical integration eventually resulted in a global food supply controlled by a handful of large corporations. These firms determine what types of crops or animals are raised, as well as how they are packaged and marketed to consumers, reducing farmer and grower independence and consumer choice. They also contribute to environmental pollution and health problems through the increased use of fertilizers and pesticides in crop farming, the use of growth hormones and other feed additives in animal husbandry, and the use of fossil fuels to power farm machinery and transport products to distant markets.

Global Vertical Integration

By the late 20th century, many vertically integrated agribusiness corporations operated on a global scale. Many of these multinational corporations are U.S.-based corporations that began as small companies in local markets and grew in part through vertical integration. The rise of multinational corporations in the agriculture industry has created an increasingly integrated global food system that has largely replaced local food systems as aspects of production and distribution are spread worldwide. International trade agreements such as the North American Free Trade Agreement (NAFTA) reducing trade barriers have also aided the rise of the global food system.

Global vertical integration of agribusiness, which places the global food supply in the hands of a few large multinational corporations, reduces the ability of countries to regulate the supply and distribution of food within their own borders. Agribusiness and increased globalization have also removed profits from the local economy, where they had traditionally been reinvested in the era of the self-sufficient family farm. Profits from most food sales now either benefit the corporation and its stockholders or are reinvested in other geographic areas. Many small, rural communities have thus lost the economic benefits traditionally derived from local agricultural production and distribution, as large multinational corporations dominate the agricultural production market and increase their control every aspect of the global food system.

See Also: Agribusiness; ConAgra; Contract Farming; Factory Farm; Horizontal Integration; Monsanto; North American Free Trade Agreement; Trade Liberalization.

Further Readings

Blank, Steven C. *The Economics of American Agriculture: Evolution and Global Development.* Armonk, NY: M.E. Sharpe, 2008.

Conkin, Paul K. *A Revolution Down on the Farm: The Transformation of American Agriculture Since 1929.* Lexington: The University Press of Kentucky, 2008.

Gardner, Bruce L. *American Agriculture in the Twentieth Century: How It Flourished and What It Cost.* Cambridge, MA: Harvard University Press, 2006.

Heffernan, William D. "Agriculture and Monopoly Capital." *Monthly Review* (July–August 1998).

Hurt, R. Douglas. *Problems of Plenty: The American Farmer in the Twentieth Century (The American Ways Series).* Chicago: Ivan R. Dee, Publisher, 2003.

Ikerd, John E. *Crisis and Opportunity: Sustainability in American Agriculture (Our Sustainable Future).* London: Bison Books, 2008.

McCullough, Ellen B., Prabhu L. Pingali, and Kostas G. Stamoulis. *The Transformation of Agri-Food Systems: Globalization, Supply Chains, and Smallholder Farms.* Sterling, VA: Earthscan, 2008.

Vogeler, Ingolf. *The Myth of the Family Farm: Agribusiness Dominance of U.S. Agriculture.* Boulder, CO: Westview Press, 1981.

Marcella Bush Trevino
Barry University

WAL-MART

Wal-Mart is the world's largest retail operation, with over $400 billion in annual sales, 4,100 stores in the United States, and 3,500 stores overseas. In 2009, Wal-Mart was the highest-volume grocer in the United States, with approximately $100 billion in sales and a 21 percent share of the grocery market. When H. Lee Scott was named president and chief executive officer in 2000, the company suffered from a maligned public image, defending itself against allegations of underpaid and underinsured workers, Clean Water Act violations, and a reputation for destroying the viability of local businesses in many localities. Over the past eight years, the company has pledged a commitment to sustainable principles, adopted new food safety standards, and entered the organic marketplace. Although detractors dismiss Wal-Mart's efforts to clean up its practices, an argument can be made for the positive effects of shifting such a large-scale supply chain to environmentally conscious principles.

Wal-Mart first introduced grocery products into its Supercenters in 1988, using its powerful supply management tactics to compete at the lowest price point. Such tactics forced other supermarket chains including Grand Union, Bruno's, and Homeland Stores into bankruptcy. Workers at competitors like Kroger, Safeway, and Albertsons worried that their wages would be reduced as a result of the lower wages paid to Wal-Mart employees. Critics questioned the safety of food sold by a company that imports massive amounts of goods from China and used pressure tactics to gain cooperation from suppliers.

In October 2004, the United States sued Wal-Mart for violating the Clean Water Act in nine states, calling for changes in store building codes. The U.S. Environmental Protection Agency fined Wal-Mart $1 million to settle charges of violations incurred while building stores in Massachusetts, New Mexico, Oklahoma, and Texas. The company's carbon footprint is still questionable because of its vast trucking network and the number of car trips to its stores daily. Various localities also complain about the tax breaks and government assistance that Wal-Mart negotiates for new building sites.

Wal-Mart Goes Green

Wal-Mart began implementing its green initiatives in 2005 and promised to have over 400 organic items in stores by the summer of 2006. The demand for organic food has grown

451

15 percent annually over the past several years. Retailers frequently charge a 30 percent to 40 percent premium for organic food over conventional items. Organic standards have typically been associated with food that is locally produced, using sustainable methods sans pesticides and fertilizers. Free-range chickens, grass-fed beef, and labor-intensive techniques of harvesting produce are among the offerings recognized by educated consumers. The giant retailer's sheer size allows it to negotiate with suppliers about how goods are produced, which could allow Wal-Mart to push manufacturers and competitors to adopt more sustainable practices and organic food. A company-wide goal is to increase the energy efficiency of producing its most energy intensive products, including those products manufactured overseas. Wal-Mart has received praise for their green initiatives, which include building environmentally friendly stores and reducing packaging on merchandise. Entering the organic food market is an approach that both counters criticism from the press and labor unions and caters to a more upscale customer who demands organic products and favors socially responsible businesses. Wal-Mart chairman S. Robson Walton serves on the board of Conservation International, and Sam Walton's grandson, Sam R. Walton, joined the board of Environmental Defense.

One of Wal-Mart's major environmental initiatives has been to have all its wild-caught fish certified by the Marine Stewardship Council. The London-based council provides consumers with the knowledge of which fisheries avoid overfishing and adopt practices that do not harm the ocean environment. Rather than eliminating fisheries that are not sustainable, Wal-Mart asked suppliers to allow fisheries up to 18 months to be in compliance with the new standards. This move put Wal-Mart in competition with Austin-based Whole Foods Market, one of the premier organic chains, which has used the Marine Stewardship Council to certify its wild-caught salmon. Suppliers are concerned that up to 20 percent of wholesalers might be forced out of the supply chain for noncompliance; however, Wal-Mart has promised to give them time to make the environmental improvements necessary to meet the certification requirements. Critics question whether the commitment to have 100 percent of its seafood certified as sustainable will cause prices to rise and wonder whether this is a greenwashing effort to avoid an outcry over the increased amount of fish needed to fill the demand. Concerns remain about whether store managers and employees have the expertise and training to understand all the factors that differentiate organic foods from conventional offerings.

An investigation by the Cornucopia Institute cited Wal-Mart for deceptive labeling of organic products, which is complicated by the complex distinctions between certified organic products and those that are only partially organic. Organic standards have been disputed since the 1990s, when activists charged that the U.S. Department of Agriculture was inconsistent in its policies, and there have been continued questions about its vigilance in maintaining them. Products labeled "natural" or "all natural" are easily confused with those that meet the organic standards. A white paper issued by the Cornucopia Institute, "Wal-Mart Rolls Out Organic Products—Market Expansion or Market Delusion?" considers whether the company's entry into the marketplace has devalued the organic label by stocking products from large-scale factory farms and developing countries.

Wal-Mart and Food Safety

In 2008, Wal-Mart shifted policies on its Great Value store-brand milk, insisting to suppliers that milk should come exclusively from cows that are not given the recombinant bovine somatotropin artificial hormone as a result of pressure from consumers seeking products free from genetically modified organisms. Wal-Mart was the first U.S. grocer to require that suppliers of produce, meat, fish, poultry, and ready-to-eat foods have their factories certified to meet the rigid standards of the Global Food Safety Initiative. This may have been partially a result of a recall of peanut butter sold under the house brand Great Value because of the presence of *Salmonella* in its product, responsible for an outbreak that sickened consumers in 41 states. The Global Food Safety Initiative requires factories to meet audit certification using one of its standards including Safe Quality Food, British Retail Consortium, International Food Standard, or equivalents like Global-GAP. Wal-Mart has promised full certification by July 2009.

Former Environmental Defense Fund president Fred Krupp credits Scott, who stepped down from his post in early 2009, with changing how Wal-Mart interfaced with the press and its critics. Scott's interest in environmentalism led him to shift store lighting to energy-efficient compact fluorescent lightbulbs and begin a path toward making Wal-Mart a leader in delivering sustainable goods to its market. Recent environmental initiatives include doubling the fuel efficiency of its truck fleet, a commitment to using 100 percent renewable energy, and creating zero waste in company stores.

Philanthropic contributions to various charities have helped to counteract the tarnished image of Wal-Mart. From providing water and food to victims of Hurricane Katrina to a partnership with Feeding America to provide 90 million pounds of food annually to families in need by the end of 2009, the corporate giant has pledged gifts to hunger-relief efforts in 32 states. In addition, Wal-Mart donated $2.5 million to Feeding America to increase warehouse capacity and purchase 20 new refrigerated trucks to transport food directly from Wal-Mart stores to food pantries, soup kitchens, and other agencies.

Wal-Mart has the leverage to make an enormous contribution to increasing the demand for organic food and to strengthening food safety standards. Only time will tell if they are up to the challenge.

See Also: Certified Organic; Food Safety; Recombinant Bovine Growth Hormone; Supermarket Chains; Supply Chain; Sustainable Fisheries Act.

Further Readings

Fishman, Charles. *The Wal-Mart Effect*. New York: Penguin, 2006.

Gogoi, Pallavi. "Are Wal-Mart's 'Organics' Organic?" *BusinessWeek* (January 18, 2007). http://www.businessweek.com/print/bsdaily/dnflash/content/jan2007/db20070117_88739 (Accessed January 2009).

National Public Radio. "Wal-Mart CEO Stepping Down After Nine Years." *Morning Edition* (January 30, 2009).

Reuters. "Wal-Mart Becomes First Nationwide U.S. Grocer to Adopt Global Food Safety Initiative" (February 4, 2008). http://www.reuters.com/article/pressRelease/idUS199047+ 04-Feb-2008+PRN20080204 (Accessed January, 2009).

Walton, Sam. *Sam Walton: Made in America*. New York: Bantam, 1993.

Stephanie Yuhas
University of Denver

WEED MANAGEMENT

A farmer (left) and an agronomist used specialized computer software to develop a carefully timed weed management strategy to counter the emergence of pigweed in a soybean field.

Source: U.S. Department of Agriculture, Agricultural Research Service/Bruce Fritz

An informal definition and subjective definition of a weed is a plant growing anywhere that it is not wanted. Sometimes plants such as flowers or vegetables stray and come up spontaneously as volunteers. Quite often for gardeners these are acceptable, even if technically these are weeds. At other times, a weed is a plant in a place where humans do not want it to grow.

Weeds are nuisances that infest areas humans cultivate including lawns, gardens, parks, farm fields, rangelands, or fish ponds. Weeds may be native plants or invasive imports. They are undesirable in lawns because they are unsightly. They are undesirable also because they rob nutrition from cultivated plants. If not cleared, they can choke gardens and fields to the point that crop yields are reduced or almost completely eliminated. In the case of horticultural plants, they can be unsightly as well as damaging. Weeds can spread plant pathogens. They can also be an irritating nuisance with thorns or prickles. Others such as poison ivy can be skin irritants. Others are poisonous if eaten by either humans or animals. Loco weed in the western areas of the United States extracts arsenic from the soil. If eaten by a horse or cow, it causes poisoning. There have been cases of large numbers of cattle or horses being sickened or killed from eating poisonous weeds. Other weeds have burrs or parts that stick to fur or skin. All of these characteristics of weeds make them undesirable where humans cultivate other plants. These types of weeds are a nuisance to hunters, domestic animals, and farmers.

Weeds from the fields are often produced by wind-blown seeds that arrive in the garden, in horticultural displays, or in vast agricultural fields. Gardeners are faced with the need to use hoes or other implements to free their gardens from weeds that otherwise will overwhelm the cultivated plants. Farmers face the same problems, only on a larger scale.

Adaptations by weedy plants give them the ability to spread widely and thickly. Often weeds spread seeds, which have a high fertility rate. If the soil has been disturbed by nature or by humans, it gives weeds the opportunity to grow abundantly. Part of the adaptation of weeds is quick growth—the source of the expression "growing like a weed." Rapid growth rates allow weeds to outgrow plants used for crops. They also have the ability to remain dormant for years. Some weeds can be dormant for 20 or 40 years. Some weeds have extensive root systems that inhibit the growth of other plants. Other species of weeds are allelopathic, producing chemicals that prevent germination or growth by other plants.

Weeds have been part of the human story since the beginning. Literature and religion contain references to weeds. William Shakespeare wrote a sonnet on weeds. The book of

Genesis records that God punished Adam for his disobedience with a curse of weeds on the ground Adam cultivated.

Weeds can be invasive species that have usually been spread globally by human actions. They are transported with crops or with fodder for animals to new places. Some invasive weeds have entered ecologies where there was little to inhibit their spread. Prickly pear cactus became a plague in Australia, crowding out other plants. It was brought under control by the introduction of a moth that feeds on it.

Some weeds are beneficial. Dandelions (*Taraxacum*) are edible and can be used in herbal medicine or made into a tea or a wine. Burdock is a weed spread widely in Asia, where it has several uses including in a soup or as a medicine. Other weeds can repel insects or some animals, add nutrients to the soil, repel fungi, or distract insects.

Most weeds are noxious because, if allowed to flourish, they will destroy crops or grazing land. Many of the plants of the western United States such as sagebrush are weeds that flourished after gazing lands were overgrazed. Sheep and cattle so weakened the stands of native grasses that weeds were able to flourish, hindering the grasses' return.

Farming practices can promote the spread of weeds. In fact, weeds are in part a result of human agricultural and range management decisions. Monocrops allow weeds opportunities to infest gardens and fields. If gardens or farms are allowed to be uncovered for seasons between plantings, the bare soil can become host to a variety of weeds. To prevent weeds, many farmers use a variety of weed management practices. Among these are cultivation, timely planting, and herbicides.

Using herbicides is a widespread practice that is at times about the only effective remedy short of intense labor—trying to physically pull out all weeds. Chemical companies have developed a large number of general and specific herbicides for specific weeds from poison ivy to pigweed. Some of these are applied preemergence, and others postemergence. Regardless of the time of application, it is necessary that spraying not pollute the soil or cause runoff that is toxic to fish and other aquatic life, nor be applied to wrongly identified weeds.

A good weed management system includes exact identification of the weeds that must be controlled. A second advantage is to understand the life cycle of the specific weeds. Another is to understand the growth habits of weeds and their potential for crop competitiveness. Another technique is to maximize the planting time and seeding rates so that the crop has the maximum opportunity to compete with the weeds. In gardens, the use of mulch also inhibits weed growth.

Weed management practices seek to control weeds in an economical manner to prevent weed pressure from interfering with crop yield. Organic farming requires a variety of weed control methods. One technique is to reduce the risk of soil exposure to weeds. The more weeds can be prevented, the less opportunity they have to compete with crops for water and nutrients. Reducing weeds also reduces the opportunity for fungal diseases to gain opportunities to infect crops.

Constant weeding is a management technique that reduces the weed seed bank. This reduces the number of future weeds. Preventing any weed from growing in a garden, vineyard, or field is important to reduce the aggregate number of weeds in future crop seasons. This requires giving attention to weeds constantly, both in and out of growing seasons.

Weed management is necessary for gardens, vineyards, farms, rangeland, fish ponds, and lakes. Water weeds can inhibit the growth of fish or can foul water and interfere with boating. Of future concern is the emergence of herbicide-resistant weeds.

See Also: Grazing; Organic Farming.

Further Readings

Buhler, Douglas. *Expanding the Context of Weed Management*. London: Taylor & Francis, 1998.

Liebman, Matt, et al. *Ecological Management of Agricultural Weeds*. Cambridge, UK: Cambridge University Press, 2007.

Naylor, Robert E. L., ed. *Weed Management Handbook*, 9th ed. New York: John Wiley & Sons, 2002.

Ross, Merrill A. and Carol A. Lembi. *Applied Weed Science: Including the Ecology and Management of Invasive Plants*. Englewood Cliffs, NJ: Prentice Hall, 1998.

Singh, Harinder P., et al, eds. *Handbook of Sustainable Weed Management*. London: Taylor & Francis, 2006.

Upadhyaya, M. K. and R. E. Blackshaw, eds. *Non-Chemical Weed Management: Principles, Concepts and Technology*. Wallingford, UK: CABI, 2007.

Walters, Charles. *Weeds: Control Without Poisons*. Austin, TX: Acres U.S.A., 1999.

Andrew Jackson Waskey
Dalton State College

WHEAT

Wheat is the major food resource for about half of the world's population. At the same time, wheat production provides for the income of millions of small farmers and agribusiness firms likewise in developed and developing countries. The resources for producing wheat can be also used for producing fuels and for maintaining natural forest and prairies. Although some consumers are able to choose among practically hundreds of bread types, others obtain their main protein intake from wheat flour. Different answers are being put forward to the questions posed by the interdependence of resources involved in food production and the incompatible use of their outcomes. The total world wheat production is 500–600 million tons, of which more than half is produced in developing countries. The main producing countries include China, the United States, India, Kazakhstan, Pakistan, Russia, Turkey, the Ukraine, Canada, and Australia, which together account for about 80 percent of total production. Estimations for the year 2010 indicate that total wheat production will increase to about 680 million tons, especially driven by the need to replenish remarkably low present world stocks. Long-term factors affecting agricultural global production, however, indicate a slight decline in the planted area and the supply in the coming 10 years. This is a definite slowdown from previous growth, which doubled production in 30 years. The most important driver of this trend is the competition of soybeans and corn that turn out more profitable alternatives for farmers than wheat in the near future.

The demand for wheat-based products has slightly declined in recent years and is expected to remain steady in contrast to the strong growth in demand for vegetable oil and protein meal from soybeans and corn. The consumption of wheat bread is pervaded by economics but also by political, aesthetic, social, symbolic, and health aspects of each society. Although wheat demand is relatively steady in terms of quantity, there are changes in the demanded quality associated with consumption trends in quality food upgrading, health concerns, and social and environmental responsibility. These changes translate to all food system participants in the form of business opportunities and coordination and innovation challenges. For example, a growing niche of organic wheat bread requires a tighter coordination between organic producer, wheat miller (not to mix organic with nonorganic wheat), and baker.

International trade of wheat is about 128 million tons. Soybeans and corn have outpaced this figure in recent years and moved wheat from its traditional lead in global agricultural commodity markets to third place. The traditional wheat-exporting countries are the United States, Canada, Australia, the European Union, and Argentina. In recent years, there has been a strong increase of exported wheat from the Black Sea countries, especially from Russia and from the Ukraine, making wheat global trade become less concentrated.

The trade flow is primarily from North to South, as developed countries account for about 85 percent of exports, whereas developing countries account for 75 percent of imports. Two main types of wheat are traded globally: milling wheat, which accounts for about 95 percent of total world wheat production, and durum wheat. The milling market is further segmented into the highest quality, demanded by developed countries (about 10 percent of the milling wheat market) and a lower quality, demanded by developing countries including Iraq, Iran, Indonesia, and Egypt. China, the largest wheat producer in the world, has been a small net exporter in recent years; it is expected, however, that production constraints will lead it to become a small net

A Pakistani farmer in the Loralai District bringing in some of his wheat harvest, which was grown from a drought-tolerant strain appropriate to local conditions.

Source: U.S. Agency for International Development

importer in the next 10 years. Brazil is the largest wheat importer but also the place where agriculture is growing the most from land expansion in the Cerrados (a large savanna environment in the center of the country).

Wheat prices have always been highly volatile, similar to other commodity prices. However, an unusual hike almost doubled agricultural commodity prices between 2005 and early 2008. The causes behind this inflation entailed a combination of mutually reinforcing production, demand, and financial factors. The main factors point to the demand growth for animal feed and biofuels, simultaneous droughts in key grain-producing regions, low stock levels for cereals and oilseeds, and rising oil prices. The context of increased speculation in commodity futures markets and the devaluation of the U.S. dollar also contributed to the inflation of wheat and other agricultural commodity prices. Commodity prices have already dropped from the hike but are mostly expected to stay above average levels of the past.

The Effect of Wheat on Economic Development

Wheat prices play a multiple role in economic development, as they simultaneously affect the food purchasing power of consumers and the income of producers, as well as global trade. Although the cost of the wheat commodity is a small part of the final retail price of wheat food products (about 10 percent), it has risen to have an important effect on retail prices. The effect on people's welfare varies according to their income, being the highest for lower-income consumers who spend a greater portion of their income on food.

After the agricultural commodity price hike, international organizations called for adjusting domestic agricultural policies so as to promote a better balance between supply and demand in international markets. A high emphasis is being put (particularly by the United Nations Food and Agriculture Organization) on the role of commodity production and prices in correcting the growing gap in living standards in countries in different stages of economic development.

Another policy focus of major wheat-exporting countries is developing business opportunities from meeting the needs of new and value-added markets for wheat. One example is the Feed Opportunities from the BioFuels Industries initiative, which joins private, public, and academic organizations to research the opportunities of improving farmers' income from the integration of livestock feeding and wheat-based ethanol production in Canada.

The Effect of Domestic Policies on Wheat Availability and Access

Wheat market prices are also highly affected by tariffs and price controls implemented by government policies. Developed countries have traditionally supported their wheat farmers' income through support prices. These policies have meant surplus wheat production from relatively higher prices in the European Union and the United States and a diminishing production and market share from relatively lower prices in developing countries. Gradual removal of wheat subsidies in developed countries is expected from negotiations at the World Trade Organization, as well as other multilateral organizations in the coming years.

In some developing countries such as Pakistan and India, support prices and input subsidies are current policies targeted to increasing per acre yield. Conversely, in other countries, including Argentina, export tariffs have been the typical domestic policy in wheat exporting. In addition to having an important role in government income, export tariffs reduce the domestic price of wheat. For example, in Argentina the ongoing tariff has been intended as a tool for decoupling the international and the domestic prices. The extent to which the policy reduced the effect of international wheat price hikes for processors and consumers is a controversial issue. Moreover, social and political conflict resulted from the policy effect on diminishing the competitiveness of domestic wheat producers from reduced prices relative to their counterparts overseas.

Support output price and input price subsidy policies in developed and developing producing countries have largely fostered intensification of industrial agriculture. The consequences of these policies have been increasing yields beyond efficiency, both production- and economic-wise, in European Union countries, for example, and poor crop rotation systems, such as wheat–rice in Indian agriculture.

The most widespread wheat production system in the world is conventional agriculture, also termed industrial agriculture, which relies on the use of high-yield varieties, chemical fertilizers, and irrigation to increase the wheat yield per hectare. This technology was widely implemented through the agricultural transformation process known as the green revolution, which radically increased wheat and other major crops' output in developing countries.

The Effect of Wheat on International Cooperation and Conflict

Wheat availability and access are involved in dilemmas stemming from the definition of property rights, production technologies, and consumption preferences over wheat resources. One is the "feed versus food versus fuels" dilemma. Wheat competes for limited production resources with other crops, mainly oilseeds and corn, which are increasing their demand as inputs for animal and biofuels production. This demand drives the supply of wheat down, as these alternatives generate higher returns for producers, with the consequence that wheat-based foods are becoming more expensive.

Another dilemma is the industrial versus more sustainable agrofood systems. Past increases in food production through industrial agriculture have been important in decreasing overall food costs with an important effect on reducing poverty in developing countries. There are, however, other views contending that industrial agriculture is not the only or even the best solution to providing food to meet increasing demand.

In relation to production technology, another dilemma is the use of biotechnology in wheat production. Unlike the widespread genetically modified (GM) soybeans and corn varieties, GM wheat is not available yet in the market, but several traits are being developed, and some are expected to be launched in the coming years. The release by the bioscience companies of GM wheat is expected to generate conflict among the food systems stakeholders. On the one hand, biotechnology crops are attractive to farmers because they simplify the operations management of conventional farming and allow increasing returns from scale economies. On the other hand, it is fully expected that buyers will be reluctant to buy GM wheat at least until regulatory approval and information and discussion on food safety and environmental impact.

A range of collaborative initiatives are underway to tackle these conflicts and develop a more sustainable approach toward food availability and access. These approaches understand agriculture as an ecosystem and weigh the interrelations of the crop with its natural, economic, and human environment. The work at international organizations including the International Center for Maize and Wheat Improvement and the Food and Agriculture Organization is incorporating this approach in wheat through several initiatives including no-till wheat production, integrated plant nutrient systems and pest management, integration of wheat and livestock feeding, and organic wheat farming.

Alternative technologies, just like mainstream ones, are the outcome of a complex social process that includes the definition of knowledge as well as the development of operating procedures and rules for implementation and operation. In practice, sustainable wheat production, as it purports a change of the dominant view, demands a revision of the institutions and policies that foster industrial agriculture.

Experiences at a small local level point to the importance of farmers' organizations and civil organizations in which producers and consumers exchange information for marketing and accessing the more sustainable product. Comprehensive studies are needed that analyze the evidence of comparative achievements and prospects of these alternative solutions under the various circumstances of real-world wheat resources. The dissemination through the media of research and experience in sustainable wheat production are crucial to understanding and fostering the development role of wheat as a fundamental food resource.

See Also: Agribusiness; Agrofood System (Agrifood); Commodity Chain; Supply Chain; Sustainable Agriculture.

Further Readings

Bruinsma, J., ed. *World Agriculture: Towards 2015/2030. An FAO Perspective.* Food and Agriculture Organization of the United Nations (2003). http://www.fao.org/docrep/004/Y3557E/Y3557E00.htm (Accessed February 2009).

Ekboir, J., ed. *CIMMYT 2000–2001 World Wheat Overview and Outlook: Developing No-Till Packages for Small-Scale Farmers.* Mexico: CIMMYT, 2002.

Organisation for Economic Co-operation and Development (OECD). "Rising Food Prices: Causes and Consequences." Working Document, Paris, 2008. http://www.oecd.org/dataoecd/54/42/40847088.pdf (Accessed February 2009).

Reynolds, M. P. and N. E. Borlaug. "Impacts of Breeding on International Collaborative Wheat Improvement." *Journal of Agricultural Science,* 144:1:3–17 (2006).

U.S. Department of Agriculture. "Agricultural Projections to 2017." OCE-2008-1, February 2008. http://www.ers.usda.gov/Publications/OCE081 (Accessed February 2009).

M. Laura Donnet
Austral University

YEOMAN FARMER

The term *yeoman* is a word dating to the 14th century that came to mean a farmer who owned the land he farmed (unlike nobles of higher rank, a yeoman was usually actively involved in the farming himself, rather than simply owning the farm and delegating its management to others). The etymology of the word is unclear. It may be related to the Middle English term *yemen*, "to care," making the yeoman a caretaker of the land, but it has generally been assumed that the word was formed by prepending "yeo" to "man," as was common throughout English then as now. Early editions of the *Oxford English Dictionary* read *yeo-* as a variant of young, and *yeoman* as "young man," reflecting the junior status of the yeoman relative to loftier nobles. Later editions of the Oxford dictionary suggested the *yeo-* could be the same as that of the English and Dutch word *yeomath*, meaning "second crop of grass," and that yeoman would then mean "an additional man." The word's exact usage has varied from era to era; in some centuries, a yeoman is simply a freeman, or a faithful assistant (like the French *valet*). In others, he is a landowner but may not necessarily farm his land. Older, non-farm-related usages of the word have been retained in phrases like the Yeomen of the Guard, a ceremonial military corps created by Henry VII in 1485 to serve as bodyguards to the British monarch, and in the use of "yeoman" as a military rank in both the British and the U.S. navies, generally indicating minor administrative duties. The connotations of the word have varied, sometimes derogatory (unlike other landowners, the yeoman gets his hands dirty), sometimes laudatory (he is a self-sufficient man who does honest work). Until the 18th century, both "yee-man" and "yo-men" were acceptable pronunciations; since then, only the latter has been used.

In the United States, the term *yeoman* was often used for farm owners who did not own slaves, and who in many cases hired them from other farmers at harvest time. In some parts of the country, yeomen were significantly poorer than other farm owners (i.e., planters), growing little more than what they needed for self-sufficiency, with the little excess sold to pay taxes and buy goods that could not be produced on the farm, which might range from milled flour or imported spices to appliances like stoves and iceboxes. The yeoman, in many ways, is the archetypal "American farmer," whose needs have fueled U.S. political debates since before the birth of the union—the hard worker who is unable to be completely self-sufficient and is forced to participate in the market to pay for obligations (taxes, debts), services (education, medicine), and goods that cannot be produced directly on the farm.

It's important to consider the age during which North America was colonized, and during which the United States was formed. Agriculture, once dismissed as the toil necessary for the lower classes to support the upper classes, was newly important in all areas of Western thought. Theology emphasized the nature of Eden as a garden and the role of Man as caretaker of the Earth; the same strains of Western philosophy that would characterize the Native American as a Noble Savage who had been kept pure by living off the land similarly saw the white farmer as honest, down-to-earth, reliable, wise, and possessed of an innate moral character. The New World was compared frequently to the Garden of Eden by writers and speakers of all the European nations who settled it. Often these comparisons were rhetorical—the Americas were untouched by the "civilized world" of the West and Middle East, having existed outside the history and memory of Western civilization, and there were large areas of completely untamed wilderness the likes of which were rarely found in Europe anymore. But more than one comparison was literal, with various religious figures claiming that the New World was the site of the real Eden (with human civilization having been relocated to other continents by the travel of Noah's Ark). This made the role of the American colonist more significant, as he was entrusted not simply with some land to be worked for his profit or the king's but with a holy land where he was meant to coexist with the natural world as Adam had before the Fall. The southern settlers in particular saw as their duty the exploration of ways they could profit from the land without unduly transforming it; the Puritans of New England were more likely to see themselves as the bringers of civilization to the savage wild lands.

By the time of the Revolutionary War, the yeoman had become the central figure of agrarian philosophy, symbolic of one-half of the debate over the role of government and the American character. Although Alexander Hamilton's Federalists called for strong centralized government and an American economic identity based on commerce, industry, and banking, the Republicans formed around Thomas Jefferson's agrarian writings, which held dear a strong local government that was a natural outgrowth of the independence of small farmers, and a portrait of America as principally a nation of farmers. Chances are that neither side would consider themselves the winner; as history unfolded, the America that developed contained elements of both. Interestingly, however, though the economy skewed more toward the Hamiltonian ideal (debates over banking and economic policy dominated American political debate from the 19th century through the New Deal and were second only to the slavery question), and although each generation included fewer and fewer American farmers, the American self-image was undeniably Jeffersonian, throughout the 19th century and arguably still today in the 21st century. Hamilton's America of Banks never provided such a strong image as the yeoman, nor did the advocates of slavery, whose characterization of planters as cavaliers in the old English model and preservers of codes of honor was at odds with much of the American identity and individualism.

This rhetorical yeoman worked hard to make the most of his land. Slavery was portrayed by abolitionists as a threat to the American work ethic, a crutch for landowners that could make them lazy as they grew fat and rich off the sweat of others. Even Jefferson, a slave owner himself, said of the practice that it had the effect of destroying both the morals and the industry of the people who embraced it.

Today, of course, although the planter is no longer around, American farming is similarly divided between smaller (often family) farms that seem closer to the yeoman ideal and large agribusinesses in which there is a sharp divide between those who own the land (and make the decisions) and those who work it; further, the fact that the bulk of agricultural subsidies goes to those large corporations creates a class of farmer whose fortunes, unlike the yeoman's, are not dependent on those of the farm.

See Also: Agrarianism; Berry, Wendell.

Further Readings

Freyfogle, Eric T. *Agrarianism and the Good Society: Land, Culture, Conflict, and Hope.* Lexington: University Press of Kentucky, 2007.

Inge, M. Thomas, ed. *Agrarianism in American Literature*. New York: Odyssey Press, 1969.

Twelve Southerners. *I'll Take My Stand: The South and the Agrarian Tradition.* Baton Rouge: Louisiana State University Press, 2006.

Bill Kte'pi
Independent Scholar

Green Food Glossary

A

Added Sugars: Sugars and syrups, such as corn syrup, that are put in foods during processing or preparation. Added sugars do not include naturally occurring sugars such as those that occur in milk and fruits.

Antibiotic-Free Protein: Poultry, pork, and lamb that does not come from animals that have been given antibiotics as a feed stimulant.

Antioxidants: Antioxidants are chemical compounds that work to protect the body from cell damage by inhibiting oxidation, rendering free radicals—molecules with one or more unpaired election—harmless. Some research indicates that this process improves general health and helps prevent degenerative diseases.

B

Biodynamic: Methods of farming that combine organic policies, including crop rotation, animal and mineral preparation, and the rhythmic influences of the solar system; originally made popular by Austrian philosopher Rudolf Steiner.

Biodynamic Agriculture: An older, and in many ways more rigorous, process than organic farming, Biodynamic Agriculture sustains soil productivity and ensures habitat protection with the use of organic pest controls, crop rotation, and the cyclical properties of nature. The process aims at creating a greener planet through self-sustaining farms and healthier soil, while producing better food that is rich in nutrients. Biodynamic is a registered trademark, owned by Demeter USA.

Bovine Growth Hormone: A synthetic hormone given to dairy cows to maximize milk production. Some evidence suggests increased risk for breast and prostate cancer in humans who drink milk from cows given this and similar hormones. Approved in 1993, today about 15 to 20 percent of U.S. dairy cows are injected with the hormone. Also known as rBST (recombinant bovine somatotropin) and rBGH (recombinant bovine growth hormone).

BPA or Bisphenol A: Recent research shows that this chemical, found in plastic beverage bottles, is believed to leach carcinogens into the liquid. Active studies may recommend bottlers make a change to the plastic used in many liquid containers.

C

CAFE Practices: Coffee and Farmer Equity (CAFE) Practices are a set of socially and environmentally responsible guidelines for producing, processing, and buying coffee. They were established in 2001 by coffee giant Starbucks in cooperation with Conservation International, a nonprofit environmental group.

Cage-Free Eggs: Cage-free eggs are laid by hens that are not kept in typical cages. This does not always mean they are organic, as conditions are not heavily regulated and cage-free operations often confine the animals by densely packing them into enclosed areas.

California Certified Organic Farmers (CCOF): An independent organization that provided the first official certifications for all stages of organic food commercialization, including farming, processing, and sale. They operate in 29 states and five foreign countries.

Carob Bean Gum, Carrageenan, and Locust Bean Gum: The most widely used and thought to be effective natural food preservatives. They are derived from plant gums and seaweeds.

Certified Agent: Certified agents audit independent organizations and organic farms to ensure that U.S. Department of Agriculture (USDA) National Organic Program (NOP) standards are being upheld.

Certified Humane Food: The Certified Humane Raised & Handled program is an inspection, certification, and labeling program for meat, poultry, egg, and dairy products from animals raised to humane care standards. The program is a voluntary, user-fee based service available to producers, processors, and transporters of animals raised for food. The USDA verifies the inspection process of the Certified Humane Raised and Handled program.

Certified Organic: Food products that meet or exceed standards set forth by the USDA NOP. Products "made with organic ingredients" include 70 percent organic ingredients and cannot contain the organic label. "Organic" products must have at least 95 percent organic ingredients and may feature the USDA organic seal. "100% Organic" is the most stringent standard but does not count water or salt.

Closed Herd: A herd of animals that lives its entire life on a farm. Outside animals can spread diseases—mad cow for example—causing many farmers to quarantine their livestock. Because herds of animals for organic products cannot be treated with antibiotics, maintaining a closed herd is essential.

Community Supported Agriculture (CSA): A subscription-based farming system in which individual shares of a farm are purchased and, on a routine basis, shareholders are provided with an assortment of in-season fruits and vegetables and other groceries. In 2008, more than 1,500 CSAs were operating nationwide, with shares costing anywhere from $20 to $900.

Composting: The act of recycling organic matter back into the Earth, creating a nutritious mixture that enriches the soil. Compost is typically made up of lawn clippings, vegetable scraps, and untreated papers. When these materials are combined, they make for an environmentally safe fertilizer.

Cover Cropping: A means of using crops that provide temporary protection for delicate seedlings as well as provide a canopy for seasonal soil protection and improvement between normal crop production periods. Used on organic farms, cover cropping is a natural way of shielding fragile plants from the weather. Cover crops are also called "green mature crops."

Crop Rotation: A system of planning in which crops vary from season to season. No single crop is planted in the same field twice in two seasons. Crops grown successively in the same field will rapidly deplete the minerals in the soil. By employing crop rotation, farmers ensure that their fields are bountiful for years longer. The USDA NOP highly recommends, and in many cases requires, that certified organic farms use crop rotation.

F

Fair Trade: Products branded with the fair trade label are produced and imported from developing nations in which workers are often subject to substandard conditions and compensation. Fair trade labeling assures consumers that farmers are paid a better-than-conventional wage and are trained in sustainable agriculture practices.

Farmers Market: Farmers markets are where local farmers gather to sell their produce or specialty goods in a specific place at a designated time. All food bought at a farmers market is probably not produced using green or organic practices, but in general, the selection of organic food is broader than at a supermarket.

Food Quality Protection Act (FQPA): Passed by the U.S. Congress in 1996, the FQPA amendments changed the way the Environmental Protection Agency regulates pesticides by requiring a new safety standard—"reasonable certainty of no harm"—that applies to all pesticides used on foods.

Free Range: A reaction to the unnatural and sometimes unethical treatment of animals used for food, especially chickens, the term *free range* refers to any product related to, or produced by, animals that live their lives primarily in outside spaces and that are permitted to graze and forage freely. Non–free range animals used for commercial food products typically spend their whole lives in a crowded, confined feeding lot before slaughter.

G

Genetically Engineered (GE): GE foods have foreign genes inserted into their genetic codes. Almost any living thing can undergo some type of genetic engineering; the most widespread use typically involves plants. For instance, certain strains of rice have been altered to be more resistant frost or flooding, saving thousands from starvation. GE practices are sometimes referred to as *bioengineering* or *biotechnology*.

Genetically Modified Organism (GMO): A plant, animal, or microorganism that is transformed by genetic engineering. Something that is the result of this process is called a "product of genetic engineering," or a "derivative of GMOs." Although GMOs, especially crops such as rice or corn, create an otherwise impossible food supply for impoverished nations worldwide, their use is considered by some in the scientific community to be dangerous. They claim that the health of the population and the environment is at risk as a result of insufficient knowledge about the safely and predictably modified plant genomes.

Green: The term that refers to products made from materials that are recycled, renewable, or otherwise environmentally friendly.

Green Tea: Tea that is made from fermented dry leaves, raising the antioxidant level significantly over black tea. A recent European study concluded that a specific compound in green tea helps prevent cancer cells from growing.

Guarana: A South American plant whose seeds contain caffeine, added to soft drinks as a stimulant.

H

Humus: The result of organic material being decomposed into a dark soil-like material that contains plant nutrients.

I

Integrated Pest Management (IPM): The use of different techniques in combination to control pests—typically insects that feed on commercial crops—with an emphasis on methods that are least harmful to the environment and most specific to the particular pest. For example, pest-resistant plant varieties, regular monitoring for pests, pesticides, natural predators of the pest, and good stand management practices may be used in combination to control or prevent particular pests.

International Federation of Organic Agriculture Movements (IFOAM): IFOAM's mission is to lead, unite, and assist the organic movement in its full diversity with a goal of worldwide adoption of ecologically, socially, and economically sound systems that are based on the principles of organic agriculture.

Irradiation: Exposure to ionizing radiation to kill any form of bacteria. Critics claim that irradiation can be used to mask poor handling of products, which leads to other kinds of contamination. Food irradiation is a synthetic process that is not allowed in organic production.

N

National Organic Program (NOP): In 1990, the U.S. Congress passed the Organic Food Production Act, which called on the USDA to establish national standards for growing, processing, and marketing organic products. NOP was established to create a system of criteria for certifying organic food by the USDA. The program unifies all organic foods under a cohesive national agenda and creates the organic seal, which appears on products that meet criteria.

National Organic Standards Board (NOSB): A government-appointed panel that advises the NOP in assisting in the development of standards for substances to be used in organic production and to advise on any other aspects of the implementation of the NOP.

Natural: Sometimes incorrectly confused with their organic cousins, natural foods do not contain additives or preservatives, but ingredients may have been grown using conventional farming methods or GE grain. Also called "all-natural."

No Preservatives: A food product that is made without ingredients used for the purpose of extending its shelf life. Typical chemical preservatives include nitrates, nitrites, butylated hydroxytoluene, and sulfites.

Northeast Organic Farming Association (NOFA): NOFA is a nonprofit organization of nearly 4,000 farmers, gardeners, and consumers working to promote healthy food, organic farming practices, and a cleaner environment. NOFA has chapters in Connecticut, Massachusetts, New Hampshire, New Jersey, New York, Rhode Island, and Vermont.

O

Organic Consumers Association (OCA): A research and action center for the organic and fair trade movements that campaigns for a myriad of environmental issues. The OCA is a proponent of labeling for GE food.

Organic Crop Improvement Association (OCIA International): This nonprofit, member-owned organization is one of the world's oldest and largest leaders in the organic certification industry. The OCIA promotes environmental protection policies and is dedicated to providing the highest quality organic certification services and access to viable global markets.

Organic Farming: Coined in 1940, the term *organic farming* is derived from the concept of the farm as a living organism. Organic farming is a form of agriculture that relies on crop rotation, compost, and environment friendly pest control and excludes synthetic fertilizer and pesticides. The IFOAM defines organic farming as "a production system that sustains the health of soils, ecosystems and people. It relies on ecological processes, biodiversity and cycles adapted to local conditions, rather than the use of inputs with adverse effects." Organic farming also excludes all types of growth or feed stimulants and any genetic modifications.

Organic Matter: Anything that is composed of a once-living organism that is capable of decay or a process of decay. Compost used in organic farming is entirely made from organic matter.

Organic Trade Association (OTA): The OTA is a membership-based business association that focuses on the organic community in North America. The OTA's mission is to promote and protect the growth of the organic industry to benefit the environment, farmers, the public, and the economy. The OTA is a member of IFOAM.

P

Persistent Toxic Chemicals: Detrimental materials, such as Styrofoam or dichlorodiphenyltrichloroethane (DDT), that remain active for a long time after their application and can be found in the environment years, and sometimes decades, after they were used.

Pesticide: A general term for chemicals used to destroy living things that interact with crops and animals that are considered pests. There are many specific forms of pesticides including insecticides, herbicides, fungicides, and rodenticides, used to eliminate the organisms for which they are named.

Q

Quality Assurance International, Inc. (QAI, Inc.): U.S. based and USDA accredited, QAI is a for-profit corporation, considered by most to be the global authority in organic certification services and has certified more than a quarter of a million organic products worldwide. QAI offers organic certification under the National Organic Program for producers, processors, private labelers, distributors, retailers, restaurants, wild crop harvesters, greenhouse, mushrooms, and facilities. QAI also offers "fiber certification" under the American Organic Standards.

R

Rainforest Alliance: A U.S.-based conservation group and certifier that sets rigorous environmental standards for coffee, cacao, and other products to protect the rainforest. Its seal ensures that farms have met environmental and social standards including biodiversity, conservation, and ecosystem protection, as well as worker protection, health care, and education for children of the farm workers.

S

Soil Association Certification Ltd.: The United Kingdom's leading organic certification organization. Their criteria is used for up to 80 percent of the organic food sold in that country.

Sustainable: In agriculture, the term *sustainable* means capable of being maintained with minimal long-term effect on the environment. Sustainable agriculture integrates three main goals: environmental stewardship, farm profitability, and prosperous farming communities. Sustainable development recognizes the need to work with living environments in a balanced manner.

Sustainable Seafood: The act of not overfishing, causing the possibility of extinction or adverse effects on a habitat.

U

United States Department of Agriculture (USDA): Established by President Lincoln, the USDA is an umbrella organization encompassing all aspects of farming production that has executive and legislative authority to ensure food safety and protect national resources. Active operating units include the NOP, Agricultural Resource Service, Food Safety and Inspection Service, Risk Management Agency, and Animal and Plant Health Inspection Service.

W

Wild-Crafted: A plant gathered in the wild in its natural habitat from a site that is not maintained under cultivation or other agricultural management.

Sources: U.S. Environmental Protection Agency and organic.org

Green Food Resource Guide

Books

Agrios, G.N. *Plant Pathology.* San Diego, CA: Academic Press Inc., 1988.

Allen, Patricia and Debra van Dusen, eds. *Global Perspectives on Agroecology and Sustainable Agricultural Systems.* IFOAM International Scientific Conference. Santa Cruz: University of California Press, 1988.

Altieri, Miguel A. *Agroecology: The Science of Sustainable Agriculture.* Boulder, CO: Westview Press, 1995.

Altieri, Miguel A. and Susanna B. Hecht. *Agroecology and Small Farm Development.* Boca Raton, FL: CRC Press, 1990.

Barrow, C. J. *Environmental Management and Development.* London: Routledge, 2005.

Borlaug, Norman E. *Norman Borlaug on World Hunger.* San Diego, CA: Bookservice International, 1997.

Borlaug, Norman, et al. *Vetiver Grass: A Thin Green Line Against Erosion.* Washington, D.C.: National Academy of Sciences, 1993.

Carroll, C. Ronald, et al. *Agroecology.* New York: McGraw-Hill, 1990.

Committee on Sustainable Agriculture and the Environment in the Humid Tropics, National Research Council. *Sustainable Agriculture and the Environment in the Humid Tropics.* Washington, D.C.: National Academy of Sciences, 1993.

Committee on the Role of Alternative Farming Methods in Modern Production Agriculture, National Research Council. *Alternative Agriculture.* Washington, D.C.: National Academy of Sciences, 1989.

Conca, Ken and Geoffrey Dabelko, eds. *Environmental Peacemaking.* Washington, D.C.: Woodrow Wilson Center, 2002.

Conway, Gordon R. *The Doubly Green Revolution: Food for All in the 21st Century.* Ithaca, NY: Cornell University Press, 1998.

Dahlberg, Kenneth A. *Beyond the Green Revolution: The Ecology and Politics of Global Agricultural Development.* New York: Plenum, 1979.

Diehl, Paul and Nils Gleditsch Petter, eds. *Environmental Conflict.* Boulder, CO: Westview Press, 2001.

Edens, Thomas C., et al. *Sustainable Agriculture and Integrated Farming Systems.* 1984 Conference Proceedings. Lansing: Michigan State University Press, 1985.

Edwards, Clive A., et al. *Sustainable Agricultural Systems*. Ankeney, IA: Soil and Water Conservation Society, 1990.

Fluck, Richard C., ed. *Energy in Farm Production*. Amsterdam: Elsevier, 1992.

Gadgil, Madhav and Ramachandra Guha. *Ecology and Equity*. London: Routledge, 1995.

Gliessman, Stephen R. *Agroecology: Researching the Ecological Basis for Sustainable Agriculture*. New York: Springer, 1990.

Global Environment Facility. *Producing Results for the Global Environment*. New York: Global Environment Facility, 2005.

Griffiths, T. and L. Robin, eds. *Ecology and Empire*. Seattle: University of Washington Press, 1997.

Kidd, Charles and David Pimentel. *Integrated Resource Management: Agroforestry for Development*. San Diego, CA: Academic Press Inc., 1992.

Lal, Rattan and B. A. Stewart. *Soil Degradation*. New York: Springer, 1990.

Mandelker, D. R. *Environment and Equity: A Regulatory Challenge*. New York: McGraw-Hill, 1981.

Manwaring, Max, ed. *Environmental Security and Global Stability*. Lanham, MD: Lexington Books, 2002.

Marsh, George. *Man and Nature*. Cambridge, MA: Harvard University Press, 1965.

McIsaac, Gregory and William R. Edwards, eds. *Sustainable Agriculture in the American Midwest: Lessons From the Past, Prospects for the Future*. Urbana: University of Illinois Press, 1994.

McLaren, Digby J. and Brian J. Skinner, eds. *Resources and World Development*. Chichester, UK: Wiley, 1987.

Meffe, G. K. and C. R. Carroll. *Principles of Conservation Biology*. Sunderland, MA: Sinauer Associates, 1994.

Mepham, T. B., et al. *Issues in Agricultural Bioethics*. Nottingham, UK: Nottingham University Press, 1995.

Merchant, Carolyn, ed. *Major Problems in American Environmental History*, 2nd edition. Boston, MA: Houghton Mifflin Co., 2005.

Metcalf, Robert Lee and Robert A. Metcalf. *Destructive and Useful Insects: Their Habits and Control*. New York: McGraw-Hill, 1993.

Myers, Norman. *Gaia: An Atlas of Planet Management*. Garden City, NY: Anchor Press/ Doubleday, 1993.

Olson, Richard K., ed. *Integrating Sustainable Agriculture, Ecology, and Environmental Policy*. Binghamton, NY: Food Products, 1992.

Perrin, Constance. *Everything in Its Place: Social Order and Land Use in America*. Princeton, NJ: Princeton University Press, 1992.

Pimentel, David, ed. *World Soil Erosion and Conservation*. Cambridge, UK: Cambridge University Press, 1993.

Pimentel, David and Charles W. Hall. *Food and Natural Resources*. San Diego, CA: Academic Press Inc., 1989.

Postel, Sandra. *Last Oasis: Facing Water Scarcity*. New York: Norton, 1992.

Schwab, Jim. *Deeper Shades of Green: The Rise of Blue-Collar and Minority Environmentalism in America*. San Francisco: Sierra Club, 1994.

Sen, Amartya K. *Poverty and Famines: An Essay on Entitlement and Deprivation*. Oxford: Oxford University Press, 1981.

Shiva, Vandana. *The Violence of the Green Revolution: Third World Agriculture, Ecology, and Politics*. London: Zed Books, 1991.

Snow, Donald. *Inside the Environmental Movement: Meeting the Leadership Challenge*. Washington, D.C.: Island, 1992.

Thurston, H. David, et al. *Slash/Mulch: How Farmers Use It and What Researchers Know About It*. Ithaca, NY: Cornell International Institute for Food, Agriculture and Development, 1994.

Treoh, Frederick and Louis Thompson. *Soils and Soil Fertility*. New York: Oxford University Press, 1993.

Turner, B. L., II, et al., eds. *The Earth as Transformed by Human Action: Global and Regional Changes in the Biosphere Over the Past 300 Years*. Cambridge, UK: Cambridge University Press, 2003.

Journals

Agriculture, Ecosystems and Environment (Elsevier Science)
Alternatives (Alternatives Inc.)
American Naturalist (Thomson Corporation)
Amicus Journal (National Resources Defense Council)

Biodiversity and Conservation (Chapman and Hall)
Biological Conservation (Elsevier Science)
BioScience (American Institute and Biological Sciences)

Conservation Biology (Blackwell Publishing)
Critical Reviews in Environmental Science and Technology (Taylor and Francis)

Ecological Economics (International Ecological Economics)
Ecologist, The (Ecosystems Ltd.)
Environment (Voyage Publications)
Environmental Action (American Chemical Society)
Environmental Science and Technology (Center for Environment and Energy Research and Studies)
Environment and Behavior (SAGE Publications)

Global Environment Politics (MIT Press)

Human and Ecological Risk Assessment (Taylor and Francis)
Human Ecology (Springer Science and Business Media)

International Journal of Sustainable Development & World Ecology (Taylor and Francis)

Journal of Agriculture and Food Chemistry (American Chemical Society)
Journal of Environmental Economics and Management (Academic Press)
Journal of Environmental Management (Academic Press)
Journal of Environment and Development (SAGE Publications)

Nature (Palgrave Macmillan)
New Scientist (Reed Business Information)

Planning (Oxford University Press)
Population and Environment (Center for Environment and Population)
Progressive (Progressive Inc.)

Science (American Association for the Advancement of Science)
Sierra (Sierra Club)
Society and Natural Resources (Routledge)

Trends in Ecology and Evolution (Oxford University Press)

Waste Age (Prism Business Media)
Whole Earth Review (Point Foundation)

Websites

Certified Humane Foods
www.certifiedhumane.com

Demeter Association
www.demeter-usa.org

Ecology and Society
www.ecologyandsociety.org

Ecorazzi
www.ecorazzi.com

Fair Trade Organization:
www.TransFairUSA.org

Food and Drug Administration
www.fda.gov

Growing Green Co-op
www.goinggreenevents.com

National Agricultural Law Center
www.nationalaglawcenter.org

National Agricultural Library
www.nal.usda.gov

National Resources Conservation Service
www.nrcs.usda.gov

*The New York Time*s Dot Earth Blog
www.dotearth.blogs.nytimes.com

U.S. Department of Agriculture
www.usda.gov

U.S. Environmental Protection Agency
www.epa.gov

U.S. Green Party
www.gp.org

Worldwatch Institute
www.worldwatch.org

Green Food Appendix

Community-Supported Agriculture

http://www.nal.usda.gov/afsic/pubs/csa/csa.shtml

This website, part of the National Agricultural Library of the U.S. Department of Agriculture, offers information on the history and current state of community-supported agriculture in the United States, a method of agricultural production and distribution in which a group of consumers invests in a local farm and receives shares of the harvested crops in return. It includes information useful to both farmers and consumers, as well as links to information about related topics including sustainable agriculture, alternative plants and crops, grazing systems and alternative livestock breeds, organic production, and ecological pest management. It also includes links to several resources that allow searching to find participating farms in a geographical area, news items about community-supported agriculture, and academic and governmental reports about the growth of community-supported agriculture.

Food Safety—From the Farm to the Fork

http://ec.europa.eu/food/food/index_en.htm

This website is part of the Health and Consumer Protection site of the European Commission. The guiding principles of the commission's food program are explained in its White Paper, downloadable in 11 languages from this site: it covers all aspects of the food chain from production through processing, storage, transport, and retail sale. Further information relating to food safety is organized into eight topics: general food law, animal nutrition, labeling and nutrition, biotechnology, novel food, chemical safety, biological safety, and official controls. A separate section is devoted to topics of current concern, including BSE (mad cow disease), genetically modified food and feed, and *Salmonella*. Many commission reports and publications are available for download from this site, as are speeches and press releases.

Genetically Modified Organisms

http://www.newscientist.com/topic/gm-food

This website, created by the London-based international weekly journal *New Scientist*, includes basic information about genetically modified (GM) organisms and articles and editorials from the magazine on different topics pertaining to genetic modification. Like the journal, the articles are intended for a general audience and are reports of scientific developments rather than peer-reviewed research. The website does not advocate for or against GM products but provides an overview of their history and current use and the differing opinions regarding them. Topics covered in the articles include GM food crops, genome smuggling, the possibility and ethics of breeding livestock that are unable to feel pain, and the potential for GM crops that could reduce greenhouse emissions. The website also contains links to organizations involved in the GM debate including the World Health Organization, the Institute of Food Science and Technology, the Royal Society, and the Friends of the Earth.

National Organic Program

http://www.ams.usda.gov/AMSv1.0/nop

This website, created and maintained by the Agricultural Marketing Service of the U.S. Department of Agriculture, explains the National Organic Program that develops and administers standards for organic agricultural products sold in the United States (including those produced abroad) and certifies agents to inspect the production and handling of products sold as organic. The website is directed primarily toward professionals in agriculture and the food industry but also contains much information of use to journalists, politicians, and consumers interested in knowing exactly what it means when a product is sold bearing the "USDA Organic" seal. Basic information about the history of the program, its purposes, and current regulations are available from the website and from downloadable fact sheets covering topics such as the process of obtaining certification and grading and labeling standards. The website also includes the national list of prohibited and allowed substances, a searchable interface to locate accredited certifying agents by region, and news items about the program.

Oldways

http://oldwayspt.org

Oldways is an international nonprofit organization that partners with scientists and academics to promote healthy food and lifestyle habits. Oldways is best known for promoting the Mediterranean Diet based on fruits, vegetables, whole grains, nuts, and fish and olive oil as a healthy alternative to the traditional American diet based on meat, dairy, and refined grains. In partnership with the Harvard School of Public Health and the World Health Organization, Oldways introduced the Mediterranean Diet Pyramid in 1993 as an guide to healthy eating that draws on the traditions of Crete, Greece, and southern Italy circa 1960 and offered it as a healthier alternative to the food pyramid that the U.S. Department of Agriculture was then promoting. Oldways has also created an Asian Diet Pyramid, Latin American Diet Pyramid, and Vegetarian Diet Pyramid that make nutritional recommendations drawing on the traditions of those cultures. Further information about the pyramids is available from the Oldways website, as is information and recommendations about other health and nutrition topics including whole grains, sugar, coffee, and seafood.

Slow Food

http://www.slowfood.com

Slow Food is an international nonprofit organization founded in 1989 to support local traditions and combat the spread of fast food around the world. The Slow Food website is available in English, Italian, French, Spanish, German, Portuguese, Russian, Japanese, and Arabic and includes links to national Slow Food websites. The site includes the Slow Food Manifesto (written by Folco Portinari in 1989) and information about the history of the slow food movement, current projects, and upcoming events. Specific focuses of the website include taste education (reawakening and training the senses to enjoy food and also encouraging local food production), defending biodiversity (including the Ark of Taste that catalogs animal breeds, fruit and vegetable varieties, and specific dishes that are in danger of disappearing), and linking producers and coproducers to encourage consumers to take an interest in where and how their food is produced. Many Slow Food publications, including newsletters, books of traditional recipes, and their annual report, are available for download from this website.

Sustainweb

http://www.sustainweb.org/about

Sustain is a British organization representing about 100 national public interest organizations that advocate food and agriculture practices that improve the welfare and health of people and animals, improve the environment, enrich society and culture, and promote equity. The website includes information about many food-related issues including healthy school food, protecting children from junk food marketing, the contribution of farming to climate change, connections between food and mental health, and information about the environmental and health impact of common food products. Many Sustain publications, including policy recommendations and reports covering everything from carrot production in the United Kingdom to the nutritional quality of food marketed to babies and young children, can be downloaded from the website, which also acts as a clearinghouse for news items relevant to healthy food and sustainable agriculture.

Sarah Boslaugh
Washington University in St. Louis

Index

Article titles and their page numbers are in **bold.**